MW01625730

The Quantum Measurement Problem

Michael Steiner   Ronald Rendell

Volume I of the Series: Progress on the Physics of Quantum Measurement

Inspire Institute

Published by Inspire Institute, Inc.
Alexandria, VA

Inspire Institute, Inc. is a 501(c)(3) nonprofit organization for scientific research

Library of Congress Control Number: 2018906297

ISBN-13: 978-1-7322910-0-3
ISBN-10: 1-732-29100-4

Printed in the United States of America
1 2 3 9 8 4 1 9 7 4

First Edition, 2018

Cover: "Schrödinger's cat - dead and alive" by Mopic/Shutterstock/227038018/Enhanced License

Contents

Acknowledgments

The authors wish to express their gratitude to the reviewers whose comments helped mold the book into one that we hope will stimulate new interest within the scientific community with the goal of resolving the measurement problem. The authors acknowledge: Dr. Steven Adler, Dr. Mario Ancona, Dr. Tom Bullock, Professor A. R. Usha Devi, Professor Michael Frey, Dr. Armen Gulian, Professor Louis Marchildon, Professor Kelvin McQueen, Dr. A. K. Rajagopal, Ms. Wendy Roseberry, Ms. Elissa Rudolph, Dr. Louis Sica, Ms. Marjorie Steiner, and Mr. Brian Whelan. The reviewers' comments and recommendations greatly improved the book's content and presentation.

Illustration Credits

We also wish to acknowledge the authors that have contributed a large body of material to the public domain. This material was highly valuable in developing the illustrations in the book. Additionally, the licensing system developed via creative commons was particularly useful in this respect.

Chapter 1

Voltmeter by Smashicons/https://www.flaticon.com/free-icon/voltmeter_778843/licensed under Flaticon basic license

Cat by Clker-Free-Vector-Images/http://pixabay.com/en/cat-cartoon-yellow-animal-domestic-30720/licensed under CC0

Radiation Symbol by Clker-Free-Vector-Images/http://pixabay.com/en/radioactive-symbols-danger-39417/licensed under CC0

Geiger counter by Rama/https://commons.wikimedia.org/wiki/File:Crocus-p1020562.jpg#filelinks/licensed under CC Attribution-Share Alike 2.0 France

Flask by Clker-Free-Vector-Images/http://pixabay.com/en/chemistry-lab-flask-experiment-306005/licensed under CC0

Table by OpenClipart Vectors/http://pixabay.com/en/desk-furniture-school-table-brown-149332/licensed under CC0

Old hammer with clipping path by Melinda Fawyer/http://www.shutterstock.com/en/pic.mhtml?id=53954512 Image ID: 53954512/licensed under Shutterstock Standard License

Chapter 2

Gears by geralt/http://pixabay.com/en/clock-time-gear-gears-blue-70189/licensed under CC0

Eye by TobiasD/http://pixabay.com/en/eye-blue-eye-eyeball-eye-closeup-369557/licensed under CC0

Neuron and glia at 60x by Christopher Meade/http://www.shutterstock.com/pic-79216378/stock-photo-neuron-x.html/licensed under Shutterstock Standard License

Brain neurons synapse by Naeblys/http://www.shutterstock.com/pic-215498779/stock-photo-brain-neurons-synapses-reasoning.html/licensed under Shutterstock Standard License

Emotions by RyanMcGuire/https://pixabay.com/en/emotions-man-happy-sad-face-adult-371238/licensed under CC0

Steam Locomotive
/https://commons.wikimedia.org/wiki/File:SAR_500_Class_Steam_Locomotive,_1953.jpeg/ Public Domain

Universe expansion by Gnixon/https://commons.wikimedia.org/wiki/File:Universe_expansion2.png/Public Domain

Mona Lisa by Leonardo da Vinci/https://commons.wikimedia.org/wiki/Category:Mona_Lisa/Public Domain

Chapter 5:

Bust of Aristotle from Ludovisi Collection
/https://commons.wikimedia.org/wiki/File:Aristotle_Altemps_Inv8575.jpg/Public Domain

Bust of an unknown Greek from Museo archeologico nazionale di Napoli
/https://commons.wikimedia.org/wiki/File:Bust_of_an_unknown_Greek_-_Museo_archeologico_nazionale_di_Napoli.jpg/Public Domain

Epicurus from Baumeister: Denkmäler des klassischen Altertums/https://commons.wikimedia.org/wiki/File:Epikur2.png/Public Domain

Ludwig Boltzmann from Uni Frankfort/https://commons.wikimedia.org/wiki/File:Boltzmann2.jpg/Public Domain

Isaac Newton from Arthur Shuster & Arthur E. Shipley: *Britain's Heritage of Science*/https://commons.wikimedia.org/wiki/File:SS-newton.jpg/Public Domain

Albert Einstein official 1921 Nobel prize in physics photograph/
https://commons.wikimedia.org/wiki/File:Albert_Einstein_(Nobel).png/Pubic Domain

Werner Heisenberg from German Federal Archive/
https://commons.wikimedia.org/wiki/File:Bundesarchiv_Bild183-R57262,_Werner_Heisenberg.jpg/Licensed under CC-BY-SA 3.0

John von Neumann from Los Alamos National Labratory/https://commons.wikimedia.org/wiki/File:JohnvonNeumann-LosAlamos.jpg/ Public Domain/ Unless otherwise indicated, this information has been authored by an employee or employees of the Los Alamos National Security, LLC (LANS), operator of the Los Alamos National Laboratory under Contract No. DE-AC52-06NA25396 with the U.S. Department of Energy. The U.S. Government has rights to use, reproduce, and distribute this information. The public may copy and use this information without charge, provided that this Notice and any statement of authorship are reproduced on all copies. Neither the Government nor LANS makes any warranty, express or implied, or assumes any liability or responsibility for the use of this information.

Flammarion engraving/https://commons.wikimedia.org/wiki/File:Flammarion.jpg/Public Domain

View from the Left Eye by Ernst Mach/In Beiträge zur Analyse der Empfindungen, 1886/Public Domain

Alpha particle and electrons from a thorium rod in a cloud chamber by Cloudylabs/https://commons.wikimedia.org/wiki/File:Alpha_particle_and_electrons_from_a_thorium_rod_in_a_cloud_chamber.jpg/Licensed under the Creative Commons Attribution-Share Alike 3.0 Unported license.

Mathematical equations formulas on tree concept by Juliann/https://www.shutterstock.com/image-vector/mathematical-equations-formulas-on-tree-concept-227079898/Licensed under Shutterstock Standard license

Antikythera Mechanism by Tilemahos Efthimiadis/https://commons.wikimedia.org/wiki/File:The_Antikythera_Mechanism_(3471171927).jpg/Licensed under Creative Commons Attribution 2.0 Generic

Gear and idler wheel model by Maxwell/https://commons.wikimedia.org/wiki/File:Maxwell_Molekularwirbel.jpg/Public Domain

View of a Skull by Leonardo da Vinci/https://commons.wikimedia.org/wiki/File:View_of_a_Skull.jpg/Public Domain

Neural circuitry of a rodent hippocampus by Santiago Ramón y Cajal/https://commons.wikimedia.org/wiki/File:CajalHippocampus.jpeg/Public Domain

Composition & size of atoms by John Dalton/https://commons.wikimedia.org/wiki/File:John_Dalton;_Composition_%26_size_of_atoms._1806-1809_Wellcome_M0004649.jpg/Creative Commons Attribution 4.0 International license.

First Solvay Conference group portrait by Benjamin Couprie/https://commons.wikimedia.org/wiki/File:1911_Solvay_conference.jpg/
Public Domain

Prototype photon counting system, c 1980s by Science Museum London/Science and Society Picture Library/ https://commons.wikimedia.org/wiki/File:Prototype_photon_counting_system,_c_1980s._(9660571969).jpg/Licensed under Creative Commons Attribution-Share Alike 2.0 Generic license.

Previously Published Work

This book includes sections that have been previously published in the article "L. Marchildon, Causal Loops and Collapse in the Transactional Interpretation of Quantum Mechanics," Physics Essays, Vol. 19 No. 3, pp. 422-429 (2006). They are reprinted with permission of Physics Essays Publication.

Preface

The quantum measurement problem evokes responses ranging from the extraordinary—as seen by the perspective of the young, dreamy-eyed student looking for something new and incredible, to responses of distain, outrage, and general fury—as seen by the old curmudgeon who believes that the fundamental principles of physics are known. In the case of the old curmudgeon, anything that cannot be predicted currently may be said to be due to the problem's substantial complexity. This may be either due to insufficiently accurate modeling or that simulations of a known accurate model are simply not currently practical due to computer technology limitations.

The question of how to approach the measurement problem, and more generally, any unsolved problem that may not be rooted in currently known theory, is addressed as one important aspect of this book. The proper scientific methodology to utilize for investigating such problems is deductive reasoning as opposed to inductive reasoning.

What is not often realized is that by its very nature, inductive reasoning can fail miserably for problems that cannot be solved using existing theory. These latter problems demand the use of exceptions to the currently known theory for their respective solutions. For these reasons, those who strongly believe that the measurement problem is strictly a philosophical problem may identify more closely with inductive reasoning than deductive reasoning. Interestingly, it appears that such thinking leads to the difficulty of having new theories accepted by the scientific community. Max Planck said [1, p. 151]:

> *A new scientific truth does not triumph by convincing its opponents and making them see the light, but rather because its opponents eventually die, and a new generation grows up familiar with it.*

This is a rather bleak statement. If true, one has to question why one would even consider working on such problems. On the other hand, if the problem is not just an academic problem, and its solution could provide new observable consequences, then the problem and its resolution would be seen in a different light. In such a case, it opens up the possibility that the solution to such a problem could have significant implications in both physics and science as a whole.

If the reader is willing to read the book cover-to-cover with an open mind, we expect that this will greatly enhance your level of understanding of the measurement problem. On the other hand, if you approach this book as a critic looking for ammunition to find some errors, you will surely succeed—in finding some errors. Deduction is a process in which one makes errors and learns from one's mistakes. As Niels Bohr had said [2],

> *An expert is a man who has made all the mistakes which can be made, in a very narrow field.*

Introduction

The quantum measurement problem is a problem in physics that has spawned many debates and strong responses from most physicists, scientists, and popular science writers. It is not immediately clear whether the problem is simply philosophical in the sense that its solution would provide no additional observational consequences beyond our current theoretical treatment of quantum mechanics. This problem is elusive to grasp and requires a substantial command of the major issues involved.

This book has been developed to assist the reader to reach one major goal regarding the measurement problem. That goal is for the reader to understand, in a precise manner, the measurement problem. In order to meet this goal, the reader will need to understand the major arguments examining whether or not the measurement problem is solely an interpretational problem or a problem with a potential solution that could provide new predictive power above and beyond the currently known quantum mechanics. These major arguments are addressed by providing arguments that:

1. Unitary Schrödinger processes can be experimentally distinguished from measurement processes.

2. Measurement is not a unitary Schrödinger process.

Chapter 1 shows by clear and concise examples how the wave and particle natures of matter and light appear to be problematic when taken together. An illustrated walk is given through logical possibilities of wave and particle models with the goal of elucidating the measurement problem. The phenomenon of unitary interference is considered as governed by Schrödinger's equation. Born's proposed rule is examined, that is, distinct from the unitary evolution inherent in Schrödinger's equation. Born's rule, which was to become the measurement postulate of von Neumann, accounts for the results of measurement. Paradoxes involving predictions of macroscopic superpositions, such as Schrödinger's Cat, are illustrated.

The characteristics of quantum state evolution as described by Schrödinger's equation or unitary evolution are presented in Chapter 2. The aspects of reversibility, entropy, and unitary evolution for interactions between quantum systems are considered. This is expanded into interactions between multiple systems, and finally the theory and predictions during large-scale interaction are considered. Concepts of the quantum locomotive are presented which are utilized in later chapters. The painting of the Mona Lisa is examined as a measurement paradox that is related to complexity and entropy and is different than the usual superposition paradox of Schrödinger's Cat.

The details of the measurement problem are further elaborated in Chapter 3. Under fairly general assumptions, contradictions appear if one attempts to explain an experimental phenomenon by assuming only a single mode of evolution; neither Schrödinger's equation nor the measurement postulate by themselves appear to be fully sufficient to explain the phenomenon of measurement. Arguments are given based on the modeling of particle-detector interactions as well as their predictions assuming unitary evolution versus conditions required for observation or measurement. The measurement problem cannot be avoided: within the conventional framework, unitary evolution can be experimentally distinguished from measurement.

With the use of the Chapter 3 development, numerous proposed scientific approaches to resolving the measurement problem are analyzed in Chapter 4. Issues of decoherence, hidden variable models, and light emission from matter are considered as well as other potential solutions of these phenomena. A major point of this book is that the measurement problem is not resolved by unitary evolution alone and must be supplemented by meeting additional requirements that we have laid out in Chapter 4 in order to be scientifically accepted. We attempted to present a balanced view of the various approaches that at this time have some chance of meeting the latter requirements. On the other hand, it is our ardent viewpoint that approaches that are claimed to have resolved the measurement problem, for which no attempt has been made to address basic requirements of scientific acceptability, should be taken with a grain of salt.

For any given problem area, it is often beneficial to familiarize oneself with the history of the problem, and the measurement problem is no exception. There exists a substantial body of history surrounding this problem, much of which is still pertinent today. In the other technical chapters, historical context that is directly relevant to particular material is generally minimized when presented alongside technical material in order to concentrate on communicating essential concepts. Chapter 5, however, is solely devoted to exploring relevant historical issues. Often overlooked and potentially significant historical issues related to the measurement problem are presented from the 5th century BCE to the present day. The reader may wonder why the detailed history is presented in the middle of the book in Chapter 5. In the particular case of the measurement problem, the authors believe it is easiest to comprehend the detailed historical scientific issues after understanding the relevant physics of the measurement problem, rather than before. The measurement problem is atypical in this regard; as an example to the contrary, one does not need to learn the minute details of thermodynamics as a prerequisite to reading and appreciating the historical developments of the automobile.

An important aspect of science that is often overlooked is the use of the methodology of scientific reasoning that is specifically appropriate for the problem under consideration. The methods of deductive versus inductive reasoning are considered in Chapter 6. The strengths and weaknesses of these methods are considered in terms of the best approach toward resolving the measurement problem.

A broad consideration of proposed approaches to solving the measurement problem is given in Chapter 7 via the partitioning of all approaches into those that are closed versus those that are open. Several examples of both closed and open approaches are given.

This book is written with the intent of communicating the measurement problem to as broad an audience as possible. However, this is not a simple task for this particular problem, which demands dedication and study. With this in mind, we have tried to develop certain key concepts in certain parts of the book with minimal mathematical content that is meant for the reader with some scientific background, while including detailed mathematical content in other areas of the book that is meant for the more technical reader.

The reader is strongly urged to visit the website theQMP.com and also youtube.com where freely available companion chapter videos are provided as well as other useful information on the quantum measurement problem. Additionally, a FAQ discussion is available at theQMP.com which is designed to aid readers with additional questions.

Less technical readers without a background in quantum theory who are familiar with undergraduate calculus and have some basic knowledge of physics should proceed with Chapter 1. In Chapter 2, we recommend concentrating on understanding the three concepts associated with the unitary characteristics of interference, reversibility, and entanglement. The mathematical developments should be omitted as well as the exercises. In Chapter 3, a test that can experimentally distinguish unitary evolution from measurement is developed. The less technical reader should not try to follow the mathematical development of the unitary versus measurement discrimination test (UMDT), but should be generally aware that such a methodology was developed in Chapter 3. In Chapter 4, the reader should become familiar with the definition of the measurement problem and the requirements for its solution. As well, many of the approaches that have been put forth and the strengths and weaknesses of these approaches can be understood without a thorough understanding of the mathematics. Approaches that are more technical such as stochastic equations should be omitted. The less technical reader should understand the backdrop of history as it pertains to the quantum measurement problem, while the mathematical historical development can be skipped. All readers should thoroughly read Chapter 6. Chapter 7 should be understood at least in terms of the division between open and closed systems and the methods that should be involved in resolving the measurement problem.

For those more technical readers with a sufficiently strong background in quantum mechanics and some background already on the measurement problem, as well as those who are interested in further research on the measurement problem, we recommend skimming Chapters 1 and 2 and concentrating on learning in Chapter 3. Chapter 4 should be read with the intent of learning why the various interpretations generally fail to address the requirements of resolving the measurement problem and understand the pros and cons of the physical approaches that have been proposed. The history that we presented is unabridged as we include important issues of consciousness and free will that are often left out in other books. The history of the measurement problem is important to understand, particularly to learn what has been accomplished and not to repeat the errors of the past. For those interested in actually making progress on the measurement problem, the scientific approaches and particularly the deductive process in Chapter 6 should be firmly understood.

After learning Chapter 7 and working through the various exercises throughout the book, the reader will be in a much better position to begin researching the measurement problem than before reading this book.

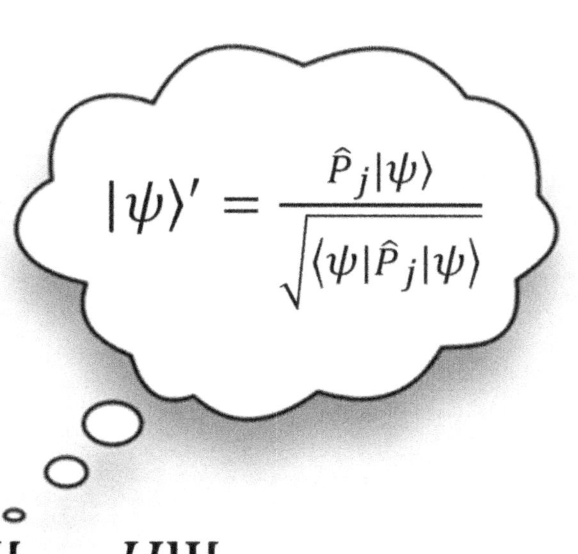

$$i\hbar\frac{\partial}{\partial t}\Psi = H\Psi$$

CHAPTER 1

Wave Particle Duality and Schrödinger's Cat

Understanding the concept of wave-particle duality is the first step towards understanding the quantum measurement problem.

The measurement problem began to be formulated early in the history of quantum mechanics. One can argue that the measurement problem is deeply rooted in the issue of wave-particle duality. Here we give a brief historical review needed for this chapter; a more in-depth historical presentation is given in Chapter 5. The issue of wave-particle duality arose with Einstein's discovery of the photoelectric effect. It had already been known by Planck's discovery in 1900 that energy was quantized. However, the mechanism of how this quantization occurred was not known at that time, and in fact it was not generally accepted that individual particles existed. At the beginning of the 20th century, many believed that there was simply a continuum and not discrete atoms.

Long before 1900, it was known that light exhibited interference. Huygens proposed early in 1678 that light behaves as a wave. Newton believed that light was corpuscular, and this was accepted by many scientists. Interestingly the concept that light is corpuscular was also put forward by Gassendi and Hobbes before Newton. However, Young in 1803 demonstrated conclusively that light exhibits interference through his double slit experiment, and many scientists abandoned the corpuscular theory as it did not adequately appear to explain the interference phenomenon. Maxwell in the 1860s developed equations that describe classical light, which can take the form of a wave equation.

Wave Properties of Light

Before the measurement problem was clearly formulated, the issue of wave-particle duality existed. In order to understand the issue of wave-particle duality, we consider the double slit experiment as shown in Figure 1.1 in which a light source shines toward a surface containing two narrow slits. In terms of wave propagation, one can consider a similar experiment in which the double slit is placed on the surface of a water tank, and the source consists of an oscillating plunger that creates waves.

Young, having studied the theory of sound propagation, argued that light should also behave as a wave phenomenon. His ideas were initially rejected by most. Later, an experiment was formulated by Augustin-Jean Fresnel, in which a circular body placed in front of the source would block the light if it consisted of particles and gave specific predictions if light were a wave. An experiment was conducted by François Arago in which a circular object was placed in the path of a light source. A particle theory predicts that there should only be a circular shadow due to the object blocking the light, whereas a wave theory predicts that light can further recombine from the edges of the object and at the center of the shadow there should be an additional bright spot due to constructive interference. Arago observed this additional spot in the center of the shadow which confirmed Fresnel's predictions for wave propagation. After this striking experimental confirmation, the theory that light propagates as a wave gained acceptance among the scientific community, and Newton's corpuscular theory was largely abandoned.

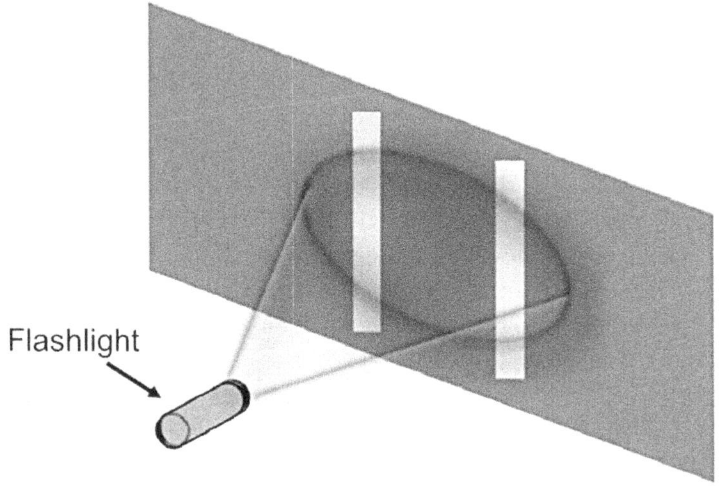

Figure 1.1: A light source shines on Young's double slit.

Young contemplated that each color of light corresponds to a particular frequency of the wave undulation or oscillation. The wave emitted from the flashlight then heads toward the two slits as shown in Figure 1.2. Once a wave enters a slit, there is an interaction between the wave and the slit. This interaction is similar to a scattering of a water wave that is initially moving in a particular direction which then enters a narrow slit. When the water wave exits the slit, it will be found to disperse in a wide range of directions. Hence a wave that enters a single slit can exit in many directions. If one considers the Point x_1 after the double slit apparatus as shown in Figure 1.3, one sees that a wave that exits the left slit can continue directly to Point x_1 but also a wave that exits the right slit could scatter and change its direction and also have an effect at Point x_1. Any other Point x_2 could likewise be considered. Any given point after the double slit apparatus will be affected by a wave that went through the left slit and a wave that went through the right slit.

Note that one might instead believe that the wave representation of two waves going toward the two slits from the flashlight is simply an incomplete model, and that

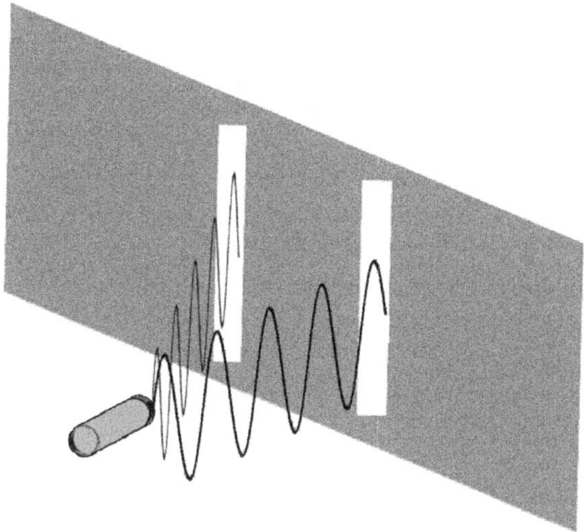

Figure 1.2: Young's double slit with light represented as a wave rather than a particle.

in reality there is but a single localized wave-particle that is simply oscillating toward only one of the two slits. As we will see, this explanation is inadequate and this is the importance of the double slit experiment. Suppose that one considers the light wave as localized but oscillating particle moving toward only one of the two slits in Figure 1.2. Such an explanation is incorrect; one must consider at any given point the contributions that could have occurred by the wave going through *both* slits. Again, let us consider Figure 1.3 where the wave at point x_1 is composed of the summation or *superposition* of waves that pass through *both* slits. Assuming $\psi_L(x_1, t_0)$ is the value of the wave at the point x_1 (x_2 can likewise be considered) at time t_0 that emerges from the left slit, and $\psi_R(x_1, t_0)$ is the value of the wave at the same point and time that emerges from the right slit, the overall wave is computed as the sum of the two waves, i.e. $\psi_L(x_1, t_0) + \psi_R(x_1, t_0)$. As the overall wave function is $\psi_L(x_1, t_0) + \psi_R(x_1, t_0)$, rather than either $\psi_L(x_1, t_0)$ or $\psi_R(x_1, t_0)$, interference is exhibited between $\psi_L(x_1, t_0)$ and $\psi_R(x_1, t_0)$ in the formation of the overall wave function. This is due to the detection probability being given by the absolute square of the overall wave function, $|\psi_L(x_1, t_0) + \psi_R(x_1, t_0)|^2 = |\psi_L(x_1, t_0)|^2 + |\psi_R(x_1, t_0)|^2 + \psi_L^*(x_1, t_0)\psi_R(x_1, t_0) + \psi_L(x_1, t_0)\psi_R^*(x_1, t_0)$, which came to be known as *Born's Rule*. Interference arises from the cross-terms and therefore it is not correct to consider the wave as a single localized particle (although as we will see in Chapter 4, Bohm's theory attempts to do just that, although by adding an additional quantum potential to the theory). Often one sees the double-slit pattern explained in terms of the placement of a screen that the light hits after it passes through the double-slit apparatus. More generally however a detector could be placed anywhere after the double slit, which is the case considered hereafter, and so a screen is not shown. Experimental results confirm that Figure 1.3 is correct, with the added interpretation that when a detector is placed at any position x_1 at time t_0 after the double slit apparatus, the detection probability is given by the square of the wave function, which is given by $|\psi_L(x_1,t_0) + \psi_R(x_1,t_0)|^2$. Interestingly an experiment reported as early as 1909 by G.I. Taylor used an attenuated light source from a gas flame to show wave interference from the shadow of a needle on photographs exposed for a period of three months. This provides evidence against the concept that light can be considered to be composed of propagating *localized* particles and in favor of the concept that light propagates as a wave that exhibits interference.

The wave in Figure 1.3 at (x_1,t_0) can be seen to be a superposition of the wave passing through both slits of the double slit apparatus. In general, for different experiments, there are many paths to any given point and infinitely many points may need to be considered.

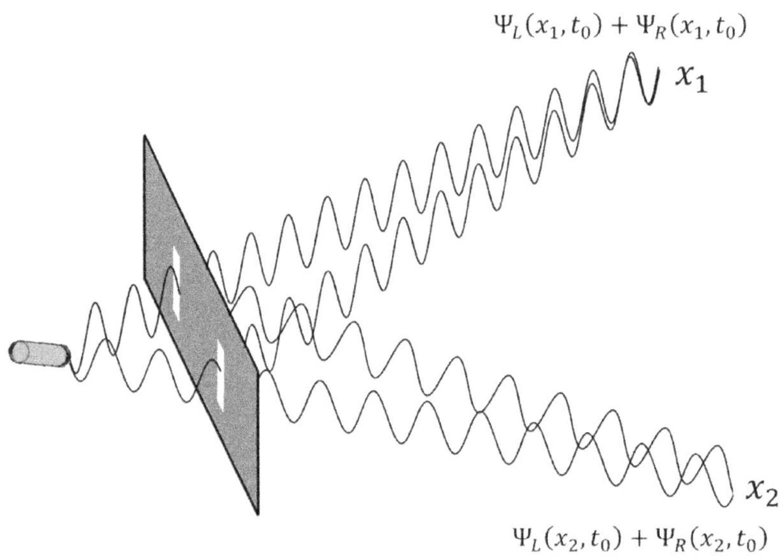

Figure 1.3: Young's double slit with output waves that are a superposition of the input wave passing through both slits.

In Figure 1.4, the full pattern (via computer simulation of the wave equation) can be seen. This pattern is similar to what would be seen if the flashlight was continually paddling water to create a water wave for a long time and a picture of the steady state amplitude of the existing wave was taken. Between the flashlight and double slit, only the direct paths to the slits are shown; the paths to the solid part of the apparatus are left out for simplicity. At any point past the double slit apparatus, a continuum is seen due to the possibility of the wave changing direction upon exiting either slit. Note the appearance of distinct peaks and nulls. These peaks and nulls are a direct consequence of the constructive and destructive interference due to summing the paths through both slits. A wave picture for describing light at least so far, provides an ample understanding to describe the phenomenon of light.

Particle Properties of Light

As has been seen, it appears from Figure 1.1 - Figure 1.4 that a wave picture of light is fully adequate to describe the situation. This was the case historically after Fresnel's experiment, for which one might expect that all would have been "said and done."

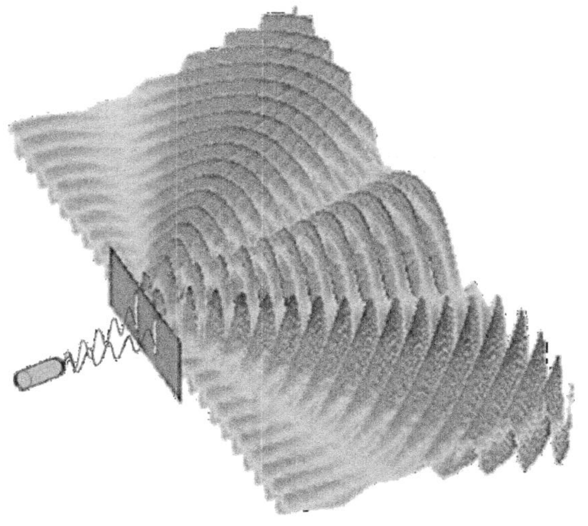

Figure 1.4: Young's double slit with a wave passing through both slits and exhibiting constructive and destructive interference.

Photoelectric Effect

When light shines on a material, it is possible under the right circumstances for electrons to be freed or ejected from the surface of a material. From Maxwell's equations one might expect that higher intensities of light would produce a proportionally higher ejection rate of electrons from the surface. But experiment showed this was not to be the case. Philipp Lenard found in an experiment in 1902 that higher intensity light of a particular frequency below a threshold did not lead to ejection of electrons. Rather, only when the frequency was increased beyond a threshold did electrons begin to be ejected, a phenomenon called the *photoelectric effect*. This was an unexpected experimental finding and was explained by Einstein in 1905.

Einstein explained the photoelectric effect by postulating that there must be individual discrete light particles or photons. The photons responsible for ionizing an atom were due to the energy of the individual particles, and not simply the amplitude of the wave. Einstein's discovery in which each photon has an energy $E = h\nu$ revived the earlier corpuscular or particle theory of light in the context of quantization.

A macroscopic light beam is composed of a large number of individual photons each of which is a packet of definite energy $E = h\nu$. The energy that is required for an electron to be liberated from an atom depends on the binding energy of the particular electron and atom. However, for a given atom, such as in a metal, the threshold required in order to see the photoelectric effect can be measured. Light that impinges on the photoelectric material with a frequency below this threshold does not lead to appreciable electron ejection; when the frequency of the light is increased beyond this

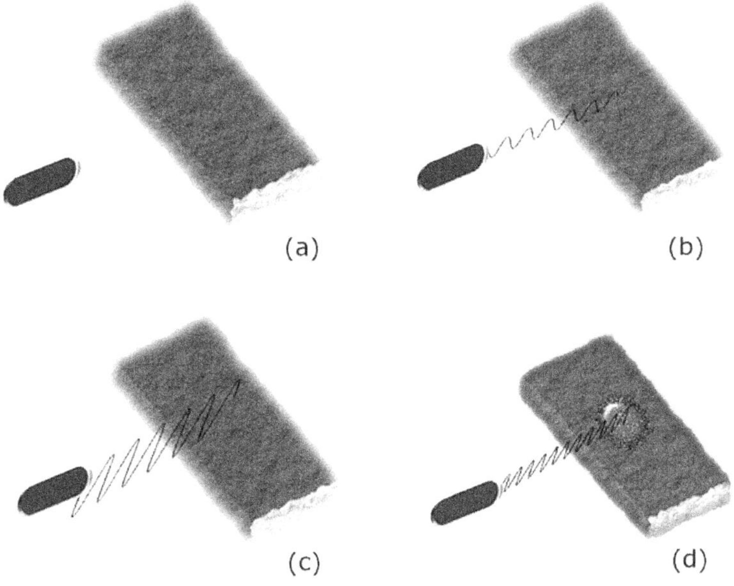

Figure 1.5: (a) Light source and metal (b) low intensity, no electrons ejected (c) high intensity, no electrons ejected (d) high frequency, electrons ejected.

threshold one can measure appreciable electron ejection.

The photoelectric effect is an important discovery in quantum mechanics. It shows that one cannot rely on the wave picture of light alone, but that light possesses an individual particle aspect. To see why this is, suppose that light was a continuum and could be represented as a wave, i.e., not composed of individual particles. Then if light were to interact with matter, it would affect the matter because of its wave-like disturbance. In Figure 1.5(a), a variable frequency and intensity light source is considered that impinges on a metal surface. The case where the frequency of the light source is lower than the binding energy of the electrons in the metal (known as the *Work Function*) is considered in Figure 1.5(b). In this case, no electrons are given off. A given amount of kinetic energy would be expected to be exchanged in such interactions depending on the conditions of the experiment. If one assumes that light is only a single composite wave, then one could expect that by increasing the amplitude of the light wave, the energy exchanged could also be continuously increased further and further until an electron is eventually ejected.

Remarkably however (other than from multiple-photon absorption that is non-linear and typically occurs at a much lower rate), this is not found experimentally when the intensity of the source is increased, as shown in Figure 1.5(c). Only in the case where the frequency of the light is increased beyond the binding energy of the electrons will the process of non-negligible photoionization begin, which is illustrated in Figure 1.5(d). Once non-negligible photoionization does begin, the rate of photoionization can be increased by increasing the intensity of the light source.

The explanation that light is solely a wave, does not appear to be adequate to explain the photoelectric effect. Einstein realized this and proposed that light also has particle properties. One might envision whether or not light could be construed on the basis of a localized particle model. Consider an alternate explanation such that a light wave is composed of small indivisible particle packets as shown in Figure 1.6. In this model, the light particles or individual photons are clustered at the peaks of the wave in a proportional manner, where the occurrence of a photon is related to the squared amplitude of the wave. This wave-particle model (WPM) may at first seem reasonable to explain wave-particle duality.

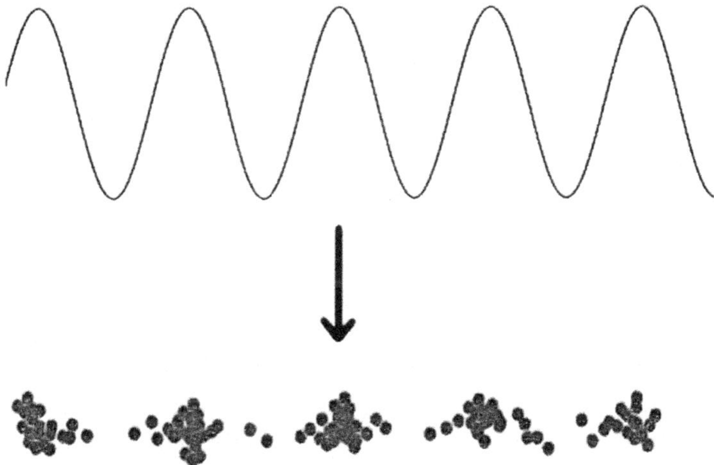

Figure 1.6: Consideration of a wave-particle model of light that is composed of particles that are clustered at the peaks of the wave.

Assume that each particle packet has an energy $E = h\nu$ where ν is the frequency of the wave. In terms of explaining the photoelectric effect, the model appears to correctly predict electron ejection only when the particles have an energy greater than the threshold of the electron binding energy. Moreover, one can see that if the particles have energy above the electron binding energy then the rate of electron ejection should be proportional to the intensity of the light since there will be more particles, which also agrees with experimental observations. So this model does appear to be reasonable for explaining the photoelectric effect. However, quantum particles were later found to have an additional intrinsic fundamental property called *entanglement* that contradicts this explanation.

Models of this type in which the wave property of light is supplemented by a localized singular particle were considered as early as 1909 by Einstein [3, p. 219]. Einstein contemplated the existence of a vector field surrounding the particles that could extend for large distances. These vector fields from many particles would sum coherently to create forces on other charged particles.

Einstein's Ghost Field

Einstein believed that the state of *N* particles should be specified via a product of states associated with each of the *N* particles and that the evolution of these particles should be specified by a product state operator. This is discussed further by Howard [3, p. 60]. A product state operator has the property of operating independently on each particle that composes the initial state. Each component or particle of a product state will add an amount toward any observable quantity that is independent of the other particles. Quantum states can be further categorized as *separable* or *entangled.* A product state is an example of a separable state.

There are several reasons why Einstein believed that particle time and space evolution should be governed by such a product state evolution operator and the initial states be product states. The primary reasons appear to be: 1) the Boltzmann entropy formula is only fully additive for multiple particles when the particles are independent [3, p. 74], 2) an entangled state composed of a Particle *A* that is separated from a Particle *B* has the property that Particle *B* does not have a physical well-defined state that is independent of a measurement on Particle *A* [3, p. 91], and 3) Einstein believed non-separable theories were non-causal [3, p. 84].

Interestingly it appears that Einstein also had more than one reason to believe the contrary: the derivation of Planck's radiation law required an assumption that particles were not mutually independent [3, p. 78] and the Bose-Einstein statistics that were later derived in 1925 also showed statistical dependence between particles that Einstein believed pointed to a "physically mysterious interaction." [3, p. 68] The entangled states might appear to be just another example of such non-independence, so one might wonder why Einstein did not accept such states. The reason may be that the lack of independence in the derivation of Planck's radiation law and Bose-Einstein statistics perhaps could still be hoped to eventually be explained by reasons other than the breakdown of the reality of local states that is seen rather poignantly in the theory of entangled states. It does appear that Einstein had an in-depth knowledge of the most significant issues regarding quantum mechanics and what was to become ultimately, the measurement problem.

In 1925 Einstein presented a lecture in which he considered that individual particles were accompanied by a ghost field that guides the particles [4, p. 222]. In Einstein's ghost field, the evolution of the particles is governed by a classical tensor product evolution. However, the problem for Einstein was that such a tensor product evolution could violate energy conservation when one particular term was considered for any single outcome.

Note that at least in this model, one is seemingly left to either accept action-at-a-distance through the existence of entangled states, or to reject energy conservation on every outcome in favor of energy conservation on-average [4, p. 226]. It does not appear that Einstein was comfortable with either of these alternatives and the ghost field was never published and faded into obscurity. The reasoning for rejection of Einstein's model has since been reconsidered in more depth by Holland [5] in regards to realistic trajectory theories.

Wave Particle Duality

The problem with both the WPM and Einstein's model is that a single photon particle that has a localized trajectory through only one of the two slits in the double slit experiment, even if it has a vector field surrounding the particle that changes causally, will generally give different predictions than a photon wave going through both slits (as we will see later in Chapter 4, Bohm's theory is a case of a local particle trajectory theory, but is supplemented with a generally *non-local* guiding field). One can still conduct a double slit experiment even when the distance between the slits is many multiples of the wavelength. Hence, one would have to believe that a particle would go through only one of the two slits, but somehow still be guided by the fact that interference is required *as if* the particle went through both slits.

One is led to the wave-particle duality problem in order to come to terms with how it can be that an individual light particle, i.e., a single photon, can interfere at different locations and thereby appear to exist as an extended or non-local wave, but also somehow cause the ionization of an atom for which the entire energy $E = h\nu$ of the photon must be transferred to the position of the atom, which can be extremely localized. In fact, the atom may be physically (orders of magnitude) smaller than the wavelength of the light.

Issues associated with wave-particle duality were being considered by Einstein, Bohr, and others prior to the development of Schrödinger's equation and Born's rule. Although many relevant experiments were being carried out during these years, experiments at the individual particle level were not plentiful and the physics of such events could be hypothesized but not easily verified by experimental evidence. It does not appear to be a simple matter to develop quantum theory without sufficient experimental evidence. In 1924 a theory by Bohr, Kramers, and Slater (BKS) was put forward in which the electromagnetic field was treated as a wave and not as a particle. BKS were not able to develop a theory that would conserve energy on all individual scattering events and at the same time retain the property that the electromagnetic field acted as a wave. BKS allowed for individual quantum transitions to violate conservation of energy, but energy would still be conserved on average. The theory was opposed by Einstein, and experiments were conducted by Bothe and Geiger and independently by Compton and Simon in 1924-1925 that examined individual scattering events. It was found that energy and momentum *are* conserved on individual scattering events. BKS theory was thus found to be incorrect and this added insight into the issue of wave-particle duality. For the most part, from 1925 through to the present there appears to be agreement on this. The framework of how this occurs however has been a hotly debated question, which seems to have begun nearly immediately upon Schrödinger's discovery of his now famous equation of quantum mechanics.

Schrödinger's Equation

In 1926, Schrödinger published an equation that would alter the path of the development of science. Schrödinger introduced a quantum wave function $\psi(r,t)$ and his equation that deterministically predicts the time evolution of $\psi(r,t)$ from the Hamiltonian of the system in the following manner:

$$i\hbar\frac{\partial}{\partial t}\psi(\mathrm{r},\mathrm{t}) = \mathrm{H}\,\psi(\mathrm{r},\mathrm{t}). \tag{1.1}$$

Schrödinger's equation would be used to help visualize physics by utilizing the wave function, $\psi(\mathrm{r},\mathrm{t})$. The equation was quickly applied to the hydrogen atom and found to successfully represent the atomic orbitals, and the energy differences corresponded to the energy of the photon emitted or absorbed when an atom made a transition between orbitals.

Schrödinger visited Bohr and Heisenberg in Copenhagen in 1926. Bohr and Schrödinger argued over whether or not Schrödinger's equation could be used alone to explain all phenomena. Bohr believed that ultimately such a deterministic and causal equation could not account for black body radiation, which historically utilized a nondeterministic or statistical approach in order to take into account individual quantum events. These issues are further elaborated in Chapter 5.

Born's Rule

Let us examine Born's rule in more detail. Consider again Young's double-slit experiment shown in Figure 1.4. In Figure 1.7 two detectors are placed after the slit in separate locations. A detector that has detected a photon will be denoted by a white detector, otherwise it will be assumed to be in its initial pre-detection state denoted by a black detector; in Figure 1.7 both detectors are in the pre-detection state and are black.

Now, suppose a single photon is incident on the double slit. According to Schrödinger's equation, a wave propagates through the slit and impinges on both

Figure 1.7: Young's double slit with detectors in their ready state as shown in black.

detectors as shown in Figure 1.8. As of yet no detection has been made. At this point the wave is physically interacting with *both* detectors. Let us assume that one of the two devices detects the photon, which is denoted by a white detector.

One might initially expect that the wave is still present—consider for example

Figure 1.8: Young's double slit with photon wave propagating via Schrödinger's equation and wave function impinging on both detectors.

Figure 1.9 (the unitary prediction involves entanglement and is discussed in Chapter 2). However, this is not the case in the Copenhagen interpretation of Bohr and Heisenberg. When a measurement is known to have been made, one invokes the measurement postulate due to Born and formalized further by von Neumann. The situation is shown in Figure 1.10.

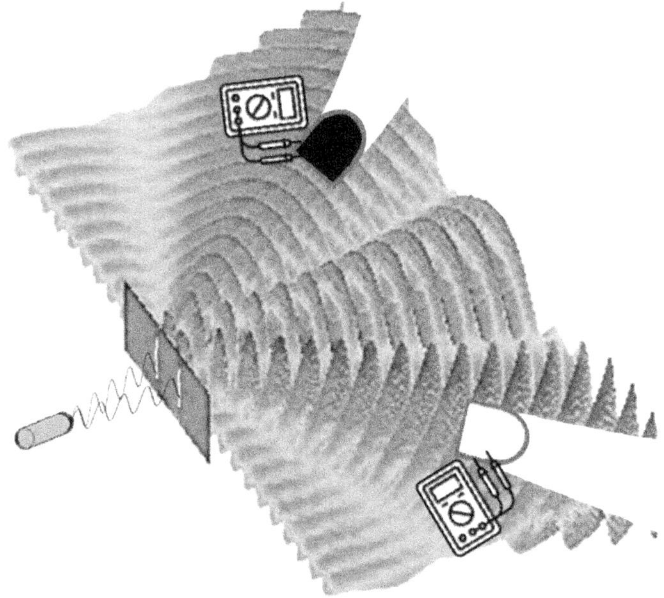

Figure 1.9: Incorrect explanation of Young's slit with a photon detection shown by white detector; wave and detection exist simultaneously.

The unitary Schrödinger wave ceases to exist after the measurement postulate is applied and the situation of Figure 1.8 is replaced with Figure 1.10, which happens when a measurement occurs in the lower detector. The replacement of Figure 1.8 with Figure 1.10 is a discontinuous change and is often referred to as a quantum jump. There was good reason historically to consider Figure 1.10 versus Figure 1.8 in order to explain measurement. At the time of Schrödinger's discovery of his equation in 1926, it had already been known for over twenty years, going back to Einstein's discovery, that light exists in quantized energy units called photons. One can see that if Figure 1.8 was considered to be correct, it would contradict the existence of such indivisible units. The detector would have to be said to have absorbed the energy in the photon (by the photoelectric effect proposed by Einstein in 1905), and yet if the wave still existed it would seemingly be able to be absorbed by another detector, which would contradict energy conservation.

So far, we have examined the case of a single detection as illustrated by the lower white detector in Figure 1.10. However, it is also possible that the upper detector detects the photon as opposed to the lower detector as shown in Figure 1.11. One sees that there are three possible outcomes of the experiment; it is possible that no detection is made, for which the result would be Figure 1.7, and as well there are two possible outcomes when detections occur—those represented by Figure 1.10 and Figure 1.11. In 1926, the question became how one mathematically describes the physics of the situation.

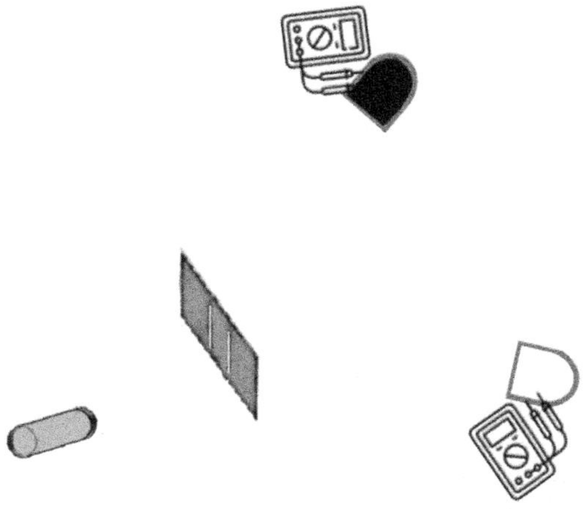

Figure 1.10: State evolution under the measurement postulate where lower detector shown in white, registers a photon.

In 1926 shortly after learning Schrödinger's equation, Born proposed the statistical nondeterministic interpretation. Born did not ask how Schrödinger's equation could be used to derive the particular wave function that occurs after a collision for individual

Figure 1.11: Second possible case of detection shown by upper white detector.

events, but rather only how probable a specific outcome of the collision would be. Born considered an electron scattering with an atom. The final electron energy would be increased by one quantum at the cost of lowering the energy of the atom. However, the direction of the electron after scattering was uncertain in Schrödinger's description and written in terms of a superposition of outcomes of the atomic unperturbed eigenstates Ψ_m that are weighted by a function $\Phi_{\mathrm{m}}(\alpha,\beta,\gamma)$ where α,β,γ specifies the scattering angles. Born proposed that the probability of any specific outcome would be found when considering the scattered wave function of the form $\sum_m \int \Phi_{\mathrm{m}}(\alpha,\beta,\gamma)\Psi_m$. The probability of finding the electron scattered with particular angles α,β,γ was proposed by Born to be proportional to $|\Phi_{\mathrm{m}}(\alpha,\beta,\gamma)|^2$.

In the case of Young's double-slit, also called an *interferometer*, one can develop a nondeterministic interpretation based on Born's rule as follows. Let the wave function that impinges on the upper detector at x_1 in Figure 1.3 be given as the superposition $\psi`_L(x_1,t_0) + \psi`_R(x_1,t_0)$ and on the lower detector as $\psi`_L(x_2,t_0) + \psi`_R(x_2,t_0)$. The initial situation shown in Figure 1.8 then jumps to either Figure 1.10 or Figure 1.11 depending on the outcome of the measurement which is determined statistically. The resulting probability of Figure 1.10 is given by the square of the wave function at (x_2, t_0), which is $|\psi`_L(x_2,t_0) + \psi`_R(x_2,t_0)|^2$. Similarly, the resulting probability of the photon being detected by the upper detector in Figure 1.11 is given by $|\psi`_L(x_1,t_0) + \psi`_R(x_1,t_0)|^2$. Upon expanding these expressions, the cross-terms produce interference between *L* and *R*. This bifurcation is shown in Figure 1.12.

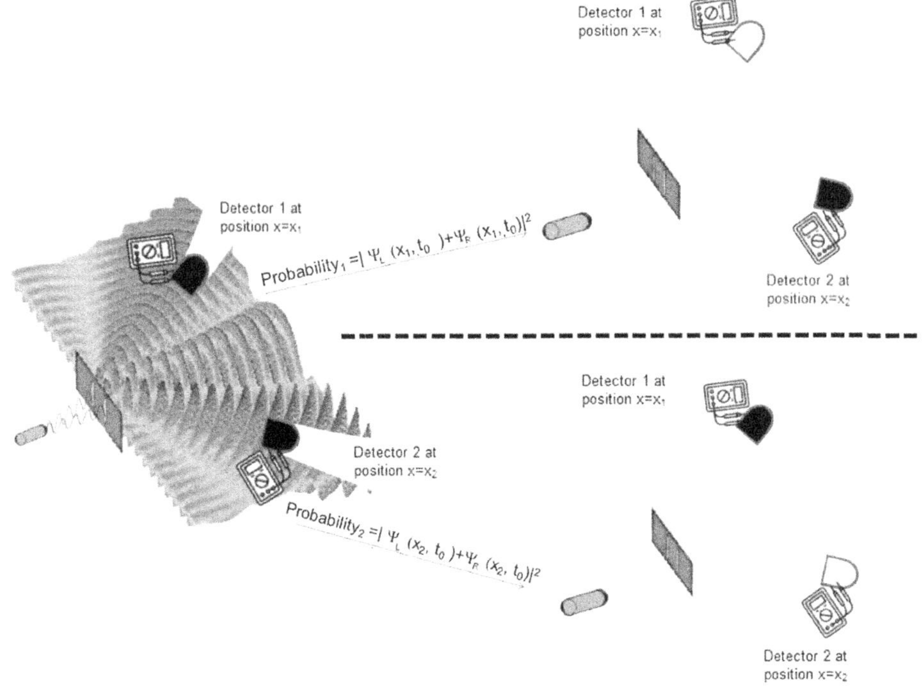

Figure 1.12: Statistical bifurcation of Born's rule.

A question that immediately arises is whether this is simply an approximation due to the lack of knowledge of the phases of the particles accumulated during the collision, or whether the underlying physics is truly nondeterministic. This issue was brought up in Born's original paper [6] and considered by Bohr, Heisenberg, and others. Von Neumann in 1927 published a series of papers that attempted to formalize the theory. Von Neumann introduced a projection postulate which represented measurement by an operator that changes the state when a measurement occurs. The projection postulate represents a discontinuous change of the state, depending on the result of the measurement. Dirac adopted the projection postulate in his 1930 textbook which became the standard presentation of quantum mechanics.

Matter versus Light

Although wave and particle duality has been considered using light, it was predicted by de Broglie that matter also has wave properties, which was subsequently verified by Davisson and Germer between 1923-1927. Hence the concept of wave-particle duality is not only applicable to light and photons but also to matter. For example, an electron could be substituted for the photon in Figure 1.1. Assuming the photon detectors are replaced by electron detectors, both wave interference as in Figure 1.4 as well as measurement as in Figure 1.12 will exhibit similar behavior.

Whether or not light or matter is utilized in a quantum experiment appears to be inconsequential to resolving the issue as to whether the projection postulate is really needed above and beyond Schrödinger's equation. A *potential* resolution of this problem was to be proposed very quickly with the discovery by Heisenberg of his uncertainty relationships.

Heisenberg's Uncertainty Relationships

Shortly after Schrödinger proposed his new equation, Heisenberg [7] developed his now famous relationships which would be used by Bohr to formulate the famous complementarity principle. The Heisenberg uncertainty principles for a quantum wave function are given in Equations (1.2):

$$\begin{aligned} \Delta x \Delta p &\geq \frac{\hbar}{2} \\ \Delta E \Delta t &\geq \frac{\hbar}{2}. \end{aligned} \tag{1.2}$$

As discussed further in Chapter 5, the time-energy uncertainty relation must be justified by a different method [8] than the position-momentum uncertainty because time does not have a well-defined quantum operator. Bohr also had a concise justification of the time-energy uncertainty relation in his Como paper where he first introduced complementarity which uses only the classical expressions for the resolving power of optical instruments [9, p. 312].

These uncertainty relationships are largely related to the well-known Fourier transform of a function *f*(*t*), which we will denote as *F*(*w*) where $w = 2\pi f$. Now one can see, from applying Einstein's result from the photoelectric effect $E = h\nu$, that *w* is

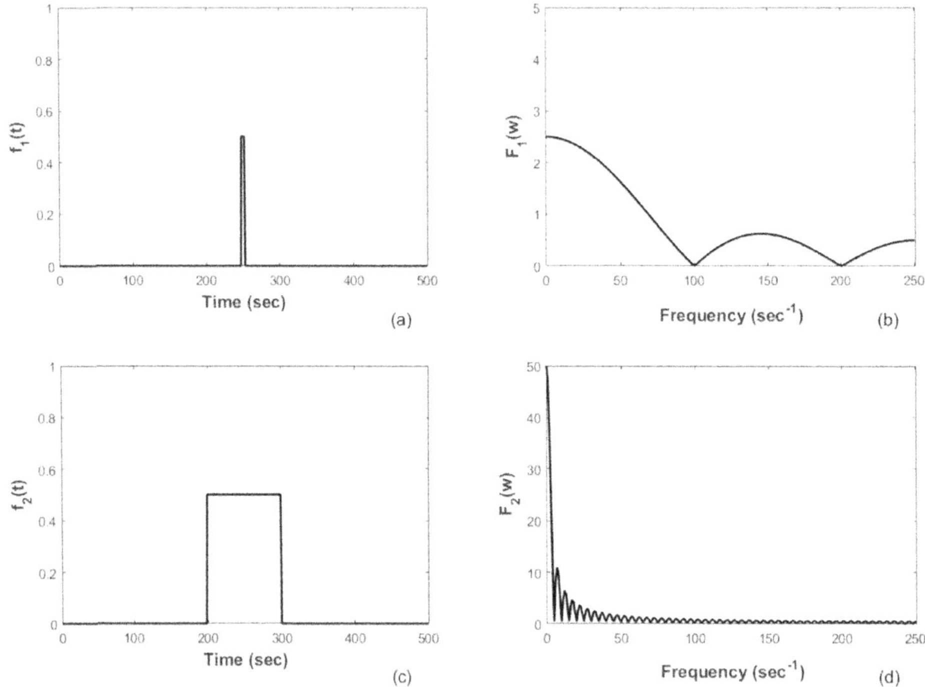

Figure 1.13: Functions of time and respective Fourier transforms: (a) time localized function $f_1(t)$; (b) Fourier transform of $f_1(t)$; (c) time uncertainty function $f_2(t)$; (d) Fourier transform of $f_2(t)$.

also a measure of the energy of a photon. For example, Figure 1.13(a) shows a highly localized time function and Figure 1.13(b) the wide Fourier transform of the function. For the case of a more delocalized square-wave time function shown in Figure 1.13(c), the Fourier transform in Figure 1.13(d) is significantly more localized in frequency when compared to Figure 1.13(b). As one further localizes a function in time, the energy becomes less certain. Variables such as time and energy that are related in this manner are called conjugate variables and are important for understanding Bohr's complementarity argument. In terms of the issue of measurement, Bohr would argue that the interaction with the device necessarily would have an effect on the particle's properties due to the Heisenberg uncertainty principle. Such a change would be in principle uncontrollable. One can consider that if a measurement device were to interact with a particle in such a manner that the particle's time when it entered the device could be known very precisely, then the particle's energy would have to have a substantial uncertainty. The gray areas shown in Figure 1.14 are valid time-energy uncertainty products, while other such products would violate the Heisenberg uncertainty principle, particularly where both the time uncertainty and energy uncertainty are small.

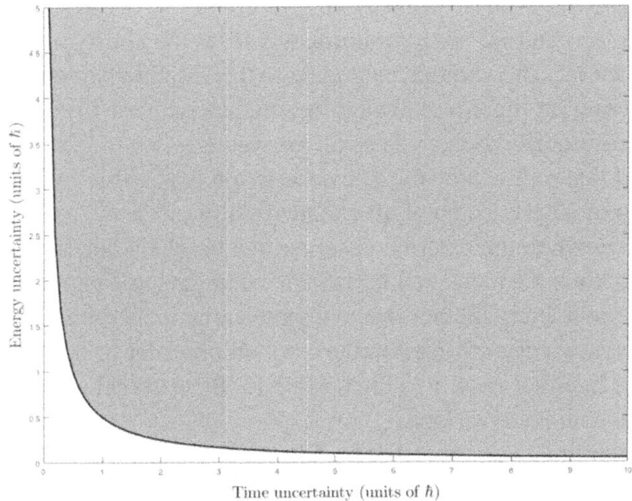

Figure 1.14: Energy versus time uncertainties as a consequence of Heisenberg's uncertainty relationships. Valid energy-time uncertainties shown in gray.

This change in uncertainty for Bohr is applicable to the inevitable consequence of requiring an interaction between the particle and device in order for the device to register a measurable quantity, such as the particle's time-of-arrival.

Consider the wave-particle light pulse in its interaction with a measurement device that will measure the position of the particle. If the position of the particle is very accurately determined, its momentum must be spread or uncertain. Or if the measurement device is such that the device can ascertain the momentum of the particle accurately, then the particle's position becomes uncertain. Now when the momentum becomes certain for a photon, then the position becomes uncertain and a plane wave results. One could then ask whether each particle has an in-principle unknowable position, but nonetheless does have an absolute position. This is an example of Einstein's objections with quantum mechanics in his debates with Bohr.

Bohr and Complementarity

Bohr's complementarity principle is one of the most important historical events regarding wave-particle duality and the measurement problem. A brief examination of complementarity is presented here and elaborated in more detail in Chapter 5.

For Bohr, the issue of whether the statistical Born rule [6] is fundamentally rooted in a physical nondeterminism or whether it is simply a mathematical approximation that is useful to derive a probability or statistical accounting was answered in the following manner. The statistical postulates are not simply useful but are ultimately necessary for an external observer for the following reasons. A physical device is required to make a measurement that necessitates interactions between system and device. This interaction is in-principle uncontrollable and thereby changes one classical variable at the expense of measuring its conjugate because of the uncertainty principle. That is, an external observer ultimately loses the ability in the process of measurement to specify the variable conjugate to the variable being measured. Hence for the external observer there will always be some incomplete knowledge in terms of specifying all measurable classical states of the system.

Bohr believed that *only* a probabilistic description is possible for the external observer in terms of a description of all classical states: no wave function evolution equation can be known to the external observer that uniquely determines the outcome of an experiment when the measured variable is conjugate to an already known variable. It does not matter whether the conjugate variable deterministically evolves with incomplete knowledge afforded to the external observer or is fundamentally nondeterministic. In either case, it is impossible for the external observer to predict with certainty the conjugate variable.

Whether 1) additional variables that are hidden from the external observer and a statistical description is required as proposed by Dirac, or 2) there is a fundamentally nondeterministic substratum, would largely become issues that did not appear to be answered by any experiment at the time of the Copenhagen interpretation. Born states at the end of his 1926 paper [6, p. 54]:

> *Ought we hope later to discover such properties (like phases of the internal atomic motions) and determine them in individual cases? Or ought we to believe that the agreement of theory and experiment—as to the impossibility of prescribing conditions for a causal evolution—is a pre-established harmony founded on the nonexistence of such conditions? I myself am inclined to give up determination in the world of atoms. But that is a philosophical question for which physical arguments are not decisive.*

Many have read these original papers and believe that the wave-particle debate is fully resolved by settling for just such a duality—that is, one can consider the phenomenon as a wave when the momentum becomes known and as a particle when the position becomes known. At this point one might tend to believe that the problem should be relegated to strictly philosophical efforts. In fact, shortly after the Copenhagen interpretation, Bohr and others appeared to have largely abandoned any efforts to model the interactions in a manner to determine both a classical variable and its conjugate.

At this stage, one may be rather content with the explanation of wave-particle duality afforded by complementarity. Suppose that the Copenhagen interpretation is correct—there is no deterministic manner to predict simultaneously a variable and its conjugate. Does this logically imply that the measurement problem is philosophical as stated by Born? Such an inference makes inductive common sense but does not follow to the deductive reasoner. In fact, later in life, Born stated [10, p. 265]:

> *"I should like only to say this: the determinism of classical physics turns out to be an illusion, created by overrating mathematico-logical concepts. It is an idol, not an ideal in scientific research and cannot, therefore, be used as an objection to the essentially nondeterministic statistical interpretation of quantum mechanics."*
>
> M. Born, Nobel Prize address, Stockholm, Sweden. ©1954 The Nobel Foundation

The issue of whether or not the problem is philosophical cannot be said to have been resolved conclusively from the issues of wave-particle duality examined so far. For that we need to dig deeper and examine the measurement problem in a different light—that of entanglement. We begin with Schrödinger's cat.

Schrödinger's Cat

In the discussion of Bohr's complementarity, Heisenberg's uncertainty principle was utilized by Bohr to show that there is always an uncontrollable uncertainty that results during measurement that broadens one variable at the expense of its conjugate, such as simultaneous measurements of position and momentum. However, there is substantially more to the measurement problem than the uncertainty principle. Understanding the entanglement predictions of Schrödinger's equation is necessary to fully comprehend the measurement problem.

Einstein was initially enthusiastic regarding the discovery of Schrödinger's equation, as he did not believe in nondeterminism. However, Schrödinger's equation can evolve product states to become entangled states which, as elaborated further in Chapter 5, Einstein similarly disliked. Einstein formulated a series of thought-experiments to illustrate that Schrödinger's equation did not appear to be the full story or that the theory was incomplete, which were sent to Schrödinger in a letter in 1935. This was originally posed as a thought-experiment in which gunpowder will spontaneously explode over a one-year period, but the time when it blows up is at an unknown random time. Furthermore, when it does blow up the explosion is so fast that one can assume the material at any time is either stable gunpowder or the state of the remnants after the explosion. Einstein was concerned that Schrödinger's equation would predict the state to be in a superposition of stable gunpowder and explosion before the one-year period. However, Einstein believed that the reality is that the actual state of the material is at any time either gunpowder or the remnants of the explosion.

Later that same year, Schrödinger published a thought-experiment similar to Einstein's in which a cat is in a box that contains a radioactive source, a Geiger counter detector, a hammer and a flask of poisonous hydrocyanic acid. When the radioactive source randomly emits a particle and the Geiger counter detects the emission, it causes a hammer to fall and break the flask, releasing the poison and killing the cat. If an observer does not open the box (assumed for times less than the mean lifetime for which the radioactive source typically takes, on average, to emit a particle), then one is forced to assume that the state of the cat inside the box is a state of a cat both alive and dead. This is further illustrated in Figure 1.15. One would have to assume by the postulates of quantum mechanics that until the box is open and the contents seen, or measured by an external observer, that the cat is in a superposition.

Figure 1.15: Schrödinger's cat, all times assumed less than the mean lifetime of the radioactive source (a) radioactive source off, cat alive, (b) radioactive source on, existence of quantum wave function, (c) quantum wave function evolves with the possibility that the hammer will fall and (d) the possibility that the hammer will break the flask causing a dead cat, but the possibility that the cat is still alive cannot be ruled out by Schrödinger's equation alone.

The issue that Einstein was most concerned with in later life is that one would not be able to predict using Schrödinger's equation when the source emits a particle that is detected by the Geiger counter, unless there are further hidden variables and a more detailed theory that exists to explain this. However, consider using the theory as proposed in the Copenhagen interpretation for which there are only the two postulates of 1) Schrödinger evolution or 2) measurement as described by Born's rule. Neither of these modes contains any such hidden variables *prima facie*, and it leads one to the conclusion that the cat is both alive and dead simultaneously. Einstein rejected this and believed the theory must therefore be incomplete.

To make matters worse for Einstein, Bohr rejected the notion that such hidden variables could ever be known to an observer, and that therefore a statistical or nondeterministic accounting was all that could ever be known in the epistemological sense. This is an important issue in a thorough understanding of Bohr's complementary principle, and is further discussed in Chapter 5. The utter insistence by Bohr of the absolute necessity of the need to resort to a statistical description could be argued to have created a dilemma for Einstein and one can begin to see why. For had Bohr accepted that there might eventually be discovered a more complete model for which entangled states did not occur, rather than the *only* possibility being a nondeterministic accounting after a measurement is made, Einstein perhaps would have rested easier knowing that such a research program would continue to uncover such a full theory. But it was not to be: Bohr never wavered on the impossibility of generally being able to predict the wave function deterministically through all stages of measurement.

Schrödinger coined the term "entanglement" and this is the primary resource currently being investigated in the theory of Quantum Computing and Quantum Information. Schrödinger took Einstein's side on the lack of completeness of quantum mechanics in disagreement with Bohr and tried in later life to develop models for which entanglement of macroscopic objects would not occur, but did not succeed.

CHAPTER 2

Characteristics of Unitary Evolution

Interference, reversibility, and entanglement are characteristics that are useful to consider toward understanding the manner in which unitary evolution differs from measurement.

Fundamental Concepts

As seen from Equation (1.1), a quantum system is characterized by the system's Hamiltonian, which is the key part of Schrödinger's equation that determines how the wave function evolves in time. The resulting time development is a deterministic *unitary* evolution and it has very particular characteristics. The Hamiltonian represents the energies of the system and its interactions and the unitary evolution is determined by these energies.

The unitary operator U corresponding to the Hamiltonian is just given by:

$$U(t) = \exp(-iHt/\hbar). \tag{2.1}$$

As discussed further in the theoretical development, Equation (2.1) is a result of *Stone's Theorem* [11], and the form of U can also be related to any symmetries that the system possesses (e.g., if the system doesn't change by rotations or translations in space) by *Wigner's Theorem* [12]. Quantum theory depicts the states of the world as evolving linearly according to a unitary operator U so that the state at a time t is given by:

$$|\psi(t)\rangle = U|\psi(0)\rangle. \tag{2.2}$$

These states are probability amplitudes made up of a space of complex unit vectors $|\psi\rangle$ (called *a Hilbert space,* a complete vector space with an inner product). Such states $|\psi\rangle$ in Hilbert space are also referred to as pure states. The imaginary number i in Equation (2.1) alerts us that quantum mechanics requires complex numbers in contrast to classical mechanics.

There can exist additional quantum states of nature that are outside the class of pure states. Quantum theory developed by von Neumann [13] included this generalization of pure states to include the set of *mixed states* [14, p. 100]. A mixed state has no representation in the set of pure states $|\psi\rangle$ in the representative Hilbert space and one must resort to the extension of pure states to the set of density matrices. A pure state $|\psi\rangle$ can also be represented by a density matrix defined by $\varrho \equiv |\psi\rangle\langle\psi|$. A mixed state is a positive density matrix for which $\mathrm{Tr}(\varrho^2) < 1$, whereas for a pure state $\mathrm{Tr}(\varrho^2) = 1$. Gleason's theorem [15] demonstrates that if the Hilbert space dimension is greater than 2, then the general expression for the expectation value of an operator A is: $\langle A\rangle = tr[A\rho]$, where ρ is the density operator. This implies that density operators are the most general description of quantum mechanical states.

The effect of a unitary operation does not depend on the state $|\psi\rangle$ to which it applies. This is a consequence of the linearity of the evolution. Quantum state unitary evolution is linear in wave function (in Chapter 7, we will also encounter quantum state evolution that is linear in density operator) because the operation satisfies:

$$U(a|\psi\rangle + b|\phi\rangle) = aU|\psi\rangle + bU|\phi\rangle. \tag{2.3}$$

This is the origin of the superposition principle in quantum mechanics, so that the linear superposition of two wave functions is also a legitimate wave function. As discussed in Chapter 1, the superposition principle in conjunction with Born's rule is the basis for understanding the double-slit experiments, Figure 1.4.

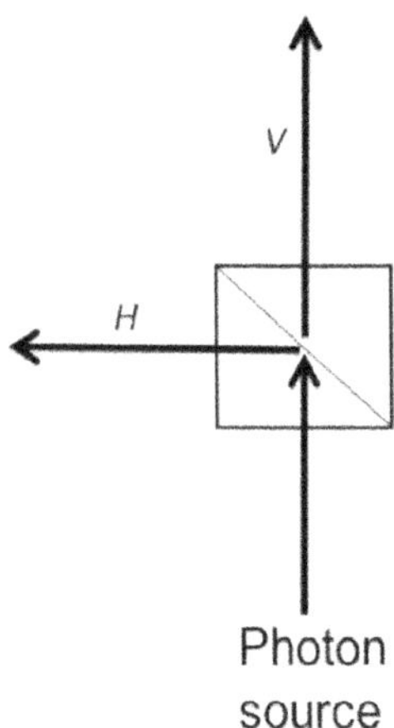

Figure 2.1: Single photon sent into a 50-50 polarizing beam splitter which allows horizontal (H) or vertical (V) polarization.

Interference

To examine the concept of interference, we consider what is known as the polarizing beam splitter for single-photons. Photons are straightforwardly generated and controlled with linear optical elements such as wave plates and beam splitters so they are convenient to use for encoding a quantum system that consists of two levels called a quantum bit or *qubit*, and as well to implement unitary transformations. A classical bit has a state of either *0* or *1*. A qubit also has two possible states $|0\rangle$ and $|1\rangle$; however, it differs from a classical bit in that it can be in states other than $|0\rangle$ and $|1\rangle$ such as the linear superposition $a|0\rangle + b|1\rangle$. This quantum property enables qubits to interfere.

A polarization qubit uses two orthogonal polarization states; e.g., horizontal polarization with the logical value H and vertical polarization with value V. A superposition of these can then be used to represent an arbitrary polarization. A coherent superposition of horizontal and vertical polarization represents a 45-degree diagonal polarization, and it will be transmitted with 100% probability by a polarizer set at that angle, Figure 2.1. However, an incoherent mixture will be transmitted with only 50% probability.

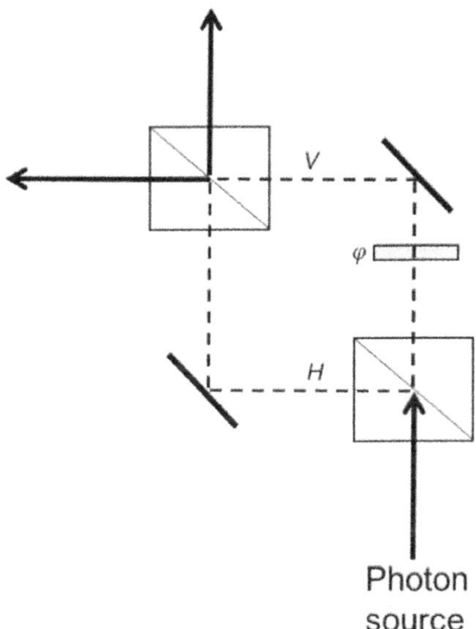

Figure 2.2: Single photon sent into an interferometer with a 50-50 polarizing beam splitter, a phase shifter, and a second beam splitter to allow interference. The mirrors are assumed sufficiently massive that there is little observable mirror recoil due to the reflection of a photon.

A polarization beam splitter, consisting of a semi-reflective mirror, transmits one polarization mode and reflects the other. A polarization rotation can be physically implemented using quarter and half-wave optical plates. For a 50-50 beam splitter with $\theta = 45$ degrees, we can use the general unitary operator for rotating a qubit given in Equation (2.16) to generate linear polarization along ±45 degrees with $\hat{n} \cdot \bar{\sigma} = \sigma_x$,

$$|+45\rangle = (|H\rangle + |V\rangle)/\sqrt{2}, |-45\rangle = (|H\rangle - |V\rangle)/\sqrt{2}. \quad (2.4)$$

Or with $\hat{n} \cdot \bar{\sigma} = \sigma_y$, generate left ($L$) and right ($R$) handed *circular polarization*

$$|L\rangle = (|H\rangle + i|V\rangle)/\sqrt{2}, |R\rangle = (|H\rangle - i|V\rangle)/\sqrt{2}. \quad (2.5)$$

A second beam splitter can also be added, along with a phase-shifter to vary the effective path-length to form a Mach-Zehnder interferometer, allowing the beams to recombine and interfere, Figure 2.2. In this case, a single photon *interferes with itself* or more precisely with the two paths the photon can take to the detector. Detectors are added to the polarizing beam splitter arrangement to make a measurement. However, photons cannot be split, therefore only one detector or the other will *click* as the photon is destroyed during the measurement process, Figure 2.3.

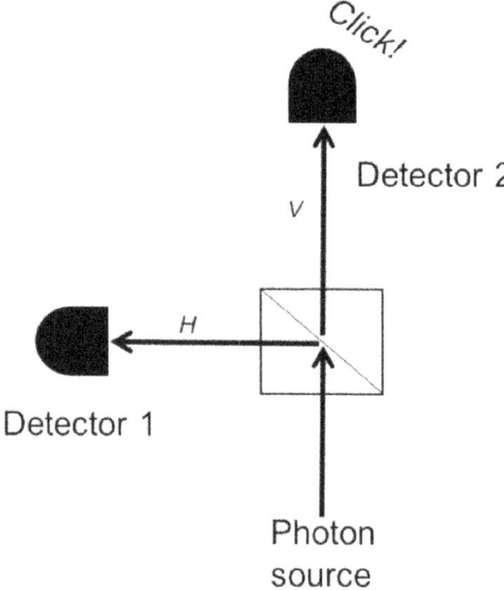

Figure 2.3: Single photon sent into a 50-50 polarizing beam splitter with click detectors that will register either horizontal (H) or vertical (V) polarization.

Reversibility

An operator that is unitary is also completely reversible. That is, suppose that at some initial time t_1 the state of the system is $|\psi(t_1)\rangle$ and a unitary operator U is found that evolves the state to $|\psi(t_2)\rangle = U|\psi(t_1)\rangle$. Then there is another unitary operation that can evolve $|\psi(t_2)\rangle$ to $|\psi(t_1)\rangle$ and completely undo the effect of the initial U. In fact, mathematically, this second unitary operation is found by the Hermitian conjugate $U^\dagger$

(found by taking the transpose and complex conjugate $U^{\dagger} \equiv (U^{*})^{T}$and which is equal to its inverse $U^{\dagger} = U^{-1}$). That is, $|\psi(t_1)\rangle = U^{\dagger}|\psi(t_2)\rangle$.

It will be shown later in Equation (2.11) of the theoretical development, that unitary evolution can be decomposed in the simple form of

$$|\psi(t)\rangle = \sum_{k} a_k(t)|\phi_k\rangle$$

where $\sum_k |a_k(t)|^2 = 1$ and for which $|\phi_k\rangle$ are fixed or stationary states called eigenstates of the Hamiltonian. The coefficients can be represented by

$$a_k(t) = a_k(0)\exp(-iE_k t/\hbar).$$

This shows that during unitary evolution, each eigenstate $|\phi_k\rangle$ undergoes a steady phase rotation that is linear in time, $\theta_k(t) = E_k t/\hbar$, with a rate proportional to the energy E_k. Therefore, unitary time evolution reduces to nothing more than simple linear progression in time, comparable to the inexorable ticking gears of a clock, Figure 2.4. Unitary evolution is akin to a simple reversible mechanical process. This characteristic steady phase rotation also results in the additional unitary properties of reversibility and preservation of information (i.e., entropy) during evolution, which

Figure 2.4: Unitary evolution reduces to the simple linear progression in time of the energy eigenstate phases comparable to the inexorable ticking gears of a clock.

will be further elaborated on in the theoretical development. Could it be possible that the click of the detector in Figure 2.3 can be described by a unitary transformation, despite its apparent lack of resemblance to the inexorable ticking gears of a clock? In Chapter 3, we will discuss the distinguishability between unitaries and measurement.

Entanglement

Multi-particle entanglement is a process of interference that occurs when two or more particles exist in a superposition. Additionally, a single particle with multiple entangled degrees of freedom also can constitute an entangled state. For example, entanglement can be created within a single particle via both the particle's internal degrees of freedom and motional degrees of freedom. We will now further illustrate examples of entangled multi-particle states.

Entanglement via Mirror Recoil

A method that can be used to create an entangled multiparticle superposition is to utilize the beam splitter and the mirrors in Figure 2.2. There are two variations on Figure 2.2 that we will consider.

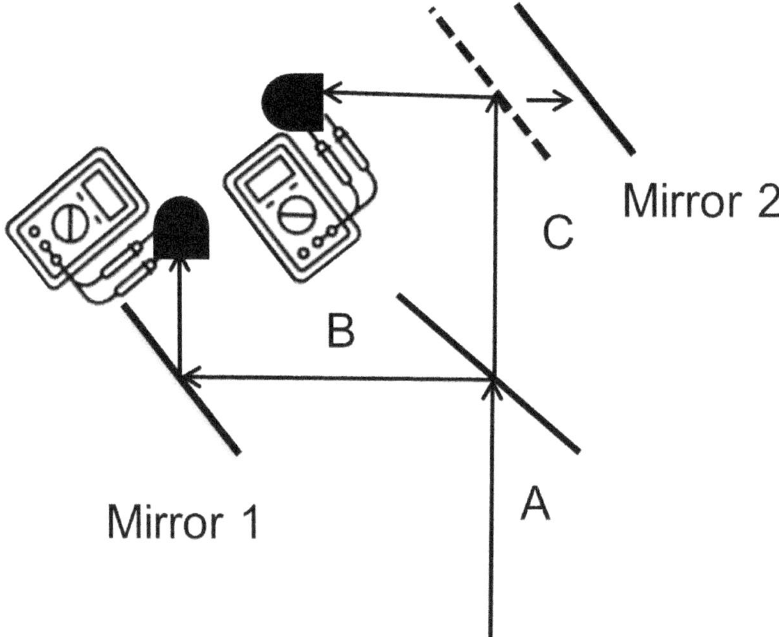

Figure 2.5: Single photon sent into a configuration of mirrors with momentum recoil of a mirror resulting in superpositions of its position.

In the first variation on creating an entangled multiparticle superposition, one could replace the heavy mirrors with a very light microscopic atomic mirror that significantly recoils when a photon is reflected by the mirror for which the situation that unitarily results is a superposition of the mirror position, as seen in Figure 2.5. Let the initial resting state of a mirror be denoted by $|\text{rest}\rangle$ and the recoiled state by $|\text{recoil}\rangle$. After being reflected by the mirrors, the state of the mirrors and the photon in

Paths B and C is given by:

$$|\Psi\rangle = \frac{1}{\sqrt{2}}\big(|\text{rest}\rangle_{\text{Mirror } B}|\text{recoil}\rangle_{\text{Mirror } C}|0\rangle_{\text{Photon,B}}|1\rangle_{\text{Photon,C}} + |\text{recoil}\rangle_{\text{Mirror } B}|\text{rest}\rangle_{\text{Mirror } C}|1\rangle_{\text{Photon,B}}|0\rangle_{\text{Photon,C}}\big).$$

As can be seen, the state $|\Psi\rangle$ involves a macroscopic superposition of a mirror. This becomes physically possible as long as the mirror is sufficiently microscopic and light so that the recoiled mirror is distinguishable from the initial state of the mirror at rest.

In the second variation of creating an entangled multiparticle superposition, we replace the output of the B and C ports of the beam splitter by a maximally entangled state of a sufficiently large number of photons that can impart a substantial momentum on not simply a microscopic mirror, but a macroscopic mirror. Such photon states, called NOON states [16] [17], are given by $|\psi\rangle = (|N\rangle|0\rangle + |0\rangle|N\rangle)/\sqrt{2}$. NOON states would be genuinely macroscopic superpositions for $N \to \infty$, though so far, they have been generated in optics or atoms only for low N [18]. When such a state of sufficiently high N impinges on mirrors, even macroscopic mirrors, the final state is entangled:

$$|\Psi\rangle = \frac{1}{\sqrt{2}}\big(|\text{rest}\rangle_{\text{Mirror } B}|\text{recoil}\rangle_{\text{Mirror } C}|0\rangle_{\text{Photon,B}}|N\rangle_{\text{Photon,C}} + |\text{recoil}\rangle_{\text{Mirror } B}|\text{rest}\rangle_{\text{Mirror } C}|N\rangle_{\text{Photon,B}}|0\rangle_{\text{Photon,C}}\big).$$

This approach can clearly be extended indefinitely via *nested interferometry* [19] allowing the superpositions to grow without limit, Figure 2.6.

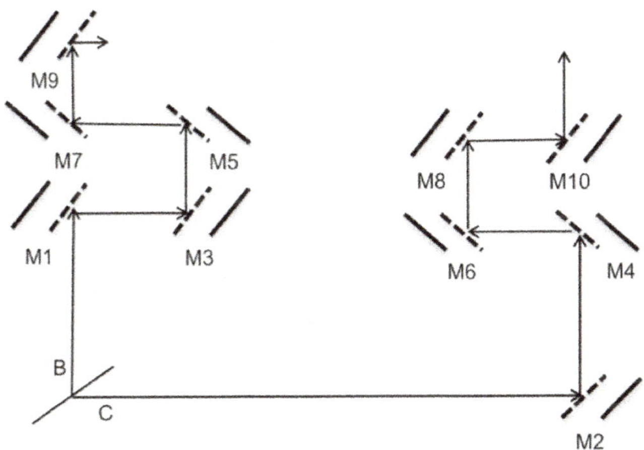

Figure 2.6: The single photon sent within a configuration of recoiling mirrors can clearly be extended indefinitely allowing the superpositions to grow without limit.

Schrödinger's Cat Without Limits

But why stop there, when the configurations of superpositions can even be extended into the physical structure of the eye of the observer [20], a superposition of one that saw the photon and one that did not, and even to the nervous system and further on to the neural correlates of consciousness in the brain, with nothing to prevent superpositions of conscious feelings, a superposition of one that was happy to see the photon and one that was not, Figure 2.7?

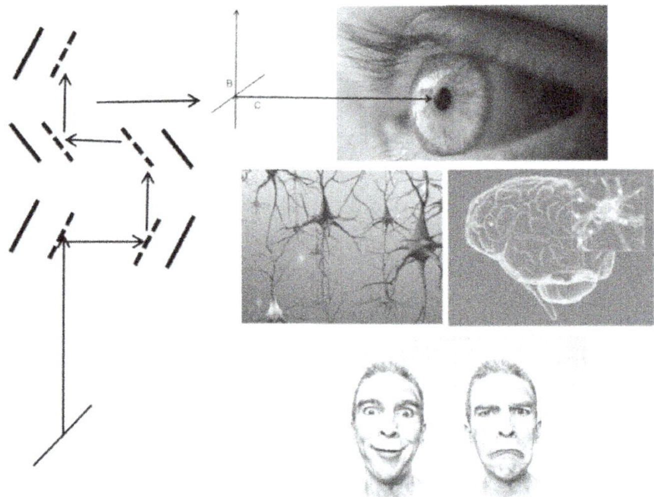

Figure 2.7: But why stop with the mirrors, when the configurations of superpositions can even be extended into the physical structure of the eye of the observer, to the nervous system, to the neural correlates of consciousness in the brain, and with nothing to prevent even superpositions of conscious feelings?

If we consider that there must be a momentum recoil for a system with finite mass, then any wave-packet will experience recoil at sufficient intensity. This would indicate that a mirror can go into superpositions of different locations. In a unitary world, there would be no boundaries to the superposition principle, any object, no matter how large, can go into a superposition of different locations [21]. One can extend the NOON photon state to an entangled matter state $|\psi\rangle = (|N\rangle|0\rangle + |0\rangle|N\rangle)/\sqrt{2}$. Consider that if N is made large enough the state could approach that of a *quantum locomotive*,

$$|\Psi\rangle = \frac{1}{\sqrt{2}}(|\text{Locomotive}\rangle|0\rangle + |0\rangle|\text{Locomotive}\rangle).$$

Such an entangled locomotive state is possible in an all-unitary theory, although it would be uncharted territory from our experience. The state might be represented as in Figure 2.8, for which the entire state of the train is traveling on the left track with the right track unoccupied, in superposition with the train on the right track with the left track unoccupied. No worries if such an extreme macroscopic object even shattered a mirror, as the superpositions unitarily would simply continue to be extended. The magnitude of its quantum of action would overwhelm that of Planck's constant and any associated decoherence would still be operating unitarily. Nothing is violated as long as there is no measurement process. However, there is no such measurement process in a unitary world, not even a single photon entering the eye of an observer allows him to fixate his location unitarily.

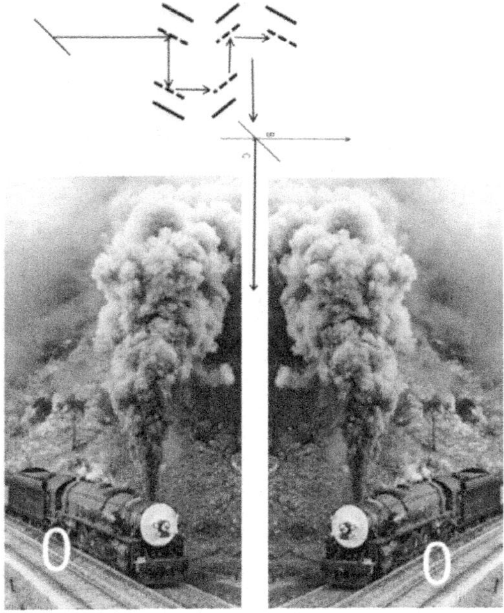

Figure 2.8: A quantum locomotive superposition via two interferometer paths represented by the empty and occupied train tracks, giving the two classical possibilities of the locomotive.

What aspects of the world around us prevent the occurrence of superpositions to qualify it as mesoscopic or macroscopic; e.g., numbers of particles or magnitudes of particular quantities? A flux qubit built from a SQUID (superconducting quantum interference device) supports superpositions of clockwise and counter-clockwise supercurrents involving billions of electrons; however, only at most a few thousand electrons are found to be distinguished when passing from one branch of the superposition to the other [22]. Interferometry has been achieved with macromolecules, beginning with diffraction of fullerenes, and since extended to higher masses in the range $10^3 - 10^{10}$ amu (where 1 amu ≈ mass of the hydrogen atom) [23] [24] [25] [26], though this is several orders of magnitude smaller than has been achieved in ultra-cold atom experiments. By applying optical pulses on a freely falling Bose-Einstein condensate of $\sim 10^5$ ultracold rubidium atoms, a superposition of wave-packets separated by a distance of 54 cm was realized [27] [28] [29]. Perfect silicon-crystal neutron quantum optical experiments have achieved a coherent massive object with a separation of 9.5 cm enclosing an area of 100 cm^2, allowing enough room for a human hand to reach through the two branches [30] [31].

Yet all of these observed phenomena are still consistent with a quantum unitary world. Will new physics appear as these experiments are scaled up still further into the world around us or is our world completely unitary? In Chapter 3, we will develop the requirements for distinguishing unitary from non-unitary processes and the implications for the quantum measurement problem.

Figure 2.9: Have all the aspects of our surroundings emerged unitarily from the initial expansion of the universe at the big bang?

Mona Lisa

A unitary world also leads to the conclusion that all aspects of our surroundings emerged unitarily from the initial evolution at the big bang. However, many aspects of what we observe would be puzzling if this were the case. Observations of the cosmic microwave background (CMB) and our local universe are consistent with a picture that the universe was initially hot and opaque, then cooled over several hundred thousand years until light decoupled, leaving a nearly isotropic CMB that embodies the afterglow of the big bang. An early inflationary phase of accelerating expansion would account for how the current universe can be nearly isotropic if the pieces of what we observe were not in causal contact when the photons were first emitted. The end result is a universe that is spatially flat at an age of $\sim$ 14 billion years since the big bang with an accelerating expansion and consisting not only of atoms (i.e., hydrogen and helium) but also cold *dark matter* and *dark energy* [32]. The early conditions of the universe on large cosmological scales some hundred thousand years after the big bang are revealed in the measured temperature distribution of the nearly uniform CMB [33]. The structure of the universe is much more complex and diverse at the smaller spatial scales we see all around us. Nevertheless, a corresponding unitary description of the universe would be described by a wave function $|\Psi\rangle$ governed by the Schrödinger equation even if the resultant probabilities for many events were not high [34].

Applied to the universe as whole, a *no-boundary* initial condition of the universe was proposed by Hartle and Hawking [35]. However, would all the delicate balances of the initial phases of the wave function of the universe remain intact to account for the world we see around us, without exceptions, not even for the seemingly historical accidents and multitude of developments streaming from our civilizations, including the inevitability of even the Mona Lisa, and Leonardo and Leonardo's mother, and the silk merchant who commissioned the Mona Lisa, etc., Figure 2.9 [36]? Unitarity would then even have led to all of the theories of the quantum measurement problem reviewed in Chapter 4, some of which predict that this same universe is non-unitary. However, these are high-level products in the hierarchy of complexity of our cultures and suggest a purposeful design and possibly that intelligence with the capability of non-unitary intention has appeared. Or alternatively, can all of this complexity be mimicked by unitary processes even if only in an emergent manner [34]? These questions form the main theme of this book.

The reader should now be familiar with the concepts of interference, reversibility, and entanglement that are inherent in unitary evolution. The reader with a less technical background should now proceed to the Introduction of Chapter 3. Although a detailed understanding of the mathematics of Chapter 3 is not a prerequisite for continuing to Chapter 4, the general results in Chapter 3 should be understood. After becoming familiar with the Introduction of Chapter 3, the less technical reader should proceed through the majority of the remaining material in Chapters 4, 5, 6, and 7, while omitting detailed technical aspects found within one section or another, which are generally not required to proceed through the remainder of the book.

Theoretical Development

Evolution in the Eigenstate Basis

An operator is unitary if its Hermitian conjugate $U^\dagger$ (found by taking the transpose and complex conjugate $U^\dagger \equiv (U^*)^T$) is equal to its inverse $U^\dagger = U^{-1}$, or

$$U^\dagger U = UU^\dagger = I \tag{2.6}$$

As a consequence, the eigenvalues of U have unit norm and take the form $\lambda_j = \exp(i\varphi_j)$. Note that if U is unitary, then so is $U^\dagger$. If U_1 and U_2 are unitaries, then so is the product $U = U_1 U_2$. The invertibility of U implies that information is preserved during the evolution. Equation (2.6) implies that the inner product of states is preserved under unitary evolution, i.e., if $|\acute{\psi}\rangle = U|\psi\rangle$ and $|\acute{\varphi}\rangle = U|\varphi\rangle$, then $\langle\acute{\psi}|\acute{\varphi}\rangle = \langle\psi|U^\dagger U|\varphi\rangle = \langle\psi|\varphi\rangle$. In particular, this implies that the net probability of finding the particle (denoted by $p(t)$) is always preserved for all time under unitary evolution.

That is,

$$p(t) = \langle\psi(t)|\psi(t)\rangle = \langle\psi(0)|U^{\dagger}U|\psi(0)\rangle = \langle\psi(0)|\psi(0)\rangle = p(0). \tag{2.7}$$

Note that Equation (2.1) implies that $U^{\dagger}(t) = U(-t)$ so that the unitary operator $U^{\dagger}$ simply reverses the time evolution from U and thus it is a time-reversible transformation. We can also see that Equations (2.1) and (2.2) are equivalent to the time-dependent Schrödinger equation since

$$i\hbar\frac{\partial U|\psi(0)\rangle}{\partial t} = HU|\psi(0)\rangle$$

or

$$i\hbar\frac{\partial|\psi(t)\rangle}{\partial t} = H|\psi(t)\rangle. \tag{2.8}$$

This is the quantum mechanical equation that has long been considered relevant for *closed* or isolated systems. In the Heisenberg representation, the time-dependence is unitarily transferred from the wave function $|\psi(t)\rangle$ to the operators of the problem such as G, which obeys

$$\frac{dG}{dt} = -\frac{i}{\hbar}[G, H]. \tag{2.9}$$

In finite dimensions and when the Hamiltonian is time-independent, the time evolution unitary operator in Equation (2.1) can be decomposed into a series of simpler pieces (this is allowed by the *spectral theorem* [37, p. 221]) as:

$$U = \sum_k e^{iE_k t}|\phi_k\rangle\langle\phi_k| \tag{2.10}$$

for any time t. The spectral theorem in infinite dimensions becomes subtler and a rigorous operational description can be problematic [37]. This simplification happens by using new states called energy eigenstates $|\phi_k\rangle$ of the Hamiltonian which have the property $H|\phi_k\rangle = E_k|\phi_k\rangle$, and the constant E_k is called the energy eigenvalue. If the state of the system is written in this eigenstate basis

$$|\psi(t)\rangle = \sum_k a_k(t)|\phi_k\rangle \tag{2.11}$$

where $\sum_k |a_k(t)|^2 = 1$, then from Equations (2.8) and (2.11), we can conclude that (see Exercise 2.1)

$$a_k(t) = a_k(0)\exp(-iE_k t/\hbar). \tag{2.12}$$

The unitary evolution is determined by the total Hamiltonian which may describe several subsystems and their mutual interactions. If the subsystems are not interacting, then the total unitary operator is simply a product of unitary operators. For example, with two non-interacting subsystems A and B, the unitary operator of the total system would be a product of local unitary operators, $U_{AB} = U_A U_B$. In this case, the subsystems evolve independently and it would be possible to average over or *trace out* the degrees of freedom of system B and still have subsystem A described by a unitary evolution. However, even in the presence of interactions, there are cases where the tracing out of subsystem degrees of freedom can still leave the evolution of the remaining subsystems unitary but instead it becomes governed by a different *effective Hamiltonian* instead of the original Hamiltonian in Equation (2.1). For example, this can occur when the timescales of different subsystems are quite disparate and the technique of *adiabatic elimination* [38] [39] can be used to eliminate the fast degrees of freedom which are adiabatically following the evolution of the slow degrees of freedom. This results in an effective Hamiltonian H_{eff} which accurately describes the unitary evolution, $U_{eff} = \exp(-iH_{eff}\mathrm{t})$, of the remaining slow degrees of freedom in this adiabatic regime. In other cases, it may be possible to exploit symmetries in the Hamiltonian to enable construction of a special unitary transformation $\widetilde{U}$ to transform the system into an effective Hamiltonian, $H_{eff} = \widetilde{U}\, H\widetilde{U}^{\dagger}$. This is yet another role for unitary transformations in quantum mechanics and is once again due to the presence of symmetries. An example is given in Exercise 2.12. The quantum evolution for more general cases when there are interactions between the subsystems is discussed in a later section.

Unitary Interference Operations

Consider a simple unitary transformation which rotates any single qubit state by the angle θ using the basis of the two orthogonal states, $|0\rangle$ and $|1\rangle$ and written as the following 2x2 matrix (Exercise 2.2):

$$U_\theta = \begin{pmatrix} \cos\theta & -\sin\theta \\ \sin\theta & \cos\theta \end{pmatrix}. \tag{2.13}$$

For $\theta = 45$ degrees, $\sin\theta = \cos\theta = 1/\sqrt{2}$, and this will rotate the original basis into another orthonormal basis given by the coherent superpositions of $|0\rangle$ and $|1\rangle$: $|+\rangle \equiv (|0\rangle + |1\rangle)/\sqrt{2}$ and $|-\rangle \equiv (|0\rangle - |1\rangle)/\sqrt{2}$,

$$U_\theta|0\rangle = |-\rangle = (|0\rangle - |1\rangle)/\sqrt{2}\,, U_\theta|1\rangle = |+\rangle = (|0\rangle + |1\rangle)/\sqrt{2}. \tag{2.14}$$

Thus, unitaries can create coherent superpositions between states. But they can also destroy coherence by destructive interference. Applying U_θ once more to Equation (2.14) cancels the coherence:

$$U_\theta U_\theta|0\rangle = -|1\rangle, U_\theta U_\theta|1\rangle = |0\rangle. \tag{2.15}$$

In fact, a unitary operator can be viewed as simply equivalent to a change of basis, since if $\{|k\rangle\}$ is an orthonormal basis, then so is $\{U|k\rangle\}$, because $(U|l\rangle)^{\dagger}U|k\rangle = \langle l|U^{\dagger}U|k\rangle = \langle l|k\rangle = \delta_{lk}$, which is the requirement for orthonormality.

For a single qubit, the most general unitary operator can be written in terms of the Pauli matrices σ_i as

$$U_{\theta} = (\cos\theta\, I \;+ i\sin\theta\;\hat{n}\cdot\bar{\sigma}\,). \tag{2.16}$$

This represents a rotation by angle θ around the direction of the unit vector $\hat{n}$ and generalizes Equation (2.13) as a convenient way for unitary manipulation of photon polarization modes.

The probability distribution for interference experiments such as the double-slit experiment is not simply the sum of the two distributions measured when only one slit is open since there is a third term from the interference between the two. Sorkin [40] has demonstrated a remarkable property of quantum mechanics that follows from this if we consider the more complicated situations of three or more slits. He showed that the probability distribution for multiple slits can always be written as sums of the possible cases where only one or two of the slits are open so that there is no interference exclusive to having more than two slits. In other words, quantum superposition and Born's rule imply that quantum interference always occurs from pairs of possible paths through the slits, and any deviation from this would indicate a failure of Born's rule [41]. The measured properties of quantum mechanics all stem from the interference of pairs of possibilities.

Mathematics of Quantum Entanglement

Consider the state ρ_{AB} of a composite system comprising just two subsystems A and B. The simplest case is a product state, $|\psi_{AB}\rangle = |\psi_A\rangle\otimes|\psi_B\rangle$ or $\rho_{AB} = |\psi_A\rangle\langle\psi_A|\otimes|\psi_B\rangle\langle\psi_B|$, which is separable or *non-entangled.* Note that in this section, the states are described explicitly using the notation of tensor products. The possibility of entanglement enters because quantum states can also be in superpositions. A simple example of a quantum entangled state is the superposition state of two qubits A and B

$$|\psi_{AB}\rangle = \tfrac{1}{\sqrt{2}}(|0_A\rangle\otimes|0_B\rangle + |1_A\rangle\otimes|1_B\rangle) \tag{2.17}$$

$$\begin{aligned}\rho_{AB} &= |\psi_{AB}\rangle\langle\psi_{AB}| \\ &= \frac{1}{2}(|0_A\rangle\langle 0_A|\otimes|0_B\rangle\langle 0_B| + |1_A\rangle\langle 1_A|\otimes|1_B\rangle\langle 1_B| \\ &+ |0_A\rangle\langle 1_A|\otimes|0_B\rangle\langle 1_B| + |1_A\rangle\langle 0_A|\otimes|1_B\rangle\langle 0_B|).\end{aligned} \tag{2.18}$$

The entanglement enters from the cross-terms $|0_A\rangle\langle 1_A|\otimes|0_B\rangle\langle 1_B| + |1_A\rangle\langle 0_A|\otimes|1_B\rangle\langle 0_B|$. Otherwise the state would be separable. More generally, a separable state takes the form of a mixture of product states and it is *entangled* if it *cannot* be written as the separable mixture of product states of the form [42]

$$\rho_{AB}^{sep} = \sum_k p_k \rho_A^k \otimes \rho_B^k = \sum_k p_k |\psi_A^k\rangle\langle\psi_A^k| \otimes |\psi_B^k\rangle\langle\psi_B^k|. \tag{2.19}$$

where $\sum_k p_k = 1$. The physical meaning of this is that it should not be possible to create entangled states just by acting on one of the subsystems individually. For example, if *local* unitaries U_A and U_B separately act on A and B respectively, no entanglement can be created,

$$\acute{\rho} = (U_A \otimes U_B)\rho(U_A \otimes U_B)^\dagger. \tag{2.20}$$

Note that the concept of a state being entangled is defined in terms of what it is not, namely Equation (2.19). This is due to the difficulty of specifying the most general entangled state. However, entanglement measures have been developed that can be applied to a range of quantum states. There are several criteria for a proper measure $E[\rho]$ for quantifying the amount of entanglement in a quantum state and several different measures have been proposed [43]. A proper measure of quantum entanglement should fulfill the following requirements [44]:

(E1) For any separable state, $E[\rho] = 0$.

(E2) The entanglement remains unchanged under local unitaries, $E[(U_A \otimes U_B)\rho(U_A \otimes U_B)^\dagger] = E[\rho]$.

(E3) Local operations and classical communication (LOCC) cannot increase the expected entanglement.

(E4) Entanglement of a pure state is equal to the von Neumann entropy of the subsystem state.

As an example, consider the two-qubit entangled pure state Equation (2.17) with density operator $\rho_{AB} = |\psi_{AB}\rangle\langle\psi_{AB}|$ and use criterion (E4) to evaluate its entanglement. The subsystem state is a simple diagonal matrix with von Neumann entropy evaluated using Equation (2.38) as (Exercise 2.3)

$$S_A = -\mathrm{Tr}(\rho_A \log \rho_A) = \log 2. \tag{2.21}$$

Therefore, the state in Equation (2.17) is maximally entangled with $E[\rho_{AB}] = S_A = \log 2$. In the literature, the von Neumann entropy is sometimes defined using the logarithm to base 2 instead, so that the maximum entanglement for two qubits is $E[\rho_{AB}] = 1$. Another entanglement measure for two qubits is the *concurrence*, which is a proper measure for both pure and mixed states. This is discussed and used in Chapter 3 as part of the test for distinguishing unitary from non-unitary processes.

Theorems of Wigner and Stone

Eugene Wigner [12] [45, p. 71] proved that all operations in quantum mechanics corresponding to symmetries must be either unitary U (e.g., rotations, translations, reflections) or anti-unitary UK (e.g., time-reversal), where K denotes complex conjugation and that these preserve probabilities. Therefore, the new states after a symmetry operation will obey the same laws of nature as the original states. The symmetry of time-translation corresponds to unitary time-evolution. *Wigner's Theorem* allows us to speak as if the pure states $|\psi\rangle$ of quantum mechanics live in Hilbert space. Technically, it is the entire set $\{e^{i\alpha}|\psi\rangle\}$ (called a *ray*) for all real phases α that describes one physical state since the phases do not affect probabilities or expectation values. However, Wigner's result means that all mappings between rays that preserve probabilities can be accomplished by either unitary or anti-unitary mappings between vectors in Hilbert space. That most of the physically significant transformations in quantum mechanics are unitary is due to Wigner's theorem.

A particular system and its interactions will determine the specific form for U. This can often be facilitated by *Stone's Theorem* [11] [37, p. 264] which implies that for every (strongly continuous) unitary operator U, there is a Hermitian operator H (so that $H = H^{\dagger}$), called the Hamiltonian, such that

$$U(t) = \exp(-iHt/\hbar). \tag{2.22}$$

The Hamiltonian represents the energies of the system and its interactions and the unitary evolution is determined by these energies. This implements the time-translation invariance symmetry of Wigner's theorem—the laws of nature do not depend on how our clocks are set. Operating with the unitary operator of Equation (2.22) simply shifts the time (Exercise 2.1)

$$U(\tau)|\psi(t)\rangle = |\psi(t+\tau)\rangle. \tag{2.23}$$

The time translation symmetry corresponds to conservation of energy, since $UHU^{\dagger} = H$. Similarly, if the laws of nature are independent of a shift in spatial translation (i.e. where we mark the origin of our coordinate system), there must exist an equivalent unitary operation. This corresponds to conservation of linear momentum, with the spatial translation unitary operator given by

$$U_P(\bar{x}) = \exp(-i\bar{P}\cdot\bar{x}), \tag{2.24}$$

and which shifts the spatial location according to (Exercise 2.1):

$$U_P(\bar{a})|\psi(\bar{x},t)\rangle = |\psi(\bar{x}+\bar{a},t)\rangle. \tag{2.25}$$

For each symmetry, there is a unitary operator to generate the corresponding transformation in terms of the complementary canonical coordinate [45, p. 69]. Symmetries can be *continuous* (rotations, translations) or *discrete* (time reversal, space-inversion or parity). The unitary operators for continuous symmetries differ infinitesimally from the identity by a Hermitian operator G called the *generator* of the corresponding symmetry

$$U = I - i\frac{\varepsilon}{\hbar}G + O(\varepsilon^2). \tag{2.26}$$

If the symmetry is conserved by the time evolution so that $UHU^{\dagger} = H$, then we can conclude from Equation (2.26) that:

$$[G, H] = 0 \tag{2.27}$$

where the *commutator* is given by $[G, H] \equiv GH - HG$. Therefore, conservation laws can be related to commutation relations of the Hamiltonian (Exercise 2.5).

In the Heisenberg representation, the time-dependence is unitarily transferred from the Schrödinger representation wave function $|\psi(t)\rangle$ to the operators of the problem such as G, which obeys

$$\frac{dG}{dt} = -\frac{i}{\hbar}[G, H]. \tag{2.28}$$

If G is a constant of the motion, then we can conclude from Equations (2.27) and (2.28) that $dG/dt = 0$. For example, if H is invariant under spatial translation, then $[P, H] = 0$ and, therefore, the momentum P is a constant of the motion, $dP/dt = 0$.

Quantum Evolution with Subsystem Interactions

In the presence of interactions between subsystems, when the degrees of freedom of some of the subsystems are traced out, the evolution of the remaining set of subsystems generally will not remain unitary, but instead becomes described by a non-unitary stochastic equation. This is the problem of open systems, which is discussed further in Chapter 7. It occurs, for example, when the system of interest may be coupled to an environment that cannot be easily described in detail. When the degrees of freedom of the environment are traced out, the unitary evolution becomes modified by stochastic terms representing the influence of the environment. An open quantum system suffers noise and decoherence due to interaction with its surroundings. This leads to a type of evolution $\mathcal{E}(\rho)$ more general than unitary evolution. The requirements on $\mathcal{E}(\rho)$ are that it be trace-preserving (since we must have $\mathrm{Tr}\rho = 1$) and *completely positive* (CP), meaning that it remains positive even when we extend the map as $\mathcal{E} \otimes I$ and apply it to a larger composite system by appending an *ancilla* environment E to our quantum system. This is required because the quantum system of interest must interact with other physical systems as necessary for a complete description of the world, and the evolution of the interacting system must still remain physically valid. However, even these more general maps $\mathcal{E}(\rho)$ can be described by unitary transformations in terms of an ancilla environment as long as it is not initially correlated with the quantum system. This is a result of *dilation theorems* due to

Naimark [46] and Stinespring [47] [48]. If E is initially in the state $|0\rangle$, a unitary operator U on the composite system can always be found so that the CP map $\mathcal{E}(\rho)$ takes the following form for all ρ,

$$\mathcal{E}(\rho) = \mathrm{Tr}_E U(\rho \otimes |0\rangle\langle 0|) U^\dagger. \tag{2.29}$$

The devotion to describing irreversible processes in terms of reversible unitary operators of a larger system has been called going to the Church of the Higher Hilbert Space. One of the questions of the Quantum Measurement Problem is whether or not Nature is a devotee to this creed and is always able take refuge in the Church of the Higher Hilbert Space.

A more general description of CP maps is the *operator-sum representation* [49]. There exist operators A_μ such that

$$\mathcal{E}(\rho) = \sum\nolimits_\mu A_\mu \rho A_\mu^\dagger. \tag{2.30}$$

where $\sum_\mu A_\mu^\dagger A_\mu = 1$. The operators A_μ are generally not unitary and their representations are not unique. An important property of quantum evolutions is that they cannot be used to signal instantaneously. If Alice and Bob are spatially separated, then a measurement by Alice cannot immediately influence Bob's state. This can be demonstrated by taking the partial trace of the overall system with respect to Alice's system, whereby it can be shown (Exercise 2.6) via a direct calculation,

$$\mathrm{Tr}_A\big(\mathcal{E}(\rho)\big) = \mathrm{Tr}_\mathrm{A}\left(\sum\nolimits_\mu (A_\mu \otimes I_B)\rho\big(A_\mu \otimes I_B\big)^\dagger\right) = \mathrm{Tr}_A(\rho). \tag{2.31}$$

Equation (2.31) implies that Bob cannot distinguish a random measurement from any actions taken by Alice. Therefore, quantum mechanics is protected against the unphysical occurrence of instantaneous signaling.

Entropy of Quantum States

The entropy of a state ρ in quantum mechanics is given by the von Neumann entropy

$$S(\rho) = -\mathrm{Tr}(\rho \log \rho) = -\sum\nolimits_i \lambda_i \log \lambda_i. \tag{2.32}$$

where the λ_i are the eigenvalues of ρ. A pure state $\rho = |\psi\rangle\langle\psi|$ has the minimal value of the entropy, $S(\rho) = 0$. This signifies that further information cannot be obtained for a pure state by repeated measurement over many copies. The entropy of a system obeying the Schrödinger equation always remains constant since the entropy only depends on the eigenvalues of ρ, which cannot change under a unitary transformation.

Thus, given a unitary transformation U and a density matrix ρ, we find

$$S(U^\dagger \rho U) = S(\rho). \tag{2.33}$$

The entropy is thus independent of time under unitary evolution and the information about a state does not change, i.e.,

$$\frac{dS}{dt} = 0. \tag{2.34}$$

The entropies of the subsystems are also of interest since they give information about quantum entanglement for pure states. Details about subsystem entropies in terms of the *Araki-Lieb Inequality* are presented next.

Subsystem Entropies and the Araki-Lieb Inequality

The *Araki-Lieb Inequality* [50] is an important result of quantum correlations. Let the composite system of A and B be given by the state ρ_{AB} and the reduced density matrices of the two sub-systems A and B be ρ_A and ρ_B. Then the total entropy and subsystems entropies are related by

$$S(\rho_A) + S(\rho_B) \geq S(\rho_{AB}) \geq |S(\rho_A) - S(\rho_B)|. \tag{2.35}$$

Note that the entropy of the total system $S(\rho_{AB})$ can be less than that of A and B considered separately. In this case, there is more information in the total system due to correlations between A and B, which may include the effect of quantum entanglement. If ρ_{AB} is a pure state, then $S(\rho_{AB}) = 0$ so that $S(\rho_A) = S(\rho_B)$ from the right-hand side of Inequality (2.35). For a bipartite pure state, the subsystem entropies are always equal no matter how disparate the two physical systems. An example is an atom in an electromagnetic field. Remarkably, as long as the combined system is in a pure state, the finite degrees of freedom of the atom and the infinite degrees of freedom of the field each must have identical entropies. Interactions between subsystems A and B result in correlations with an additional measure of information called their mutual information $I(A{:}B)$ [44], the information about A gained by learning about B or vice-versa,

$$I(A{:}B) = S(\rho_A) + S(\rho_B) - S(\rho_{AB}). \tag{2.36}$$

In the special case where A and B are independent, the entropies of the subsystems simply add, as is also the case for classical systems.

$$S(\rho_A) + S(\rho_B) = S(\rho_A \otimes \rho_B). \tag{2.37}$$

As an example, consider the two-qubit entangled pure state Equation (2.17) with density operator $\rho_{AB} = |\psi_{AB}\rangle\langle\psi_{AB}|$ and use criterion (E4) to evaluate its entanglement. The von Neumann entropy of a pure state is zero so that $S(\rho_{AB}) = 0$. However, the entanglement of ρ_{AB} is given by the von Neumann entropy of each of the subsystems ρ_A and ρ_B. The entropies of these are identical as we know from the Araki-Lieb Inequality (2.35), $S(\rho_A) = S(\rho_B)$, so we only need to calculate the reduced state for one of the subsystems, for example (Exercise 2.7)

$$\rho_A = \mathrm{Tr}_B(\rho_{AB}) = \frac{I_2}{2}. \tag{2.38}$$

where I_2 is the 2x2 identity matrix. This indicates that, although the total state ρ_{AB} is pure, each of the subsystems is a *maximally mixed* or chaotic state. The state in Equation (2.17) is maximally entangled with $E[\rho_{AB}] = S_A = \log 2$. Therefore, it has the maximum possible entropy and maximum entanglement.

Two Polarization-Entangled Photons

We have discussed using polarizing beam splitters to create single qubit states involving the horizontal *H* and vertical *V* polarization degrees of freedom. Polarizing beam splitters can also be used to create two-qubit entangled states similar to Equation (2.17) as were found to be maximally entangled in the previous section. These are created by inputting two photons into the beam splitter as in Figure (2.10). The two photons can be incident into the same port or into different ports. With two photons now present, the boson statistics of the photons also have to be take into account. The

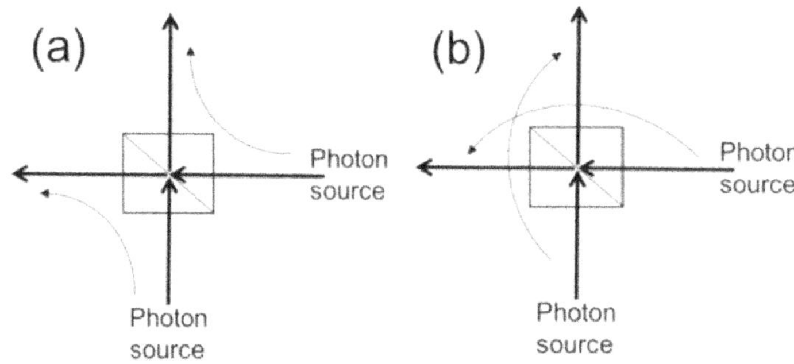

Figure 2.10: Two photons sent into a 50-50 polarizing beam splitter for the cases (a) both are reflected, (b) both are transmitted.

photon has two polarization modes H and V as well as two spatial degrees of freedom a and b, so that the vector describing the two-photon state is given by

$$\Phi = \begin{pmatrix} a_{\mathrm{H}} \\ b_{\mathrm{H}} \\ \mathrm{a}_{\mathrm{V}} \\ \mathrm{b}_{\mathrm{V}} \end{pmatrix} \tag{2.39}$$

where a_i and b_i denote the spatial degrees of freedom associated with polarization i.

Since photons are bosons, the total state for the combined polarization and spatial degrees of freedom must be symmetric under permutation of the particles. Using the basis in Equation (2.39), the polarization response from a polarizing beam splitter is identical to that of the controlled-not or CNOT gate used in quantum computing [51] (Exercise 2.8):

$$\mathrm{CNOT} = \begin{pmatrix} 1 & 0 & 0 & 0 \\ 0 & 1 & 0 & 0 \\ 0 & 0 & 0 & 1 \\ 0 & 0 & 1 & 0 \end{pmatrix} \tag{2.40}$$

The CNOT gate is a convenient unitary transformation that has the ability to produce two entangled qubits. It's customary to consider one of the qubit inputs to the CNOT as the *control qubit* C and the other as the *target qubit* T. The CNOT unitary transformation functions by flipping the target qubit depending on the state of the control qubit. Thus, it is a unitary operator that changes one qubit conditional on the state of the other. Figure 2.11 shows the standard circuit diagram representing the CNOT and its operation for the possible two-qubit inputs.

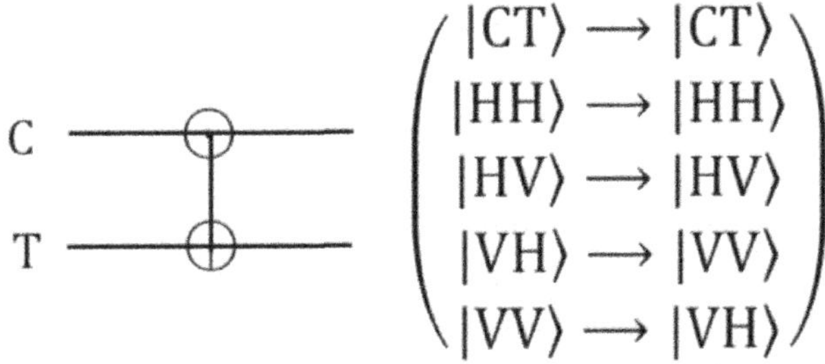

Figure 2.11: Quantum Controlled-Not or CNOT two-qubit unitary gate which flips the target state (T) only if the control state (C) is 1.

As an example, consider the non-entangled polarization state $|\mathrm{CT}\rangle_{in} = (|\mathrm{H}\rangle + |\mathrm{V}\rangle)\otimes|\mathrm{H}\rangle/\sqrt{2}$ as input to the CNOT. From the table in Figure 2.11 the output is found to be the maximally entangled state (Exercise 2.9):

$$|\mathrm{CT}\rangle_{\mathrm{out}} = \mathrm{CNOT}\left(\tfrac{1}{\sqrt{2}}(|H\rangle + |V\rangle)\otimes|H\rangle\right) = \tfrac{1}{\sqrt{2}}(|H\rangle\otimes|H\rangle + |V\rangle\otimes|V\rangle). \tag{2.41}$$

The output in Equation (2.41) is one of the *Bell-states*, the commonly used basis in quantum optics consisting of the four maximally entangled states (Exercise 2.10) as shown in Equation (2.42). Of these four, only the $|\Psi^-\rangle$ is anti-symmetric in interchange of the two photons. For an overall symmetric wave function required by boson statistics, the $|\Psi^-\rangle$ would have to be accompanied by an anti-symmetric spatial wave-function. The other Bell states are symmetric in polarizations and would require a symmetric spatial component. These possibilities for the two-photon statistics have been verified in two-photon interference experiments using polarizing beam splitters

[51]. Of particular interest is the behavior called *bunching* which is a consequence of the bosonic statistics and can occur for two photons each incident in two different ports and was found to be maximally entangled in the previous section.

$$
\begin{aligned}
|\Psi^+\rangle &= \frac{1}{\sqrt{2}}(|H\rangle\otimes|V\rangle + |V\rangle\otimes|H\rangle) \\
|\Psi^-\rangle &= \frac{1}{\sqrt{2}}(|H\rangle\otimes|V\rangle - |V\rangle\otimes|H\rangle) \\
|\Phi^+\rangle &= \frac{1}{\sqrt{2}}(|H\rangle\otimes|H\rangle + |V\rangle\otimes|V\rangle) \\
|\Phi^-\rangle &= \frac{1}{\sqrt{2}}(|H\rangle\otimes|H\rangle - |V\rangle\otimes|V\rangle).
\end{aligned} \tag{2.42}
$$

These are created by inputting two photons into the beam splitter as in Figure 2.10 (a). If the photons arrive simultaneously, they become indistinguishable and produce a destructive interference called the *Hong-Ou-Mandel* dip in the measured photon coincidence rate [52]. Methods for implementing polarization entangled photons using spontaneous parametric down-conversion (SPDC) are discussed in Appendix 2.A.

Information Flow in Entangled Composite Systems

The unitary properties of reversibility and constant entropy during evolution impose constraints on what tasks can be accomplished by unitaries and in what way they are able to direct information within entangled composite systems. This provides insight into the capabilities and limitations of unitaries. Consider an example involving three systems *A*, *B*, and *C*, which may be coupled by a variety of unitary interactions. Although the system is completely governed by reversible unitary evolution, let us examine whether configurations are possible which allow only one-way information transfer between subsystems. Information transfer means, for example, that there

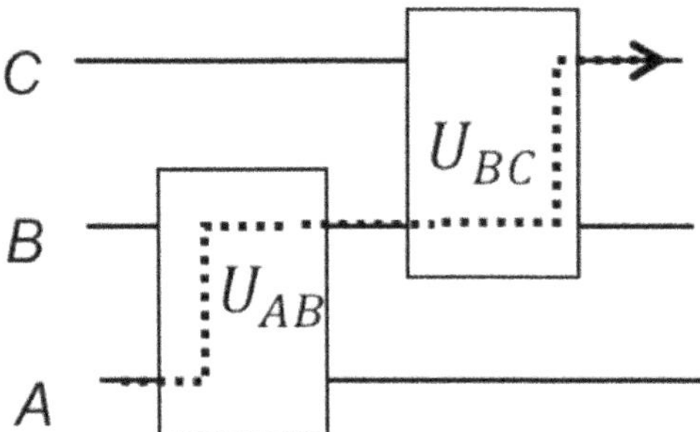

Figure 2.12: Configuration of the only possible unitary evolution of system *ABC* not allowing information flow from *C* to *A* [After Schumacher et al. [53]].

would be no information flow from *C* to *A* if the final state of *A* depended on the initial state of *AB* alone. The only possibility for this occurring within a unitary transformation is the configuration in Figure 2.12 as proved by Schumacher and

Westmoreland [53]. This has local interactions of the form $U_{ABC} = (I_A \otimes U_{BC})(U_{AB} \otimes I_C)$ which allows information to flow from A to B to C but prohibits any information from C to A. It is the local sequencing of interactions—first AB and then BC—that allows local restrictions in flow but still maintains the complete reversibility of the entire unitary operator U_{ABC} as it must (Exercise 2.11).

A unitary transformation must resort to local sequencing of interactions to shuttle and restrict information flow because linearity implies that certain unitary transformations are impossible. A prominent example is the *no-cloning theorem* [54]: no unitary transformation can copy an unknown quantum state $|\psi\rangle$. If this was possible, we could construct the following transformation for unknown states $|\psi_1\rangle$ and $|\psi_2\rangle$ [14, p. 532]

$$U(|\psi_1\rangle \otimes |c\rangle) = |\psi_1\rangle \otimes |\psi_1\rangle \quad (2.43)$$

$$U(|\psi_2\rangle \otimes |c\rangle) = |\psi_2\rangle \otimes |\psi_2\rangle. \quad (2.44)$$

where $|c\rangle$ is the initial state of the cloning machine. Taking the inner product between Equations (2.43) and (2.44) yields $\langle\psi_2|\psi_1\rangle = \langle\psi_2|\psi_1\rangle^2$, which only has the solutions $\langle\psi_2|\psi_1\rangle = 0$ or 1. Therefore unitary cloning can only be done for orthogonal states and not for a general quantum state. The limits to cloning have also been quantified and there are by now also a variety of other *no-go theorems,* which restrict what can be accomplished by unitaries, including the quantum no-deletion theorem [55], no-broadcasting [56], and no-unconditionally secure quantum bit commitment [57] [58].

The root of the problem for the no-cloning and related theorems is the necessary restriction to having only local sequencing of information flow in unitaries similar to those illustrated in Figure 2.12. Such restrictions also occur in the functioning of quantum gates such as the CNOT gate which carries out photonic entanglement in the polarizing beam splitters [53]. The CNOT cannot simply have a global exchange of information between the control and target qubits as it would then violate no-cloning. Instead, the information flow is locally directional from control to target and this limits what a CNOT or other unitary gates can achieve. However, it should be emphasized that local information flow from A to C need not imply a causal relation between A and C. Because of the linearity of unitary transformations, it is possible to form a superposition of two gates, one of which allows flow from A to C and the other from C to A. Such a composite gate of quantum-causal superposition would have the remarkable property of having no causal order between A and C, since signals would be sent in each order simultaneously [59]. In an experiment using polarized photons in such a superposition of causal orders, it has even been demonstrated that the photon can be measured without revealing the causal order [60]. This was done by embedding the result of the measurement into the photon itself but, then delaying the extraction of any information about the measurement result until after the photon goes through the entire circuit. This allows a coherent measurement at different times which erases the causal ordering information. The properties of unitary transformations thus allow the possibility of having physical processes without causal ordering, and this is also consistent with the inherent no-signaling property of quantum mechanics in Equation (2.31).

Appendix 2.A

Entanglement by SPDC

Currently, the most prominent method for generating entangled photon pairs is spontaneous parametric down-conversion (SPDC) [61] [62] [63], in which one photon from a pump laser beam is spontaneously converted into two photons, which emerge

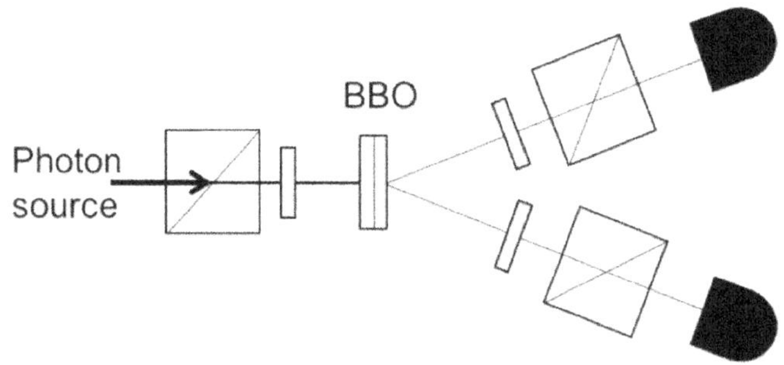

Figure 2.13: Single photon sent into BBO crystal producing entangled photons via spontaneous parametric down conversion [After Kwiat et al. [61]].

simultaneously from a pumped nonlinear crystal such as beta-barium-borate (BBO), Figure 2.13.

Nonlinear interaction of the pump laser pulse with the crystal can split the high frequency pump photon into two lower-frequency photons, called the *signal* and *idler*, and the three photons are constrained to satisfy energy conservation and phase-matching conditions. Generally, the outgoing photons are not collinear. In this method, detection of one photon can also indicate that the second photon has been generated. The down-conversion is Type-I if the signal and idler photons have

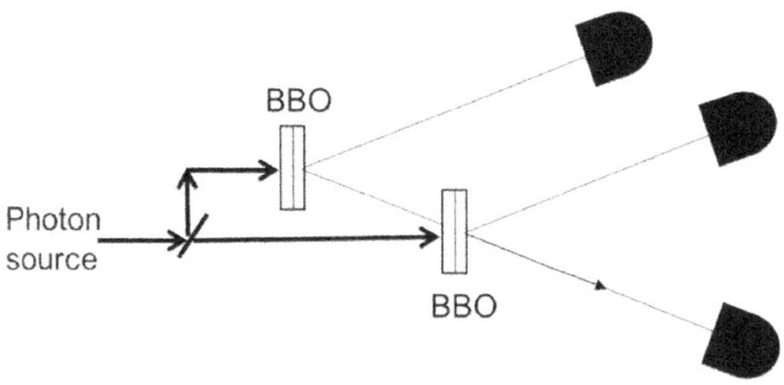

Figure 2.14: Single photon sent into BBO crystals where overlapping paths remove which-crystal information resulting in entanglement [After Krenn et al. [67]].

identical polarizations and Type-II if they have orthogonal polarizations and it is possible to prepare any of the four polarization Bell-states [64]. Many other variations are possible with SPDC, such as the method of path identification of Zou-Wang-Mandel [65] [66] [67] in which paths from two pumped BBO's are overlapped thereby removing the which-BBO-crystal information resulting in entanglement, Figure 2.14.

Exercises

2.1 Derive the energy phase rotation, Equation (2.12).

2.2 Verify that Equation (2.13) is a unitary transformation.

2.3 Calculate the von Neumann entropy, Equation (2.21), of the pure state Equation (2.17).

2.4 Prove the time-translation Equation (2.23) and the spatial translation Equation (2.25).

2.5 Suppose the operator G of Equation (2.27) obeys the eigenvalue equation $G|g\rangle = g|g\rangle$. Show that the time evolved eigenstate $|g,t\rangle \equiv \exp(-iHt/\hbar)|g\rangle$ is also an eigenstate of G with the same eigenvalue g.

2.6 Derive the no-signaling condition, Equation (2.31).

2.7 Calculate the subsystem state ρ_A , Equation (2.38).

2.8 Verify that CNOT in Equation (2.40) is a unitary transformation.

2.9 Verify the CNOT polarizing beam splitter output in Equation (2.41).

2.10 Find initial states for CNOT that generate each of the Bell states of Equation (2.42).

2.11 Demonstrate that local information flow from C to A in Figure 2.12 is not possible.

2.12 Consider the Hamiltonian $H_{int} = \Delta X_3 + g(X_+ + X_-)$ where g is a coupling constant, Δ is a detuning between frequencies of different subsystems, and the operators obey the commutation relations

$$[X_3, X_\pm] = \pm X_\pm \,, \; [X_+, X_-] = P(X_3)$$

and $P(X_3)$ is an arbitrary polynomial function of the diagonal operator X_3. Use the following unitary transformation to obtain the effective diagonal Hamiltonian for the case of weak coupling $g/\Delta \ll 1$ [68].

$$\widetilde{U} = \exp\left[\frac{g}{\Delta}(X_+ - X_-)\right]$$

$$H_{eff} = \widetilde{U} H_{int} \widetilde{U}^\dagger = \Delta X_3 + \frac{g^2}{\Delta} P(X_3) + (\text{terms higher order in } g/\Delta).$$

Hint: use the Baker-Campbell-Hausdorff expansion,

$e^A B e^{-A} = B + [A, B] + \frac{1}{2!}[A, [A, B]] + \cdots$]. In this limit, the time evolution of the system can be described by $U_{eff} = \exp(-iH_{eff}t)$.

CHAPTER 3

Interpretation or Existence of a Non-Unitary Process

The measurement problem cannot be avoided: within the conventional framework, unitary evolution can be experimentally distinguished from measurement.

Introduction

In Chapter 1, the concept of wave-particle duality was explored. It was seen that light propagates as a wave and light can exhibit wave interference properties, but light also exists as individual quantized chunks of energy. The requirement that light is quantized in individual units originally came from Einstein's theoretical quantization predictions for light quanta. This led to the concept of wave-particle duality whereby light appears to propagate as a wave while at the same time existing in quantized individual energy units or photons. Wave-particle duality is an important concept, but by itself does not explain the measurement problem.

Einstein's 1905 proposal for the light quantum or photon was initially confirmed by experiments on the photoelectric effect by Millikan (1916) [69] and by electron-light scattering experiments by Compton (1923) [70], and Bothe and Geiger (1925) [71]. The postulate of the photon was made on the basis that the amount of energy flowing during a characteristic coherence time is small compared to $h\nu$ and that the detection process appears discrete [72]. However, the photoelectric effect could also be described semi-classically by assuming classical waves and quantized detector atoms [73] [74]. The photon concept was finally demonstrated definitively by photon-anti-correlation experiments in the 1970-80s [72] [75] [76]. In 1974, Clauser demonstrated the 'whole photon or nothing' property that photons do not split at a half-silvered mirror and that true single photons are not measured at both detectors simultaneously [75]. Thus, there exists the experimental results that once light of a single photon is detected, it is no longer found within the double slit experimental apparatus. Such experimental evidence indicates that although light may exist in quantized energies while at the same time exhibiting wave-like properties (wave-particle duality), the detection or *measurement* of the photon appears to be a process that results in some destruction of the wave-like properties of the photon within the double-slit apparatus.

In Chapter 2 characteristics of unitary evolution were explored in relation to the measurement problem. Light interferes as a wave if and only if which-way information about the path taken is not accessible. It was seen that unitary evolution generally produces entanglement and shows a difference in the state that is expected under unitary evolution when compared with the case of measurement. From experimental evidence, it is known for the case of measurement that a product state results. In the case of unitary evolution, we expect a superposition state. However, the arguments that were given in Chapter 2 did not rigorously establish that there was necessarily a real measurement problem in the aspects of wave-particle duality. That is, how do we know for general detector-photon interaction (other than a mirror) that the measurement problem doesn't go away in the sense that the expectation of a superposition under unitary evolution could be wrong and that the quantum state does become a product state. Or even if a superposition does exist, why are we assured that the particular superposition is not a sufficiently good approximation to a product state when the number of particles and interactions composing a detector becomes macroscopically large?

There have been a number of arguments in the literature indicating that there is a measurement problem (see for example Bassi et al. [77], Leggett [78], Gisin [79]), but the argument originated by the authors in this chapter is based on a constructive approach to show that there will generally be observable differences between the predictions of unitary evolution and measurement. We believe that for many, an explicit construction lends itself to a fuller understanding and appreciation of the measurement problem than an abstract argument. This construction will be utilized substantially in Chapter 4 in order to analyze various approaches that have previously been proposed to resolve the measurement problem.

In this chapter, the following will be shown:

- There is a "measurement problem" for any given configuration of particles that forms the detector
    - The argument often put forward that "somehow" the measurement problem goes away for macroscopically large detectors is conclusively invalidated.
    - Unitary evolution does not describe the quantum state evolution that occurs during measurement.

Hamiltonian Description

In Chapter 2, the characteristics of unitary evolution were developed using thought experiments with mirrors and beam splitters. Quantum mechanics employs the Hamiltonian description to describe wave function evolution under Schrödinger's equation. We will assume a spin system initially to represent the detector, although the results will be shown to be generalizable to any system of particles representing the detector.

A beam splitter will be extensively utilized throughout this chapter in the process of developing a unitary versus measurement discrimination test (UMDT). Unless otherwise noted, the beam splitter will be assumed to be a polarizing beam splitter such that the action of the beam splitter depends on the polarization of the input. The states $|1_H\rangle$ and $|1_V\rangle$ denote single photon Fock states of horizontal and vertical polarization respectively, relative to the alignment of the surface of the beam splitter.

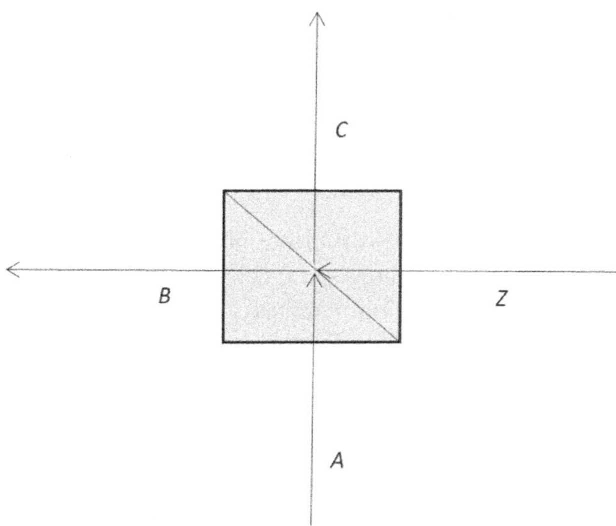

Figure 3.1: 4-port beam splitter, inputs A and Z.

A beam splitter is a four-port device as illustrated in Figure 3.1 for which there are two input ports *A* and *Z*, and in the case of a polarizing beam splitter, two possible polarization states for each port. Unless otherwise noted, it will be assumed that the input to Port *Z* is the vacuum state, denoted $|0\rangle_Z$ so that any (pure state) input to Port *A* is a linear combination of the kets $\{|1_H\rangle_A, |1_V\rangle_A\}$. For such a configuration, assuming the beam splitter transmits horizontal polarization and reflects vertical polarization on either Port *A* or Port *Z* [80, p. 142], then $|1_H\rangle_A$ evolves to $|1_H\rangle_C$, $|1_V\rangle_A$ evolves to $|1_V\rangle_B$, $|1_H\rangle_Z$ evolves to $|1_H\rangle_B$, and $|1_V\rangle_Z$ evolves to $|1_V\rangle_C$.

Consider a photon that can enter a polarizing beam splitter at Port *A* and for which the two output ports *B* and *C* are directed toward two photon detectors as illustrated in

Figure 3.2. If the state $(|1_H\rangle_A + |1_V\rangle_A)/\sqrt{2}$ is input into the beam splitter, the output is given by $(|0\rangle_B|1_H\rangle_C + |1_V\rangle_B|0\rangle_C)/\sqrt{2}$, or in short-hand dropping the vacuum states this state is often written as $(|1_H\rangle_C + |1_V\rangle_B)/\sqrt{2}$. The same input photon is predicted to exist in a superposition of both output ports.

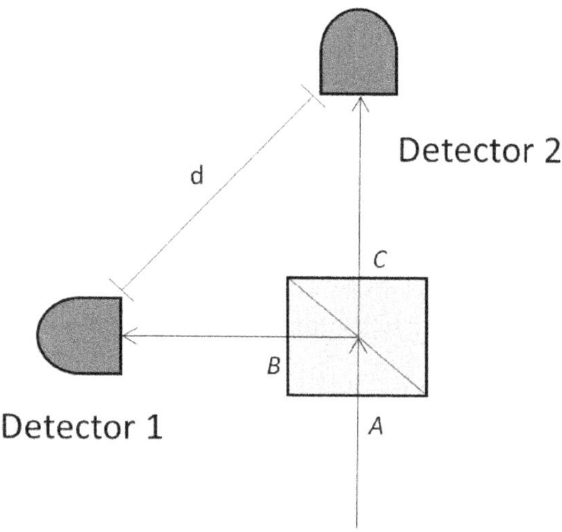

Figure 3.2: Detector configuration for a single photon impinging on beam splitter at A.

Suppose that the detectors are 100% efficient, meaning that if the photon is in the state $|1_V\rangle_C$ then Detector 1 will always register a detection of the photon whereas Detector 2 will never register a detection. Similarly, if the photon is in the state $|1_H\rangle_C$, Detector 2 will always register a detection of the photon whereas Detector 1 will never register a detection. Both Detector 1 and Detector 2 are composed of atomic particles (which might also be in a state of condensed matter) and are subject to a Hamiltonian description.

The photon Hamiltonian (after the polarizing beam splitter) will be assumed to be composed of the two field modes B and C so that $H_F = H_B \otimes H_C$. Detector 1 is assumed to be composed of n_1 spin ½ particles (the arguments given in this chapter are independent of the spin and can be extended to arbitrary spin particles) and has the self-interaction Hamiltonian H_1. Similarly, Detector 2 is composed of n_2 spin ½ particles and has the self-interaction Hamiltonian H_2. Consider the general class of Hamiltonians given by

$$\begin{aligned} H = H_F \otimes I \otimes I + I \otimes H_1 \otimes I + I \otimes I \otimes H_2 + H_{int} \\ H_{int} = H_{F,1} + H_{F,2}. \end{aligned} \tag{3.1}$$

$H_{F,1}$ denotes the interaction Hamiltonian between the photon and Detector 1 while $H_{F,2}$ is the interaction Hamiltonian between the photon and Detector 2. The Hilbert space of the input photon is denoted $\mathcal{H}_A$, similarly $\mathcal{H}_B \otimes \mathcal{H}_C$, $\mathcal{H}_1$, and $\mathcal{H}_2$ denote the Hilbert spaces of the output photon state from the polarizing beam splitter and the

Hilbert spaces of Detector 1 and 2 respectively. Furthermore, $\mathcal{D}(\mathcal{H}_A) = 2, \mathcal{D}(\mathcal{H}_B) = 2, \mathcal{D}(\mathcal{H}_C) = 2, \mathcal{D}(\mathcal{H}_1) = n_1, \mathcal{D}(\mathcal{H}_2) = n_2$, where $\mathcal{D}(\mathcal{H})$ denotes the dimension of the Hilbert space $\mathcal{H}$.

Bell's Inequality

The theory of entanglement is paramount in the development and understanding of the measurement problem. As will be further discussed in Chapter 4, the problem of the unitary prediction of entanglement versus the expectation that such entanglement is not present in local classical physics, was understood by Einstein rather earlier on historically than is generally realized. With the discovery by Bell [81] in 1964 of inequalities that are satisfied by any local hidden variable model and yet violated by quantum mechanical predictions, the issues of entanglement and nonlocality became clear. Experiments beginning in 1972 and refined over subsequent years by Clauser et al. [82], Aspect et al. [83], Ou et al. [84], and Shih et al. [85] ultimately led to the larger physics community's acceptance of entanglement and later, with proposed applications and theory of entanglement, to the emergence of the field of Quantum Information. The remaining loopholes that have remained regarding nonlocality due to detector inefficiency, etc., have recently been essentially closed [86] [87].

One might have expected then, with entanglement being understood and accepted by many, that the Quantum Measurement Problem, which deals with the dichotomy between entanglement and separable systems would be clearly understood and appreciated by most. This is not the case—the great majority of experiments (there are exceptions but often these systems violate the continuous Gaussian bound rather than Bell's inequality) in quantum information deal with the dichotomy between the entanglement of very few particles versus separable systems. On the other hand, the measurement problem requires one to confront the issue of the entanglement of detectors which are often considered to be macroscopic bodies.

The two particles that are emitted, impinge on (separate) beam splitters as illustrated in Figure 3.3. In the case when the source emits photons that are entangled in polarization, the beam splitter is assumed to be a polarizing beam splitter. After the polarizing beam splitter there are two detectors A and B corresponding to the two possible non-reflected and reflected outputs of the polarizing beam splitter. If the photon emitted toward Detector A is detected in the reflected path, A is assigned the value -1, otherwise A=1. Similarly, if the photon emitted toward Detector B is detected in the reflected path, B is assigned the value -1, otherwise B=1.

Each polarizing beam splitter can be set at an arbitrary angle θ that would allow an input horizontal polarization state $|1_H\rangle$ to always pass through un-reflected when $\theta = 0$ but has an increasing probability of being reflected as θ is increased until an incoming $|1_H\rangle$ is completely reflected when $\theta = 90°$.

Over multiple trials of the experiment, the statistic $\hat{C}$ is defined as the average of the product of the outcomes of A and B and is a measure of the statistical correlation between random variables A and B. For a given set of beam splitter parameters θ_A, θ_B

the correlation computed is denoted $\hat{C}(\theta_A, \theta_B)$.

When the Source S emits particles, one can consider the possibility of a hidden variable parameter defined by $\lambda \in \mathbb{R}$ where $\mathbb{R}$ denotes the real numbers such that the detection result at A is a function of only λ and θ_A while the detection result at B is a function of only λ and θ_B. In this case note that $A(\lambda, \theta_A)$ cannot be a function of the parameter setting θ_B. Such a model is defined to be a local hidden variable (LHV) model. The Clauser-Horne-Shimony-Holt (CHSH) inequality [88] is a more widely used extension of Bell's original inequality [81] that is easier to test experimentally. The CHSH inequality consists of computing the sample $\hat{C}(\theta_A, \theta_B)$ for four sets of experimental configurations with parameters $\{(\theta_A, \theta_B), (\acute{\theta}_A, \theta_B), (\theta_A, \acute{\theta}_B), (\acute{\theta}_A, \acute{\theta}_B)\}$.

For each experimental configuration, one computes $\hat{C}$ over the series of experimental trials. As shown in [88], the CHSH inequality provides that any LHV model will satisfy

$$S = \left|\hat{C}(\theta_A, \theta_B) + \hat{C}(\acute{\theta}_A, \theta_B) + \hat{C}(\theta_A, \acute{\theta}_B) - \hat{C}(\acute{\theta}_A, \acute{\theta}_B)\right| \leq 2. \tag{3.2}$$

This inequality is also satisfied by every separable quantum state. However, it may be violated by any pure entangled quantum state up to a maximum possible violation given by $S = 2\sqrt{2}$, known as the Cirel'son bound [89]. Violation of the local CHSH inequality is a sufficient condition for detection of entanglement between the two parties, regardless of the system and details of implementation. For every pure entangled state, a set of local measurements can be found so that the resulting correlations will violate the CHSH inequality [90] [91]. For mixed states, the relation between entanglement and nonlocality is subtler since any associated nonlocality sometimes cannot be revealed using only direct measurements [92].

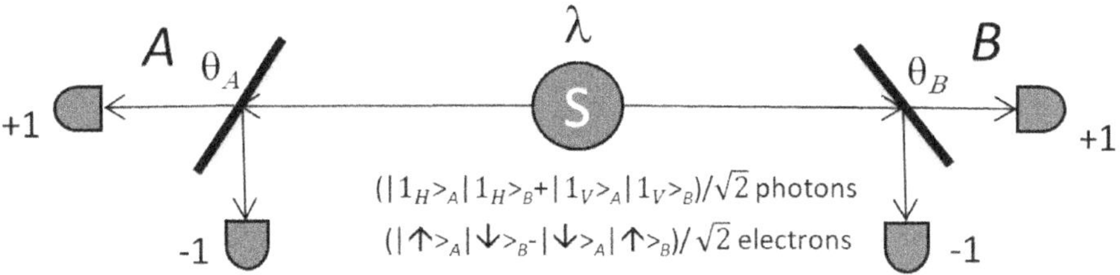

Figure 3.3 Experimental configuration for testing Bell's Inequality.

As an example of a violation of the CHSH inequality at the Cirel'son bound, consider the maximally entangled input state $(|1_H\rangle_A|1_H\rangle_B + |1_V\rangle_A|1_V\rangle_B)/\sqrt{2}$. The quantum mechanical prediction can be shown to be given by $C(\theta_A, \theta_B) = \cos(2(\theta_A - \theta_B))$ where we define $C(\theta_A, \theta_B) \equiv E(\hat{C}(\theta_A, \theta_B))$ and denote the expected value of a random variable x by $E(x)$. Then this maximally entangled input state and the particular measurement settings separated by successive 22.5-degree angles given by

$$\{(0°, 22.5°), (0°, 67.5°), (45°, 22.5°), (45°, 67.5°)\} \tag{3.3}$$

produce a correlation at Cirel'son's bound $S = 2\sqrt{2} \approx 2.8284$, which significantly violates Equation (3.2). There have now been numerous experimental confirmations of violations of Bell's inequality, closing existing "loopholes" provided by additional assumptions that previous experimenters were forced to make due to technological constraints (such as low detection efficiency for which hidden variable models still exist, even though the correlation S exceeds Bell's inequality) [93] [86] [94]. The preponderance of evidence indicates that Nature is non-local and consistent with quantum mechanics.

Separable quantum states are restricted by the CHSH bound $S_{\text{sep}} \leq 2$ consistent with LHV models while entangled quantum states can achieve the larger Cirel'son bound $S_{\text{ent}} \leq 2\sqrt{2}$. In order to achieve the maximal violation for entangled quantum states, both pairs of local observables must be chosen to be anti-commuting [91]. Remarkably, the case of anti-commuting observables also affects the case of separable quantum states, resulting in a strengthened CHSH bound of $S_{\text{sep}} \leq \sqrt{2}$, considerably smaller than the CHSH bound of $S_{\text{sep}} \leq 2$ [95] [96]. As will be seen in this chapter, the bound of $\sqrt{2}$ can be achieved by pure product states for which there is no quantum entanglement. The dichotomy between unitary and non-unitary processes can be quantified in terms of the contrast between the corresponding limits of $S_{\text{product}} = \sqrt{2}$ and $S_{\text{entangled}} = 2\sqrt{2}$ in Bell measurements.

Assumptions

There are several assumptions that are utilized in developing the theoretical basis to analyze the major issue of this chapter—whether the measurement problem has a solution with potentially new observational consequences or is a philosophical problem that only requires an interpretation, for which no new observational consequences are possible. In order to address this and related questions, it cannot be taken for granted that the detectors in Figure 3.2 or Figure 3.3 actually generate the statistics that are observed under Born's rule. In the consideration of the measurement problem (as opposed to the problem of nonlocality), the particles that compose a particular "detector" will be initially referred to as a device. Only after it is established via experimental test for which the operations will be specified in this chapter, will it be considered to be a bona fide detector.

One can separate the two devices in Figure 3.2 via an arbitrary large distance d, and one could in principle place a shutter in front of the photon emitter to limit the uncertainty of the time of emission to some Δt_e. As long as $d > 2c(\Delta t_e + \tau_l)$ where τ_l is the latency time of the devices, there can be no direct interaction between the devices such that the state of one device affects the individual statistics of the other device without violating relativistic constraints of no-signaling. Another distinct issue is whether there can be indirect interaction between separated systems and the correlation properties such as entanglement are affected. The potential existence of mechanisms that can affect the entanglement or correlation in separated systems in a manner that is faster-than-light is quite important for the issue of whether or not the measurement problem is an interpretational problem or whether the measurement problem is a real problem with observational consequences. It has been found under unitary evolution that interactions exist, such that the correlation properties between two separated systems can be affected faster than $d/\Delta t_e$ [97] [98] [99]. Although remarkable, such effects are predicted to be significantly reduced when the devices are sufficiently separated [98]. As the separation d between the devices can be made arbitrarily large so that the effects of such correlation can be made arbitrarily small, one can safely ignore such effects so long as any correlation that is under consideration also does not decrease toward zero with increasing separation; otherwise further analysis would be needed. The Hamiltonian is for now approximated in the form of Equation (3.1), but this issue will be revisited later to ensure that such an approximation is valid.

Moreover, the dimension of the detector Hilbert space could conceivably grow as a detection is made and the detector affects other particles in the local area. However, any growth in the dimension of the final detector state must be bounded because there is a well-defined distance that any time-like signals can propagate. Adler in [100] shows that only environmental particles that are within a distance $c\tau_l$, where c is the speed of light and τ_l is the measurement latency time, can causally affect the experimental outcome without violating no-signaling. Hence, to simplify matters, it will be assumed that the initial states of the detectors includes not only the initial particles that compose the detector but also the particles composing the environment surrounding the detector within a time-like effect between the time the photon wave function begins impinging upon the device and time τ_l, for which there can be potential direct interaction $\mathcal{D}(\mathcal{H}_i) = N$. Hence N remains the same throughout the interaction.

The initial ready state of Device i is initially assumed to be a pure state and is denoted by $|\psi_r^0\rangle_i \in \mathcal{H}_i$, i=*1,2*, $\mathcal{D}(\mathcal{H}_i) = N$. Later in this chapter, initial device states that are mixed will also be considered. The two devices are assumed to be capable of being initialized by an experimenter, and we assume that they are initialized in a product state given by $|\psi_r^0\rangle_1 \otimes |\psi_r^0\rangle_2$. The assumption that one can consider the device states to be initialized in this manner is important, and this assumption will be considered later in Chapter 4. The initial state of the photon sent into the polarizing beam splitter is assumed to be given by an arbitrary pure state $\sqrt{a}|1_V\rangle_A + \sqrt{1-a}|1_H\rangle_A$, where $a \in \mathbb{C}$, $|a| \leq 1$, $\mathbb{C}$ denotes the field of complex numbers and $|a|$ denotes the absolute value of a.

A single photon is assumed to obey quantum electrodynamics and to propagate unitarily. Given that the input to the polarizing beam splitter is $|1_V\rangle_A$, the output of the polarizing beam splitter is given by $|1_V\rangle_B|0\rangle_C$ and the photon will always be found in Port B when measured. Similarly, when the input is $|1_H\rangle_A$, the output of the polarizing beam splitter is given by $|0\rangle_B|1_H\rangle_C$, and the photon will always be found in Port C if measured. Assuming the time of photon emission is (approximately) known, it will be assumed that the start-time of the devices is synchronized to start in a short window just before the photons impinge on the devices. If the entire wave function of the photon interacts locally with the device at the start-time, it is assumed that the device will always evolve to its final state, whether the evolution is via a unitary or non-unitary process.

Suppose that in both paths there is only the vacuum state $|0\rangle$. After the interaction time, the initial ready device state $|\psi_r^0\rangle_i$ is assumed to change to a read-out state (or remains the same) that is indicative of no measurement and is denoted $|\psi^0\rangle_i, i = 1,2$. Consider the case that a photon with state $|1_V\rangle_B$ is inserted in Path B. Assuming the device always completely absorbs the photon, which will be referred to as 100% absorption, the state of Device 1 evolves to its readout state indicating that the Device has detected a photon, that will be denoted $|\psi^1\rangle_1 \in \mathcal{H}_1$. Similarly, if a photon with state $|1_H\rangle_C$ is inserted solely in Path C and Device 2 is 100% absorbing, the final state of Device 2 is denoted as $|\psi^1\rangle_2 \in \mathcal{H}_2$.

For macroscopic devices for which copies can be made, the states $|\psi^0\rangle_i$ and $|\psi^1\rangle_i$ $(i = 1,2)$ are fully distinguishable classically. Assuming that the two states are fully distinguishable and classical copies can be made, it follows by the no-cloning theorem [54] that such initial and final quantum device states must be orthogonal. However, such orthogonality is not guaranteed, and devices for which the device states $|\psi^0\rangle_i$ and $|\psi^1\rangle_i$ are not orthogonal, will be referred to as being less than 100% efficient. In order to consider such cases in more generality, we will include operations on the local electromagnetic field surrounding the devices as well as the devices themselves, denoted by System $1'$ and System $2'$.

An important assumption that is inherently made is that it is possible at least in principle to implement a unitary operation and perform a measurement on System $1'$ and System $2'$. However, a logical possibility is that it is impossible to implement one or more of the required operations because the conditions under which a measurement occurs prohibits such operations from being successfully implemented. This is an interesting possibility and is further examined in Chapter 4.

The method of state evolution of the devices is at this point unspecified as only an initial ready state and final detection state given a photon input are assumed; this is important as the device evolution can be specified either by unitary or non-unitary dynamics.

Hypothesis Tests

At this point, one might consider what Bell's inequality has to do with the measurement problem. Although it is generally accepted as a refutation of local hidden variables as regards the action of entangled states, the case for its application to the measurement problem has yet to be justified. In the measurement problem, it is desirable to know whether or not the particles composing a particular device are actually a bona fide detector in the sense of corresponding to the expected observations given by Born's rule. In this case, it is desired to determine if the particles that compose a device are indeed conducting a measurement in a sense that is 1) different than Schrödinger unitary evolution or 2) whether the particles composing the device are acting unitarily and Born's rule is simply a good approximation to the unitarily predicted evolution. There are two related questions to be considered. Firstly, is unitary evolution of devices sufficient to describe the expected results that are seen via Born's rule in the limit that the number of particles composing the device becomes large? Suppose the answer to this first question is found to be that unitary evolution is not sufficient to describe device evolution as an approximation with large numbers of particles. This leads to the second issue which is how to distinguish whether or not a particular device is acting unitarily.

That is, it is desired to utilize Bell's inequality to answer questions that will be put in the form of hypothesis tests:

Hypothesis Test 3.1

H_0: Schrödinger's equation describes the evolution of all particles.
H_1: Schrödinger's equation does not describe the evolution of all particles.

Hypothesis Test 3.2

H_0: Schrödinger unitary evolution of a device as the number of particles increases, is a valid description for the bifurcation rule observed under Born's rule.

H_1: There are cases whereby the quantum state of a device that evolves via Schrödinger unitary evolution does not approach, in the limit of large number of device particles, the expected bifurcation rule observed under Born's rule.

Note that these two hypothesis tests are related but not the same. Consider the possibility in the first hypothesis test that a device is found to evolve non-unitarily. The case of H_0 under Hypothesis Test 3.2, i.e., that Schrödinger unitary evolution with a sufficient number of particles for the description of Born's rule, is not logically ruled out. On the other hand, suppose it is found in Hypothesis Test 3.2 that there are cases whereby Schrödinger unitary evolution of a device with sufficiently large number of particles does not approach the expected observed bifurcation rule observed under Born's rule. This would either imply that a device is not evolving via Schrödinger's equation in Hypothesis Test 3.1 given the initial assumptions, or the device is evolving according to Schrödinger's equation and the initial assumptions are wrong. That the device is evolving via Schrödinger's equation and the initial assumptions are incorrect will be further considered in Chapter 4.

In considering these questions, the experimental configuration seen in Figure 3.2 is different than in Figure 3.3. In experiments testing Bell's inequality, the source emits two entangled particles, and there is a total of four detectors. On the other hand, in Figure 3.2 there is only a single particle emitted and there are only two detectors. In order to develop the theory in parallel with a Bell-like experiment, the configuration seen in Figure 3.4 will be assumed, whereby two additional ancilla qubits have been added.

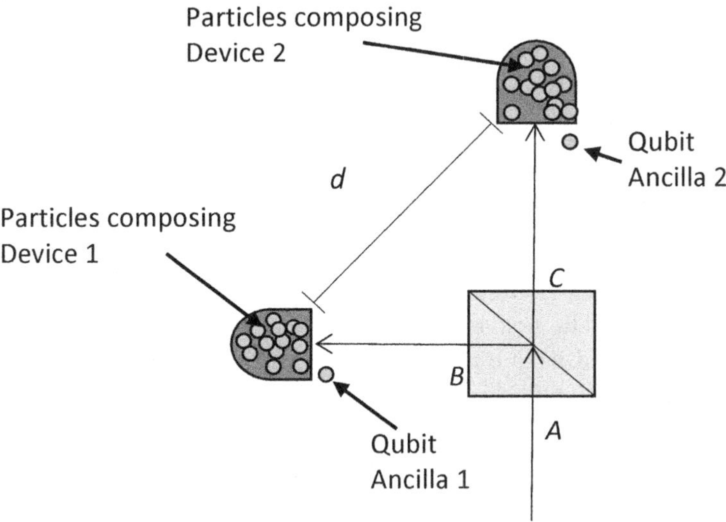

Figure 3.4: Experimental configuration for testing the measurement problem.

The goal is to specify in three steps the operations that will distinguish the Hypotheses:

- Step 1: Provide the Schrödinger unitary evolution that predicts the evolution between a single photon and the two Devices.
- Step 2: A Schrödinger unitary operation is specified that transfers the entanglement that exists between the two Devices (plus a local electromagnetic field if the devices are inefficient) into entanglement between the two ancilla qubits.
- Step 3: Perform a Bell measurement on the two qubits.

It will be seen later that, with the single configuration in Figure 3.4, both Hypothesis Tests 3.1 and 3.2 can be tested experimentally subject to our assumptions.

The secondary measurements are made on the particles that compose the device and on the electromagnetic field locally around the detector. The reason that the secondary measurement includes the electromagnetic field around the detector is that when one considers detectors via simple Hamiltonian models, it may be the case under the hypothesis of Schrödinger evolution that the photon that was absorbed by the device is re-emitted into the electromagnetic field. In order to develop the theory in the most general manner when re-emission is possible, it is required that a secondary measurement is made both on the particles composing the device and on the local electromagnetic field surrounding each device. In this manner, any device Hamiltonian, simple or complex, can be fully considered. On the other hand, if a sufficiently detailed model of a device is used such that the probability of re-emission is small and can be ignored, then it is a good approximation that the electromagnetic field will be found in the vacuum state after a sufficiently long interaction time. In such cases, the electromagnetic vacuum state is the same for all the terms in the analysis and can be factored out, and secondary measurements are only needed on the particles that compose the individual devices.

As can be seen in Figure 3.4, the measurement problem is now cast in a manner similar to Bell's inequality, with the exceptions that 1) the particles that compose the device under question are not *a priori* assumed to be a bona fide detection device, and 2) a secondary measurement takes place on the particles that compose a given device and the local electromagnetic field around each device. The answers to Hypothesis Tests 3.1 or 3.2 are desired. That is, we don't know whether the device, if evolving via Schrödinger's equation, would be a sufficiently good approximation to Born's law for all purposes, nor do we yet know if the device evolves according to Schrödinger's equation or otherwise. Step 1 in our development is to specify a Schrödinger unitary operation on the particles that compose the devices and surrounding electromagnetic field shown in Figure 3.4 in a manner to transfer the entanglement to the two ancilla particles. After that is completed, the last step is to conduct a Bell experiment on the two ancillae.

The analysis to follow, as well as a geometric picture of the Schrödinger unitary evolution, relies on a theorem known as the Schmidt decomposition.

Schmidt Decomposition

The following theorems regarding Schmidt decomposition will be utilized in the analysis to follow.

> *Theorem 3.1:* Consider two quantum systems 1 and 2 with pure states $|\psi\rangle_1 \in \mathcal{H}_1$, $|\chi\rangle_2 \in \mathcal{H}_2$ where $\mathcal{H}_1$ and $\mathcal{H}_2$ denote the Hilbert space of states of systems 1 and 2 respectively. Let $|\psi\rangle$ be a pure state of the composite system 1-2, $|\psi\rangle \in \mathcal{H}_{1,2}$ where $\mathcal{H}_{1,2} \equiv \mathcal{H}_1 \otimes \mathcal{H}_2$. There exist orthonormal states $|e_i\rangle \in \mathcal{H}_1$ and $|f_i\rangle \in \mathcal{H}_2$ whereby
>
> 1. $|\psi\rangle = \sum_{i=1}^{N} \sqrt{v_i}\, |e_i\rangle \otimes |f_i\rangle$
> 2. v_i are non-negative and satisfy $\sum_i v_i = 1$.
> 3. N is defined as the Schmidt rank.
>
> *Theorem 3.2:* Suppose $|\psi\rangle \in \mathcal{H}_{1,2}$, $|\phi\rangle \in \mathcal{H}_{1,2}$ are pure states of the joint system 1-2 that have the same Schmidt coefficients. There exist local unitary transformations X, Y acting on systems 1 and 2 respectively such that $|\psi\rangle = X \otimes Y\, |\phi\rangle$.

More information on Schmidt decomposition can be found in [14]. In order to gain further insight into the measurement problem, consider the visualization of the major issues involved.

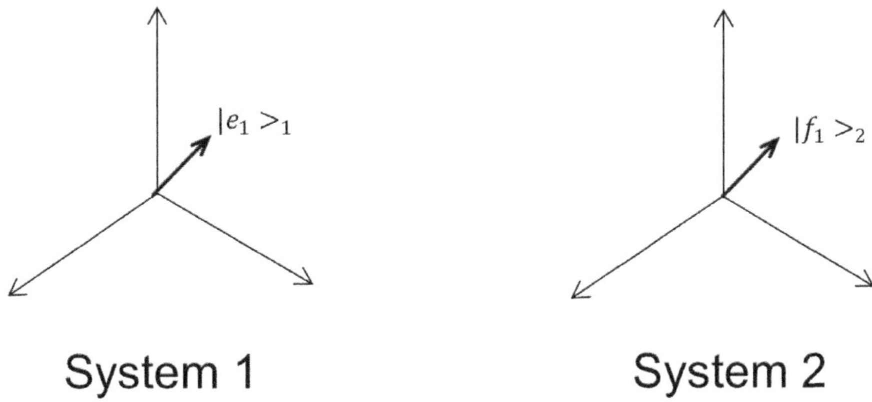

Figure 3.5: Geometric visualization of a product bipartite state with Schmidt vectors $|e_1\rangle_1$ and $|f_1\rangle_2$.

Geometry of Entanglement

Consider an arbitrary product pure state where $|\psi\rangle = |e_1\rangle_1 \otimes |f_1\rangle_2$. This is a single product state hence by Theorem 3.1, has Schmidt rank 1. Consider the geometry of this state represented by two vectors representing the kets $|e_1\rangle_1$ and $|f_1\rangle_2$ illustrated in 3-dimensional space for visualization purposes in Figure 3.5.

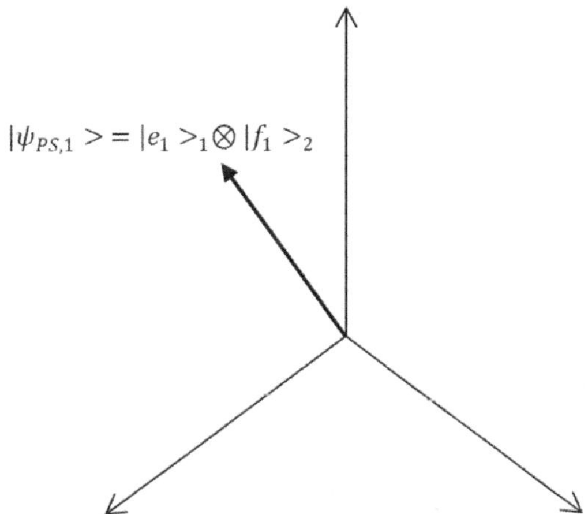

Figure 3.6: Vector representing a product state of two systems.

This can also be looked at as a single vector in the joint Hilbert space composing System 1-System 2, $\mathcal{H}_{1,2}$ as shown in Figure 3.6. In reality, the vector will exist in a higher than 3-dimensional space.

In an entangled system of particles in a pure state, the quantum state cannot be decomposed as a single product state. In Figure 3.7 an entangled bipartite system is illustrated that is composed of the superposition of two product states, i.e.,

$$|\psi\rangle = \sqrt{a}|e_1\rangle_1 \otimes |f_1\rangle_2 + \sqrt{1-a}|e_2\rangle_1 \otimes |f_2\rangle_2 \text{ where } a \in \mathbb{C}, |a| \leq 1.$$

The dotted line between the vectors is shown to be weighted by the coefficients of each respective product state. There is a 90° angle between the orthogonal states $|e_1\rangle_1$ and $|e_2\rangle_1$ and similarly between $|f_1\rangle_2$ and $|f_2\rangle_2$. This entangled state can also be visualized as a superposition of $|\psi_{PS,1}\rangle = |e_1\rangle_1 \otimes |f_1\rangle_2$ and $|\psi_{PS,2}\rangle = |e_2\rangle_1 \otimes |f_2\rangle_2$, where the subscript PS denotes product state. In this case, $|\psi\rangle = \sqrt{a}|\psi_{PS,1}\rangle + \sqrt{1-a}|\psi_{PS,2}\rangle$.

This is illustrated in Figure 3.8. The degree of the superposition $\sqrt{a}$ related to the entanglement, can be seen to simply correspond geometrically to the angle between the entangled state $|\psi\rangle$ and the product state $|\psi_{PS,1}\rangle$ via $\theta = \text{acos}(\sqrt{a})$. The three vectors $|\psi_{PS,1}\rangle$, $|\psi_{PS,2}\rangle$, and $|\psi\rangle$ can be seen to all lie in a single plane determined by any two of these vectors.

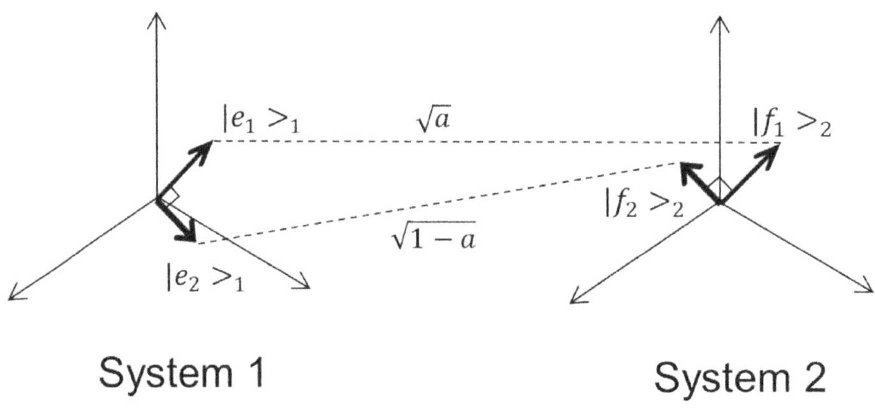

Figure 3.7: Geometric visualization of the Schmidt vectors composing an entangled bipartite system.

Geometry of the Measurement Problem

Note that state evolution via Schrödinger's equation is a requirement in H_1 in both Tests 3.1 and 3.2. In this respect, first we will examine what occurs during Schrödinger unitary evolution in the interaction of the photon with the particles composing the devices, and later what occurs when a measurement has occurred. Consider an initial photon state in Figure 3.4 that is input into the polarizing beam splitter given by

$$\sqrt{a}|1_V\rangle_A + \sqrt{1-a}|1_H\rangle_A,$$

where $a \in \mathbb{C}$, $|a| \leq 1$. This leads to the initial photon state after the polarizing beam splitter (PBS)

$$|\psi_{\text{photon,PBS}}\rangle = \sqrt{a}|1_V\rangle_B \otimes |0\rangle_C + \sqrt{1-a}|0\rangle_B \otimes |1_H\rangle_C \tag{3.4}$$

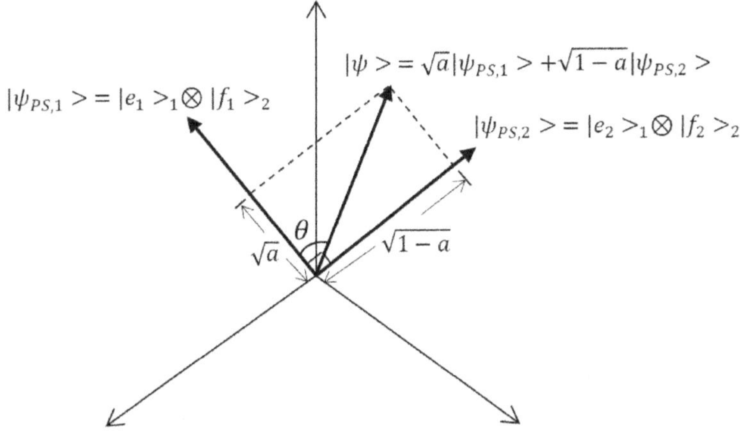

Figure 3.8: Geometry of an entanglement superposition state composed of two product states.

Note that a valid set of Schmidt vectors of the initial photon set can be directly determined via $|e_1\rangle_1 = |1_V\rangle_C$, $|e_2\rangle_1 = |0_H\rangle_B$, $|f_1\rangle_2 = |0_H\rangle_C$, $|f_2\rangle_2 = |1_V\rangle_C$. As all of these vectors are elements of Hilbert space $\mathcal{H}_{\text{photon}}$, with dimension $\mathcal{D}(\mathcal{H}_{\text{photon}}) = 2$, the geometry of this photon state is such that the vectors lie in within a natural basis of Euclidean space as seen in Figure 3.9.

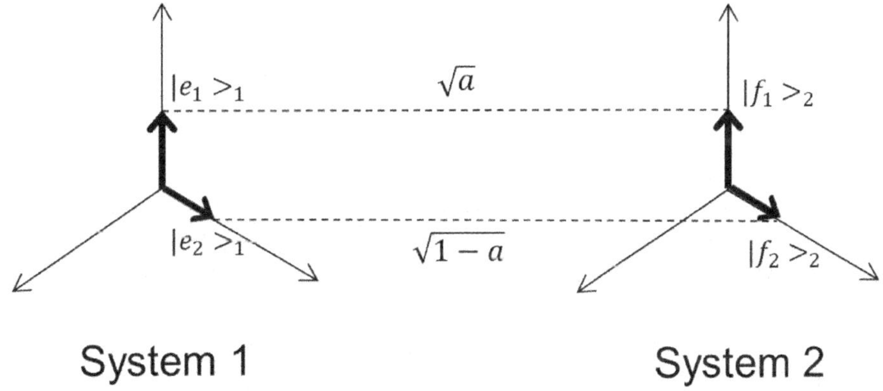

Figure 3.9: Geometry of entangled initial photon.

Combining this with the initial state of the two devices and the two ancilla qubits in Figure 3.4 gives an initial state that will be found to be conveniently ordered as

$$|\psi_{\text{init}}\rangle = \sqrt{a}|0\rangle_{A_1} \otimes |1_V\rangle_B \otimes |\psi_r^0\rangle_1 \otimes |0\rangle_{A_2} \otimes |0_H\rangle_C \otimes |\psi_r^0\rangle_2 + (1-\sqrt{a})|0\rangle_{A_1} \otimes |0_H\rangle_B \otimes |\psi_r^0\rangle_1 \otimes |0\rangle_{A_2} \otimes |1_H\rangle_C \otimes |\psi_r^0\rangle_2) \tag{3.5}$$

where $|\psi_r^0\rangle_i \in \mathcal{H}_i$, i=1,2, $\mathcal{D}(\mathcal{H}_i) = N$, $|0\rangle_{A_1} \in \mathcal{H}_{A_1}$, $|0\rangle_{A2} \subset \mathcal{H}_{A2}$, $\mathcal{D}(\mathcal{H}_{A_1}) = \mathcal{D}(\mathcal{H}_{A2}) = 2$.

We have included in $|\psi_{init}\rangle$ the state space of the two ancilla qubits that are depicted in Figure 3.4. Ancilla 1 and the B photon modes have been inserted adjacent to Device 1, while Ancilla 2 and the C photon modes have been inserted adjacent to Device 2. The composite system composed of the state of the photon in the B port and Device 1 will be referred to as System $1'$. System $2'$ denotes the state of the photon in the C port and Device 2. The ancillae are assumed initially uncoupled with Systems $1'$ and $2'$ and remain in the same state during the Schrödinger unitary evolution whereby the photon interacts with the two devices. For notational simplicity, these constant states of the ancillae will not be written but are assumed present, and will be inserted later when the ancillae are directly utilized, resulting in

$$|\psi_{\text{init}}\rangle = \sqrt{a}|e_1^0\rangle_{1'} \otimes |f_1^0\rangle_{2'} + \sqrt{1-a}|e_2^0\rangle_{1'} \otimes |f_2^0\rangle_{2'} \tag{3.6}$$

where

$$\begin{aligned} |e_1^0\rangle_{1'} &= |1_V\rangle_B \otimes |\psi_r^0\rangle_1 \\ |e_2^0\rangle_{1'} &= |0\rangle_B \otimes |\psi_r^0\rangle_1 \\ |f_1^0\rangle_{2'} &= |0\rangle_C \otimes |\psi_r^0\rangle_2 \\ |f_2^0\rangle_{2'} &= |1_H\rangle_C \otimes |\psi_r^0\rangle_2. \end{aligned}$$

The Hamiltonian is assumed to lack interaction terms between Device 1 and 2, likewise there are no photon interaction terms between Arm B and Arm C in Figure 3.4 locally at Device 1 and 2 respectively. It is assumed under Hypothesis H_0 of Hypothesis Test 3.1 that the state evolution follows Schrödinger's equation. Due to the lack of interaction between Systems $1'$ and $2'$, the Hamiltonian consists of interaction terms only between the individual particles composing a device (which is assumed to include all time-like particles within a local environment) and between the individual devices and the respective local arms of the photon. The Schrödinger evolution in Step 1, therefore, takes the form of a tensor product operating on Systems $1'$ and $2'$ with W and V Schrödinger unitary. Hence

$$U = e^{-iHt} = (I \otimes W) \otimes (I \otimes V) \tag{3.7}$$

and $|\psi_{fin}\rangle = (I \otimes W) \otimes (I \otimes V)\ |\psi_{init}\rangle$. The ancillae are multiplied by the identity operator and are unaffected by U. Substituting Equation (3.6), the initial state under

Schrödinger unitary evolution evolves to

$$
\begin{aligned}
|\psi_{\text{fin},1}\rangle &= \sqrt{a}W|e_1^0\rangle_{1'} \otimes V|f_1^0\rangle_{2'} + \sqrt{1-a}W|e_2^0\rangle_{1'} \otimes V|f_2^0\rangle_{2'} \\
|\psi_{\text{fin},1}\rangle &= \sqrt{a}|e_1^1\rangle_{1'} \otimes |f_1^1\rangle_{2'} + \sqrt{1-a}|e_2^1\rangle_{1'} \otimes |f_2^1\rangle_{2'},
\end{aligned} \tag{3.8}
$$

where the ancillae have again been dropped from the notation for simplicity,

$$|e_i^1\rangle_{1'} = W|e_i^0\rangle_{1'}, |f_i^1\rangle_{2'} = V|f_i^0\rangle_{2'}, i = 1,2.$$

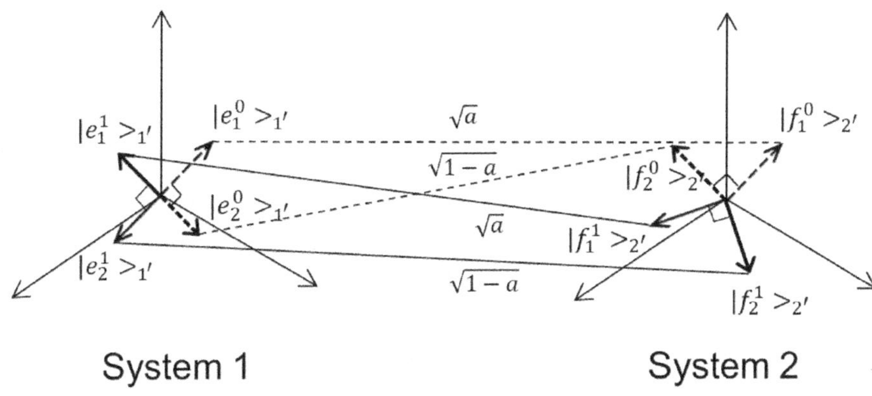

Figure 3.10: Geometry of the Schmidt states in the quantum measurement problem.

The geometry of the initial states and final states of Figure 3.4 is illustrated in Figure 3.10 in terms of the Schmidt decomposition; the ancillae are not affected by the Schrödinger unitary and are not shown. The two orthogonal vectors $\{|e_1^0\rangle_{1'}, |e_2^0\rangle_{1'}\}$ are seen to be rotated into $\{|e_1^1\rangle_{1'}, |e_2^1\rangle_{1'}\}$ and similarly $\{|f_1^0\rangle_{1'}, |f_2^0\rangle_{1'}\}$ are rotated into $\{|f_1^1\rangle_{1'}, |f_2^1\rangle_{1'}\}$. The geometry of orthogonality is maintained as well as the relationships of the respective coefficients in the superposition given by $\sqrt{a}$ and $\sqrt{1-a}$.

One can also examine the geometry of the Schrödinger unitary evolution of Figure 3.4 in a manner similar to Figure 3.8. Shown in Figure 3.11 is the initial state geometry of the superposition in terms of the product states.

The final entangled state under Schrödinger unitary evolution is illustrated in Figure 3.12. A comparison of Figure 3.11 with Figure 3.12 shows that the geometry between the vectors is the same; the two figures are related by rotation. If we assume the devices absorb the photon with probability, one which will be defined as a 100% absorbing device, it follows

$$
\begin{aligned}
|e_1^1\rangle_{1'} &= |0\rangle_B \otimes |\psi^1\rangle_1 \\
|e_2^1\rangle_{1'} &= |0\rangle_B \otimes |\psi^0\rangle_1 \\
|f_1^1\rangle_{2'} &= |0\rangle_C \otimes |\psi^0\rangle_2 \\
|f_2^1\rangle_{2'} &= |0\rangle_C \otimes |\psi^1\rangle_2.
\end{aligned}
$$

One can then factor out the electromagnetic field in the final Schrödinger predicted state in Step 1, resulting in:

$$|\psi_{\text{fin},1}\rangle = |0\rangle \otimes (\sqrt{a}\,|\psi^1\rangle_1 \otimes |\psi^0\rangle_2 + \sqrt{1-a}|\psi^0\rangle_1 \otimes |\psi^1\rangle_2. \tag{3.9}$$

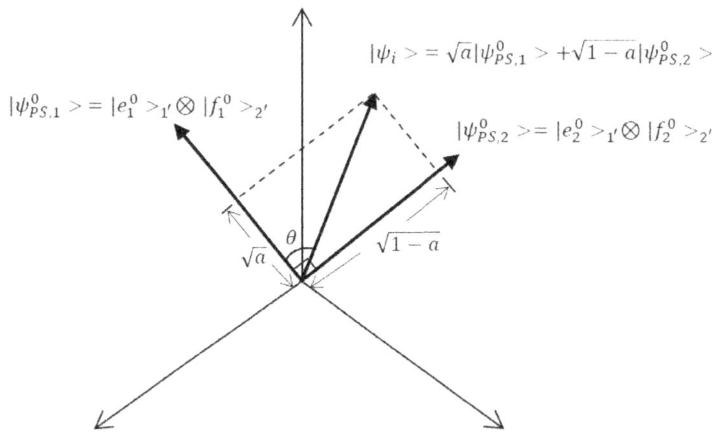

Figure 3.11 Geometry of the initial entangled superposition state composed of two product states in the quantum measurement problem.

We will see from this latter expression that under Schrödinger unitary evolution, the entanglement that originally existed within the photon field has been swapped into entanglement between the initial and final device states. This also aids in explaining

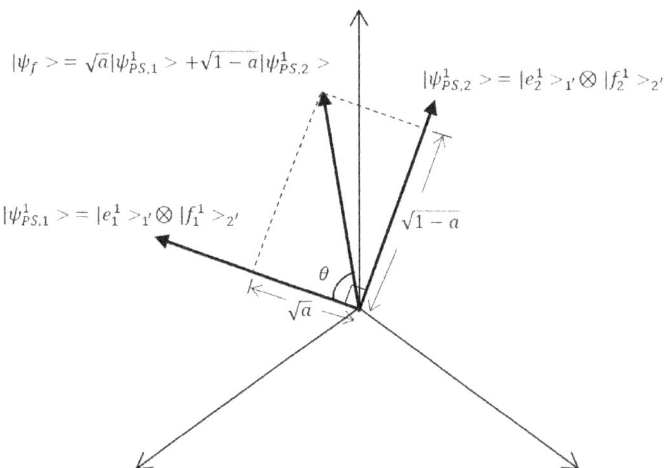

Figure 3.12 Geometry of the final entangled superposition state in the quantum measurement problem.

why Figure 3.11 and Figure 3.12 are both 2-dimensional planes.

On the other hand, if the devices do not operate in the sense that a photon was either not absorbed 100% of the time by the device or has been re-emitted with some probability by the device, then the Schrödinger unitarily predicted final state can be an entangled state of field-device. One might consider a solution via post-selection such that only final states are further analyzed when the particular case in which no photon is found in the local electromagnetic field surrounding each device. However, it can be shown (Exercise 3.20) that such post-selection can still be problematic in that it is not guaranteed that the final states are orthogonal, i.e. $|\psi^0\rangle_i \perp |\psi^1\rangle_i$ (although there may exist initial states for which this orthogonality does occur). In order to consider the problem for all devices in full generality for arbitrary initial states, and particularly to illustrate the theory for specific device-particle models, each device and its local electromagnetic field as previously defined by Systems $1'$ and $2'$ will be considered. Unitarily it can be shown (Exercise 3.1) that these local joint systems do provide the desired orthogonality.

Specification of Operations

Equation (3.8) will be the starting point for the Step 2 operation. In Step 2 local unitary operations will be applied that evolve Equation (3.8) to a state for which the entanglement that was present between the two devices is swapped to create entanglement between the two ancilla qubits. Step 3 is to perform a Bell measurement on the two qubits. At this point, the two ancilla qubits are included in the description of $|\psi_{\text{fin}}\rangle$ resulting in,

$$\begin{aligned} |\psi_{fin,1}\rangle &= \sqrt{a}|0\rangle_{A_1} \otimes |e_1^1\rangle_{1'} \otimes |0\rangle_{A_2} \otimes |f_1^1\rangle_{2'} + \\ &\sqrt{1-a}|0\rangle_{A_1} \otimes |e_2^1\rangle_{1'} \otimes |0\rangle_{A_2} \otimes |f_2^1\rangle_{2'}. \end{aligned} \quad (3.10)$$

The operations that are desired in Figure 3.4 are such that they will enable the discrimination between the hypotheses in Tests 3.1 and 3.2. Firstly, we will develop the procedure needed for addressing Test 3.2; it will be seen later that the same procedure also suffices to enable the discrimination of the hypotheses in Test 3.1.

A reasonable question is whether or not the entanglement that is unitarily predicted between System $1'$ and $2'$ can necessarily be transferred to the qubits in Step 2. Note that as $U = (I \otimes W) \otimes (I \otimes V)$ in Step 1, one could simply choose the Step 2 operation to be $U^{-1} = (I \otimes W^{-1}) \otimes (I \otimes V^{-1})$, which reverses the original interaction and transfers the entanglement back to the photon field. Continue by interacting the local fields B and C in a manner that transfers the entanglement between the single photon and the qubits similar to the approach taken in [101]. This shows that it is always possible to find a local Step 2 operation, by reversing the original interaction and transferring the photon superposition into two-qubit entanglement.

However, there are several reasons to consider if it is possible to transfer the entanglement without reversing the original detector-field unitary operation. For example, one might object to such inverse methodology as being at the very least extremely difficult as it requires the implementation of a new Hamiltonian that accomplishes the complete time-reversal of the naturally occurring process of the original interaction. Another important reason is that it is desirable to work towards implementable techniques that can discriminate unitary evolution from measurement. As well, there are theories that will be examined in Chapter 4 that maintain that the measurement problem is resolved by certain naturally occurring unitary processes that ultimately cannot be reversed. With these reasons in mind, we will not allow the consideration of Step 2 unitary operations that require the time-reversal of the detector. If there are terms in the device state after Step 1 for which the device has evolved from an initial state to a final state (or in the case of a mixed initial device state from an initial eigenstate of the device to a final state), the only operations that will be considered allow the device to remain in such states or evolve further in time via the original unitary.

The final vectors in Figure 3.12 lie in a 2-dimensional plane. These vectors are typically in a high dimensional subspace determined by the particles that compose the detectors. We begin by showing there exists a local unitary matrix that rotates these vectors into any two natural Euclidean basis vectors, as any two such basis vectors also form a two-dimensional space, which can contain the plane of the final vectors. This operation will be implemented between the particles that compose each device and the respective local qubit, as shown in Figure 3.13.

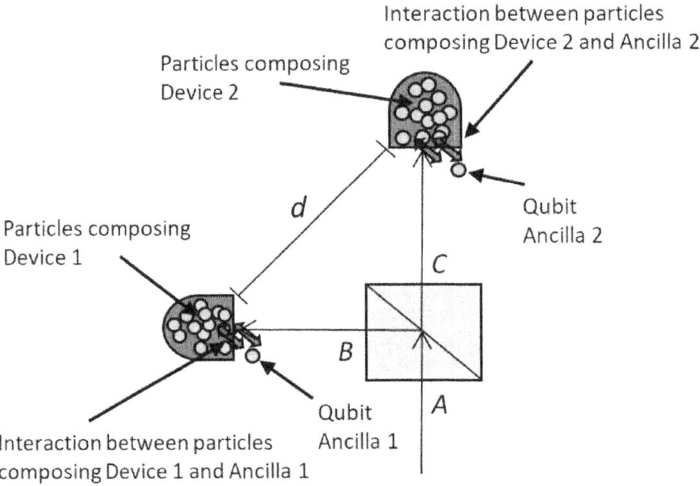

Figure 3.13 Operation of interaction between the particle composing each device and the respective ancilla qubits and local electromagnetic field.

Define

$$T_1 = (|1\rangle_{A_1} \otimes |e_1^1\rangle_{1'})(\langle 0|_{A_1} \otimes {}_{1'}\langle e_1^1|) + (|0\rangle_{A_1} \otimes |e_1^1\rangle_{1'})(\langle 0|_{A_1} \otimes {}_{1'}\langle e_2^1|)$$
$$T_2 = (|1\rangle_{A_2} \otimes |f_2^1\rangle_{2'})(\langle 0|_{A_2} \otimes {}_{2'}\langle f_2^1|) - (|0\rangle_{A_2} \otimes |f_2^1\rangle\rangle_{2'})(\langle 0|_{A_2} \otimes {}_{2'}\langle f_1^1|) \tag{3.11}$$

where $|0\rangle_{A1}$, $|1\rangle_{A1}$ are natural basis vectors of the ancilla qubit (a second ancilla qubit is similarly required for System $2'$),

$$|0\rangle_{A_1} = \begin{pmatrix}1\\0\end{pmatrix}, \quad |1\rangle_{A2} = \begin{pmatrix}0\\1\end{pmatrix}.$$

As it was assumed that both System 1 and 2 have dimension N, T_1 and T_2 are $2N$ by $2N$ matrices with 2 non-zero rows. T_1 will have the effect of mapping $|0\rangle_{A_1} \otimes |e_1^1\rangle_{1'}$ to $|1\rangle_{A_1} \otimes |e_1^1\rangle_{1'}$ and $|0\rangle_{A_1} \otimes |e_2^1\rangle_{1'}$ to $|0\rangle_{A_1} \otimes |e_1^1\rangle_{1'}$, where $|1\rangle_1 \in \mathcal{H}_1$. Note that this operation has similar functionality as a controlled-not operation commonly

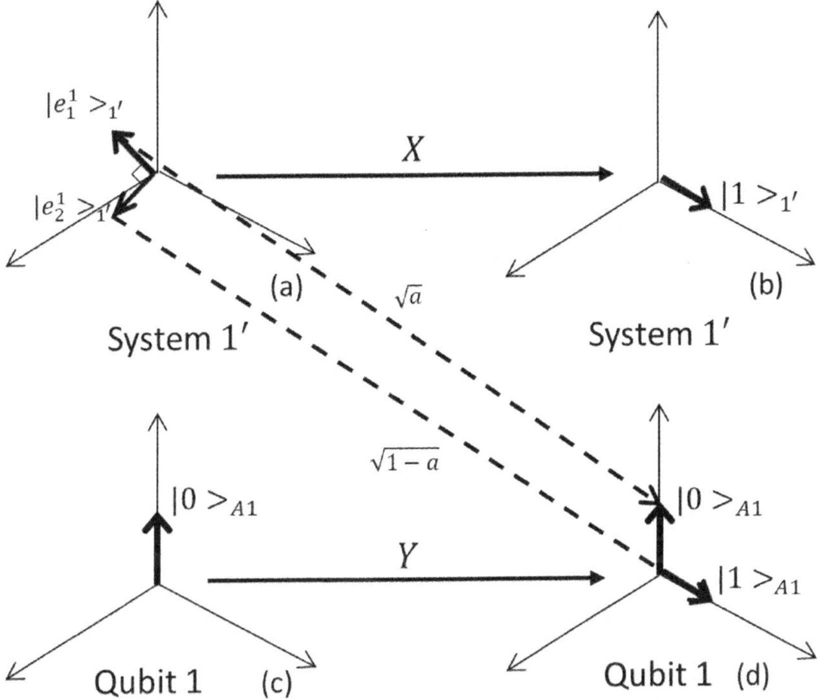

Figure 3.14: The unitarily evolved Schmidt vectors in Systems $1'$ are shown in Figures (a), (b) operated on by X causing the initial ancilla qubit 1 in Figure (c) to evolve to Figure (d) in a manner that retains the coefficients $\sqrt{a}, \sqrt{1-a}$ of the initial state of System$1'$ in Figure (a). A similar diagram is found for System $2'$ via the transformation Y_t.

encountered in quantum computing, although applied to a device rather than a qubit.

Consider the specification of a unitary matrix X as an extension of T_1 and Y as an extension of T_2 (the X and Y extensions do not need to be identical). Let X have the same two non-zero rows as T_1 and Y the same two non-zero rows as T_2. The rows of X will be orthonormal if and only if $|e_1^1\rangle_{1'}$ is orthogonal to $|e_1^2\rangle_{1'}$. Note from the previous discussion regarding efficient detection, such orthogonality is a reasonable assumption for the device state of many realistic detectors; however, there exist models for which orthogonality is not guaranteed, which is why we extended Systems 1 and 2 to

Systems 1′ and 2′ respectively. Assuming these rows are orthonormal, one can complete X (and similarly Y) by adding $2N-2$ rows such that all rows form a complete basis of the $2N$ dimensional linear vector space. As the rows form a complete orthonormal basis and the matrices are square, X and Y are also unitary. Note that we chose to map both $|e_1^1\rangle_{1'}$ and $|e_1^2\rangle_{1'}$ to $|e_1^1\rangle_{1'}$ in the Hilbert space $\mathcal{H}_1$. However, it can be shown for the case when the initial detector state is pure (Exercise 3.3) that an arbitrary state in $\mathcal{H}_1$ could have been chosen (and similarly for System 2). Conditions that the chosen final states must meet when the initial detector state is mixed are examined in Exercise 3.14

Let X be such a $2N$ by $2N$ extension of T_1 and Y an extension of T_2 (the X and Y extensions do not need to be identical). The geometrical effect of X is to transfer the quantum state coefficients of the vectors $\{|e_1^1\rangle, |e_2^1\rangle\}$ into the coefficients of the natural vectors of the ancilla qubit A_1, as seen in Figure 3.14. Step 2 is to apply $X \otimes Y$ to Equation (3.10) resulting in

$$|\psi_{fin,2}\rangle \equiv (X \otimes Y)\, |\psi_{fin,1}\rangle \tag{3.12}$$

$$|\psi_{\text{fin},2}\rangle = \sqrt{a}|1\rangle_{A_1} \otimes |e_1^1\rangle_{1'} \otimes |0\rangle_{A_2} \otimes |f_2^1\rangle_{2'} + \sqrt{1-a}|0\rangle_{A_1} \otimes |e_1^1\rangle_{1'} \otimes |1\rangle_{A_2} \otimes |f_2^1\rangle_{2'} \,.$$

This can be rewritten by swapping the second and third systems:

$$= \sqrt{a}|1\rangle_{A_1} \otimes |0\rangle_{A_2} \otimes |e_1^1\rangle_{1'} \otimes |f_2^1\rangle_{2'} + \sqrt{1-a}|0\rangle_{A_1} \otimes |1\rangle_{A_2} \otimes |e_1^1\rangle_{1'} \otimes |f_2^1\rangle_{2'}$$

for which the state of the detector particles can be factored out:

$$= \{\sqrt{a}|1\rangle_{A_1} \otimes |0\rangle_{A_2} - \sqrt{1-a}|0\rangle_{A_1} \otimes |1\rangle_{A_2}\} \otimes |e_1^1\rangle_{1'} \otimes |f_2^1\rangle_{2'}, \tag{3.13}$$

which includes our desired 2-qubit entangled state (in the first two systems).

Now that Step 2 has been completed, a standard Bell experiment is required to be implemented on the two ancillae for completion of Step 3 (utilized in the quantification of the Bell inequality), as shown in Figure 3.15.

Note that the unitary operations that are utilized in Step 2 are not of the form of the time-reversal of the unitary that occurred in Step 1. Hence it has been demonstrated that the entanglement can be unitarily transferred without the impediment of requiring a large-scale time-reversal, which was a desired property of the operation for the reasons discussed previously. We will refer to the use of Figure 3.4 along with device Hamiltonian and the Step 1, 2, and 3 operations that have now been fully specified as a *unitary versus measurement discrimination test* (UMDT).

Specific Device-Particle Modeling

In Equation (3.1), a general class of Hamiltonians was considered for which a UMDT can be defined. Now, a specific Hamiltonian model of device-particle is considered both to illustrate how to apply these operations and also to understand how these operations address Hypothesis Tests 3.1 and 3.2.

Given that a single photon has impinged on the two devices as in Figure 3.4 and assuming the Step 1 Schrödinger unitary evolution in Hypothesis H_0 is common in both Hypothesis Tests 3.1 and 3.2, the state will evolve to Equation (3.9). Step 2 is to apply the local unitary X to Device 1 and Y to Device 2 that swaps the unitarily predicted entanglement between System $1'$ and System $2'$, into entanglement between the two ancilla qubits A_1 and A_2. Step 3 is to conduct a Bell measurement on these two qubits.

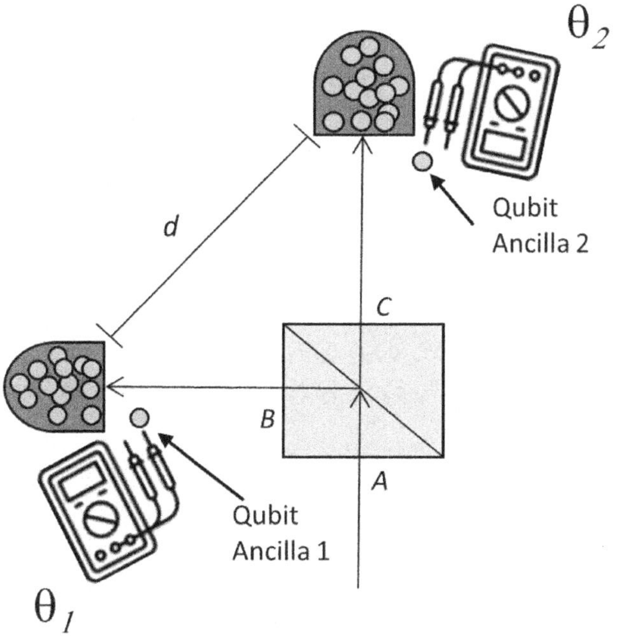

Figure 3.15: Step 3: Bell measurement on Ancilla Qubits.

Consider first a model for which each device is composed of two spin ½ particles with states that lie in a 4-dimensional Hilbert space. The relevant electromagnetic field in Figure 3.4 can be specified by the B and C photon modes and the two polarization modes for which the states lie in a 4-dimensional Hilbert space.

The Hamiltonian then is given by,

$$H = H_F \otimes I_4 \otimes I_4 + I_4 \otimes H_1 \otimes I_4 + I_4 \otimes I_4 \otimes H_2 + H_{int} \qquad (3.14)$$

where I_j is the j by j identity matrix. Denote the spin ½ operators in terms of the Pauli matrices and raising and lowering operators as

$$S_x = \frac{1}{2}\sigma_x = \frac{1}{2}\begin{pmatrix} 0 & 1 \\ 1 & 0 \end{pmatrix}, S_y = \frac{1}{2}\sigma_y = \frac{1}{2}\begin{pmatrix} 0 & -i \\ i & 0 \end{pmatrix}, S_z = \frac{1}{2}\sigma_z = \frac{1}{2}\begin{pmatrix} 1 & 0 \\ 0 & -1 \end{pmatrix}$$

$$S_+ = \begin{pmatrix} 0 & 0 \\ 1 & 0 \end{pmatrix}, S_- = \begin{pmatrix} 0 & 1 \\ 0 & 0 \end{pmatrix}.$$

We consider a two-spin Heisenberg model for the Hamiltonian of each device

$$H_i = \Omega_{1,i}S_z \otimes I_2 + \Omega_{2,i}I_2 \otimes \sigma_z + J_{x,i}S_x \otimes S_x + J_{y,i}S_y \otimes S_y + J_{z,i}S_z \otimes S_z. \quad (3.15)$$

Note that interactions between the particles composing each device have been included via $\{J_{x,i}, J_{y,i}, J_{z,i}\}$. The relevant electromagnetic field free Hamiltonian is given by

$$H_F = w a_B{}^{\dagger} a_B \otimes I_2 + w I_2 \otimes a_C{}^{\dagger} a_C,$$

where $a_B{}^{\dagger}$ (a_B) is an operator that creates (annihilates) a photon of vertical polarization in the B Port of the beam splitter shown in Figure 3.4 and $a_C{}^{\dagger}(a_C)$ creates (annihilates) a photon of horizontal polarization in the C Port of the beam splitter. The cases of a horizontally polarized photon in the B output port or vertical polarized photon in the C Port can be included if desired; however, as the input to the beam splitter in the Z Port of Figure 3.1 is assumed to be the vacuum state, such states cannot occur via Schrödinger's equation.

Consider the following field-particle interaction model,

$$H_{F,1} = c_1(a_B \otimes I_2 \otimes S_+ \otimes I_2 \otimes I_4 + a_B{}^{\dagger} \otimes I_2 \otimes S_- \otimes I_2 \otimes I_4)$$
$$H_{F,2} = c_2(I_2 \otimes a_c \otimes I_4 \otimes S_+ \otimes I_2 + I_2 \otimes a_C{}^{\dagger} \otimes I_4 \otimes S_- \otimes I_2)$$

for which $H_{int} = H_{F,1} + H_{F,2}$. These interaction terms have the effect of interacting the photon mode that impinges on a device with the first of the two spin particles that compose the device. However, as the two spins that compose that device can themselves interact via the coefficients $\{J_{x,i}, J_{y,i}, J_{z,i}\}$ in Equation (3.15) a photon that impinges on a device can affect both particles.

Let the initial state of the ancillae, EM field, Device 1, Device 2 be

$$|\psi_{\text{init}}\rangle = \sqrt{a}|0\rangle_{A_1} \otimes |1_V\rangle_B \otimes |\psi_r^0\rangle_1 \otimes |0\rangle_{A_2} \otimes |0_H\rangle_C \otimes |\psi_r^0\rangle_2 + \sqrt{1-a}|0\rangle_{A_1} \otimes |0_H\rangle_B \otimes |\psi_r^0\rangle_1 \otimes |0\rangle_{A_2} \otimes |1_H\rangle_C \otimes |\psi_r^0\rangle_2 \quad (3.16)$$

where $|\psi_r^0\rangle_i \in \mathcal{H}_i$, $i=1,2$, $\mathcal{D}(\mathcal{H}_i) = 2$,

$$|0\rangle \equiv \binom{1}{0}, |1\rangle \equiv \binom{0}{1}.$$

Consider the initial device states of $|\psi_r^0\rangle_1 = |0\rangle \otimes |0\rangle$, $|\psi_r^0\rangle_2 = |0\rangle \otimes |1\rangle$. The Schrödinger unitary evolution operator to be applied to $|\psi_{\text{init}}\rangle$ is given by

$$U(t) = e^{-iHt} = (I \otimes W) \otimes (I \otimes V).$$

For a fixed Schrödinger unitary evolution time $t = \tau_e$, the initial state is evolved in Step 1 to $|\psi_{\text{fin},1}\rangle = U(\tau_e)|\psi_{\text{init}}\rangle$ or

$$|\psi_{\text{fin},1}\rangle = \sqrt{a}\,|0\rangle_{A_1} \otimes |e_1^1\rangle_{1'} \otimes |0\rangle_{A_2} \otimes |f_1^1\rangle_{2'} + \sqrt{1-a}|0\rangle_{A_1} \otimes |e_2^1\rangle_{1'} \otimes |0\rangle_{A_2} \otimes |f_2^1\rangle_{2'}.$$

The operators $T_1$and T_2 of Step 2 are used to transfer the entanglement between Systems $1'$ and $2'$ to entanglement between the two ancilla qubits:

$$T_1 = (|1\rangle_{A_1} \otimes |e_1^1\rangle_{1'})(\langle 0|_{A_1} \otimes_{1'} \langle e_1^1|) + (|0\rangle_{A_1} \otimes |e_1^1\rangle_{1'})(\langle 0|_{A_1} \otimes_{1'} \langle e_2^1|)$$
$$T_2 = (|1\rangle_{A_2} \otimes |f_2^1\rangle_{2'})(\langle 0|_{A2} \otimes_{2'} \langle f_2^1|) - (|0\rangle_{A_2} \otimes |f_2^1\rangle_{2'})(\langle 0|_{A_2} \otimes_{2'} \langle f_1^1|).$$

The vectors $|e_i^1\rangle_{1'}, |f_i^1\rangle_{1'}, i = 1{,}2$ can be computed from Equation (3.8) and for this model are:

$$|e_1^1\rangle_{1'} = W|1_V\rangle_B \otimes |0\rangle \otimes |0\rangle,\ |e_2^1\rangle_{1'} = W|0_V\rangle_B \otimes |0\rangle \otimes |0\rangle$$
$$|f_1^1\rangle_{2'} = V\,|0_H\rangle_C \otimes |0\rangle \otimes |0\rangle,\ |f_2^1\rangle_{2'} = V\,|1_H\rangle_C \otimes |0\rangle \otimes |0\rangle.$$

Upon extending T_1 and T_2 to unitary matrices via the procedure previously shown, i.e., adding orthogonal rows to complete the basis, we arrive at X and Y. Applying the unitary $X \otimes Y$ to $|\psi_{\text{fin}}\rangle$ will transfer the state in the two ancilla qubits to become entangled at the expense of the loss of entanglement between Systems $1'$ and $2'$.

In Step 3, a Bell experiment is performed on the two ancilla qubits. Steps 1 and 2 were simulated for $t = .4$ using the Hamiltonian parameters $w = 2$, $\{\Omega_{1,1,}\Omega_{2,1,,}J_{x,1}, J_{y,1}, J_{z,1}\} = \{.5, .2, .1, .3, .5\}$ and $\{\Omega_{1,2,}\Omega_{2,2,,}J_{x,2}, J_{y,2}, J_{z,2}\} = \{.3, .4, .7, .2, .6\}$. Two initial device states were chosen as $|\psi_r^0\rangle_1 = |0\rangle \otimes |0\rangle, |\psi_r^0\rangle_2 = |0\rangle \otimes |1\rangle$. Results of the Bell experiment are given in terms of the CHSH sum in Figure 3.16 for both the Step 1 Schrödinger unitary prediction for *t=.4* and under an assumption that the photon takes a definite path via either the *B* or *C* Port. One sees that the CHSH sum is always greater than the case of a known photon path, for any initial superposition of the photon. Furthermore, the CHSH sum is the same independent of the initial state, the time of the unitary evolution, the number of particles composing the devices, the interactions within the device, the interactions between the photon and particles composing the device. In short, the argument given is rigorous.

Since the entanglement has been transferred to the two ancilla qubits, the unitary prediction can also be calculated analytically using a two-qubit version of the CHSH inequality [102]

$$S = \left|\hat{C}(a,b) + \hat{C}(a,\acute{b}) + \hat{C}(\acute{a},b) - \hat{C}(\acute{a},\acute{b})\right| \leq 2, \tag{3.17}$$

where $a, \acute{a}$ and $b, \acute{b}$ are the two-valued variables (± 1) for the first and second qubits respectively. Quantum mechanically, the correlation between a and b is given in terms of the system density matrix ρ and the Hermitian operators $\hat{a}$ and $\hat{b}$ corresponding to a and b:

$$\hat{C}(a,b) = \mathrm{Tr}\big(\rho(\hat{a}\otimes\hat{b})\big). \tag{3.18}$$

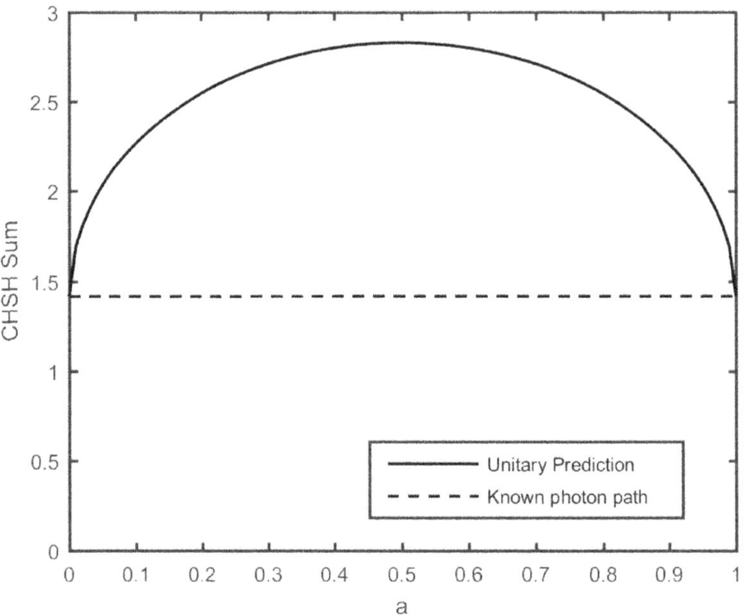

Figure 3.16: CHSH Sum versus degree of photon superposition, Unitary CHSH sum is $2\sqrt{2}\ at\ a = .5$ whereas for known photon path of either $|\psi^1\rangle_1 \otimes |\psi^0\rangle_2$ or $|\psi^0\rangle_1 \otimes |\psi^1\rangle_2$ the CHSH sum is $\sqrt{2}$ as seen in the dotted curve, independent of a.

The two-qubit measurement settings corresponding to the polarization measurement settings of Equation (3.3) can be shown to be given by

$$a = \sigma_z, \qquad \acute{a} = \sigma_x \tag{3.19}$$

$$b = \frac{\sigma_z+\sigma_x}{\sqrt{2}}, \ \acute{b} = \frac{\sigma_z-\sigma_x}{\sqrt{2}}. \tag{3.20}$$

Using the entangled pure state

$$|\psi\rangle = \sqrt{a}|00\rangle + \sqrt{1-a}|11\rangle, \tag{3.21}$$

the CHSH sum for the two ancillary qubits is found from the above equations to be

$$S = \sqrt{2}(1 + 2\sqrt{a(1-a)}). \tag{3.22}$$

As expected, Equation (3.22) is the same as the simulation of the unitary prediction in

Figure 3.16. The entanglement between two qubits can be quantified in terms of Wootters' *concurrence* [103] which, for the pure state of Equation (3.21) is given by

$$\mathbb{C} = 2\sqrt{a(1-a)} \tag{3.23}$$

so that the CHSH sum can be rewritten as

$$S = \sqrt{2}(1 + \mathbb{C}). \tag{3.24}$$

The CHSH sum for the known photon path is

$$S_{Cl} = \sqrt{2}. \tag{3.25}$$

Therefore, at this measurement setting, the fractional difference between the unitary and known photon path CHSH sums is exactly given by the entanglement of the state in the form of the concurrence

$$\frac{S_U - S_{Cl}}{S_{Cl}} = \mathbb{C}. \tag{3.26}$$

The CHSH sum in Equation (3.24) also results using a mixed state ρ with the same concurrence $\mathbb{C}(\rho) = 2\sqrt{a(1-a)}$. This can be seen from the result [103] that a mixed state of two qubits can be decomposed into a convex sum of pure states, all with concurrence equal to the concurrence of the mixed state

$$\rho = \sum_i c_i |\psi_i\rangle\langle\psi_i| \tag{3.27}$$

with $\sum_i c_i = 1$. Since each $|\psi_i\rangle$ is unitarily equivalent to a Schmidt decomposition of the form of Equation (3.21), the properties of the trace in Equation (3.18) imply that the mixed state CHSH sum is identical to that of the pure state result in Equation (3.24) at this measurement setting

$$S_\rho = \sqrt{2}(1 + \mathbb{C}(\rho)) = \sqrt{2}\left(1 + 2\sqrt{a(1-a)}\right). \tag{3.28}$$

Interpretation or Existence

Let us set $a = .5$ (which is indicative of a maximally entangled state) in order to examine the impact of our development, and assume the devices are 100% efficient. Assume in the Hypothesis H_0 in Hypothesis Test 3.1, that all systems evolve via Schrödinger's equation. After the initial Schrödinger unitary evolution of photon and devices, the subsequent operation of Step 3 is performed while assuming H_0 also proceeds unitarily. Now we know that the unitary prediction is to entangle the two-qubits, and hence Step 3 of the UMDT will result in a CHSH sum $2\sqrt{2}$ under H_0.

On the other hand, suppose that Device 1 is a system that we *know* is a bona fide efficient measurement device. In Schrödinger unitary theory the state after Step 1 is given in Equation (3.9)

$$|\psi_{\text{fin}}\rangle = |0\rangle \otimes (\sqrt{a}\,|\psi^1\rangle_1 \otimes |\psi^0\rangle_2 + \sqrt{1-a}|\psi^0\rangle_1 \otimes |\psi^1\rangle_2).$$

The state is predicted via Schrödinger's equation to be in a superposition of two terms. The first is that Device 1 absorbed the photon in Path B and evolved to a new readout state $|\psi^1\rangle_1$ while Device 1 evolved to state $|\psi^0\rangle_2$ indicative of no absorption. The second term is that Device 2 absorbed the photon in Path C and evolved to a new readout state $|\psi^1\rangle_2$ while Device 1 evolved to state $|\psi^0\rangle_1$ indicative of no absorption. On the other hand, we know that if a photon is sent via only Path B that the only unitary possibility, as the devices are assumed 100% efficient, is to evolve to $|\psi^1\rangle_1 \otimes |\psi^0\rangle_2$, and we know that if a photon is sent via only Path C that the final state that is predicted unitarily is $|\psi^0\rangle_1 \otimes |\psi^1\rangle_2$. Hence, we know the final states of the devices under Hypothesis H_0 when no superposition occurs. If one assumes that unitary evolution suffices to explain all processes including measurement, and that when a measurement occurs the final state exists in a definite pointer state of either $|\psi^1\rangle_1 \otimes |\psi^0\rangle_2$ or $|\psi^0\rangle_1 \otimes |\psi^1\rangle_2$ then one needs to examine whether there are any observational consequences if indeed the final state of the two devices ends up after Step 1 in the Schrödinger unitarily predicted state of Equation (3.9) or ends up in either $|\psi^1\rangle_1 \otimes |\psi^0\rangle_2$ or $|\psi^0\rangle_1 \otimes |\psi^1\rangle_2$.

The case when both Device 1 and Device 2 are in a state that corresponds to what would have been the case if the photon took either the upper or lower path results in a tensor product state of devices $|\psi^1\rangle_1 \otimes |\psi^0\rangle_2$ or $|\psi^0\rangle_1 \otimes |\psi^1\rangle_2$. One finds the UMDT CHSH sum corresponds to taking a 2-qubit state that is in a tensor product, and results in $\sqrt{2}$ as seen in the dotted curve in Figure 3.16. On the other hand, under H_0 the CHSH sum can be computed as $\sqrt{2}(1+\mathbb{C})$.

An important issue is whether or not the CHSH sum changes if one increases the number of particles in the device. Note that we first generally developed theory that generates the Step 1 and 2 operations for an arbitrary Hamiltonian. Suppose that one changes the Hamiltonian from the two-particle example to three particles and computes the results. The results are the same, $\sqrt{2}(1+\mathbb{C})$ for Schrödinger unitary and $\sqrt{2}$ for the case when the final state is a definite device pointer state of either $|\psi^1\rangle_1 \otimes |\psi^0\rangle_2$ or $|\psi^0\rangle_1 \otimes |\psi^1\rangle_2$. Suppose one changes the Hamiltonian to 100000 particles. It would take a long time to simulate on a computer, but one would obtain the same result.

The UMDT CHSH sum is independent of the number of particles. The result is independent of the operators: for example, one can include creation and annihilation operators common in quantum optics. The result is independent of the local time-like environment surrounding the detector that could possibly be directly affected by the interaction. The result is also independent of the type of particles that compose the device as they can be either bosons or fermions or composite particles. And the result is independent of local interactions between the particles composing the device; these can be completely arbitrary.

One might take exception to the fact that the initial states of the device were always pure states, and in fact there is the possibility of mixed state initial states. How are we assured that the same results will occur for such mixed initial states? This is a valid objection; the theory will be extended to mixed states shortly and shown that the same result occurs for initially mixed device states.

In this manner, the UMDT operations specified are indeed capable of resolving Hypothesis Test 3.1. Theoretically our construction shows that Hypothesis H_0 is false and H_1 is correct; given our assumptions, one *must* include an evolution process that diverges from Schrödinger's equation if one expects to be consistent with the experimental facts that measurements do occur and are statistically correlated with the degree of superposition *a*.

Now if it were the case that the substitution of either of these states were such that it did not provide any observational differences in this setup or others, the problem can rightly be deemed to be strictly interpretational and a matter that could be argued is best left as philosophy. However, as there does appear that there are observational differences, then the problem cannot be assumed *a priori* to be a strictly interpretational issue.

Let us now turn to Hypothesis Test 3.2 which examines whether or not Schrödinger unitary evolution is a valid description for Born's bifurcation rule. Born's rule predicts that when a measurement occurs in one of the two devices, there is a probabilistic bifurcation of the quantum state. In the case that we are considering, Born's rule predicts that the photon is found with probability a in Path B and $(1 - a)$ in Path C. This is consistent with either the state $|\psi^1\rangle_1 \otimes |\psi^0\rangle_2$ whereby Device 1 is in a final readout state and the Device 2 is in a final state indicating no measurement, or whereby the state is $|\psi^0\rangle_1 \otimes |\psi^1\rangle_2$ whereby Device 1 indicates no measurement and Device 2 is in a final state indicating a measurement. But either case results in the CHSH sum of $\sqrt{2}$. Now if one assumes only Schrödinger unitary evolution under H_0 if Hypothesis Test 3.2 and then applies Step 1 and Step 2, the result is predicted to be $\sqrt{2}(1 + \mathbb{C})$. As these predictions are the same for all numbers of particles that compose the device, it *cannot* be that Schrödinger's equation is a valid description of the physics as the number of particles of the device increases: Hypothesis H_0 of Hypothesis Test 3.2 is false.

Extension to Mixed States

In the prior development, it was assumed that the initial ready state of Device *i* is the pure state $|\psi_r^0\rangle_i$, *i=1,2*, $|\psi_r^0\rangle_i \in \mathcal{H}_i$, *i=1,2*, $\mathcal{D}(\mathcal{H}_i) = N$. The joint state of Device 1 and 2 was assumed to be a tensor product state given by $|\psi_r^0\rangle_1 \otimes |\psi_r^0\rangle_2$. The UMDT procedure developed previously is herein extended to the case of mixed states.

The initial ready state of Device *i* will be assumed to be described by the density matrix $\rho^0_{r,i}$, $\mathcal{D}(\mathcal{H}_i) = N$. Given a Hilbert space $\mathcal{H}$, a density matrix ρ sometimes referred to as a density operator $\rho: \mathcal{H} \to \mathcal{H}$, is a bounded positive semi-definite Hermitian unity trace operator. The set of density matrices, denoted $\mathfrak{M}(\mathcal{H})$, is a convex set with extremal elements that are pure states.

A Hermitian positive semi-definite matrix $\rho^0_{r,i} \in \mathfrak{M}(\mathcal{H})$ can be written as

$$\rho^0_{r,i} = \sum_{j=1}^{R_i} \lambda^i_j |\psi^0_{r,j}\rangle_i \, {}_i\langle \psi^0_{r,j}|$$

where $|\psi^0_{r,j}\rangle_i \in \mathcal{H}_i$,*i=1,2* are any set of orthogonal eigenstates of the density matrix with respective eigenstates $\lambda^i_j \geq 0$. The rank R_i of the density matrix denoted $\mathcal{R}(\rho^0_{r,i})$ is equal to unity in the case that the state is pure and is strictly greater than 1 in the case that the matrix is mixed.

The initial photon state after the PBS will be assumed as before to be a pure state given as

$$|\psi_{\text{photon,PBS}}\rangle = \sqrt{a}|1_V\rangle_B \otimes |0\rangle_C + \sqrt{1-a}|0\rangle_B \otimes |1_H\rangle_C,$$

where $a \in \mathbb{C}$, $|a| \leq 1$ is the degree of superposition. The density operator of the photon initial state is represented as

$$\rho_{\text{photon,PBS}} = |\psi_{\text{photon,PBS}}\rangle \langle \psi_{\text{photon,PBS}}|.$$

The initial state of the Ancilla, Photon, Device 1, and Device 2 can be written given as

$$\tilde{\rho}_{\text{init}} = \rho^0_{A_1,A_2} \otimes \rho_{\text{photon,PBS}} \otimes \rho^0_{r,1} \otimes \rho^0_{r,2}. \tag{3.29}$$

where $\rho^0_{A_1,A_2} = \rho^0_{A1} \otimes \rho^0_{A2}$ is the initial joint density matrix of Ancilla 1 and 2. This can be re-written in a manner similar to the case of pure states whereby Ancilla 1 and the *B* photon modes are adjacent to Device 1 while Ancilla 2 and *C* photon modes have been inserted adjacent to Device 2, which will be denoted by ρ_{init}. As in the case of the pure state treatment, the composite system of the joint state of the photon in the B port and Device 1 will be referred to as System $1'$ while System $2'$ denotes the joint state of the photon in the C port and Device 2. Let $U^{(1)}_{\text{sw}}$ denote the unitary operator that swaps Mode C with Device 1 so that System $1'$ and $2'$ are adjacent and denote $U^{(2)}_{\text{sw}}$ the unitary operator that changes the position of Ancilla 2 so that it is inserted between System $1'$ and $2'$. In this case,

$$\rho_{init} = U^{(3)}_{\text{sw}} \tilde{\rho}_{\text{init}} U^{(3)\,\prime}_{\text{sw}}$$

where $U^{(3)}_{\text{sw}} = U^{(2)}_{\text{sw}} U^{(1)}_{\text{sw}}$. Consider the unitary evolution of the initial state of Equation

(3.29) followed by the Step 1 unitary evolution to be specified via the matrices X and Y, which is then followed by the Step 2 Bell experiment which consists of measuring a set of observables. The expected value one of any of the observables $\mathcal{O}$ is given by,

$$\mathcal{E}(\rho_{\text{init}},\mathcal{O}) = \text{Tr}(\mathcal{O}U_c(\rho^{(0)}_{A_1,A_2} \otimes \rho_{\text{photon,PBS}} \otimes \sum_{j=1}^{R_1} \lambda^i_j |\psi^0_{r,j}\rangle_1 {}_1\langle\psi^0_{r,j}| \otimes \rho^0_{r,2})U_c{}')$$

where $U_c \equiv (X \otimes Y)U^{(3)}_{\text{sw}}$. Denoting the initial density matrix

$$\rho_{\text{init}}(j) \equiv \rho_{\text{photon,PBS}} \otimes |\psi^0_{r,j}\rangle_1 {}_1\langle\psi^0_{r,j}| \otimes \rho^0_{r,2},$$

it is found by completing Exercises 3.4-3.6 that $\mathcal{E}(\rho_{\text{init}},\mathcal{O})$ is completely characterized by $\mathcal{E}(\rho_{\text{init}}(j),\mathcal{O}), \forall$ j.

As we already have developed a methodology of specifying T_1 when both Device 1 and 2 are each pure states, let us for now hold j constant and assume that Device 1 is initialized to $|\psi^0_{r,j}\rangle_1 {}_1\langle\psi^0_{r,j}|$ and Device 2 initialized to a pure state. Consider then the following extension to the methodology of specifying T_1 for mixed states, for which after Step 1, $\mathcal{E}(\rho_{\text{init}}(j),\mathcal{O})$ has transferred the entanglement that existed between System $1'$ and System $2'$ to entanglement of the two qubits. The theory will be completed if a linear operator $T_1 \otimes T_2$ can be specified that achieves the desired entanglement transfer between System $1'$ and System $2'$ into entanglement of the two ancilla, for *all* of Device 1's eigenstates $|\psi^0_{r,j}\rangle_1$. With this goal in mind, define

$$|e^0_{1,j,}\rangle_{1'} = |1_V\rangle_B \otimes |\psi^0_{r,j}\rangle_1$$
$$|e^0_{2,j}\rangle_{1'} = |0\rangle_B \otimes |\psi^0_{r,j}\rangle_1$$

and $|e^1_{i,j}\rangle_{1'} = W|e^0_{i,j}\rangle_{1'}$, $i = 1{,}2$ as the unitarily evolved states after interaction of Photon B mode with Device 1. The solution for T_1 when Device 1 is initially in a pure state has previously been found and can be written (with a slight change to the notation)

$$T_{1,j} \equiv (|1\rangle_{A_1} \otimes |e^1_{1,j}\rangle_{1'})(\langle 0|_{A_1} \otimes {}_{1'}\langle e^1_{1,j}|) + (|0\rangle_{A_1} \otimes |e^1_{1,j}\rangle_{1'})(\langle 0|_{A_1} \otimes {}_{1'}\langle e^1_{2,j}|).$$

A linear map is needed that maps $|0\rangle_{A_1} \otimes |e^1_{1,j}\rangle_{1'}$ to $|1\rangle_{A_1} \otimes |e^1_{1,j}\rangle_{1'}$ and $|0\rangle_{A_1} \otimes |e^1_{2,j}\rangle_{1'}$ to $|0\rangle_{A_1} \otimes |e^1_{1,j}\rangle_{1'}$. Hence the mapping for the case of pure states must be extended to hold not only for any single j, but for all j. Consider the ansatz that the form of T_1 is the sum of these $T_{1,j}$, i.e.,

$$T_1 = \sum_{j=1}^{R} T_{1,j}. \tag{3.30}$$

Note that T_1 is a linear operator because the sum of linear operators is itself a linear operator. Given a similar construction for T_2, $T_1 \otimes T_2$ achieves the desired transfer from the entanglement between System $1'$ and System $2'$ to entanglement between the two Ancilla, see Exercise 3.8-3.9. Furthermore, T_1 and T_2 can be extended to unitary mappings X and Y, see Exercises 3.10-3.12.

So far, a procedure has been developed to handle any initial state that is of the form whereby one of the two Devices is initialized to a mixed state while the other device is initialized to a pure state. To complete the argument, the latter result is extended in which Device 1 and Device 2 can both be initialized to arbitrary mixed states, see Exercise 3.13. Hence, we have specified a UMDT for arbitrary initial device states.

Partial Density Matrices

Consider a bipartite composite system in a tensor product state given by $\rho = \rho_1 \otimes \rho_2$. We denote the operation of partial trace $\mathrm{Tr}_j\rho$ whereby System j has been traced from ρ in a manner that $\mathrm{Tr}_1\rho = \rho_2$, $\mathrm{Tr}_2\rho = \rho_1$. Partial trace is a unique linear operator with these properties. It is useful to investigate how the density matrices of the various systems are affected by the 3-Step operations; herein the partial densities of Systems $1'$ and $2'$ are found after each of the three steps. Defining,

$$\begin{aligned} \rho_{\mathrm{init},1'} &\equiv \mathrm{Tr}_{2'}\rho_{\mathrm{init}} \\ \rho_{\mathrm{init},2'} &\equiv \mathrm{Tr}_{1'}\rho_{\mathrm{init}} \\ \rho_{\mathrm{photon},B} &\equiv \mathrm{Tr}_{\mathrm{C}}|\psi_{\mathrm{photon,PBS}}\rangle\langle\psi_{\mathrm{photon,PBS}}| \\ \rho_{\mathrm{photon},C} &\equiv \mathrm{Tr}_{\mathrm{B}}|\psi_{\mathrm{photon,PBS}}\rangle\langle\psi_{\mathrm{photon,PBS}}|, \end{aligned}$$

it can be shown (Exercise 3.17)

$$\rho_{\mathrm{init},1'} = \rho_{\mathrm{photon},B} \otimes \rho_{r,1}^{(0)} \tag{3.31}$$

and similarly,

$$\rho_{\mathrm{init},2'} = \rho_{\mathrm{photon},C} \otimes \rho_{r,2}^{(0)}.$$

Consider the density matrix representation of the states $|e_1^0\rangle_{1'}$ and $|e_2^0\rangle_{1'}$ given by

$$\begin{aligned} \rho_{1,1'}^{(0)} &\equiv |1\rangle_{B\,B}\langle 1| \otimes \rho_{r,1}^{(0)} \\ \rho_{2,1'}^{(0)} &\equiv |0\rangle_{B\,B}\langle 0| \otimes \rho_{r,1}^{(0)}. \end{aligned}$$

The Schrödinger's equation unitary evolution of ρ_{init} is given by

$$\rho_{\mathrm{fin},1} = (I \otimes W \otimes I \otimes V)\rho_{\mathrm{init}}\,(I \otimes W \otimes I \otimes V)',$$

where the identity matrices apply to the ancilla qubits. It follows that the density matrix representations of the evolves states $|e_1^1\rangle_{1'}$ and $|e_2^1\rangle_{1'}$ can be written

$$\rho_{1,1'}^{(1)} = W\rho_{1,1'}^{(0)}W'$$
$$\rho_{2,1'}^{(1)} = W\rho_{2,1'}^{(0)}W'.$$

It can be shown (Exercise 3.18) after Step 1, the density matrix of System $1'$ denoted $\rho_{1'}^{(1)}$ is

$$\rho_{1'}^{(1)} = a\rho_{1,1'}^{(1)} + (1-a)\rho_{2,1'}^{(1)}. \quad (3.32)$$

Step 2 is applied, for which the resulting density matrix is given by

$$\rho_{\text{fin},2} \equiv (X \otimes Y)\rho_{\text{fin},1}(X \otimes Y)'.$$

Now, for the case of pure states, X maps both $\rho_{1,1'}^{(1)}$ and $\rho_{2,1'}^{(1)}$ to $\rho_{1,1'}^{(1)}$. Hence the density matrix of System $1'$ denoted $\rho_{1'}^{(2)}$ becomes

$$\rho_{1'}^{(2)} = \text{Tr}_{A_1}X(|0\rangle_{A_1\,A_1}\langle 0| \otimes \rho_{1'}^{(1)})X'$$
$$\rho_{1'}^{(2)} = \rho_{1,1'}^{(1)}.$$

After Step 2, the density matrix of Qubit 1 is

$$\rho_{A_1}^{(2)} = \text{Tr}_{1'}X(\rho_{A_1}^{(1)} \otimes \rho_{1'}^{(1)})X',$$

which is found to be

$$\rho_{A_1}^{(2)} = a|0\rangle_{A_1\,A_1}\langle 0| + (1-a)|1\rangle_{A_1\,A_1}\langle 1|.$$

Step 3 is the application of the observables representing Bell's experiment. Each of these observables can be written as a (different) tensor product $\mathcal{O} = \mathcal{O}_1 \otimes \mathcal{O}_2$ that is applied to the two ancilla qubits. Step 3 yields the expected value $\mathcal{E}(\rho_{\text{init}}, \mathcal{O}) = \text{Tr}\left((\mathcal{O}_1 \otimes \mathcal{O}_2)\rho_{A_1,A_2}\right)$, where ρ_{A_1,A_2} is the joint state of the composite system of the two ancillae after Step 2.

Specific Device-Particle Modeling

The new UMDT construction applicable to mixed states can now be applied to a specific Device-Particle Model. We will consider the same Hamiltonian of the devices given by Equation (3.14) as in the previous case when the devices were initialized to a pure state. Consider the initial density matrices of the two spins that compose Device

1 and Device 2:

$$\rho_{r,1}^{(0)} = \begin{pmatrix} 1/2 & 0 & 0 & 0 \\ 0 & 1/4 & 0 & 0 \\ 0 & 0 & 1/4 & 0 \\ 0 & 0 & 0 & 0 \end{pmatrix}$$

$$\rho_{r,2}^{(0)} = \begin{pmatrix} 1/4 & 0 & 0 & 0 \\ 0 & 3/8 & 3/8 & 0 \\ 0 & 3/8 & 3/8 & 0 \\ 0 & 0 & 0 & 0 \end{pmatrix}.$$

Note that $\rho_{r,1}^{(0)}$ is a rank 3 matrix with a repeated eigenvalue of ¼ and can be shown to be an entangled two-spin state. There is a 2-dimensional subspace that is spanned by two orthogonal eigenstates corresponding to the eigenvalue 1/4. Consider the following choice of orthogonal eigenstates of $\rho_{r,1}^{(0)}$:

$$|\psi_{r,1}^0\rangle_1 = \begin{pmatrix} 1 \\ 0 \\ 0 \\ 0 \end{pmatrix}, |\psi_{r,2}^0\rangle_1 = \begin{pmatrix} 0 \\ 1 \\ 0 \\ 0 \end{pmatrix}, |\psi_{r,3}^0\rangle_1 = \begin{pmatrix} 0 \\ 0 \\ 1 \\ 0 \end{pmatrix},$$

$\rho_{r,2}^{(0)}$ is a rank 2 matrix with no repeated eigenvalues. Consider the following choice of orthogonal eigenstates of $\rho_{r,2}^{(0)}$:

$$|\psi_{r,1}^0\rangle_2 = \begin{pmatrix} 1 \\ 0 \\ 0 \\ 0 \end{pmatrix}, |\psi_{r,2}^0\rangle_2 = \frac{1}{\sqrt{2}}\begin{pmatrix} 0 \\ 1 \\ 1 \\ 0 \end{pmatrix}.$$

Now construct for both $j = 1,2$

$$\begin{aligned} |e_{1,j,}^0\rangle_{1'} &= |1_V\rangle_B \otimes |\psi_{r,j}^0\rangle_1 \\ |e_{2,j}^0\rangle_{1'} &= |0\rangle_B \otimes |\psi_{r,j}^0\rangle_1 \\ |f_{1,j}^0\rangle_{2'} &= |0\rangle_C \otimes |\psi_{r,j}^0\rangle_2 \\ |f_{2,j}^0\rangle_{2'} &= |1_H\rangle_C \otimes |\psi_{r,j}^0\rangle_2 \end{aligned}$$

which are unitarily evolved via Schrödinger's equation in Step 1 to $|e_{i,j}^1\rangle_{1'} = W|e_{i,j}^0\rangle_{1'}$ and $|f_{i,j}^1\rangle_{2'} = V|f_{i,j}^0\rangle_{2'}$, for $i = 1,2, j = 1,2$. In order to specify the Step 2 operation, consider the composite Ancilla 1-System $1'$. The two matrices $T_{1,1}$ and $T_{1,2}$ are computed via

$$T_{1,j} = (|1\rangle_{A_1} \otimes |e_{1,j}^1\rangle_{1'})(\langle 0|_{A_1} \otimes {}_{1'}\langle e_{1,j}^1|) + (|0\rangle_{A_1} \otimes |e_{1,j}^1\rangle_{1'})(\langle 0|_{A_1} \otimes {}_{1'}\langle e_{2,j}^1|)$$

and the sum is formed, $T_1 = T_{1,1} + T_{1,2}$, which can be extended to a unitary operation

X with use of the singular value decomposition (Exercises 3.10-3.12). Utilizing the same procedure applied to the composite Ancilla 2-System 2′, a unitary operation Y is found and $X \otimes Y$ is applied to complete Step 2.

Once Step 2 is completed, the ancilla qubits are extracted from the overall state and a Bell (CHSH) experiment is performed. The computation for the initial state above was found to result via computer simulation in Figure 3.17 for the Hamiltonian with parameters considered previously, i.e. $t = .4$ using the Hamiltonian parameters $w = 2$, $\{\Omega_{1,1,}\Omega_{2,1,,}J_{x,1},J_{y,1},J_{z,1}\} = \{.5,.2,.1,.3,.5\}$ and $\{\Omega_{1,2,}\Omega_{2,2,,}J_{x,2},J_{y,2},J_{z,2}\} = \{.3,.4,.7,.2,.6\}$. As expected from the mixed-state two-qubit CHSH sum result in Equation (3.28), it can be seen that this result in Figure 3.17 is identical to Figure 3.16 which is the case when the devices are initialized to pure states. Moreover, the unitary operation that is utilized in Step 2 is not of the form that requires a time-reversal of the unitary that occurred in Step 1.

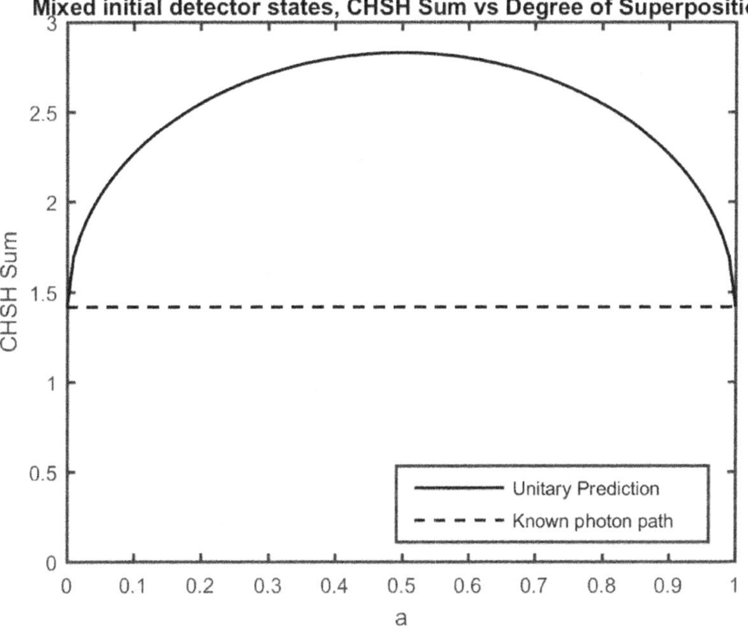

Figure 3.17 CHSH Sum versus degree of photon superposition when initial device states are mixed.

Entanglement in the Measurement Problem

In the field of Quantum Information, an important application is the computation of entanglement measures. Understanding the entanglement prediction under Schrödinger's equation is quintessential to achieve a full understanding of the measurement problem.

In the UMDT, a Bell experiment has been applied as Step 3 in the construction to discriminate Schrödinger's unitary evolution from the quantum states required under measurement. The CHSH sum was quantified by the entanglement between System 1′ and System 2′ in terms of the concurrence. After Step 1 it can be seen from Equation (3.32) that $\mathbb{C}(\rho_{\text{fin},1}) = 2\sqrt{a(1-a)}$. And since it is known that local unitarily evolution does not change entanglement for any entanglement monotone [104] and the concurrence is an entanglement monotone, it follows $\mathbb{C}(\rho_{\text{fin},2}) = 2\sqrt{a(1-a)}$. In the case whereby the devices are bonafide measurement devices, then System 1′ and System 2′ are represented in a product state or a measurement state, in which case $\mathbb{C}(\rho_{\text{fin},2}) = 0$.

One might consider what relationship the CHSH inequality has to the issue of nonlocality in the measurement setup that has been proposed. The reader should note that the CHSH inequality so far has been used herein as a tool to demonstrate that the case of measurement is distinguishable from the case of unitary evolution. That is, the argument presented requires only that there be a difference in the CHSH sum between the cases of measurement and unitary evolution, and the issue of whether or not there exists a local hidden variable model for a particular photon superposition state has not been utilized. Nevertheless, one can consider the issue of nonlocality as it relates to the operations that have been proposed. Note that for the case of measurement, whereby the devices result in a tensor product state, the CHSH inequality obtains the value $\sqrt{2}$, which is the upper bound for the strengthed CHSH inequality $S \leq \sqrt{2}$ for separable states. If one considers the tensor product measurement result as resulting from a hidden local variable such that a photon is emitted from the B Port of the beam splitter with probability a and the C port with probability 1-a, then this can be seen to represent a local hidden variable model for the case of measurement. On the other hand, for the unitarily predicted cases whereby the degree of superposition a is such that the CHSH inequality is less than 2, one cannot conclude from the current demonstration whether or not there is a local hidden variable that can describe the results of follow-up measurements. There is, however, a different set of measurements proposed in a generalized Bell inequality in [105] that could be used to demonstrate nonlocality after Step 2 when the state is not maximally entangled. This or other considerations might also be useful in developing procedures that provide maximal distinguishability between measurement and unitary evolution, particularly if the degree of superposition of the photon is less than maximally entangled.

Exercises

3.1 Assume the setup as in Figure 3.4 and a bipartite system that consists of the Mode B photon interacting with Device 1 for which Device 1 is initialized to an arbitrary pure state, denoted $|\psi_i\rangle$. Consider the two initial states of System 1′ given by $|m_0\rangle = |1\rangle_B \otimes |\psi_i\rangle$. and $|m_1\rangle = |0\rangle_B \otimes |\psi_i\rangle$ where $|1\rangle_B$ denotes the presence of a photon in Mode B and $|0\rangle_B$ denotes the vacuum state. Given any unitary operator that evolves System 1′, show that the state after evolving $|m_0\rangle$ remains orthogonal to the state after evolving $|m_1\rangle$.. More generally, show that unitary evolution maintains the geometry between the two system states, hence maintains the properties of orthogonality and non-orthogonality.

3.2 Assume in Exercise 3.1 that the initial photon state is $|1_V\rangle_B$ and after unitary evolution and interaction with Device 1, the final photon state is $|0\rangle$. Is the initial device state necessarily orthogonal to the final device state?

3.3 Assume initially pure states of the devices in the setup of Figure 3.13 and that the state $|0\rangle_{A_1} \otimes |e_1^1\rangle_{1'}$ is mapped to $|1\rangle_{A_1} \otimes |e_1^1\rangle_{1'}$ and $|0\rangle_{A_1} \otimes |e_2^1\rangle_{1'}$ is mapped to $|0\rangle_{A_1} \otimes |e_1^1\rangle_{1'}$ in Step 2 (and similarly for System 2). In this exercise, the final state of System $1'$ is generalized from $|e_1^1\rangle_{1'}$ to an arbitrary state $|\psi_f^1\rangle$:

a. Given arbitrary pure states $|\psi_f^1\rangle$ and $|\psi_f^2\rangle$, construct unitary operators U_N^1 and U_N^2 such that $|\psi_f^1\rangle_{1'} = U_N^1|e_1^1\rangle_{1'}$ and $|\psi_f^2\rangle_{2'} = U_N^2|e_1^1\rangle_{2'}$.

b. Let $X_t \equiv (I_2 \otimes U_N^1)X, Y_t \equiv (I_2 \otimes U_N^2)Y$, where $|\psi_f^1\rangle = U_N^1|e_1^1\rangle_1$, and $|\psi_f^2\rangle = U_N^2|e_1^1\rangle_2$ and apply X_t and Y_t in place of Equation X and Y in Equation (3.12) to show

$$|\psi_{\text{fin},2}\rangle = \{\sqrt{a}|1\rangle_{A_1} \otimes |0\rangle_{A_2} - \sqrt{1-a}|0\rangle_{A_1} \otimes |1\rangle_{A_2}\} \otimes |\psi_f^1\rangle_{1'} \otimes |\psi_f^2\rangle_{2'}.$$

c. Conclude that there is a large class of unitary operators that achieves entanglement transfer for pure states.

3.4 Show that that the expected value of any Observable $\mathcal{O}$ of a system in an initially mixed state ρ is fully characterized by knowledge of the expected value of Observable $\mathcal{O}$ when the initial state is any eigenstate of the density matrix. That is, show that $\mathcal{E}(\rho, \mathcal{O}) \equiv \text{Tr}(\mathcal{O}\rho)$ can be written in terms of $\mathcal{E}(|\psi_j\rangle\langle\psi_j|, \mathcal{O})$, where $\rho = \sum_{j=1}^{R} \lambda_j |\psi_j\rangle\langle\psi_j|$.

3.5 Let $\rho = \rho_1 \otimes \rho_2 \otimes \rho_3$. Show that $\mathcal{E}(\rho, \mathcal{O}) \equiv \text{Tr}(\mathcal{O}\rho)$ can be written in terms of $\mathcal{E}(\rho_1 \otimes |\psi_j\rangle\langle\psi_j| \otimes \rho_3, \mathcal{O})$, where $\rho_2 = \sum_{j=1}^{R} \lambda_j |\psi_j\rangle\langle\psi_j|$.

3.6 Suppose that ρ in Exercise 3.5 undergoes any unitary evolution U prior to the measurement of Observable $\mathcal{O}$.

a. Show that $\mathcal{E}(\rho, \mathcal{O})$ can be written in terms of

$\mathcal{E}(\rho_1 \otimes |\psi_j\rangle\langle\psi_j| \otimes \rho_3, \mathcal{O})$, where $\rho_2 = \sum_{j=1}^{R} \lambda_j |\psi_j\rangle\langle\psi_j|$.

b. Find suitable $\rho_1, |\psi_j\rangle$, and ρ_3 to prove using the setup of Figure 3.13

$$\mathcal{E}(\tilde{\rho}_{\text{init}}, \mathcal{O}) = \text{Tr}(\mathcal{O}U_c(\rho_{\text{anc}}^0 \otimes \rho_{\text{photon,PBS}} \otimes \sum_{j=1}^{R_1} \lambda_j^i |\psi_{r,j}^0\rangle_1 {}_1\langle\psi_{r,j}^0| \otimes \rho_{r,2}^0)U_c{}^\dagger)$$

is completely characterized by $\mathcal{E}(\tilde{\rho}_{\text{init}}(j), \mathcal{O})$, where

$$\tilde{\rho}_{\text{init}}(j) = \rho_{\text{photon,PBS}} \otimes |\psi_{r,j}^0\rangle_1\,{}_1\langle\psi_{r,j}^0| \otimes \rho_{r,2}^0.$$

3.7 Assume a system where initial state ρ undergoes any unitary evolution U prior to the measurement of Observable $\mathcal{O}$. Show that $\mathcal{E}(\rho,\mathcal{O})$ is the same as the Heisenberg representation of measuring the evolved Observable $\mathcal{O}' = U^{\dagger}\mathcal{O}\,U$ when the system has initial state ρ but does not undergo unitary evolution prior to measuring $\mathcal{O}'$.

3.8 Consider an initial state of Field-System 1′-System 2′ assuming the setup in Figure 3.13 is given by

$$\tilde{\rho}_{\text{init}}(j) = \rho_{\text{photon,PBS}} \otimes |\psi_{r,j}^0\rangle_1\,{}_1\langle\psi_{r,j}^0| \otimes \rho_{r,2}^0.$$

Let $T_1 = \sum_{j=1}^{R_1} T_{1,j}$ where

$$T_{1,j} = (|1\rangle_{A_1} \otimes |e_{1,j}^1\rangle_{1'})(\langle 0|_{A_1} \otimes {}_{1'}\langle e_{1,j}^1|) + (|0\rangle_{A_1} \otimes |e_{1,j}^1\rangle_{1'})(\langle 0|_{A_1} \otimes {}_{1'}\langle e_{2,j}^1|)$$

$$|e_{1,j,}^0\rangle_{1'} = |1_V\rangle_B \otimes |\psi_{r,j}^0\rangle_1$$
$$|e_{2,h}^0\rangle_{1'} = |0\rangle_B \otimes |\psi_{r,j}^0\rangle_1.$$

and $|e_{i,j}^1\rangle_{1'} = W|e_{i,j}^0\rangle_{1'}$, $i = 1,2$ are the unitarily evolved states after interaction of Photon B mode with Device 1. Show for all $j = k$,

$$T_{1,j}(|0\rangle_{A_1} \otimes |e_{1,k}^1\rangle_{1'}) = |1\rangle_{A_1} \otimes |e_{1,j}^1\rangle_{1'}$$
$$T_{1,j}(|0\rangle_{A_1} \otimes |e_{2,k}^1\rangle_{1'}) = |0\rangle_{A_1} \otimes |e_{1,j}^1\rangle_{1'}$$

and that for all $j \neq k$,

$$T_{1,j}\,(|0\rangle_{A_1} \otimes |e_{1,k}^1\rangle_{1'}) = 0$$
$$T_{1,j}\,(|0\rangle_{A_1} \otimes |e_{2,k}^1\rangle_{1'}) = 0,$$

and that therefore $|0\rangle_{A_1} \otimes |e_{1,k}^1\rangle_{1'} \in \mathfrak{N}(T_{1,j})$ and $|0\rangle_{A_1} \otimes |e_{2,k}^1\rangle_{1'} \in \mathfrak{N}(T_{1,j})$ where $\mathfrak{N}(A)$ denotes the null space of an arbitrary matrix A.

3.9 Assume an initial state in the setup of Figure 3.13 is of the form $\rho_{\text{init}}(j,k) = \rho_{\text{photon,PBS}} \otimes |\psi_{r,j}^0\rangle_1\,{}_1\langle\psi_{r,j}^0| \otimes |\psi_{r,k}^0\rangle_2\,{}_2\langle\psi_{r,k}^0|$, and that $T_1 = \sum_{j=1}^{R_1} T_{1,j}$ is derived as in Exercise 3.8 and similarly $T_2 = \sum_{k=1}^{R_2} T_{2,k}$ using the same procedure for System 2′. Show that $(T_1 \otimes T_2)$ maps $|0\rangle_{A_1} \otimes |e_{1,j}^1\rangle_{1'}$ to $|1\rangle_{A_1} \otimes |e_{1,j}^1\rangle_{1'}$ and $|0\rangle_{A_1} \otimes |e_{2,j}^1\rangle_{1'}$ to $|0\rangle_{A_1} \otimes |e_{1,j}^1\rangle_{1'}$ for all $j = 1,2,\ldots,R_1$ in System 1′, and similarly $|0\rangle_{A_2} \otimes |e_{1,k}^1\rangle_{2'}$ to $|1\rangle_{A_2} \otimes |e_{2,k}^1\rangle_{2'}$ and $|0\rangle_{A_2} \otimes |e_{2,k}^1\rangle_{2'}$ to $|0\rangle_{A_2} \otimes |e_{2,k}^1\rangle_{2'}$ for all $k = 1,2,\ldots,R_2$.

3.10 Given a singular value decomposition (SVD) of a density matrix $\rho = USV^{\dagger}$, with U, V unitary and S a diagonal matrix containing the singular values of ρ, show that there are r singular values all unity where $r = \mathcal{R}(\rho)$ (the rank of ρ) and the remaining diagonal elements are zero.

3.11 Consider the setup of Figure 3.13. Given a SVD of $T_{1,j} = USV^{\dagger}$, show that $W = UV^{\dagger}$ is a unitary matrix that maintains the desired mappings of $T_{1,j}$ for the initial state $\tilde{\rho}_{\text{init}}(j,k) = \rho_{\text{photon,PBS}} \otimes |\psi_{r,j}^{0}\rangle_{1\,1}\langle\psi_{r,j}^{0}| \otimes |\psi_{r,k}^{0}\rangle_{2\,2}\langle\psi_{r,k}^{0}|$, i.e. W maps $|0\rangle_{A_1} \otimes |e_{1,j}^{1}\rangle_{1'}$ to $|1\rangle_{A_1} \otimes |e_{1,j}^{1}\rangle_{1'}$ and $|0\rangle_{A_1} \otimes |e_{2,j}^{1}\rangle_{1'}$ to $|0\rangle_{A_1} \otimes |e_{1,j}^{1}\rangle_{1'}$.

3.12 Consider the setup of Figure 3.13. Assuming $T_1 = \sum_{j=1}^{R_1} T_{1,j}$ and given a SVD of $T_1 = USV^{\dagger}$, show that $W = UV^{\dagger}$ is a unitary matrix that maintains the desired mappings of $T_{1,j}$ for all $j = 1,2,\dots,R_1$.

3.13 Consider the setup of Figure 3.13. Assume $T_1 = \sum_{j=1}^{R_1} T_{1,j}$ and $T_2 = \sum_{k=1}^{R_2} T_{2,k}$ are computed for an initial state of the form $\rho_{\text{init}} = \rho_{\text{photon,PBS}} \otimes \rho_1 \otimes \rho_2$ where ρ_1 is the initial state of Device 1 and ρ_2 is the initial state of Device 2. Show when both ρ_1 and ρ_2 are mixed states that the unitary operators $W \otimes V$ continue to achieve the desired transfer of the entanglement between System $1'$ and System $2'$ to entanglement between the ancillae.

3.14 In Exercise 3.3, it was shown that it is not a necessary condition that the final state of System $1'$ be mapped to $|e_1^1\rangle_{1'}$ but rather can be arbitrarily chosen as $|\psi_f^1\rangle$. For the case of an initially mixed System $1'$ density matrix, it has been demonstrated in Exercise 3.11 that a unitary matrix W can be found that maps $|0\rangle_{A_1} \otimes |e_{1,j}^{1}\rangle_{1'}$ to $|1\rangle_{A_1} \otimes |e_{1,j}^{1}\rangle_{1'}$ and $|0\rangle_{A_1} \otimes |e_{2,j}^{1}\rangle_{1'}$ to $|0\rangle_{A_1} \otimes |e_{1,j}^{1}\rangle_{1'}$, hence the final state of System $1'$ is dependent on the eigenvector j $|e_{1,j}^{1}\rangle_{1'}$. Can the procedure be extended in the same manner as in Exercise 3.3 so that the final state of System $1'$ can be chosen arbitrarily but fixed i.e. $|\psi_f^1\rangle$ independent of j?

3.15 Consider the setup of Figure 3.13. Can ρ_{init} always be written in the form

$$\rho_{A1}^{(0)} \otimes \rho_{\text{photon},B} \otimes \sum_{j=1}^{R_1} \lambda_j^1 |\psi_{r,j}^{0}\rangle_{1\,1}\langle\psi_{r,j}^{0}| \otimes \rho_{A2}^{(0)} \otimes \rho_{\text{photon},C} \otimes \sum_{j=1}^{R_2} \lambda_j^2 |\psi_{r,j}^{0}\rangle_{2\,2}\langle\psi_{r,j}^{0}|$$

where $\rho_{\text{photon},B} \equiv \text{Tr}_{\text{C}}|\psi_{\text{photon},PBS}\rangle\langle\psi_{\text{photon},PBS}|$ and $\rho_{\text{photon},C} \equiv \text{Tr}_{\text{B}}|\psi_{\text{photon,PBS}}\rangle\langle\psi_{\text{photon,PBS}}|$?

3.16 Consider the computation of $\mathcal{E}(\rho_{\{A_1,A_2\}},\mathcal{O}) = \text{Tr}\left((\mathcal{O}_1 \otimes \mathcal{O}_2)\,\rho_{\text{anc}}\right)$ where ρ_{anc} is the joint state of the composite system of two ancillae in Figure 3.13. Given $\rho_{A_1} \equiv \text{Tr}_{A_2}\rho_{\text{anc}}$, $\rho_{A_2} \equiv \text{Tr}_{A_1}\rho_{\text{anc}}$, does $\text{Tr}\left((\mathcal{O}_1 \otimes \mathcal{O}_2)\,\rho_{\text{anc}}\right) = \text{Tr}\left(\mathcal{O}_1\rho_{A_1} \otimes \mathcal{O}_2\,\rho_{A_2}\right)$?

3.17 Given that $\rho_{\text{init},1'} \equiv \text{Tr}_{2'}\rho_{\text{init}}$ and $\rho_{\text{photon},B} \equiv \text{Tr}_{C}|\psi_{\text{photon,PBS}}\rangle\langle\psi_{\text{photon,PBS}}|$, derive Equation (3.31): $\rho_{\text{init},1'} = \rho_{\text{photon},B} \otimes \rho_{r,1}^{(0)}$.

3.18 Show that after Step 1 is completed $\rho_{1'}^{(1)}$ (the density matrix of System $1'$) satisfies Equation (3.32): $\rho_{1'}^{(1)} = a\rho_{1,1'}^{(1)} + (1-a)\rho_{2,1'}^{(1)}$.

3.19 Consider the setup of Figure 3.13. Given that the rank of the initial density matrix $\rho_{r,1}^0$ of Device 1 is equal to R_1, what is the rank of the density matrix of System 1′ after Step 1 is completed?

3.20 Consider the setup of Figure 3.13. Suppose that after Step 1 is completed, that additional photon detectors are placed in the local area completely surrounding each device that can be used to determine if a photon has been re-emitted by the device. Suppose it is found that no photons exist in either the B or C Mode. The case whereby one post-selects upon finding no photons is related to the effect of negative result measurement, also commonly referred to as interaction-free measurement [106] [107]:

a. Describe the set of possible outcomes via a positive operator value measurement.

b. Provide an example via computer simulation or theoretically of an initial device state $|\psi_r^0\rangle_1$ and a Hamiltonian whereby orthogonal pure device states $|\psi^0\rangle_1 \perp |\psi^1\rangle_1$ result after post-selecting the no-photon case.

c. Provide an example via computer simulation or theoretically of an initial pure device state whereby $|\psi^0\rangle_1$ is not orthogonal to $|\psi^1\rangle_1$ after post-selecting the no-photon case.

d. Is this result consistent with Exercise 3.1 assuming the evolution were strictly unitary? If all evolution is unitary wouldn't it be the case that $|\psi^0\rangle_1 \perp |\psi^1\rangle_1$ in an actual experiment if it was determined from interaction-free measurement that the photon was necessarily absorbed into the device?

3.21 Assume a photon in the Mode A impinges entirely on a device. A device is defined as 100% absorbing if during the interaction time 1) the device absorbs the photon 100% of the time when the state of Mode A is $|1\rangle_A$, and 2) does not re-emit photons. A device is defined as 100% efficient if, given a photon absorption, the final device state $\rho_1^{(1)}$ is always fully distinguishable from the final device state $\rho_0^{(1)}$ obtained when the state of Photon Mode A is $|0\rangle_A$:

a. Is 100% absorption a necessary and/or a sufficient condition for 100% efficiency under unitary evolution?

b. Given that the final device states are the pure states $|\psi^0\rangle_1$, $|\psi^1\rangle_1$, is orthogonality of the final device states a necessary and/or sufficient c condition for 100% efficiency under unitary evolution?

3.22 Assume that the final device states in Exercise 3.21 can be mixed states. Provide necessary and sufficient conditions on $\rho_0^{(1)}$ and $\rho_1^{(1)}$ so that the device is 100% efficient under unitary evolution (hint: use results from [108]).

3.23 Suppose after Step 1 is completed in the setup of Figure 3.13, a POVM is applied to both Device 1 and 2. Given that the concurrence $\mathbb{C}$ is an entanglement monotone, use the properties of entanglement monotones from [109] to conclude that the concurrence on-average after the POVM must be less than or equal to the concurrence before the POVM. Conclude that the concurrence on-average after the POVM of Exercise 3.20 is applied must remain the same or decrease.

3.24 Suppose that Device 1 in the setup of Figure 3.13 is described by a Hilbert space of dimension N and is 100% absorbing. Using the results from prior exercises, show that a necessary condition that Device 1 is 100% efficient is that $\mathcal{R}(\rho_0^{(1)}) + \mathcal{R}(\rho_1^{(1)}) \leq N$.

3.25 Suppose that a measurement device is utilized in the laboratory which is macroscopic and for which the device readout state when the device detects a particle is macroscopically different than when it does not detect a particle. Furthermore, the device is sensitive in the sense that it always absorbs the particle if it impinges fully on the detector, and does not re-emit photons.

a. Using the results from Exercise 3.24, conclude that the initial state of the device could not have been a fully mixed state.

b. Suppose that the device is characterized experimentally to be strictly less than 100% efficient. Can it still be concluded that the initial state could not have been a fully mixed state?

3.26 Suppose that a device has an efficiency of $\eta = 100\% - \mathcal{E}$, where $\mathcal{E} > 0$ is small. Assume after Step 1 that the density matrices $\rho_1^{(1)}, \rho_0^{(1)}$ are full rank and have the same eigenstates. Show by counter-example that the set of eigenvalues of $\rho_1^{(1)}$ and $\rho_0^{(1)}$ are constrained by η, that is, they cannot be arbitrary positive values that sum to unity (an interesting extension would be to investigate the constraint on the set of eigenvalues when the eigenstates of $\rho_1^{(1)}$ and $\rho_0^{(1)}$ can be different).

3.27 In the methodology in the text developed to handle initial mixed device states applicable to the setup in Figure 3.13, the operations T_1 and T_2 computed in Equation (3.30) are used to derived unitary operators that are applied to System 1′ and System 2′ respectively. Suppose that one were constrained to apply local unitary operations only to System 1 and 2, i.e. to the devices themselves that does not include the local electromagnetic field surrounding each device. Given that the device is 100% absorbing, do you expect that a procedure for arbitrary mixed initial states of the devices in the form $\rho_{r,1}^{(0)} \otimes \rho_{r,2}^{(0)}$ can be developed that would give the same results as found in Figure 3.17? If not, can you find sufficient conditions on $\mathcal{R}(\rho_{r,1}^{(0)})$ and $\mathcal{R}(\rho_{r,2}^{(0)})$ so that a procedure can be found?

3.28 Are the device conditions of 100% absorption and 100% efficiency sufficient in order for a device to be considered to be a bona fide measurement device?

CHAPTER 4

Discerning Approaches

Humpty Dumpty sat on a wall
Humpty Dumpty had a great fall.
All of Schrödinger's minions were put into action
But Rosenfeld's many pages were just a distraction.
Scully, Englert, and Schwinger strived to gain traction
And even Aharonov and Feynman could get no satisfaction.
All of Schrödinger's successors tried mightily in vain
But couldn't put Humpty Dumpty together again.

The Incompleteness of Quantum Mechanics

Both measurement and unitarity form the basis of Bohr's complementarity, and the Copenhagen Interpretation in which there are dual descriptions in Nature. Whether or not the measurement postulate plus Schrödinger unitarity are sufficient aspects in the full description of Nature is related to the question of the incompleteness of quantum mechanics.

One impact of the measurement problem is that the measurement postulate of von Neumann cannot follow from the second Schrödinger unitary postulate. Historically this was anticipated rather quickly. In fact, the inadequacy of a completely causal deterministic description had been argued by Bohr shortly after Schrödinger's equation was formulated in 1926 as will be further examined in Chapter 5. Born's rule was then proposed and found to agree with the results of experiment and was incorporated by von Neumann in 1932 into the two postulates of quantum mechanics [13]. Moreover, this methodology is what is used today in nearly every paper on quantum mechanics. Why then is the measurement problem so subtle to comprehend? Because the arguments that had been put forward by von Neumann and others are not operational methods that can distinguish Schrödinger unitary evolution from measurement. The arguments have not been fully convincing and one is often left to wonder what would happen in the limit as the number of particles composing the measurement device increases. One is left in the dark as to how to conduct an experiment that conclusively shows that the difference does not diminish with the size of the devices or the type of interactions or the types of particles, etc. In order to develop an operational method, it is necessary to show that there is a predicted experimental difference that does not approach zero as the number of particles grows, between measurement and Schrödinger unitary evolution, as has been demonstrated in Chapter 3. But that there is an experimentally observed difference between product states and entangled states was not generally known until after 1964 by Bell [110] and only for two qubits in a pure state.

Einstein believed that Quantum Mechanics is incomplete: "Is the Moon there when nobody looks," he asks. Why would Einstein say such things? The reason should be clear by now. Schrödinger unitary evolution makes predictions of entangled states whereas product states are predicted under measurement. It has come to light in research by Howard [3] [111], which is discussed in Chapter 5, that Einstein understood the difference between product states and entangled states well before most. He believed that physical processes should be described by separable product states. And yet, Schrödinger unitary evolution does not make this prediction. Something does not seem quite right. Einstein's arguments with Bohr regarding the simultaneous existence of non-commuting quantities were faulty. However, the existence of simultaneous non-commuting quantities is not necessary to make Einstein's incompleteness argument. What would have sufficed is what we have demonstrated in Chapter 3. The argument that has been presented does not require the simultaneous existence of non-commuting quantities. Unfortunately, the detailed theory of entanglement (which was to become developed in the area of Quantum Information) required to make this argument was not known at the time of the Bohr-Einstein debates.

Definition of the Measurement Problem

Having completed the development of a test within the formalism of quantum mechanics that allows Schrödinger unitary evolution to be distinguished from evolution that occurs during measurement and considered the incompleteness of the current formalism and the requirements to improve upon the current formalism, the point has been reached whereby the measurement problem can be well defined.

Definition of the Quantum Measurement Problem:

> *The measurement and unitary postulates of quantum mechanics, when applied to particles that include a bona fide measurement device, are inconsistent in that two contrary predictions arise. The joint system and device including the local (time-like) environment evolve via Schrödinger's equation to an entangled state that is in principle experimentally distinguishable within the current formalism from the case of measurement for which there is evolution to a single system-device-local environment state that is a product state. Such*

distinguishability is independent of the Hamiltonian, parameters of the device, and on interaction with the local environment surrounding the device.

The fact that the two evolution modes are, within the same formalism, in principle experimentally distinguishable, is why the measurement problem is a real problem that can be both theoretically and experimentally investigated using scientific methodology. This situation can be summarized by the dictum:

To the extent there is entanglement, there is no measurement.

Requirements on Solutions

Given a clear definition of the measurement problem and the significance that a solution could have, one should consider what it would take for a proposed solution to be considered as a correct solution to the Quantum Measurement Problem. The following is an initial step to define the requirements that an acceptable theory should meet. There are two categories of potential solutions. A Category 1 Theory addresses the following requirements:

R1.1: The theory predicts the specific physical conditions under which measurement occurs and the physical conditions under which Schrödinger unitary evolution occurs.

R1.2: R1.1 is validated by experiment.

R1.3: Given a specific set of physical conditions under which a measurement has occurred (R1.1), the theory predicts:

R1.3.1 the potential system outcomes that might occur under the specific physical conditions assumed and their relative frequency of occurrence.

R1.3.2 for each potential system occurrence in R1.3.1, the potential state evolution(s) of the system and accompanying potential state evolution(s) of the measuring device (plus local environment, if needed).

R1.4: R1.3 is validated by experiment.

Requirement R1.1 requires specifying the micro-macro divide that is needed to resolve the measurement problem. Note that both R1.3 and R1.4 might appear to be met by invoking the measurement postulate of von Neumann. However, the prescription must be based on the underlying phenomena. The Hermitian operator that defines the measurement in von Neumann's theory is assumed known. However, R1.3 goes beyond this in the sense that such a Hermitian operator is not assumed known but would be derived from physics of the proposed theory.

A second class of theories, denoted as Category 2, make the claim that one or more of the requirements above are impossible to achieve by virtue that the conditions under which a measurement occurs prohibit one or more of the requirements from being met. This is a rather interesting logical possibility. In such a case, there would need to be an in-principle failure of the ability to experimentally investigate the measurement problem via techniques that, under the current quantum formalism, are in principle capable of discriminating Schrödinger unitary evolution from non-Schrödinger evolution such as the operations developed in Chapter 3. However, a Category 2 theory still must include non-Schrödinger or non-unitary measurement operations, in order to resolve the measurement problem. That is, a Category 2 theory cannot be entirely a Schrödinger unitary solution.

An assumption was made in Category 2 that the operations proposed to discriminate unitary evolution from measurement are in-principle impossible. On the other hand, if the conditions under which a measurement occurs prohibit these discriminating operations, then it is plausible that some or all of R1.1-R1.4 cannot be fulfilled. Therefore, it is proposed that a Category 2 theory meet the following requirements before being accepted:

R2.1: The theory provides a theoretical rationale that the conditions under which a measurement occurs prohibits the experimental discrimination of Schrödinger unitary evolution from non-Schrödinger evolution via UMDT tests.

R2.2: The theory provides a restricted set of implementable unitary and measurement quantum operations. The conditions are specified for which unitary operations are applicable, and the conditions for which measurement operations are applicable, subject to the limitations of R2.1.

R2.3: Experimental evidence confirms that the operations that are outside the proposed restricted set of quantum operations indeed fail to be implementable and provides compelling evidence for the proposed theory.

As an example of a theory that could fall into this category, consider a theoretical measurement model that shows to actually discriminate unitary evolution from non-unitary evolution, two non-commuting observables of the device would be needed to be known simultaneously, which is impossible. And that the model proposes a non-unitary collapse mechanism that occurs in relation to the conditions for which discrimination between unitary evolution and non-unitary evolution becomes impossible. Moreover, suppose it is found via experiments that indeed such operations fail to be capable of being implemented successfully because there does appear to be such an inherent limitation in the process of measurement. Then such a theoretical model and a negative experimental demonstration might logically be capable of ruling out theories that purport to meet one or more of the R1.1-R1.4. On the other hand, such a demonstration would imply that there exists a restricted set of implementable operations.

If a solution is found for a Category 1 theory that meets R1.1-R1.4, then the current formalism of quantum mechanics would be augmented to explain the conditions under which measurement occurs as well as a prescription for the measurement observable in von Neumann's theory. In such a case Einstein would be correct in his thesis that Quantum Mechanics in its current form is incomplete. If a resolution is found for a Category 2 theory that meet R2.1-R2.3, then it is a logical possibility that the current two postulates of quantum mechanics have predictive power, but with an amended structure. The set of implementable operations would need to be restricted. The amended structure of a Category 2 theory would include the conditions under which the unitary postulate can be applied versus when the measurement postulate would be applied, subject to the limitations of R2.1. In this case, the Category 2 theory and its new restrictions would be critical in order to complete quantum mechanics. In either case, it appears that Einstein is correct in the following: quantum mechanics in its current form cannot at this time be considered to be complete.

What if a given theory has not yet met either R1.1-R1.4 or R2.1-R2.3? Then such a theory in our view should be considered in a deductive investigation as a scientific hypothesis if there is some evidence that supports the theory, or a conjecture or speculation otherwise regarding the resolution of the measurement problem.

Resolution of the Quantum Measurement Problem

> *The Quantum Measurement Problem shall be resolved when the current theory of quantum mechanics is amended via either a Category 1 theory that meets Requirements R1.1-R1.4 or a Category 2 theory that meets Requirements R2.1-R2.3.*

Philosophy and the Measurement Problem

There are a substantial number of philosophical concepts that are relevant to the themes of the measurement problem as well as the ultimate resolution of the measurement problem. Philosophers often deal with issues of *Ontology* and *Epistemology*. Ontology is the philosophical study of the nature of being, becoming, existence or reality as well as the basic categories of being and their relations [112]. The measurement problem deals with the currently unknown problem of how an entangled state becomes a product state that is demanded under measurement. This is an ontological problem. As will be explained in more detail in Chapter 5, Einstein did not consider entangled states an adequate description of reality whereas he did consider product states an adequate description of reality. For Einstein, the measurement problem would have been a major issue in ontology. Furthermore, philosophers often categorize how a given theory explains Nature in terms of ontology.

Another important concept of philosophy that is relevant to measurement is the term "Epistemology." Epistemology is a branch of philosophy that studies the nature of knowledge, justification, and the rationality of belief [113]. Bohr considers that measurement is a means by which knowledge comes about. Bohr wrote at length on issues of epistemology and its relationship to complementarity. A resolution of the measurement problem that goes beyond the current quantum theory could, therefore, have a major impact on epistemology.

Concepts of free will, determinism, and consciousness are also studied at length in philosophy. Issues of the physics of free will, determinism, and consciousness were examined by Aristotle, Descartes, and many others. It was believed by many physicists that these issues are related to the measurement problem. Hence, the resolution of the measurement problem may be expected to have major implications on these questions.

Unitarianism traditionally is the belief in a single God. Unitarianism is also a philosophy that rejects principles of dualism such as mind and matter. We will use the word in the sense of those that reject principles of dualism and furthermore believe that the motion of particles in Nature is fundamentally described by a single law. Unitarians in this sense often believe that all phenomena emerge from the evolution described by a unitary theory.

What a Solution to the Measurement Problem is Not

Assume Measurement has Occurred

There are a number of proposed solutions that do indeed explain Born's rule very well, *given* that a measurement has occurred. The measurement postulate is a postulate that tells one how to proceed with state evolution, *given* that a measurement has already occurred and *given* it has occurred via a *given* Hermitian observable.

If a theory provides an explanation of how something came about *a posteriori* via a measurement, then has the measurement problem been resolved? If both Device 1 and 2 are bona fide measurement devices for which it is known that a measurement occurs, then one can at that point throw away the unitarily predicted state and substitute the state of the result predicted from the measurement postulate.

Let us examine what such an *a posteriori* theory would tell us regarding our Chapter 3 demonstration that distinguishes unitary evolution from measurement. Suppose that in our Device 1 and 2 we perform numerous experiments in which we slowly increased the number of atoms that compose the device. Perhaps when two particles are used, we find experimentally that Step 1 is verified to be evolving via Schrödinger's equation as the follow-up Step 2 and 3 give a correlation that agrees with the Bell correlation for maximally entangled particles. As an example to illustrate the logical issues of an *a posteriori* theory, suppose that 200 atoms are added in a particular configuration and after Step 1, it was suddenly found that the result is not commensurate with Schrödinger evolution. An *a posteriori* theory could substitute at this point the tensor product measurement state for the entangled state predicted under Schrödinger evolution. But such a substitution in an *a posteriori* theory is missing everything that is vital to the solution of the measurement problem. It does not tell us why this particular configuration produced measurement. It does not give us necessary and sufficient conditions of the physical phenomena under which non-Schrödinger evolution occurs versus when Schrödinger evolution occurs. It does not provide us with answers to Requirement 1.3 that the theory gives a prescription based on the underlying phenomena, be it nondeterministic or deterministic, for the evolution or change of states when a measurement occurs. An *a posteriori* prescription is based on knowing a measurement occurs and not based on the underlying phenomena. An *a posteriori* theory is able to satisfy Requirement 1.4, because it is based on what has already occurred. Hence the reader should understand that a satisfactory solution to the measurement problem must go well beyond simply providing an explanation of the predictions consistent with von Neumann's measurement postulate, i.e., given *a posteriori* knowledge that a measurement has occurred.

An *a posteriori* theory is nothing more than another interpretation, with no more predictive power then von Neumann's theory. Due to the results of Chapter 3, it is impossible for the Schrödinger predicted entangled state to be taken as the measurement state. Because of this, one will find interpretations that invariably have a step whereby Schrödinger's predicted state is thrown out and the measurement state is substituted. *A posteriori* theories seem to be most often put forward by individuals who strongly believe everything evolves according to Schrödinger's equation and have tried but failed to directly solve the measurement problem by showing an exact correspondence between measurement and Schrödinger unitarity. One should be keen to check how the theorist of an *a posteriori* justifies this substitution. Sometimes one finds justification via pejoratives such as "Of Course," or "Obviously" added for good measure in an attempt to ward off those who would question the validity of such a substitution.

Many philosophers define the measurement problem precisely as an *a posteriori* problem. For example, Lewis [114] defines the measurement problem as:

> *"The measurement problem, in a nutshell, is the problem that at the end of a measurement, there is nothing in standard quantum mechanics that represents the determinate outcome of the measurement" ... "A minimal condition for solving the measurement problem, then, is that a theory provides an explanation of our determinate measurement results."*
>
>

At first glance, this seems like a reasonable definition of the measurement problem. However, on closer examination the definition assumes that the current *a posteriori* theory is satisfactory, and a minimal condition to resolve the measurement problem is some explanation of the determinate measurement results. As many or most philosophers appear (from our reading) to subscribe to Lewis's definition of the measurement problem, we will throughout the remainder of the book refer to such *a posteriori* definitions as the *Philosophers' Definition* of the measurement problem. The philosophers' definition is certainly a convenient definition for philosophers, as numerous philosophical theories can be proposed and analyzed using philosophical categorizations and philosophical analysis with little or no risk of being scientifically falsified—but only under the condition for which one has been *given* that a measurement has occurred.

The issues of who-what-when-where-and-why a measurement occurs are largely taken off the table by the use of the philosophers' definition, and yet such issues are primary fundamental physical issues that need to be addressed in the quantum measurement problem. Furthermore, our discussion in Chapter 3 shows why the *a posteriori* philosophers' definition is not sufficient to encompass the issues of incompleteness of current quantum mechanics that Einstein and others have voiced regarding the conflict between measurement and unitary evolution.

The definition that we propose requires that the issues of entanglement versus product state be addressed by the development of either a Category 1 or Category 2 physical theory and via physical experimentation that provides verification of the related requirements that have been elaborated on. As such a theory must address the physical reasons as to the who-what-when-where-and-why a measurement occurs, we will henceforth refer to our definition as the *Physical Definition* of the measurement problem

Necessary and Sufficient Conditions for Measurement

Consider a slightly modified *quantum locomotive* (originally presented in Chapter 2) for which the locomotive impinges on one of two doors, yours and your neighbor's. Suppose $|n_1\rangle_{D_1}$ represents the number of particles impinging on your door, and $|n_2\rangle_{D_2}$ on your neighbor's door. Assuming N is sufficiently large that its effect would be to smash through your door as if a locomotive is coming toward you, then the superposition state $(|N\rangle_{D_1}|0\rangle_{D_2} + |0\rangle_{D_1}|N\rangle_{D_2})/\sqrt{2}$ can be seen to represent a quantum locomotive.

If the state has just smashed through your door and you are aware that it is coming toward you rather fast, we would contend that this is a sufficient condition for measurement. On the other hand, a quantum locomotive smashing through your door is not a necessary condition for measurement. One might be under the impression that if one can specify a necessary and sufficient condition under which measurement occurs, then this would resolve the measurement problem.

Specifying a necessary and sufficient condition may or may not lead to fully resolving the measurement problem. For argument's sake, consider the condition of *irreversibility*. Suppose that it were found that true irreversibility is both necessary and sufficient for measurement to occur. In such a case, irreversibility can be considered to be a condition that is logically equivalent to the condition under which measurement occurs. But unless one can explain the physical conditions under which such true irreversibility occurs, one still has significant work to perform. On the other hand, suppose a theory predicts that a necessary and sufficient condition for measurement is the interaction of a photon with an atomic sample of .1-.2 mole of a particular category of molecules at a temperature of $20° - 25°$ Celsius and a pressure of 10-15 psi. Furthermore, suppose that the theory is experimentally verified. Then such a condition would certainly be more specific and relevant to meeting the conditions under which measurement occurs, compared to the use of an equivalent condition which occurs under completely unknown physical conditions. In the case of irreversibility, one has succeeded in replacing the unknown physical conditions under which measurement occurs with the unknown physical conditions associated with when irreversibility occurs. Generating necessary and sufficient conditions for measurement is an important tool; however, substituting one unknown condition for another unknown condition is hardly substantial progress regarding the measurement problem. In order to resolve the measurement problem via the use of equivalent conditions, one needs to substitute the unknown necessary and sufficient conditions under which measurement occurs via an equivalent but known and understood necessary and sufficient condition. This situation is comparable to Feynman's discussion of the physical meaning of force in Newton's Second Law, $F = ma$, in which he gives the example of attempting to explain the observation of an object changing its position by introducing the term "gorce," defined as that which causes an object's change in position [115, pp. (12-1)-(12-2)]. No predictions whatsoever can be made from such a definition. Instead, the physical meaning of force in $F = ma$ comes from relating force to specific independent underlying properties. Otherwise, $F = ma$ is an incomplete law. Similarly, a theory of the measurement problem must relate the act of measurement to underlying physical phenomena as in R1.3.

For All Practical Purposes, FAPP

Bell coined the phrase in the paper [116], "for all practical purposes, FAPP." Bell believed that the measurement postulate, with its replacement of the quantum predicted unitary state with the results obtained through measurement was (as far as he knew) acceptable for all practical purposes but did not appear to be acceptable in a theoretical manner. Bell [116] states:

> *Ordinary Quantum Mechanics (as far as I know) is just fine for all practical purposes.*
>
> Reprinted from Physics World, Against Measurement by J. S. Bell, Vol. 3, p. 33, 1990 with permission from the Institute of Physics.

If ordinary quantum mechanics were correct and no alteration was ever needed, then the measurement problem would be an issue strictly for interpretation and might be an interesting problem in philosophy, but hardly worthy of substantial scientific investigation. What we can state from our developments so far, is that ordinary quantum mechanics cannot be considered to be complete at this time, because the measurement problem implies a contradiction that can be further investigated both experimentally via the UMDT of Chapter 3, as well as theoretically. A solution of the measurement problem, that is, of Category 1 would require a substantial revision of the current formalism. A solution that is of Category 2 would still require some augmentation of the current formalism. In either case, it does not appear that the current formalism can stand on its own.

It should also be noted that the general intent of Bell's paper [116], which is sometimes quoted in an effort to justify the philosopher's definition of the measurement problem, is actually recommending that an effort be made to go further than the current two postulates. Later Bell states for example, "Is it not good to know what follows from what, even if it is not really necessary FAPP?" and looks at several different theories.

External Orthogonalization

A unitary methodology of external orthogonalization that is within the Hamiltonian formulation is presented, which will be useful in analyzing numerous approaches that have been proposed to resolve the measurement problem. Consider the Hamiltonian consisting of terms of the system to be measured H_S, a set of external particles with Hamiltonian H_D with an orthonormal set of eigenstates $|\Psi_r\rangle$ and an interaction Hamiltonian H_{int} between the system and the set of external particles. The external particles could, for example, compose a device or be part of the local environment surrounding the system and are assumed initialized to state $|\Psi_0\rangle$. The interaction Hamiltonian will be constructed in a manner that if a system is in the r th eigenstate of H_S denoted $|\phi_r\rangle$ then the external set of particles will make a unitary transition from $|\Psi_0\rangle$ toward populating the r th eigenstate of the external particle set $|\Psi_r\rangle$.

This method will be referred to as *external orthogonalization* because each system eigenstate $|\phi_r\rangle$ is paired with a corresponding external state $|\Psi_r\rangle$, with the property that the $|\Psi_r\rangle$ form an orthonormal set. A Hamiltonian that accomplishes this is given by

$$H = H_S \otimes I + I \otimes H_D + H_{int} \tag{4.1}$$

$$H_S = \sum_r E_S |\phi_r\rangle\langle\phi_r|$$

$$H_D = \sum_r E_P |\Psi_r\rangle\langle\Psi_r|$$

$$H_{int} = \sum_r (|\phi_r\rangle \otimes |\Psi_0\rangle)(\langle\phi_r| \otimes \langle\Psi_r|) + (|\phi_r\rangle \otimes |\Psi_r\rangle)(\langle\phi_r| \otimes \langle\Psi_0|).$$

In order to illustrate this concept, it is assumed that the energies are the same E_S for all $|\phi_r\rangle$ and as well the pointer states $|\Psi_r\rangle$ all have energy E_P. If the system is allowed to evolve for a time $t = \pi/2$, the unitary evolution operator

$$U\left(\frac{\pi}{2}\right) = e^{-\frac{iH\pi}{2}}$$

performs the desired operation. If the system is initially in the superposition given by

$$|\psi_S\rangle = \sum_r c_r|\phi_r\rangle$$

then $U(\pi/2)(|\psi_S\rangle \otimes |\Psi_0\rangle)$ is given by

$$\sum_r c_r e^{i\theta_r}|\phi_r\rangle \otimes |\Psi_r\rangle. \qquad (4.2)$$

This method will be referred to as external orthogonalization because each system eigenstate $|\phi_r\rangle$ is paired with a corresponding external state $|\Psi_r\rangle$, with the property that the $|\Psi_r\rangle$ are an orthogonal set.

If the energies E_S and E_P are the same for all $|\phi_r\rangle$ and $|\Psi_r\rangle$, the phase θ_r is independent of r. The term $e^{i\theta_r}$ factors out and, since the overall phase of the state has no observable consequences, the term $e^{i\theta_r}$ can be dropped from the expression, resulting in:

$$\sum_r c_r|\phi_r\rangle \otimes |\Psi_r\rangle.$$

One can trace the external particles from the overall state and compute the partial density matrix of the system. This is seen to be the diagonal matrix:

$$\begin{pmatrix} |c_1|^2 & \cdots & 0 \\ \vdots & \ddots & \vdots \\ 0 & \cdots & |c_N|^2 \end{pmatrix}. \qquad (4.3)$$

Interpretations of Quantum Mechanics

Various approaches have been proposed to resolve the measurement problem that are loosely referred to as *interpretations*. An interpretation is generally proposed with the idea that it will agree with the discovery of Schrödinger's equation as well as Born's rule which were to become the two postulates of quantum mechanics. However, Born's rule is an *a posteriori* rule, i.e., it assumes a measurement has already occurred and in a known measurement basis. As we will see, however, there is no "interpretation" that fully addresses the requirements R1.1-R1.4 or R2.1-R2.3 for which measurement occurs.

Copenhagen Interpretation

Arguably the most well-known interpretation of quantum mechanics is the *Copenhagen Interpretation.* The Copenhagen interpretation has a remarkable history, and important details will be discussed later in Chapter 5. The Copenhagen interpretation is essentially that a deterministic description such as Schrödinger's equation cannot suffice to describe Planck's quantum of action that Bohr considered to be rooted in a process that, at best, can be modeled nondeterministically or statistically. Hence a deterministic description necessitates the augmentation by a nondeterministic statistical description in order to provide a full accounting of phenomenon. Such dual methods that are employed in the Copenhagen interpretation represent the principle of complementarity. Complementarity has also been applied to many other situations in Bohr's writings.

The reason that Bohr argued that the measurement process could at best be modeled by an external observer statistically was that he believed that the process of measurement requires an interaction between a device and a particle, which could be quite unpredictable and uncontrollable under certain circumstances. He considered that a measurement device generally possessed the qualities of classicality in terms of its possible readouts. Consider a device that is designed to measure momentum and provides two readout positions that indicates a range of distinct momentum values. If a particle is input with a momentum that corresponds precisely to one of the momentum readout values, then there is no uncontrollable interaction and the device directly reads out the momentum. On the other hand, suppose that a particle is input with a well-defined position. Then the particle will have a large uncertainty in its momentum and will exist as a superposition of numerous momentum states. Because of the Heisenberg uncertainty relationship, Bohr would expect that the device would still read-out one out of the many possible momentum values because of the impossibility of knowing the position and momentum simultaneously, which in turn causes the interaction to become uncontrollable and unpredictable. Hence the outcome can, from an external observer's perspective, at best be modeled probabilistically or statistically. Interestingly, position/momentum non-commutativity is also a form of complementarity. Hence in this example, position/momentum complementarity begets wave/particle complementarity.

Essentially because of this viewpoint, Bohr did not believe that it made sense to go further into a detailed deterministic description of the process of measurement. This is rather unfortunate, but the era of Quantum Information where entanglement plays a central role, was largely begun in a very significant manner only over the last twenty years.

Rosenfeld's Solution

A unitary methodology was presented by Rosenfeld in 1965 [117] as an exposition of a measurement theory by Daneri, Loiniger, and Prosperi, based on the ergodic theorem [118]. This method uses a large number of device states and Rosenfeld believed that in the limit of large numbers of particles his method would be sufficient to explain measurement. Rosenfeld considers an atomic system initialized to state $|\varphi_0\rangle = \sum_r c_r|\varphi_r\rangle$ and a device initialized to state $|\Psi_0\rangle$. Two devices are consequently used to measure two different observables *A, B* with eigenstates given by $|\varphi_r\rangle$ and $|\chi_s\rangle$ respectively. There is a short time τ when the device interacts with the system, such that if the initial state of the system is $|\varphi_r\rangle$ the final state of system-device is $\alpha_r|\varphi_r\rangle\otimes|\Psi_r\rangle$, whereby the device is in the pointer state $|\Psi_r\rangle$ that occurs unitarily when the system input is in state $|\varphi_r\rangle$. The approximation $\alpha_r \approx 1$ is made as the interaction is assumed to be sufficiently short. After time τ, the atomic system and device are assumed non-interacting and evolve via the tensor product unitary $U_S(t) \otimes U_D(t)$. The Schrödinger predicted unitary evolution when the device is set to measure observable *A* is given by

$$\sum_r c_r\,|\varphi_r\rangle \otimes |\Psi_r\rangle.$$

After a time t the system-device evolves to

$$\sum_r c_r\,U_S(t)|\varphi_r\rangle \otimes U_D(t)|\Psi_r\rangle. \tag{4.4}$$

The system is then followed by a measurement of Observable *B* via interaction with a second measurement device, initialized to $|\Psi_0\rangle$, that evolves to $|\chi_s\rangle\otimes|\Psi_r\rangle$ when the system input is in state $|\chi_s\rangle$. The probability of finding the quantum state $|\chi_s\rangle$ is on average given by use of the Born and/or von Neumann measurement postulate

$$\sum_r |c_r|^2\langle\chi_s|U_S(t)\varphi_r\rangle^2. \tag{4.5}$$

Rosenfeld states that since this expression only depends on the relative probabilities $|c_r|^2$ of the eigenstates $|\varphi_r\rangle$, but not on the phases of the complex coefficients c_r, this is referred to as the "reduction" of the state $|\varphi_0\rangle$. That is, if no measurement of *A* had been performed, the probability of finding $|\chi_s\rangle$ is given by

$$|\sum_r c_r\,\langle\chi_s|U_S(t)\varphi_r\rangle|^2 \tag{4.6}$$

which differs from Equation (4.5) by the cross terms:

$$\sum_{r\neq r'} c_r\,c_{r'}^*\langle\chi_s|U_S(t)\varphi_r\rangle\langle\chi_s|U_S(t)\varphi_{r'}\rangle^*. \tag{4.7}$$

Rosenfeld claims that the effect of Measurement *A* is the disappearance of the terms in Equation (4.7) and that is formally denoted as the "reduction" of the initial state.

Rosenfeld desires to impose an ergodicity condition on the measurement device for which Equation (4.7) is claimed to go to zero when the device is macroscopic. In this case, if Measurement *A* is followed by Measurement *B*, the probability of finding $|\chi_s\rangle$

will correspond correctly to Equation (4.5). The argument Rosenfeld presents is given in Appendix 4.A.

External orthogonalization will also be useful in discerning many approaches that have been proposed to resolve the measurement problem.

External Orthogonalization and Rosenfeld

Rather than use the method of ergodicity that is statistical and outside of a rigorous Hamiltonian formulation that has been the focus of this book, it will be shown that external orthogonalization is a unitary solution within the Hamiltonian formulation, that can also be used to derive Equation (4.5).

Consider the use of two unitary devices denoted *X* and *Y* that are used to implement external orthogonalization in order to determine if such devices result in the outcome that Rosenfeld desires under measurement, Equation (4.5). In the first measurement *X* of Rosenfeld, let us set the eigenstates of the system required by external orthogonalization equal to the eigenstates of Observable *A* in Rosenfeld's proposal, i.e., $|\phi_r\rangle|\varphi_r\rangle$ and whereby the corresponding unitary evolution is denoted $U_{S,X}(t)$. The second measurement is specified by using external orthogonalization with $|\phi_r\rangle = |\chi_r\rangle$, the eigenstates of Observable *B*, and the corresponding unitary evolution is denoted $U_{S,Y}(t)$. In both cases the external particles are initialized to $|\Psi_0\rangle$. Given the initial state $|\psi_S\rangle = \sum_r c_r|\phi_r\rangle$, the resulting state of external orthogonalization after unitary evolution is

$$|\psi_{S,X}^{(1)}\rangle \equiv U_{S,X}(\pi/2)(|\psi_S\rangle \otimes |\Psi_0\rangle)$$

$$|\psi_{S,X}^{(1)}\rangle = \sum_r c_r|\varphi_r\rangle \otimes |\Psi_r\rangle.$$

Rosenfeld then proposes that the system and Device *X* are separated or non-interacting and the system is allowed to unitarily evolve via its self-Hamiltonian. In this case, consider the partial density matrix of the system. The system is in a mixed state, $\rho_S^{(1)} \equiv \mathrm{Tr}_X|\psi_{S,X}^{(1)}\rangle\langle\psi_{S,X}^{(1)}|$ which can be computed as the diagonal matrix

$$\rho_S^{(1)} = \begin{pmatrix} |c_1|^2 & \cdots & 0 \\ \vdots & \ddots & \vdots \\ 0 & \cdots & |c_N|^2 \end{pmatrix}$$

where *N* is the number of eigenvectors of Observable *A*. After the system and Device *X* are separated, the system is evolved unitarily for a time $t = t_U$. Denoting

$$U_S(t) = e^{-\frac{iH_S t}{2}}$$

the resulting state of the system becomes $\rho_S^{(2)} \equiv U_S(t_U)\rho_S^{(1)}U_S(t_U)^\dagger$. This will leave the density matrix unchanged i.e., $\rho_S^{(2)} = \rho_S^{(1)}$ as $\rho_S^{(1)}$ is diagonal in the self-Hamiltonian H_A. The third and final step is to interact the system (specified by density matrix $\rho_S^{(2)}$) with

Device Y initialized to the density matrix $\rho_{D,0} \equiv |\Psi_0\rangle\langle\Psi_0|$. The density matrix of System-Device Y is then given by

$$\rho_{S,Y}^{(3)} \equiv U_{S,Y}(\pi/2)(\rho_S^{(2)} \otimes \rho_{D,0})U_{S,Y}(\pi/2)^\dagger .$$

The system after interacting with Device Y is given by $\rho_S^{(2)} \equiv \mathrm{Tr}_Y|\psi_{S,Y}^{(1)}\rangle\langle\psi_{S,Y}^{(1)}|$, which can be computed as

$$\rho_S^{(2)} = \begin{pmatrix} \sum_r |c_r|^2\langle\chi_1|\varphi_1\rangle^2 & \cdots & 0 \\ \vdots & \ddots & \vdots \\ 0 & \cdots & \sum_r |c_r|^2\langle\chi_N|\varphi_N\rangle^2 \end{pmatrix},$$

and is in agreement with Rosenfeld's Equation (4.5). Hence it has been shown that unitary evolution via external orthogonalization in the manner that Rosenfeld considered, is fully consistent with Born's rule and von Neumann's measurement postulate insofar as results on the system are measured.

External Orthogonalization and UMDT

If the measurement problem could be resolved by developing a unitary approach that results in Rosenfeld's Equation (4.5), then external orthogonalization would indeed be a satisfactory resolution to the measurement problem. If this were the case, then external orthogonalization should also be able to satisfactorily resolve the dilemma for which the Chapter 3 UMDT was specifically developed to address. Let us then examine how well Rosenfeld's criterion and external orthogonalization fare in comparison with the UMDT.

Suppose that the external orthogonalization Hamiltonian of Equation (4.1) is utilized in the Chapter 3 development for the detectors shown in Figure 3.4. Consider the case for which both devices are identical and are two-level systems in which $|\Psi_0\rangle$ that will change its state to $|\Psi_1\rangle$ upon absorbing a photon. Setting $|\phi_0\rangle = |0\rangle$ and $|\phi_1\rangle = |1\rangle$, a Hamiltonian applicable to Figure 3.4 is given by:

$$H = H_F \otimes I \otimes I + I \otimes H_1 \otimes I + I \otimes I \otimes H_2 + H_{int}$$

$$H_F = w{a_B}^\dagger a_B \otimes I_2 + wI_2 \otimes {a_C}^\dagger a_C,$$

$$H_i = |0\rangle\langle 0| + 2|1\rangle\langle 1|$$

$$H_{F,1} = a_B \otimes I_2 \otimes \sigma_+ \otimes I_2 + {a_B}^\dagger \otimes I_2 \otimes \sigma_- \otimes I_2$$

$$H_{F,2} = I_2 \otimes a_C \otimes I_2 \otimes \sigma_+ + I_2 \otimes {a_C}^\dagger \otimes I_2 \otimes \sigma_-$$

$$H_{int} = H_{F,1} + H_{F,2}.$$

In this model, the atomic system that is considered to be measured in Rosenfeld's paper has been replaced by a photon in order to consider its application using the

Chapter 3 development. Both H_1 and H_2 are device Hamiltonians that are each a two-level system. The field is described by a harmonic oscillator of mode energy $w = 1$, and the interaction is such that when a single photon is annihilated the device resonantly undergoes a raising operation and when the device falls from state $|1\rangle$ to $|0\rangle$ a resonant photon is added to the mode. The application of this particular external orthogonalization Hamiltonian can be seen to be a 100% absorbing model for interaction times $\pi/2$ (and modulus $2\,\pi$).

The initial state of the device i, denoted $|\psi_r^0\rangle_i$, unitarily evolves to a read-out state that is indicative of no measurement and, adopting the notation in Chapter 3, is denoted $|\psi^0\rangle_i$, $i = 1, 2$. Consider the case that a photon with state $|1_V\rangle_B$ is inserted in Path B. After time $\pi/2$, the state of Device 1 evolves to its read-out state indicating that the Device has detected a photon, denoted $|\psi^1\rangle_1$. Similarly, if a photon with state $|1_H\rangle_C$ is inserted solely in Path C, the final state of Device 2 after time $\pi/2$ is denoted as $|\psi^1\rangle_2$.

In the case of Rosenfeld whereby a single device is used in order to detect the system, the initial system state was assumed to be a pure state given by $|\psi_S\rangle = \sum_r c_r\,|\varphi_r\rangle$. However, in the Chapter 3 development there are now two detectors that interact with the electromagnetic field in separate locations. However, for the case needed to be considered in Figure 3.4, the initial photon state given by

$$|\psi_{photon,PBS}\rangle = \sqrt{a}|1_V\rangle_B \otimes |0\rangle_C + \sqrt{1-a}\,|0\rangle_B \otimes |1_H\rangle_C$$

results in a mixed photon state for both photon Mode B as well as photon Mode C. In the case of Figure 3.4 there are two detectors after the initial Step 1 unitary evolution and the partial density matrices of the devices are found by Equation (3.32) for which $\rho_{1'}^{(1)} = a\rho_{1,1'}^{(1)} + (1-a)\rho_{2,1'}^{(1)}$ and $\rho_{1,1'}^{(1)}$ is the Schrödinger evolved density matrix of the device assuming a photon impinges on the device and $\rho_{2,1'}^{(1)}$ is the Schrödinger evolved density matrix of the device with no photon impinging on the device. Assuming for simplicity that the initial states of the devices are pure states, then $\rho_{1,1'}^{(1)}$ and $\rho_{2,1'}^{(1)}$ are pure orthogonal states. With this, if one considers $\rho_{1'}^{(1)}$ transformed via change of basis using the final orthogonal states $\rho_{1,1'}^{(1)}$ and $\rho_{2,1'}^{(1)}$, the result is

$$\begin{pmatrix} a & 0 \\ 0 & 1-a \end{pmatrix} \tag{4.8}$$

and $\rho_{2'}^{(1)}$ can similarly be expressed in the basis of the final orthogonal states $\rho_{1,2'}^{(1)}$ and $\rho_{2,2'}^{(1)}$ resulting in

$$\begin{pmatrix} 1-a & 0 \\ 0 & a \end{pmatrix}. \tag{4.9}$$

Hence the devices after Step 1 are in a mixed state for which the diagonal elements correspond exactly to what is required in the measurement postulate and Born's law. If the solution to the measurement problem only required that one show how one could derive the probabilities of measurement via unitary evolution, then Step 1 would suffice and the measurement problem would be solved by Step 1. However, as we have defined the measurement problem, the problem is related to the unitary prediction of entanglement versus a product state of the device which is expected under measurement. This is brought out in the UMDT operations that distinguish unitary evolution from measurement, and there remain Steps 2 and 3. If one applies Step 2 and Step 3 to the two devices, the result is shown in Figure 3.16. Although the partial density matrix of the devices after Step 1 by themselves are in states that might appear to resolve the measurement problem as understood by Rosenfeld and many others, one sees in Figure 3.16 that the devices when acting unitarily have the Bell measurement result that is always greater than the result when measurement occurs for any non-zero superposition parameter a.

Let us now ask how well external orthogonalization solves the measurement problem by examining the requirements that we have set forward for a scientifically acceptable solution. It has been demonstrated using the UMDT of Chapter 3, that reproducing Rosenfeld's Equation (4.5) is not sufficient to resolve the measurement problem. The measurement problem is related to the prediction of entanglement versus the product state that is demanded in any theoretical treatment. That Schrödinger's equation predicts entanglement was pointed out quite early in 1935 by Schrödinger himself [119].

In order to show that the measurement problem can be solved unitarily, one would need to develop a unitary methodology that gives the Born predictions of measurement and for which the device plus system are in a tensor product state versus an entangled state. The fact that the partial density matrices of the devices are in completely mixed states, when the overall state of the devices is a pure state, can be shown to result in an entangled state (see Exercise 4.4). Hence the idea that a unitary operation can resolve the measurement problem by diagonalizing the density matrix of the devices in a manner related to Born's law, such that it removes first order coherence, is not correct. On the contrary, the fact that the partial density matrices are mixed and non-coherent when the overall joint matrix of the two density matrices is pure, only points to the transfer of the first order photon coherence that originally existed between the two paths into second order coherence of entanglement between the two qubits via the UMDT in Step 1 and 2. Such entangled states are contrary to the states that result during measurement, which are well-defined product states. The unitary prediction of entanglement is contrary to the notion of a single well-defined system-device state that occurs under the measurement process.

Has Rosenfeld addressed Requirement R1.1? There are not explicit conditions that have been specified for a measurement to occur. Rosenfeld has not addressed R1.1 as he contends that measurement is a consequence of Schrödinger unitary evolution. And if the measurement problem could be resolved by showing that Equation (4.5) converges to Equation (4.6), then indeed Rosenfeld could have resolved the measurement problem.

Environmental Decoherence

Environmental decoherence is the interaction of a coherent system with the external particles in the environment, in a manner that destroys the system coherence. External orthogonalization can be used to model decoherence in which the Hamiltonian of Equation (4.1) is utilized such that H_D represents the environmental particles that are external to the system.

Assume that the system initially has coherence which can be represented by the off-diagonal terms of the density matrix of the system. Let us assume that there are two degrees of freedom of the system, and that the system is initially in the state $|\psi_S\rangle = \sqrt{a}|\phi_1\rangle + \sqrt{1-a}|\phi_2\rangle$, for which the density matrix of the system denoted $\rho^{(0)}$ is given by

$$\rho^{(0)} = \begin{pmatrix} a & \sqrt{a(1-a)} \\ \sqrt{a(1-a)} & 1-a \end{pmatrix}.$$

For any value of a not equal to either 0 or 1 one can see that the system possesses a non-zero off-diagonal coherence term. After the system interacts with the environment, and undergoes external orthogonalization, the final state of system plus environment is given by Equation (4.2) where the $|\Psi_r\rangle$ are orthogonal states of the environment. Such states can be considered in environmental decoherence theory as environmental *pointer* states, similar to a device read-out state that results for the corresponding system state $|\phi_r\rangle$. Now if one computes the partial density matrix of the system after environment decoherence occurs, denoted by $\rho^{(1)}$, it is found via Equation (4.3) to result in

$$\rho^{(1)} = \begin{pmatrix} a & 0 \\ 0 & 1-a \end{pmatrix}.$$

At this point, one might expect that the measurement problem is resolved because the system coherence is completely eliminated. That is, one can interpret the system to be in either state $|\phi_1\rangle$ with probability a and in state $|\phi_2\rangle$ with probability $1-a$; that it is only our ignorance of the actual state that prevents us from knowing the correct state. And this is quite correct if the system is isolated from its environment and either follow-up unitary evolution or measurements are performed on the system alone.

However, even after separating the system from its environment, the system remains entangled with the environment under unitary evolution. The case of entanglement versus the case for which the system is definitely either in $|\phi_1\rangle$ or $|\phi_2\rangle$ can in fact be differentiated via a UMDT. Under unitary evolution and environmental decoherence, the original first-order off-diagonal system coherence $\sqrt{a(1-a)}$ does indeed vanish, but instead becomes second-order coherence or entanglement between the system and environment. Simply because one transferred first-order coherence into entanglement does not resolve the measurement problem one iota. However, one might be inclined to proclaim to have resolved the measurement problem because when the system is isolated all follow-up operations performed on the system-only do indeed produce results consistent in a manner that the environment did measure the system in the environmentally determined pointer basis $\{|\Psi_r\rangle\}$. But this requires ignoring the unitarily predicted entanglement between system and environment which still can, in principle, be differentiated from measurement via a UMDT.

Environmental decoherence predicts a pointer basis but does not tell us the conditions for which measurement will occur. Consider a system composed of an electron with up or down spin that is moving in an inhomogeneous magnetic field. Suppose that the environment consists of one or more electrons which can interact with and scatter the first electron. It is well known that the unitary solution of interacting electrons can produce entangled electron states. Helium can be found in states such as the material parahelium, which is found when the two electrons are in an entangled singlet state. Another possibility is that the state of the two electrons is in the triplet entangled state for which a different material results, orthohelium. Scattering of electrons is generally modeled unitarily for example using Feynman diagrams, so one would be hard pressed to produce real theoretical or experimental evidence that would prove that a few electrons interacting with a single spin electron is sufficient to produce measurement. That is, if UMDT were performed in most such cases, we expect that the interactions are largely unitary and the results would agree with unitary theory. Although evidence would suggest that elementary electron scattering is unitary in most cases, one has to be careful not to leap to the final conclusion that this implies that all interactions are unitary. Such logically incorrect conclusions are often drawn by the quick-witted inductionist.

Moreover, consider the replacement of the devices in the UMDT of Chapter 3 by the local environment. Suppose that one were to speculate that environmental decoherence with some particular characteristic time or characteristic environmental size, resolved the measurement problem. As we can make the distance d between the two devices as large as desired in Figure 3.4, we can include both longer-and-longer interaction times and/or larger-and-larger local environments with increasing d and preclude interaction between the two respective local environments. Hence a UMDT can be designed in principle to investigate experimentally whether or not this is the case. As to date, there is substantial experimental evidence that indicates that minimal environmental interaction does not result in measurement. If at some point the environment can be considered a measurement device, then experimental investigation using UMDT would be needed to specify the conditions for which measurement would occur under such an environmental decoherence theory. Furthermore, these conditions would necessitate a substantial revision of quantum mechanics.

Decoherence by Stipulation

Bub claims in [120, p. 20]:

> *...we can take the decoherence pointer as definite by stipulation, and that decoherence then guarantees the objectivity of the macroworld, which resolves the measurement problem without resorting to*

Copenhagen or neo-Copenhagen instrumentalism.

Bub appears to be attempting to shift the definition of the measurement problem from the physical definition to the philosophical definition by taking the decoherence pointer basis as definite by stipulation.

Although Bub did not propose that decoherence by stipulation resolves the physical measurement problem it is still a useful exercise to analyze. To see that this does not resolve the physical measurement problem as we have defined it, let us consider several cases. In Case 1, we consider a system of particles that for the time interval $[0, t_1]$ is unitarily predicted to evolve in a manner that provides a decoherence pointer, which is equivalent to the process of external orthogonalization. Now suppose that such a system is predicted to continue to decohere in the same pointer basis without any further interaction in the time interval $[t_1, t_3]$, and that this could be continued indefinitely if desired. Bub's approach is that this is sufficient to guarantee objectivity of the state that existed at time t_1. Consider now Case 2 in which the same exact set of particles in the same state as Case 1 at time t_1 can interact with a new set of particles in a future time interval $[t_2, t_3]$. It is entirely possible that the unitary predicted state for this new set of particles effects an interaction between the pointer states such that the new pointer states that emerge are different from what they were in time interval $[0, t_1]$, and it is these new states that continue on to represent the new pointer basis. Hence if one applied Bub's methodology, in Case 1 the state after time t_1 represents an objective state, and in Case 2, the state after time t_2 is also an objective state.

Now, we apply the UMDT test of Chapter 3 to Cases 1 and 2 for which we will consider three possible outcomes—Case 3, 4, and 5. Case 3 is defined as the particles from Case 1 for which Step 2 and Step 3 of the UMDT test are applied at time t_3 and the CHSH value is $2\sqrt{2}$. Case 4 is defined as the particles from Case 1 for which Step 2 and Step 3 of the UMDT test are applied at time t_3, and the CHSH value is $\sqrt{2}$. Case 5 is defined as the particles from Case 2 for which Step 2 and Step 3 of the UMDT test are applied at time t_3, and the CHSH value is $\sqrt{2}$. If Case 3 is found to occur, then similarly in Case 2 the CHSH value if a Bell experiment were made at time t_1 would be expected to be $2\sqrt{2}$ as these are the same particles in the same state as in Case 1. However, suppose Case 5 is found to occur. Then we can conclude from the UMDT test that after time t_1 such a set of particles composing the device did not constitute a bona fide measurement of the photon while in Case 2 that the interaction with the set of the old plus the new particles does constitute a bona fide measurement of the photon in the interval $[0, t_3]$.

It can be seen from this example that UMDT tests provides us with a manner to gather true experimental evidence for the conditions under which a device is a bona fide measurement device. Bub's methodology would claim in Case 1 after time t_1 that the state is a classical state such as a cat being alive or dead because the state continued to decohere in that basis. On the other hand, suppose the UMDT test (in this example) shows that these particles are still in a superposition and that cat is neither alive nor dead at time t_1, while the follow-up test at t_3 shows measurement. In this case the correct decoherence pointer would be that given by the former case at time t_3. We have demonstrated the possibility that Bub's methodology, applied to the measurement problem as we have defined it, could give ambiguous results at time t_1, because of the *mere existence* of the possibility that unitarily one can always in the future interact particles in a manner to change the pointer basis that results from decoherence.

One can further ask whether or not Bub's methodology constitutes a necessary condition and/or a sufficient condition for measurement. Let us consider first the condition of sufficiency with the prior Cases 1 and 2 that have already been defined. In Case 1 Bub's methodology would claim that the state after time t_1 is an objective state such as a cat being alive or dead. Yet suppose that when the particles are subject to a UMDT is Case 3 with the result of the CHSH $2\sqrt{2}$ with the same particles as in Case 1, we would conclude that the state after time t_1 was in an entangled state and not in a product state. From this argument, Bub's methodology cannot be considered to be proven to be a sufficient test for a set of particles being a bona fide measurement device. Now, suppose that we applied the UMDT test at time t_3 to the particles from Case 1 and Case 4 results. Then we can conclude that the particles composing the device in Case 1 are a bona fide measurement device and the state after time t_1 is an objective state. But suppose that none of these experiments are performed except Case 5. Then we can conclude that the entire configuration of the original particles plus the new particles constitutes a bona fide device. Furthermore, as the original particles were interacted with new particles in Case 2 so that the original particles are not a decoherence pointer state, Bub's criterion would not apply to the original set of particles. Yet if we then conducted the experiment in Case 4 we would know that the original set of particles constitutes a bona fide measurement device. Because of these counterexamples that have not been ruled out by experiment, Bub's criterion cannot be considered to be a necessary condition for measurement at this time.

Bub's condition at this time has not been proven to be either a necessary or a sufficient condition for measurement in the manner that we have defined the measurement problem. Now one might go beyond Bub's formalism and claim that it is ultimately impossible to interact or reverse decoherence, even in a unitary theory. In this case such a theory could be classified as a Category 2 theory. However, there are systems that can undergo the process of external orthogonalization, for which a measurement does not occur. For example, suppose that a photon interacts with a single atom in a cavity, under conditions that have been utilized in the work of Haroche [121]. There is substantial experimental evidence that such interactions (with the experimental parameters considered in [121]) are unitary and that entanglement would be maintained if one conducted the UMDT test on such systems. Hence if such Category 2 arguments are to be made, they demand much more theoretical and experimental evidence as to what constitutes Category 2 irreversible decoherence.

In the paper [3] Bell states regarding the Landau and Lifshitz formulation (which is similar to the orthodox von Neumann quantum mechanics):

And the Landau and Lifshitz formulation, with vaguely defined wave function collapse, when used with good taste and discretion, is adequate FAPP. It remains that the theory is ambiguous in principle, about exactly when and exactly how the collapse occurs, about what is microscopic and what is macroscopic, what quantum and what classical. We are allowed to ask: is such ambiguity dictated by experimental facts? Or could theoretical physicists do better if they tried harder?

When Bell wrote his paper [116] in 1990, Quantum Information had not seriously begun other than a handful of entanglement experiments, and some expected in the early days of quantum information that mesoscopic entangled multipartite states might be impossible to observe due to the theoretically expected complete loss of entanglement when even a single particle of a maximally entangled state is lost through decoherence. However, mesoscopic experiments have proven such thoughts ill conceived. In light of quantum information technologies continuing to improve [122] [123] [124] and with each passing year progressively larger and larger systems being coherently manipulated [125] [26] [126] [127] [27] [128] [129], invoking environmental decoherence by stipulation in this day and age is not a satisfactory argument for resolution of the physical measurement problem. One would need to theoretically and experimentally investigate what conditions of environmental decoherence are bona fide measurements in order to begin to address the requirements of resolving the physical measurement problem. Unfortunately, Bell died at the age of 62. Had Bell been alive to witness the explosion in entanglement related areas, one has to wonder if he would not have reconsidered the possibility of experimentally investigating the issue of the measurement problem, rather than considering the conventional theory correct—FAPP.

Church of the Higher Hilbert Space

Some Unitarians respond that measurement can always be explained via unitary evolution by the system-plus-device interacting with a larger system. Here two types of unitary theories are examined, 1) unitary theory defined solely by Schrödinger's equation and 2) unitary theory that is necessarily different than that predicted by Schrödinger's equation. Generally speaking, Unitarians that belong to the *Church of the Higher Hilbert Space* invoke a form of Stinespring's dilation theorem.

Von Neumann's original theory [13] was extended to positive operator valued measurement theory by Kraus et al. [49]. Suppose a measurement on a system defined on the Hilbert space $\mathcal{H}_A$ has k possible outcomes, and for which there exists a matrix M_k corresponding to the k th outcome that maps any state $\psi \in \mathcal{H}_A$ to $\psi' = M_k\,\psi / ||M_k\,\psi||$. For each outcome k, there exists an associated Hermitian operator of the form $E_k \equiv {M_k}^\dagger M_k$ with the properties,

$$\sum E_k = I$$

$$\langle\psi|E_k|\psi\rangle \geq 0 \;\forall\; |\psi\rangle \in \mathcal{H}_A,$$

the latter property ensures positivity of the operators. The set $\{E_k\}$ are defined as the elements of the positive operator valued measurement (POVM). When the k th result occurs, the density operator ϱ is transformed according to

$$\hat{\varrho} = \frac{M_k \varrho M_k^{\dagger}}{P(k)}$$

with probability $p_k = \mathrm{Tr}\left[M_k \varrho M_k^{\dagger}\right] = \mathrm{Tr}[\varrho E_k]$. Note that the set $\{E_k\}$ may not be sufficient to uniquely define the transformation rule of the density matrix as there could be multiple decompositions. Hence it is desirable in the study of the measurement problem to define the set M_k as well. This is further examined as well as various classes of POVMs encountered in Chapter 7. An average density matrix can be found by averaging over all measurement results. This is often referred to as a Kraus decomposition or operator-sum form:

$$T(\varrho) = \sum M_k \varrho M_k^{\dagger}. \tag{4.10}$$

A completely positive map is a map that remains positive when $T(\varrho)$ is extended to operate on a larger system $\mathcal{H}_A \otimes \mathcal{H}_B$ via $(T \otimes I)\varrho_{A,B}$ where $\varrho_{A,B}$ is a density matrix on $\mathcal{H}_A \otimes \mathcal{H}_B$. It can be shown that T as defined by Equation (4.10) is a completely positive and trace-preserving linear map T. Not all positive maps are completely positive, see Exercise 4.9.

Now, suppose one has made a measurement on a system. Then there must exist a completely positive trace-preserving map T that corresponds to the transformation of ϱ on average. In order to proceed let us consider the following dilation Theorem due to Stinespring [130, p. 22]

> Theorem 4.1: *Given a completely positive, trace-preserving map T between states on a finite-dimensional Hilbert space $\mathcal{H}_A$, there exists an ancilla A defined in an external Hilbert space $\mathcal{H}_B$ and a unitary operation U_S acting on $\mathcal{H}_A \otimes \mathcal{H}_B$ such that $T(\varrho) = Tr_B\left(U_S(\varrho \otimes |0\rangle\langle 0|)U_S{'}\right)$.*

Stinespring's dilation theorem shows that given a measurement that is characterized on average by the operator T, one can always find a unitary operator U and an additional Hilbert space $\mathcal{H}_B$ such that the result of any measurement is the same, on average, as what would be found if the system interacts with an external ancilla via U_S. This implies that T cannot be distinguished on average from that expected under the measurement postulate. Let us consider particle evolution that is evolving via the two cases of 1) Schrödinger unitary evolution and 2) unitary evolution other than

Schrödinger's equation.

Schrödinger Unitary

First, the case whereby the unitary is found via Schrödinger's equation is considered. Consider again our UMDT of Chapter 3. It was assumed in Chapter 3 that one can separate the two devices in Figure 3.4 via an arbitrary large distance d, and one could in principle place a shutter in front of the photon emitter to limit the uncertainty of the time of emission to some Δt_e. Assuming $d > 2c(\Delta t_e + \tau_l)$ where τ_l is the latency time of the devices, there can be no direct interaction between the devices such that the state of one device affects the individual statistics of the other device, without violating relativistic constraints of no-signaling. With two devices in the initial state $|\psi_r^0\rangle_1 \otimes |\psi_r^0\rangle_2$ one needs to implement

$$T(\cdot) = \mathrm{Tr}_{A_1}\big(U_S\big(|\psi_{\text{photon,PBS}}\rangle\langle|\psi_{\text{photon,PBS}}| \otimes |\psi_r^0\rangle_{1\,1}\langle\psi_r^0| \otimes |\psi_r^0\rangle_{2\,2}\langle\psi_r^0| \otimes |0\rangle\langle 0|\big)U_S{}^{\dagger}\big)$$

where the ancilla is initially in the state $|0\rangle$. However, one cannot implement an arbitrary unitary U that is required under the Stinespring dilation for two separated detectors as direct interaction between the two detectors is not allowed. The best that one can do without violating no-signaling is to implement two Stinespring dilations in which two ancillae are used: A_1 appended to the Hilbert space of Device 1, and A_2 appended to the Hilbert space of Device 2 in the following manner:

$$T_1(\cdot) = \mathrm{Tr}_{A_1}\big(U_{S,1}\big(|\psi_{\text{photon,PBS}}\rangle\langle|\psi_{\text{photon,PBS}}| \otimes |\psi_r^0\rangle_{1\,1}\langle\psi_r^0| \otimes |0\rangle_{A_1\,A_1}\langle 0|\big)U_{S,1}{}^{\dagger}\big)$$

$$T_2(\cdot) = \mathrm{Tr}_{A_2}\big(U_{S,2}\big(|\psi_{\text{photon,PBS}}\rangle\langle|\psi_{\text{photon,PBS}}| \otimes |\psi_r^0\rangle_{2\,2}\langle\psi_r^0| \otimes |0\rangle_{A_2\,A_2}\langle 0|\big)U_{S,2}{}^{\dagger}\big).$$

The following are necessary conditions so that the operations $\{T_1, T_2\}$ are not distinguishable on average from that expected under the measurement postulate: 1) given the photon initial state $|\psi_{\text{photon,PBS}}\rangle = \sqrt{a}|1_V\rangle_B \otimes |0\rangle_C + \sqrt{1-a}\,|0\rangle_B \otimes |1_H\rangle_C$ the state of Device 1′ must be on average

$$T_1(\cdot) = a\rho_{1,1'}^{(1)} + (1-a)\rho_{2,1'}^{(1)}, \tag{4.11}$$

2) for Device 2′ on average

$$T_2(\cdot) = (1-a)\rho_{1,2'}^{(1)} + a\rho_{2,2'}^{(1)} \tag{4.12}$$

and 3) the joint state of Device 1′ and Device 2′ must either be in the product state $\rho_{1,1'}^{(1)} \otimes \rho_{2,2'}^{(1)}$ or the product state $\rho_{2,1'}^{(1)} \otimes \rho_{1,2'}^{(1)}$ which, on average, is the separable density matrix given by:

$$a\rho^{(1)}_{1,1'} \otimes \rho^{(1)}_{2,2'} + (1-a)\rho^{(1)}_{2,1'} \otimes \rho^{(1)}_{1,2'}. \tag{4.13}$$

Now, if indeed there is a unitary that is able to interact the detector with an ancilla within the time which can have a causal effect on the measurement outcome, then it must be within a distance of $c(\Delta t_e + \tau_l)$. However, in the construction of the Chapter 3's UMDT, we included in the definition of Device $1'$ and Device $2'$, all particles that are contained within such a causal distance. Hence the ancilla are necessarily part of Device $1'$ and Device $2'$ as we have defined them. Let us consider the main result of Chapter 3. Consider any local Hamiltonian that interacts Mode B and Device 1 and Mode C and Device 2 shown in Figure 3.4. We have proven that entanglement always is predicted under unitary evolution and the CHSH value is $2\sqrt{2}$. Now, the definition of $T_1(\cdot)$ and $T_2(\cdot)$ are unitary, and one can therefore find a corresponding local Hamiltonian that can be used in the Chapter 3 UMDT. The CHSH value will be maintained at $2\sqrt{2}$ with the use of $T_1(\cdot)$ and $T_2(\cdot)$. Hence, it is impossible that the condition that is required in Equation (4.13) is met and the use of the Stinespring dilation fails to resolve the measurement problem.

Note that if one is able to ignore the ancilla, proposals that use such dilations do not appear to be disprovable. For example, consider the proposal by Ozawa in the paper [131] based on operational quantum mechanics. A similar methodology has been proposed by Ozawa based on a dilation to resolve the measurement problem. Ozawa's operational theory is applied to a single detection device and it is shown that, on average, the same density matrix results when using a dilation as would be required under measurement. And for a single measurement device, there is no theoretical or experimental methodology that can discriminate a system for which states occur with a given probability versus a system that always occurs in the same average density matrix. That is, consider an orthogonal decomposition of a mixed state ρ given by

$$\rho = \sum \alpha_i \rho_i$$

where $\alpha_i > 0$ and the ρ_i are orthogonal in the sense that $\mathrm{Tr}({\rho_i}^{\dagger}\rho_j) = 0$. Suppose that a set of particles is prepared in one of two ways by Alice. In the first manner, the particles are always prepared in the state given by $\rho = \sum \alpha_i \rho_i$. In the second manner, the particles are prepared by generating a random variable that corresponds to selection of one of the states ρ_i with probability α_i. In the second manner, each time the particles are prepared, they are in one of the orthogonal states ρ_i. If Alice then gave the particles to Bob without specifying whether or not the particles were prepared via the first method or second method, Bob cannot distinguish the two cases. Furthermore, Alice can give Bob not only a single set of particles, but can provide Bob with many sets of particles so that all cases are either prepared by an average ρ or by selecting a particular ρ_i with probability α_i and Bob still cannot distinguish the two cases.

As long as one is able to ignore the ancilla in the Stinespring dilation, Ozawa's solution cannot be disproven either theoretically or experimentally. When one considers two separated devices as in Figure 3.4 it is possible in principle to detect the entanglement if the interaction is unitary and for which all particles including any ancilla that are within a time-like effect are included in the description, a contradiction with experiment is possible. In such a case, one cannot ignore the ancilla in Stinespring's dilation. This answers a question that was posed in Chapter 2 and leads to the dictum,

> *There is no salvation from the Measurement Problem in the Church of the Higher Hilbert Space.*

Non-Schrödinger Unitary

Having seen that unitary Schrödinger theories that correspond to a particular fixed Hamiltonian cannot possibly be valid explanations of measurement, let us consider unitary theories that are proposed for which arbitrary unitary operations are invoked. Consider the paper by Chirabella et al. [132]. The authors state that they provide a complete derivation of finite dimensional quantum theory from only operational principles.

One might then examine whether or not such operational principles provides a resolution to the measurement problem. The authors develop an operational-probabilistic framework (OPF) based on circuit transformations. The OPF uses a circuit combined with probability theory. The circuits can have classical outcomes, but the theory is used to compute probability distributions for circuit outcomes when the circuits are interconnected.

A circuit or test in the OPF represents a physical device such as a beamsplitter or a photon counter. Each device has an input system and an output with capital letters. The tests have one of several classical outcomes, represented by an index i ∈ X for which each outcome corresponds to a possible event. The set $\{C_i\}$ of all events from A to B is defined as the set of transformations from A to B, Transf(A,B). Preparation tests are used to define states represented as $|\varrho_i)_B$ as tests with no input (no-input we denote as $\mathcal{I}$). Effects or observations represented as $(a_j|_A$ are defined as tests with no-output. The set of states is denoted St(A)=Transf($\mathcal{I}$,A) and the set of effects Eff(A)=Transf($A, \mathcal{I}$). The authors show that that composition of a preparation test $|\varrho_i)_A$ with an observation test $(a_j|_A$ gives rise to the joint probability $p_{i,k} = (a_j|\varrho_i)$.

Chirabella et al. in [132] introduce a *purification principle*. A purification of the state $\varrho \in \mathrm{St}_1(A)$ is a pure state $|\Psi_\rho\rangle$ of some composite system AB with the property that ϱ is the marginal of $|\Psi_\rho\rangle$. The purification principle is based on showing that any particular ϱ that exists is equivalent to a unitary operation plus a measurement on the ancilla on a larger Hilbert space with the possible addition of a deterministic effect e_B. The use of a pure state Ψ_ϱ on a larger Hilbert space for which a subsystem has the state ϱ, is termed purification. One can show that purifications exist for any state ϱ and can be accomplished via a unitary theory, for example, via the use of the Stinespring dilation [130, p. 22]. As long as the ancilla or additional environment is included in the Chapter 3 UMDT, the CHSH result is $2\sqrt{2}$ under unitary evolution. However, if the system is in a definite state, then the CHSH result (including the additional environment or ancilla required to accomplish the purification) is $\sqrt{2}$.

Chirabella et al state in [132]:

> *... purification implies the possibility of simulating any irreversible process through a reversible interaction of the system with an environment that is finally discarded.*

Indeed, one is capable of simulating an irreversible process, but only when the environment is discarded. If the environment were included, there would be a reversible process and the entanglement would be maintained in the Chapter 3 UMDT test. The point is that when the additional environment is included, the CHSH result must be $\sqrt{2}$ when a measurement occurs, but will always be $2\sqrt{2}$ under unitary evolution. One cannot simply ignore the additional particles required to accomplish the purification in the UMDT test of Chapter 3, because they are well defined and limited in number by the condition $d > c(\Delta t_e + \tau_l)$. In terms of the measurement process, the authors state:

> *The purification principle expresses a law of conservation of information, stating that at least in principle, irreversibility can always be reduced to the lack of control over an environment. More precisely, the purification principle is equivalent to the statement that every irreversible process can be simulated in an essentially unique way by a reversible interaction of the system with an environment, which is initially in a pure state. This statement can also be extended to include the case of measurement processes...*

Solutions such as this are dispersing the entanglement into the environment to try and force a unitary solution. But even if the entanglement disperses into the environment—it hasn't gone away and no measurement can be said to have occurred unitarily. The solution proposed in [132] violates our dictum:

> *To the extent there is entanglement, there is no measurement.*

Another result of OPF [132] is for any two states ϕ_0 and ψ_0 there exists a unitary transformation $U \in \text{Trans(A, B)}$ such that $U\phi_0 = \psi_0$. This is correct; there always exists a unitary transformation that maps one state to another. Consider again the Chapter 3 UMDT and for simplicity that the two devices are 100% absorbing and initialized to the pure initial states $|\psi_r^0\rangle_1$ and $|\psi_r^0\rangle_2$. Device 1 (Device 2) evolves to $|\psi^1\rangle_1$ ($|\psi^1\rangle_2$) when a photon is detected and to $|\psi^0\rangle_1$ ($|\psi^0\rangle_2$) when no photon is detected.

Suppose that $U_1|\psi_r^0\rangle_1=|\psi^1\rangle_1$ and $U_2|\psi_r^0\rangle_1 = |\psi^0\rangle_1$. Then there are reversible transformations that will take the initial state of Device 1 (and similarly for Device 2) to either the state of Device 1 representing a detection or the state of the device representing no-detection. If one knows which device detects the photon, then one could choose the correct unitary that corresponds to the action that occurred during measurement.

In [132] a unitary description of measurement is proposed. While the authors are correct in that such a choice of unitary operations can be done *a posteriori*—such a specification requires knowledge of which device detected the photon. Such *a posteriori* information of which device detected the photon, demands information regarding the *a priori* statistics of the degree of superposition of the photon, which is parameterized by the variable *a*. A methodology of applying either U_1 or U_2 depending on *a posteriori* information is not a valid solution to the problem, because prior to the measurement occurring, there is only a fixed Hamiltonian with a fixed unitary evolution. Schrödinger unitary evolution is determined *a priori* to measurement. Initially, when the photon is launched through the beam splitter, neither the Hamiltonian of Device *1* nor Device *2* can be a function of *a* without violating causality. The issue becomes whether or not such an *a priori* Hamiltonian can adapt itself to become a function of the degree of superposition *a* by interacting with the photon, that has a state that is a function of *a*. In order for this to occur, it must be possible locally at Device *1* and Device *2* for the devices themselves to be able to determine the superposition *a*, and then modify themselves in a manner to account for the evolution. However, such a device would need to be able to extract the degree of superposition *a*, but this would violate the no-cloning theorem [54] for which it is impossible to build any device that can reliably determine *a*.

The operations that would be needed by OPF to specify these unitary operators would, if Schrödinger's equation were obeyed, require the devices in the measurement problem setup to have either a time-varying Hamiltonian that depends on the degree of superposition *a,* or an environment that is a function of *a*, i.e., via a purification or Stinespring dilation. Again, this is impossible without violating causality. It does appear to be possible to develop a set of reversible unitary operations *a posteriori* after knowing the measurement results, which is consistent with a statistical or a non-unitary evolution. Papers occasionally appear in the literature that do exactly this as an approach to resolving the measurement problem. A recent example for which the authors have defined a unitary operator as a function of the measured state can be seen in Equation 11 of [133] by Cruikshank and Jacobs. These same authors' purport, "If a physical measurement-based process is designed to produce a single outcome, then even the measurement at the end is unnecessary." Such a solution may be satisfactory to resolve the philosopher's version of the measurement problem, but it does not resolve our version: such unitary operators cannot follow from Schrödinger's evolution that requires a Hamiltonian to be specified prior to the act of measurement.

One can't simply throw around unitary operators here and there and claim that Nature is describable by a reversible *a priori* theory. Simply because there exists a set of unitary operators that can describe a given evolution, given the *a posteriori* final state, does not imply that the same quantum mechanical evolution follows from the unitary evolution *a priori* that is predicted by Schrödinger's equation with a time invariant and causal Hamiltonian for which the *a posteriori* final state is unknown.

Unitary theory combined with Kraus operators or measurement can be described *a posteriori* by reversible unitary quantum operations. This may be quite correct, but such unitary operations provide no *a priori* predictive power, which is a basic requirement of any scientific theory and are not derivable from Schrödinger's equation within a causal Hamiltonian formulation.

Consistent Histories

Consistent histories is a theory that provides a means of generating potential histories of the evolution of a closed system in a Hilbert space $\mathcal{H}$. The closed system may include not only a system that is being measured, but also the device. We will describe a version of consistent histories for which the initial state at time t_0 is a pure state given by $|\psi_0\rangle$ and evolves from time t_A to time t_B via the unitary operator $U(t_A, t_B)$, except at a set of times denoted $\{t_1, t_2, \cdots, t_N\}$, for which the state undergoes projection, and N is the number of projections.

There are often more than one set of potential histories. Potential histories will be indexed by $k = 1,2,\cdots$. One can define a sample space $\mathbb{S}_k$ that contains elements, each of which is a set of projections that might occur at the times $\{t_1, t_2, \cdots, t_N\}$. A quantum sample space $\mathbb{S}_k$ consists of a set of histories that are consistent in a manner that a probability can be assigned to each history such that the probabilities of each element or individual history, sum to unity. One can see that if the elements of $\mathbb{S}_k$are disjoint or mutually exclusive and also if the union of the elements is the entire sample space, then one is guaranteed that the sum of the element probabilities in $\mathbb{S}_k$ will be unity. Sufficient conditions under which the elements of $\mathbb{S}_k$ are mutually exclusive will be found.

Suppose the initial state is given by $|\psi_0\rangle$ and that for time t_i for histories belonging to $\mathbb{S}_k$, the set of possible projections are denoted by $P_1(k, t_i), P_2(k, t_i), \cdots$. A set of projectors called histories $\{Y_{\alpha_1}(\mathrm{k}), Y_{\alpha_2}(\mathrm{k}), \cdots\}$ are defined at various times as

$$Y_{\alpha_j}(k) = P_0 \otimes P_{\alpha_{1,j}}(k, t_1) \otimes P_{\alpha_{2,j}}(k, t_2) \otimes \cdots \otimes P_{\alpha_{N,j}}(k, t_N)$$

where $\alpha_j = \left(\alpha_{1,j}, \alpha_{2,j}, \cdots, \alpha_{N,j}\right)$, and $P_0 \equiv |\psi_0\rangle\langle\psi_0|$. Now we can define the quantum sample history space $\mathbb{S}_k = \left\{Y_{\alpha_1}(k), Y_{\alpha_2}(k), \cdots, Y_{\alpha_N}(k)\right\}$ as consisting of the set of events that form a Boolean algebra with the properties

$$\sum_i Y_{\alpha_i}(k) = I$$
$$Y_{\alpha_i}(k) Y_{\alpha_j}(k) = 0, \;\; \forall i \neq j. \tag{4.14}$$

The events are assumed [134, p. 57] to be mutually orthogonal via Equation (4.14). As the events are mutually orthogonal, they also commute and the events of $\mathbb{S}_k$ are a Boolean algebra for which the conjunction and disjunction operations $\wedge$ and $\vee$ respectively are well-defined.

Events defined at times $\{t_1, t_2, \cdots, t_N\}$ in a given experiment should be consistent in the sense that a probability can be assigned to any particular history and the histories within $\mathbb{S}_k$ are disjoint. This later condition is not the same as the mutual orthogonal condition of Equation (4.14). A chain operator $K(Y_{\alpha_j}(k))$ is defined as

$$K\left(Y_{\alpha_j}(k)\right) = P_{\alpha_{N,j}}(k, t_N)U(t_N, t_{N-1})P_{\alpha_{N-1,j}} \cdots P_{\alpha_{2,j}}(k, t_2)P_{\alpha_{1,j}}(k, t_1)U(t_1, t_0)P_0.$$

The probability that the history $Y_{\alpha_j}(k)$ occurs is given by

$$p(Y_{\alpha_j}(k)) = \mathrm{Tr}\left[K\left(Y_{\alpha_j}(k)\right)^{\dagger} K\left(Y_{\alpha_j}(k)\right)\right].$$

A sufficient condition [135] [136] that a given sample space $\mathbb{S}_k$ represents a set of consistent histories (that are mutually exclusive and for which the probabilities of the elements sum to unity) is that the different histories are mutually exclusive in the sense that

$$\mathrm{Tr}\left[K\left(Y_{\alpha_j}(k)\right)^{\dagger} K\left(Y_{\alpha_l}(k)\right)\right] = 0, \quad \forall j \neq l. \tag{4.15}$$

Note that this condition restricts within each k the set of histories (see the related Exercises 4.6-4.8). When the individual histories within a given set of consistent histories satisfy Equation (4.15) the probability of any two histories will be additive via

$$p\left(Y_{\alpha_j}(k) \vee Y_{\alpha_l}(k)\right) = p\left(Y_{\alpha_j}(k)\right) + p\left(Y_{\alpha_l}(k)\right), \qquad \forall j \neq l,$$

or

$$p\left(Y_{\alpha_j}(k) \wedge Y_{\alpha_l}(k)\right) = 0 \ \forall j \neq l.$$

A history may or may not be Schrödinger unitary. For a given problem with a Hamiltonian H if there was only a valid consistent history that was Schrödinger unitary, then the measurement problem would be solved by consistent history theory. However, as has been raised by Dowker and Kent [137], this is not the case in light of the non-unique consistent history sets that can be defined when $k > 2$:

> *Our own view is that, because of the overwhelming variety of consistent sets and the strength of the condition that a proposition be includable in every consistent extension of the actual history, it will not be possible to resolve this particular problem without going*

beyond the formalism of consistent histories.

There is an alternative which cuts through all these problems. It is to accept, once and for all, that quantum theory is not sufficient to describe the world, and that it should be augmented by a further axiom which takes the form of a selection principle. The consistent histories formalism has taught us that there are infinitely many incompatible descriptions of the world within quantum mechanics. Perhaps some simple criterion can be found to pick out one of these descriptions, by selecting one particular consistent set.

Griffiths considers this criticism by Dowker and Kent in [138]. He states

Dowker and Kent conclude that the consistent history approach to quantum theory lacks predictive power, and for this reason is not satisfactory as a fundamental scientific theory. ... it is the fact that the consistent history approach, as a fundamental theory, treats all frameworks or consistent families "democratically", and provides no criterion to select out one in particular. In other words, there is no "law of nature" which specifies the framework.

Griffiths concedes that

... consistent histories, as a fundamental theory of nature, does not single out a particular framework.

Griffiths gives an example of consistent histories regarding Dowker and Kent's criticism which is similar to what we have already been examining in the UMDT of Chapter 3. That is, a photon is considered that enters a beam splitter at *A* at time t_0 in Figure 3.4 and exits at time t_1 in one of two output paths *B* and *C*. In each path *B* and *C* there is a detector. The initial joint state of the photon plus Device 1 plus Device 2 is given by $|\Psi_0\rangle$ and the initial ready state of Device i is assumed to be a pure state and is denoted by $|\psi_r^0\rangle_i$, i=1,2, $|\psi_r^0\rangle_i \in \mathcal{H}_i$, i=1,2, $\mathcal{D}(\mathcal{H}_i) = N$. For simplicity, it will be assumed that the detectors are 100% absorbing, that is the device always completely absorbs the photon. Suppose that in both paths there is only the vacuum state $|0\rangle$. At time t_1 it is assumed that the initial ready device state $|\psi_r^0\rangle_i$ changes to a read-out state that is indicative of no measurement and is denoted $|\psi^0(t_1)\rangle_i, i = 1{,}2$, for which it changes to $|\psi^0(t_2)\rangle_i, i = 1{,}2$. Consider the case that a photon with state $|1_V\rangle_B$ is inserted in Path B. The state of Device 1 evolves to its readout state indicating that the device has detected a photon, which will be denoted $|\psi^1\rangle_1 \in \mathcal{H}_1$. Similarly, if a photon with state $|1_H\rangle_C$ is inserted solely in Path *C* and Device 2 is 100% absorbing, the final state of Device 2 is denoted as $|\psi^1(t_2)\rangle_2 \in \mathcal{H}_2$.

Consider the initial local state $(|1_V\rangle_A + |1_H\rangle_A)/\sqrt{2}$ of the photon in mode *A* at time t_0.

This state then propagates through the beam splitter and evolves to the state $|\psi_{photon,PBS}\rangle = (|1_V\rangle_B \otimes |0\rangle_C + |0\rangle_B \otimes |1_H\rangle_C)/\sqrt{2}$ at time t_1 which represents an equal superposition of output paths B and C. Let $|\Psi_0\rangle = |a\rangle \otimes |\psi_r^0\rangle_1 \otimes |\psi_r^0\rangle_2$. Denote the projection operator of an arbitrary state $|\psi\rangle$ by $\mathcal{P}(|\psi\rangle) \equiv |\psi\rangle\langle\psi|$. We define $|\psi_{fin,1}\rangle$ consistent with previous notation as:

$$|\psi_{fin,1}\rangle = |0\rangle \otimes (|\psi^1(t_2)\rangle_1 \otimes |\psi^0(t_2)\rangle_2 + |\psi^0(t_2)\rangle_1 \otimes |\psi^1(t_2)\rangle_2)/\sqrt{2}.$$

Griffiths shows two sets of histories that are individually consistent. One set (denoted by $k = 1$) can be found via the following four histories:

$$Y_{\alpha_1}(1) = \left\{\mathcal{P}(|\Psi_0\rangle), \mathcal{P}\left(|\psi_{photon,PBS}\rangle \otimes |\psi^1(t_1)\rangle_1 \otimes |\psi^1(t_1)\rangle_2\right), \mathcal{P}(|\psi_{fin,1}\rangle)\right\}$$
$$Y_{\alpha_2}(1) = \left\{\mathcal{P}(|\Psi_0\rangle), \mathcal{P}\left(|\psi_{photon,PBS}\rangle \otimes |\psi^0(t_1)\rangle_1 \otimes |\psi^0(t_1)\rangle_2\right), I - \mathcal{P}(|\psi_{fin,1}\rangle)\right\}$$
$$Y_{\alpha_3}(1) = \left\{\mathcal{P}(|\Psi_0\rangle), I - \mathcal{P}\left(|\psi_{photon,PBS}\rangle \otimes |\psi^0(t_1)\rangle_1 \otimes |\psi^0(t_1)\rangle_2\right), \mathcal{P}\left(|\psi_{fin,1}\rangle\right)\right\}$$
$$Y_{\alpha_4}(1) = \left\{\mathcal{P}(|\Psi_0\rangle), I - \mathcal{P}\left(|\psi_{photon,PBS}\rangle \otimes |\psi^0(t_1)\rangle_1 \otimes |\psi^0(t_1)\rangle_2\right), I - \mathcal{P}(|\psi_{fin,1}\rangle)\right\}.$$

However,

$$p\left(Y_{\alpha_2}(1)\right) = p\left(Y_{\alpha_3}(1)\right) = p\left(Y_{\alpha_4}(1)\right) = 0$$

where $p(Y_\alpha)$ denotes the probability of history Y_α occurring. Hence for $k = 1$ there is only one non-zero probability history given by $Y_{\alpha_1}(1)$. For the set $k = 2$ the following two sets of non-zero probability histories (along with a correspond set of zero probability histories) are found:

$$Y_{\alpha_1}(2) = \begin{Bmatrix} \mathcal{P}(|\Psi_0\rangle), \mathcal{P}\left(|\psi_{photon,PBS}\rangle \otimes |\psi^0(t_1)\rangle_1 \otimes |\psi^0(t_1)\rangle_2\right), \\ \mathcal{P}(|0\rangle \otimes |\psi^1(t_2)\rangle_1 \otimes |\psi^0(t_2)\rangle_2) \end{Bmatrix}$$

$$Y_{\alpha_2}(2) = \begin{Bmatrix} \mathcal{P}(|\Psi_0\rangle), \mathcal{P}\left(|\psi_{photon,PBS}\rangle \otimes |\psi^0(t_1)\rangle_1 \otimes |\psi^0(t_1)\rangle_2\right), \\ \mathcal{P}(|0\rangle \otimes |\psi^0(t_2)\rangle_1 \otimes |\psi^1(t_2)\rangle_2) \end{Bmatrix}.$$

Note that the set of histories for $k = 2$ are mutually exclusive in the sense that

$$p\left(Y_{\alpha_j}(2) \cup Y_{\alpha_l}(2)\right) = p\left(Y_{\alpha_j}(2)\right) + p\left(Y_{\alpha_l}(2)\right), \qquad \forall j \neq l.$$

However, it can be the case that

$$p\left(Y_{\alpha_j}(1) \cup Y_{\alpha_l}(2)\right) \neq p\left(Y_{\alpha_j}(1)\right) + p\left(Y_{\alpha_l}(2)\right).$$

In this sense, the two sets of histories $k = 1$ and $k = 2$ are termed incompatible. A sufficient condition for the incompatibility is that the projectors at a given time do not commute when chosen from two different sets of histories.

Griffiths states in [138]:

> *There are many (in fact, an uncountably infinite number of) other frameworks which could have been employed to discuss the same situation. The fact that the same physical system can be discussed using many different frameworks gives rise to the problem of choice: how does one decide which framework is appropriate for describing what goes on in this closed quantum system?*

The discrimination of the incompatible sets of histories is required and incumbent upon the theory to provide. A Category 1 theory would provide such discrimination in terms of meeting Requirement R1.3 (or Requirement R2.2 for a Category 2 theory). In terms of using actual data from measurements to determine the physical history, Griffiths states [138]:

> *Checking the predictions given by different incompatible frameworks always involves alternative experimental arrangements... In any case, just as there is no single "correct" choice of consistent family for describing the system, there is no single arrangement of apparatus which can be used to verify the predictions obtained using different families.*

We disagree with this statement. The UMDT of Chapter 3 is a single experimental arrangement that one can use to determine via the CHSH sum whether or not the history came from either the set corresponding to $k = 1$ or the set corresponding to $k = 2$.

Note that the use of multiple potential histories would not necessarily be a major problem if one addresses the philosopher's measurement problem, which is an *a posteriori* problem given that a measurement has occurred in a given basis. This is because if one is given the measurement basis and given the measurement times, then one can determine the histories that are consistent with any given experiment and rule out incompatible histories. This would render the measurement problem an interpretational problem as opposed to the physical measurement problem that we have defined. Regarding this issue, Griffiths states [138]

> *The endless discussions about how to interpret the state $|S\rangle$ and resolve the corresponding "measurement" problem have been rightly criticized by Bell.*

Presumably, the "endless discussions" that Griffiths refers to include Einstein's objections, Schrödinger's cat paper, and the substantial body of work that concentrates on the issue of the lack of a product state prediction under unitary evolution. The physical measurement problem would appear to be irrelevant to Griffiths, as he is of the viewpoint that current quantum mechanics is correct FAPP. This viewpoint leads directly to the rejection of the physical measurement problem in favor of the philosophers' definition of the measurement problem.

On the other hand, consistent histories is a rather interesting development that does appear to go beyond the Copenhagen interpretation in providing sets of consistent histories. As noted by Dowker and Kent [137] consistent histories may be useful:

> *The virtues of the consistent histories approach are worth reasserting. We have a natural mathematical criterion which is empirically supported and which identifies the physical content of quantum theory to be propositions about particular collections of events encoded in the consistent sets. This gives a framework in which attempts to set out a quantum theory of the universe, and the problems inherent in this idea, can be sensibly discussed.... Many attempts have been made to find natural mathematical structures and interpretational postulates that allow one to use quantum theory to make statements about physics that go beyond the Copenhagen bounds. In our view, the consistent histories formalism is one of the most significant developments.*

Consistent histories as developed by Griffiths does indeed narrow down the set of possible histories as histories via requiring Equation (4.15) to be fulfilled. Hence Category 1 or 2 theories that violate this rule must be excluded. Consistent histories is certainly a step in the right direction regarding the physical measurement problem as we agree that it narrows down the potential sets of histories that need to be considered, in the sense that a theory that resolves the measurement problem must produce a history that is within one of the sets of consistent histories. However, there are a large number of potential histories that remain after imposing Equation (4.15) and without any additional assumptions, consistent histories is simply silent on further narrowing down to a single consistent set.

Many-Worlds Interpretation

The relative state formulation was put forward by Hugh Everett, III in 1957 [139] and popularized by DeWitt as a many-worlds interpretation (MWI) [140]. The formulation is based on applying the Schrödinger equation to all systems. Everett [139, p. 457] considers a machine with a sensor that contains a memory as a sufficient condition for an observer:

> *As models for observers we can, if we wish, consider automatically functioning machines, possessing sensory apparatus and coupled to recording devices capable of registering past sensory data and*

machine configurations.

Everett agrees that a superposition of system plus apparatus will generally occur in a manner that is similar to that shown in external orthogonalization resulting in Equation (4.2) reproduced below:

$$\sum_r c_r e^{i\theta_r} \, |\phi_r\rangle \otimes |\Psi_r\rangle.$$

Consider an observer to be a system with memory $|\Psi_r\rangle$ and the phenomenon being observed $|\phi_r\rangle$. Equation (4.2) is interpreted by Everett to be a superposition whereby the individual terms have a characteristic such that each term is an observer with a memory that is correlated to a particular phenomenon of a corresponding relative state $|\phi_r\rangle$. Everett proposes that there are many branches that simultaneously exist, whereby in each branch there exists an observer memory that is correlated to the corresponding relative state [139]:

> *Nevertheless, there is a representation in terms of a superposition, each element of which contains a definite observer state and a corresponding system state. Thus with each succeeding observation (or interaction), the observer state "branches" into a number of different states. Each branch represents a different outcome of the measurement and the corresponding eigenstate for the object-system state. All branches exist simultaneously in the superposition after any given sequence of observations.*

Everett in his original paper expected that such branching processes for the typical observer are equivalent to the Born statistical rule (this issue will be revisited shortly in the subsection *MWI and Born's Rule*) [139]:

> *In conclusion, the continuous evolution of the state function of a composite system with time gives a complete mathematical model for processes that involve an idealized observer. When interaction occurs, the result of the evolution in time is a superposition of states, each element of which assigns a different state to the memory of the observer. Judged by the state of the memory in almost all of the observer states, the probabilistic conclusion of the usual "external observation" formulation of quantum theory are valid. In other words, pure Process 2 wave mechanics, without any initial probability assertions, leads to all the probability concepts of the familiar formalism.*

Moreover, Everett does not see a need to destroy the superposition terms:

> *From the viewpoint of the theory, all elements of a superposition (all "branches") are "actual", none any more "real" than the rest. It is unnecessary to suppose that all but one are somehow destroyed, since all separate elements of a superposition individually obey the wave function equation with complete indifference to the presence or absence ("actuality" or not) of any other elements.*

Everett in [139] considers an observer that will make a number of sequential measurements which generate a particular outcome sequence in the memory of the observer. To illustrate in detail Everett's procedure, we consider the sequential measurement of two spin ½ particles in Figure 4.1. The *i* th, *i*=1,2 spin ½ particle is initially in the state given by:

$$\sqrt{a}|\uparrow\rangle_i + \sqrt{1-a}|\downarrow\rangle_i.$$

Before the *j* th branching process occurs, the state of the *k* th distinct reality or universe will be denoted $|\psi_{u_{j,k}}\rangle$. The initial state of the two particles in the first universe is denoted by

$$|\psi_{u_{1,1}}\rangle = \left(\sqrt{a}|\uparrow\rangle_1 + \sqrt{1-a}|\downarrow\rangle_1\right)\left(\sqrt{a}|\uparrow\rangle_2 + \sqrt{1-a}|\downarrow\rangle_2\right)$$

where it assumed that the multiplication between the two particles is shorthand for the tensor product. An observation or measurement on the first spin particle will be made by using device M_1 and on the second spin particle by using a second device M_2. The observation can be taken to be complete in MWI when it is registered in the memory of the device. Unitarily, the observer-particle is in a superposition of Equation (4.2) for which there are two terms. One term is for which M_1 is in the state having registered spin up, represented as $|\uparrow\rangle_{M_1}$ and for which the spin ½ particle is also found in the state $|\uparrow\rangle_1$. The second term is for which M_1 is in the state having registered spin down, represented as $|\downarrow\rangle_{M_1}$ and for which the spin ½ particle is found in the state $|\downarrow\rangle_1$. Although there are two terms in the complete wave function, Everett claims that each individual term is a reality caused by a branching process in which there are now two universes. In Reality or Universe 1, the spin ½ particle is measured up and there is a separate reality or Universe 2 in which the spin ½ particle is measured down. After the first branching process occurs, the state of the device-spin 1-spin 2 in the first reality is given by

$$|\psi_{u_{2,1}}\rangle = |\uparrow\rangle_{M_1} |\uparrow\rangle_1\left(\sqrt{a}|\uparrow\rangle_2 + \sqrt{1-a}|\downarrow\rangle_2\right)$$

and in the second universe a spin down is observed:

$$|\psi_{u_{2,2}}\rangle = |\downarrow\rangle_{M_1} |\downarrow\rangle_1\left(\sqrt{a}|\uparrow\rangle_2 + \sqrt{1-a}|\downarrow\rangle_2\right).$$

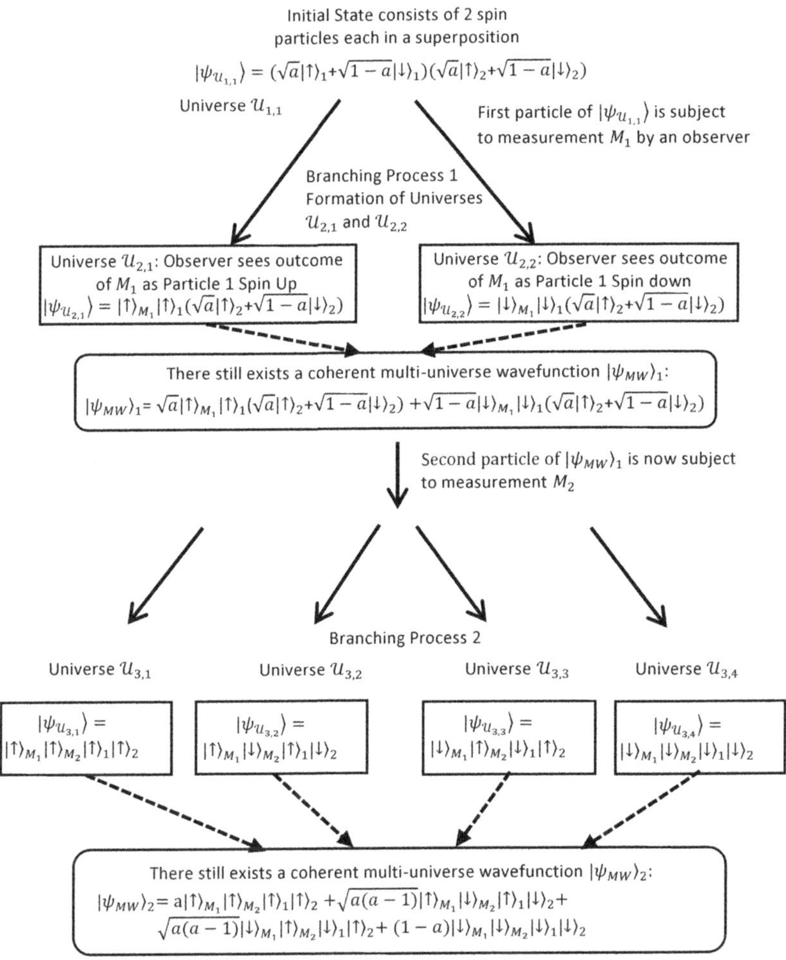

Figure 4.1: Universe formation in the Multiple Worlds Interpretation illustrated via sequential measurement of two spin ½ particles each initially in a superposition.

In addition, there still exists a fully coherent wave function that agrees with Schrödinger's equation given by:

$$|\psi_{MW_1}\rangle = \sqrt{a}|\uparrow\rangle_{M_1}\,|\uparrow\rangle_1\big(\sqrt{a}|\uparrow\rangle_2 + \sqrt{1-a}|\downarrow\rangle_2\big) \\ + \sqrt{1-a}|\downarrow\rangle_{M_1}|\downarrow\rangle_1\big(\sqrt{a}|\uparrow\rangle_2 + \sqrt{1-a}|\downarrow\rangle_2\big).$$

The individual terms such as $|\psi_{\mathcal{U}_{2,1}}\rangle$ are considered by Everett to be relative states, all of which make up the overall absolute state $|\psi_{MW_1}\rangle$. An observer that has a particular relative state is termed a relative observer. Now, consider the case when the second particle of $|\psi_{MW_1}\rangle$ is now subject to a measurement M_2. This results in the further branching of the wave function for which there are four possible outcomes as shown in Figure 4.1, which are:

$$
\begin{aligned}
|\psi_{\mathcal{U}_{3,1}}\rangle &= |\uparrow\rangle_{M_1}|\uparrow\rangle_{M_2}|\uparrow\rangle_1|\uparrow\rangle_2 \\
|\psi_{\mathcal{U}_{3,2}}\rangle &= |\uparrow\rangle_{M_1}|\downarrow\rangle_{M_2}|\uparrow\rangle_1|\downarrow\rangle_2 \\
|\psi_{\mathcal{U}_{3,3}}\rangle &= |\downarrow\rangle_{M_1}|\uparrow\rangle_{M_2}|\downarrow\rangle_1|\uparrow\rangle_2 \\
|\psi_{\mathcal{U}_{3,4}}\rangle &= |\downarrow\rangle_{M_1}|\downarrow\rangle_{M_2}|\downarrow\rangle_1|\downarrow\rangle_2.
\end{aligned}
$$

Still, in MWI there continues to exist a coherent wave function that is given by:

$$
|\psi_{MW}\rangle_2 = \mathrm{a}|\uparrow\rangle_{M_1}|\uparrow\rangle_{M_2}|\uparrow\rangle_1|\uparrow\rangle_2 + \sqrt{a(a-1)}|\uparrow\rangle_{M_1}|\downarrow\rangle_{M_2}|\uparrow\rangle_1|\downarrow\rangle_2 + \\
\sqrt{a(a-1)}|\downarrow\rangle_{M_1}|\uparrow\rangle_{M_2}|\downarrow\rangle_1|\uparrow\rangle_2 + (1-a)|\downarrow\rangle_{M_1}|\downarrow\rangle_{M_2}|\downarrow\rangle_1|\downarrow\rangle_2.
$$

Everett's MWI and Born's Rule

Now there remains the issue of the mechanics of universe splitting. Essentially how does one branch from $|\psi_{MW_1}\rangle$ to either $|\psi_{\mathcal{U}_{2,1}}\rangle$ or $|\psi_{\mathcal{U}_{2,2}}\rangle$? This issue is important in terms of how the probability interpretation in Everett's procedure can be fully reconciled with Born's rule. For a thorough discussion on this issue, the reader is referred to Barrett [141]. One problem with Everett's version of MWI is that Everett desired to derive Born's law as a consequence of deterministic pure wave function dynamics. Everett's theory proposes that if a measurement of an observable with N possible outcomes is made, there will exist only N branches of the universe for which each outcome occurs. These branches or worlds all simultaneously exist and are independent of the coefficients of the superposition which are needed to derive the Born law. Since the branches are independent of the coefficients of the superposition, it is not clear how Born's law can come about. Everett added an *additional* requirement that branches are to be measured by a mathematical measure that is equivalent to the coefficient squared of the corresponding term of the superposition. This at once gives the Born rule for the typical observer. Now, the probability of occurrence of Universe $\mathcal{U}_{3,1}$ after the second branch in Figure 4.1 is given by $\mathrm{P}(\mathcal{U}_{3,1}) = a^2$, and as well probabilities of the remaining universes now correspond with Born's law. However, as all branches physically exist in Everett's theory, it has been argued by Graham [140, p. 236] that the rationale of such a measure is inappropriate, and rather a methodology in which the universes are measured by a simple count should be applied. Deutsch also makes the point in that Everett derives the Born rule by depending on a probabilistic mechanism, which he was hoping to avoid. To rectify this possibility, Deutsch [142] added the following axiom to Everett's theory:

> *The world consists of a continuously infinite-measured set of universes*

Furthermore, Deutsch stipulates that when a superposition occurs as in Equation (4.2) the set of universes then form a disjoint set of N elements, each set of universes corresponding to a given outcome. The size (or measure) of the set of universes that correspond to the r th outcome is $|c_r|^2$. In this manner, one can ensure that outcomes occur with the correct probability. Even with this modification, MWI as a Category 1 theory does not provide the conditions for which the Chapter 3 UMDT experiment would verify a product state that is demanded by the measurement postulate.

Analysis of MWI with UMDT

Let us now critically examine this theory via the UMDT of Chapter 3. First suppose that the devices that are observers consist of quantum controlled-not (CNOT) gates that flip a bit in the device should the device interact with a photon. Basic quantum logic devices have long been the subject of experimental investigation and have already been realized in a fully unitary manner, for example, [143] [144] [145] [146]. Such devices clearly satisfy Everett's requirement for an observer as quantum CNOT gates that are automatically functioning machines that possess sensory apparatus and are coupled to recording devices capable of registering past sensory data via the flipping of the ancilla qubit. Everett expects that there is a branching that would occur for which the two devices are in a product state, "with complete indifference to the presence of absence of any other elements." If the quantum CNOT gates were in a product state, the CHSH result would be $\sqrt{2}$. However, this can be tested and will almost certainly be found to be false. As quantum CNOT gates are unitary, the value would be expected to be $2\sqrt{2}$.

One might ask if there are other types of memories that act non-unitarily. It may be, however it was also found that when a photon was slowed and stored in a slow light medium, entanglement was found to be maintained and not destroyed [147]. Hence it does not appear that Everett's requirement of automatic functioning machines possessing sensory apparatus and are coupled to recording devices capable of registering past sensory data, can be considered to be a necessary condition for a bona fide measurement device.

A problem with Everett's theory in solving the measurement problem as we have defined it, is that Everett concludes that there is a "splitting" process of which no observer will be aware:

> *This total lack of effect of one branch on another also implies that no observer will ever be aware of any "splitting" process.*

This splitting process is what many refer to as a *many minds* theory. That is, the act of observation demands many observers that possess a memory, to exist simultaneously with no observer aware of the other observers. The problem is that entanglement between the two devices in the UMDT can still be experimentally distinguished from a product state of the two devices within the current formalism. If Everett is to be in full agreement with unitary theory, the result of the UMDT CHSH must always be

$2\sqrt{2}$. It would have to then come as a surprise if particular configurations of particles were found for the detectors for which the UMDT CHSH is only $\sqrt{2}$. And we already know that memory is not a necessary condition for the UMDT CHSH to yield the value of $\sqrt{2}$. Hence there does not appear to be any particularly helpful resolution of the physical measurement problem in Everett's theory so far. On the other hand, Everett states in [148, p. 98]:

> *The irreversibility of the measuring process is therefore, within our framework, simply a subjective manifestation reflecting the fact that in observation processes the state of the observer is transformed into a superposition of observer states, each element of which describes an observer who is irrevocably cut off from the remaining elements. While it is conceivable that some outside agency could reverse the total wave function, such a change cannot be brought about by any observer which is represented by a single element of a superposition, since he is entirely powerless to have any influence on any other elements.*

In this sense, Everett seems to be claiming that the superposition terms do exist, but one is powerless to have any influence. If Everett claimed that it is impossible to reverse the total wave function, then Everett's theory would be a Category 2 theory.

The theory of Bell's inequality, which showed a difference between two-qubit product states and entangled states, was not developed until 1964 and was not known to Everett in 1957. As well, the development of a unitary CNOT gate was not begun until the 1990s. At some point, when a bona fide measurement device is utilized, the UMDT Chapter 3 test is expected to show a value of $\sqrt{2}$ within the current formalism if the resolution to the measurement problem is a Category 1 theory. On the other hand, a unitary CNOT gate would be expected to give the CHSH sum of $2\sqrt{2}$. Everett's formulation falls significantly short of predicting the physical conditions under which a configuration of particles constitutes a bona fide measurement device, which is a requirement of any Category 1 theory.

Decoherence

Zurek proposed that environmental decoherence can be utilized along with MWI. Note that once one has a composite state of two particles, it is always possible to rewrite the superposition in another basis for which it becomes a superposition. Hence the actual basis for which measurement occurs is called the basis selection problem. The basis selection problem is required for Category 1 theories via Requirement 1.3. Zurek states [149]:

> *If the quantum laws are universally valid, very nonclassical Schrödinger cat–like states should be commonplace for an apparatus that measures a quantum system and, indeed, for run-of-the-mill*

> *macroscopic systems in general Everett and other followers of the MWI philosophy tried to occasionally bypass this question by insisting that one should only discuss correlations. Correlations are indeed at the heart of the problem, but it is not enough to explain how to compute them; for that, quantum formalism is straightforward enough. What is needed instead is an explanation of why some states retain correlations, but most of them do not, in spite of the arbitrariness in basis selection that is implied.*

Zurek proposes that the pointer basis results largely from the interaction Hamiltonian formulation of device and environment. *Einselection*, a decoherence-imposed selection of preferred pointer states that remain stable in the presence of the environment, establishes stable states in which the apparatus can exist and for which the system is diagonal [149] :

> *Only the einselected pointer states will persist for long enough to retain useful (stable) correlations with—say— the memories of the observers, or, more generally, with other stable states. By contrast, their superpositions will degrade into mixtures that are diagonal in the pointer basis.*

Zurek further claims in [149] that environmental decoherence and pointer states provides the missing elements for defining the branches of Everett:

> *Decoherence and einselection are, however, rapidly becoming a part of a standard lore. Where expected, they deliver classical states, and —as we have seen above— guard against violations of the correspondence principle. The answers that emerge may not be to everyone's liking, and do not really discriminate between the Copenhagen Interpretation and the Many Worlds approach. Rather, they fit within either mold, effectively providing the missing elements —delineating the quantum-classical border postulated by Bohr (decoherence time fast or slow compared to the dynamical timescales on the two sides of the "border"), and supplying the scheme for defining distinct branches required by Everett (overlap of the branches is eliminated by decoherence).*

Zurek's environmental decoherence does indeed augment MWI by the proposal of addressing the basis selection problem of Requirement R1.3. A stable pointer basis found via the interaction Hamiltonian may very well be related to the measurement problem. This is similar to the theory of external orthogonalization and environmental decoherence which has already been presented. Zurek's theory at this time can be considered a hypothesis (as it has not been experimentally verified) of how Requirement R1.3 is met via the pointer basis selection for which MWI branches.

However, it is not a sufficiently developed theoretical solution to Requirement R1.1 of the measurement problem because it is very possible that a system undergoes external orthogonalization and yet one finds that the result of the Chapter 3 CHSH test is $2\sqrt{2}$. Only when the CHSH test is $\sqrt{2}$ can a measurement be said to have occurred in a Category 1 theory. That is, a pointer basis could be established for example with the example of a CNOT gate, yet it is known that such a gate is not a bona fide measurement device. Hence environmental decoherence does not appear to be a sufficient condition for measurement in regards to Requirement R1.1. Precise conditions under which either a device or a device plus its environment constitute a bona fide measurement device must be proposed theoretically and confirmed experimentally in order to meet Requirement R1.1.

Although we would agree that "*decoherence is rapidly becoming a part of a standard lore,*" we argue that this is due to a general lack of understanding in the scientific community of the measurement problem and its requirements for solution, for which it is the intent of this book to rectify. Decoherence certainly can occur and is commonplace in unitary theory. Decoherence does provide a working hypothesis in addressing Requirement R1.3. However, decoherence as proposed by Zurek fails as a sufficient condition to meet Requirement R1.1. Zurek believes that decoherence is a necessary condition for Requirement R1.1 [150]

> "*The environment induces, in effect, a superselection rule that prevents certain superpositions from being observed. Only states that survive this process become classical.*"
>
>

However, this claim is a statement with no scientific proof; environmental decoherence may or may not be a necessary condition. Certainly, it would be significant if it were rigorously established to be a necessary condition for Requirement R1.1. However, further theory and substantial experimentation would be needed to validate this claim.

Although there have been other proposed modifications to Everett's initial proposal, there is no theoretical nor experimental evidence at this time that we are aware of that establishes any of these MWIs as either a necessary condition or a sufficient condition for measurement. However, if a new theory is found which provides a physical reason that measurements are forming and this can be predicted via a new theory, then whether or not there are multiple universes and/or minds formed seems inconsequential and an interpretational issue that has no bearing on predictability.

Bohm's Theory

Bohm's theory is one of the several early attempts to come to grips with the measurement postulate that fall within the class of no-collapse theories in which it is desired to explain measurement without use of the measurement "collapse" postulate. In Bohm's theory, the Schrödinger wave function does not represent the particles' actual trajectory, but rather all possible trajectories that could occur. Bohm uses the Schrödinger wave function to derive a quantum potential which acts on and guides the particle with a force and for which the particle's velocity can be derived.

Bohm's theory is unique in that it includes aspects of Schrödinger's equation and also incorporates a spatially localized trajectory that the particle is said to follow. Consider Schrödinger's equation given by

$$i\hbar\frac{\partial\psi}{\partial t} = \left(-\frac{\hbar^2}{2m}\nabla^2 + V\right)\psi. \tag{4.16}$$

The wave function is generally complex and can be written as $\psi = Re^{iS/\hbar}$, where R is a real time varying function. The probability of measurement under Born's interpretation is $|\psi|^2$ which is a function of R. Substituting $\psi = Re^{iS/\hbar}$ into Equation (4.16) and taking the real part of both sides yields

$$-\frac{\partial S}{\partial t} = \frac{(\nabla S)^2}{2m} + V + \frac{\hbar}{2m}\frac{\nabla^2 R}{R}. \tag{4.17}$$

When $\hbar = 0$ this is a classical Hamilton-Jacobi equation where S is the action. If one considers a modified potential

$$\bar{V} = V + \frac{\hbar}{2m}\frac{\nabla^2 R}{R}$$

then a solution to Equation (4.17) would also constitute a solution to the original Schrödinger equation. In such a case, the velocity of the particle can be derived from the Hamilton-Jacobi equation as $v(x,t) = \nabla S(x,t)/m$. The right-hand side of Equation (4.17) is only a function of the action S and the potential V. Equation (4.17) can also be interpreted as a modified Hamiltonian-Jacobi equation where S is the action resulting from a classical potential V and a newly added quantum potential Q

$$Q = -\frac{\hbar}{2m}\frac{\nabla^2 R}{R}. \tag{4.18}$$

Note that the quantum potential is a function of R which is related to the Born probability of measurement. This is a rather interesting result by Bohm and at this stage indeed shows promise for resolving the measurement problem. One can derive from the quantum potential Q a force $F = -\nabla Q$ for which the particles are guided by the quantum potential which is in addition to the force on the particle by the potential V. The particle will have a well-defined position at all times but will experience a force as a function of the quantum potential. The quantum potential is in general non-local. Bohm's theory is an example of a non-local hidden variables theory.

This theory is of the type where you can "have your cake and eat it too." Particles travel in a classical manner yet are guided by a non-classical wave function so that the results of measurement would seem to occur in a natural manner. Suppose a theory exists such as Bohm's theory for which particles always take particular trajectories that interact with one and only one detector. By imposing a quantum potential that now guides particles in a local manner, one might expect that the measurement problem can be resolved. It is indeed true that the particles under Bohm's theory can be thought of as guided by a force that is a function of the potential V and also R in a manner that the particle obeys Born's rule. The particles require an initial position as if chosen from the probabilistic density function $|\psi|^2$ and also requires a probabilistic initial velocity that is a function of the wave function and current density. This then might appear to be a methodology for which the measurement postulate is not actually needed as the particles already are taking classical paths. In this case some incorrectly conclude that, simply because such a theory exists, the measurement problem can be relegated to an interpretational problem and is not a problem that could have any significant consequences upon its solution.

The main objective of this chapter is for the reader to further learn how to discern correct approaches from incorrect approaches regarding the measurement problem. The devil is in the details, and because of the subtleties involved, Bohm's theory presents an excellent opportunity to analyze. With this in mind, it is interesting to note that in the 1984 paper by Bohm [151], in a section entitled "The Quantum Mechanical Process of Measurement" where he considers the measurement problem, Bohm divides the measurement process into two stages.

> Stage (i): Separation of possible states of the quantum system through an in-principle reversible interaction with an apparatus
>
> Stage (ii): Registration of an actual result of the measurement in a further irreversible interaction.
>
>

Bohm [151] gives an example of an in-principle reversible interaction, as the interaction of the angular momentum of an electron in an inhomogeneous magnetic field. That is, the electron might be in a superposition of different spin states for which the paths through the magnetic field would separate the electron. Bohm states that the packets of the electron can fully separate. However, Bohm does agree there is a problem at this point to apply the measurement postulate because it is possible to recombine the packets together. He states, "It is therefore clear that we have not yet explained the irrevocability of the experimental results..." Bohm goes on to explain that it is only in the second stage for which there is an irreversible registration that there is a result of measurement. That the new quantum state can be said to occur, "when the reversal is overwhelming improbable." Bohm goes on to justify this on the basis of irreversibility found in macroscopic systems via thermodynamics. Bohm adds a further requirement to his theory: he requires at this point that when the

> *... quantum potential of the whole system is calculated, the apparatus wave function alone will guarantee that the packet not corresponding to the actual result can be left out of the discussion. Clearly it is the irreversibility of the process of registration that further guarantees that this conclusion will continue to be valid in the future.... We are thus led to the conclusion that the packets not containing this particle now correspond to inactive information. This is essentially what happens in classical solutions involving probability distributions. In such situations, after a new observation is made, we simply discard the previous probability function.*

One might agree that irreversibility is related to the solution to the measurement problem as Bohm suggests, but this is not what modern papers concentrate on when utilizing Bohm's theory to explain measurement. On the contrary, modern authors generally focus mostly on the theory without the two-step process as a resolution of the measurement problem. In fact, the two-step process that Bohm proposes is either left out entirely of the discussion by the proponents of Bohm's theory for resolving the measurement problem, or it is buried in a few lines in a manner that few would have suspected could play such a huge role. Consider Holland [5, p. 349] in which the two-stage process by Bohm is considered:

> *The complexity of the interaction gives it the appearance of being irreversible but the underlying dynamical law bringing about the wave function remains the reversible Schrödinger equation. That the process so described is not actually irreversible may be considered a weak point in our demonstration.*

Using a reversible process to demonstrate an irreversible process does appear to be a rather weak point. Holland isn't finished however. Holland makes the statement regarding the Stage 2 requirement of Bohm:

> *As we have presented it, the theory of measurement may be understood without first answering fundamental questions regarding the nature of irreversibility.*

If Holland's statement is correct, one might conclude that the measurement problem is also resolved by Bohm's theory without the Stage 2 requirement of Bohm. The latter statement by Holland refers to, "the theory of measurement." The current von Neumann theory of measurement is an *a posteriori* theory of measurement, i.e., given that a measurement has been made, the results must agree with Born's rule. In the sense that Bohm's theory is an *a posteriori* theory, it does indeed reproduce the results of measurement and resolves the philosophers' measurement problem. However, it does not yet meet the requirements of resolving the physical measurement problem.

In addition to Holland there are others that believe that Bohm's theory resolves the measurement problem [152] [153] [114] [154]. Dürr et al. states [155]:

> ... *Bohmian mechanics makes the same predictions as does orthodox quantum theory for the results of any experiment—for example, a measurement of momentum or of a spin component—provided we assume a random distribution for the configuration of the system and apparatus at the beginning of the experiment given by* $|\psi(q)|^2$*..."* *Bohmian mechanics is a counterexample to all of the claims to the effect that a deterministic theory cannot account for quantum randomness in the familiar statistical mechanical way, as arising from averaging over ignorance.*
>
> Bohmian Mechanics and Quantum Theory: An Appraisal," J. T. Cushing, A. Fine and S. Goldstein, Eds., Springer, 1996, pp. 21-44, D. Dürr, S. Goldstein and N. Zanghi, ©1996 Kluwer Academic Publishers. With permission of Springer.

Holland's and Dürr's statements are correct in the sense that Bohm's theory provides a deterministic model that is consistent with the results of the measurement postulate *given* that a measurement has occurred. Hence, they can rest knowing that they have a solution to the philosophers' measurement problem. But as has been stated, von Neumann quantum theory is an *a posteriori* theory and the measurement problem as we define it, requires one to provide a scientifically validated theory that explains the dichotomy between the entanglement predictions of Schrödinger's equation versus the product state prediction *without* pre-assuming that measurement has occurred. While it is true that Bohm's theory is a solution to the philosophical measurement problem, at best it can be considered only one out of many approaches regarding the physical measurement problem. Bohm's theory does not come close to having been scientifically validated to meet the requirements that we have set in resolving the physical measurement problem.

Historically one might consider that there are two "measurement problems." Whereupon Schrödinger developed his new equation in 1926 and was invited to Bohr's home, Bohr argued with Schrödinger that a deterministic equation would not suffice and that the quantum of action of Planck demanded a non-causal or nondeterministic element to explain various phenomena. Born's probabilistic rule was developed shortly thereafter to explain the results of measurement. This satisfied Bohr and became part of the Copenhagen interpretation. Let us suppose that historically after Born's rule was proposed, that Schrödinger had available Bohm's theory. Then Schrödinger could argue with Bohr that Bohm's theory without irreversibility does reproduce the results of measurement in a deterministic manner and that Born's rule is only needed due to the ignorance of unknown initial conditions. Bohmian mechanics is a deterministic counterexample that shows Bohmian mechanics reproduces both unitary evolution and the measurement postulate, for which the measurement postulate is a formulation of what to do given that a measurement has occurred. If one defines the measurement problem in a manner that can be resolved via a theory that demonstrates in a deterministic manner Born's rule *a posteriori* given that a measurement occurs while maintaining unitary evolution, then Bohmians can go to sleep at night content they have solved the measurement problem. And it has been said that ignorance is bliss. But it also has been said that bliss is ignorance.

In 1935, Schrödinger published two important papers [156] [119]. Schrödinger states in [119]

> *As soon as the systems begin to influence each other, the combined function ceases to be a product and moreover does not again divide up, after they have again become separated, into factors that can be assigned individually to the systems. Thus one disposes provisionally (until the entanglement is resolved by an actual observation of) of only a common description of the two in that space of higher dimension.*

Schrödinger is acknowledging that his equation predicts that an initial product state of photon and a system become entangled and cannot be represented as a product state after separating. Schrödinger in the same paper also provides his famous example of such an entangled superposition even if one were to include a living cat as one of the systems. Furthermore, the phrase "is provisional until resolved by an actual observation," is a statement that the entanglement is resolved only when a measurement occurs. The physical measurement problem, as we have defined it, is consistent with this latter issue, that is determining the conditions under which measurement occurs for which the Schrödinger entangled wave function is formally replaced by a product state. The physical measurement problem is not trivially resolved via assumption that a measurement has already occurred; rather we demand that a solution to the measurement problem provide the conditions under which a measurement occurs and the theoretical basis for such conditions.

Bohm's theory does reproduce Born's rule as an *a posteriori* theory. But let us now analyze Bohm's theory without irreversibility to see if does resolve the measurement problem, which requires one to dig deeper without skirting major issues. That Bohmian mechanics has little to offer in the sense that we have defined the measurement problem can be seen when Bohmians explain when they will change the wave function into a new wave function that is indicative of a measurement.

In this regard, consider the treatment of the measurement process by Bohm's theory in Orioles and Mompart [153, pp. 95-99]. The system is specified by a trajectory $\vec{x}_S$ and the apparatus by trajectory $\vec{x}_A$. The possible outcomes of the measurement process correspond to one of the possible eigenvalues g of a Hermitian operator $\hat{G}$ that satisfy the equation $\hat{G}\psi_g(\vec{x}_S) = g\psi_g(\vec{x}_S)$. The system wave function can be decomposed via the eigenstates of $\hat{G}$,

$$\psi_S(\vec{x}_S, t) = \sum_g c_g(t)\psi_g(\vec{x}_S).$$

When measuring the eigenvalue g_a the wave function $\psi_S(\vec{x}_S, t)$ needs to collapse to $\psi_{g_a}(\vec{x}_S, t)$. In addition to the entire apparatus, the pointer of the apparatus is represented by $\vec{x}_P(t)$. When the quantum system is in an eigenstate $\psi_g(\vec{x}_S)$, the pointer is in a region S_g, i.e. $\vec{x}_P(t) \in S_g$. A requirement is added that the pointer positions $\vec{x}_P(t)$ of the apparatus are disjoint in the sense that if g_1 and g_2 are different eigenstates of $\hat{G}$ then the pointer state trajectories are disjoint, i.e. $S_{g_1} \cap S_{g_2} = \emptyset$. A total wave function $\Phi_T(\vec{x}_S, \vec{x}_A, t)$ can then be defined that is decomposed over the possible outcomes of the experiment via the eigenvalues of $\hat{G}$,

$$\Phi_T(\vec{x}_S, \vec{x}_A, t) = \sum_g c_g(t)\, f_g(\vec{x}_A, t)\, \psi_g(\vec{x}_S). \tag{4.19}$$

with the property that $f_{g_1}(\vec{x}_A, t) \cap f_{g_2}(\vec{x}_A, t) = 0$. Now, the Bohmian process of measurement is as follows. An initial trajectory is chosen $\{\vec{x}_S(0), \vec{x}_A(0)\}$ commensurate with a probabilistic outcome of the initial wave function positions of the system and apparatus. Such a trajectory will evolve with the total wave function, and during the measurement, the total trajectory $\{\vec{x}_S(t), \vec{x}_A(t)\}$ will correspond to only a single term of Equation (4.19) which corresponds to $g = g_a$. It is stated [153, p. 97]

> *Thus, the pointer positions will be situated in $\vec{x}_P[t] \in S_{g_a}$ and we will conclude with certainty that the eigenvalue of the quantum system is g_a. In addition, the subsequent evolution of this trajectory can be computed from $f_{g_a}(\vec{x}_A, t)\psi_{g_a}(\vec{x}_S)$ alone. In other words, we do not need the entire wave function $\sum_g c_g(t)\, f_g(\vec{x}_A, t)\, \psi_g(\vec{x}_S)$ because the particle velocity can be computed from $f_{g_a}(\vec{x}_A, t)\psi_{g_a}(\vec{x}_S)$. The rest ... are empty waves that do not overlap with $f_{g_a}(\vec{x}_A, t)\psi_{g_a}(\vec{x}_S)$ so that they have no effect on the velocity of the Bohmian particle. This is the simple explanation of how the complicated orthodox collapse is interpreted....*

So far so good! If we could simply stop here, the authors would have indeed resolved the physical measurement problem. However, one cannot stop here and the authors must add an additional requirement [153, p. 98]

> *Finally, we want to enlarge the explanation of the role played by the empty waves belonging to the wave packet of ... [Equation (4.19) in this text] which do not contain the particle. In principle, one can argue that such empty waves can evolve and, in later times, overlap with the original wave that contains the particle. If we are interested in doing subsequent (i.e., two times) quantum measurements, a good measuring apparatus has to avoid these spurious overlaps. This can be understood as an additional condition for qualifying our measuring apparatus as a good apparatus.*

This statement appears designed to enforce Bohm's Stage 2 irreversibility. Bohm states in [151]:

> *Clearly it is the irreversibility of the process of registration that further guarantees that this conclusion will continue to be valid in the future.*

That is, the packets that do not contain the particle will be guaranteed to remain empty packets as long as the process of registration is irreversible. Let us consider this additional condition in the context of the UMDT of Chapter 3, which was specifically designed to determine what a given measurement theory does and does not accomplish. For simplicity, we assume that the devices are 100% absorbing in the sense defined in Chapter 3. Suppose that the photon after interaction with Device 1 and 2 in Step 1 are in states such that the state for which no measurement occurs does not overlap with the state for which measurement occurs. There are two cases to consider after Step 1. In Case 1 the Devices are bona fide measurement devices in which case the evolved Device 1 state and Device 2 state are in a product state. In Case 2 the interaction between the photon and devices is unitary in which case the final state of Device 1 and Device 2 are in an entangled state. Step 2 is then employed for which the two qubits will remain unentangled under Case 1 and in Case 2 the entanglement in the Devices will be transferred to the two qubits. In Step 3, a Bell experiment is conducted and the CHSH sum is computed. In Case 1, the CHSH will give a value of $\sqrt{2}$ and in Case 2 the CHSH will give a value of $2\sqrt{2}$. The additional Orioles-Mompart condition would imply that a bona fide measurement has only occurred when there is never subsequent overlap between the cases of absorption and non-absorption. If the angles of the Bell experiment were chosen that were 0 degrees or 180 degrees such that the results could be inferred with probability unity from the non-overlapped decomposition after Step 1, then there would be no repercussions in performing the Bell experiment and Bohm's theory will suffice to explain Steps 2 and 3. However, the angles chosen in the Bell experiment that is required in Step 3 are such that the Bell measurement do indeed mix or overlap the empty waves that existed after Step 1. Hence in the case of Bohm's theory with the additional Orioles-Mompart condition designed to enforce irreversibility, Step 3 will always be predicted to yield a CHSH sum of $2\sqrt{2}$ and thus the states would still be entangled.

Suppose that one of the other quantum measurement theories is found to be correct, for example the theory of spontaneous localization. And that such localization occurs after some amount of interaction in the first of the two-time measurements. If so, then the actual quantum state is a product state after the first of the two-time measurements. And after the particles undergo Step 2 and Step 3, the CHSH sum would be $\sqrt{2}$. So, a theory that specifies physical conditions under which measurement occurs such as a spontaneous localization theory, if experimentally found to be correct, would falsify Bohm's theory as stated by Orioles-Mompart which demands that entanglement continues to persist until such time as there are no further overlaps in the various superposition terms. Bohm's theory is predicated on unitary evolution and it is possible that a non-unitary theory that makes specific predictions and meets the requirements R1.1-R1.3 can falsify Bohm's theory, but only if such a theory is ultimately found to be correct.

One might argue that the additional Orioles-Mompart condition can be dropped or is superfluous. However, Bohm in his paper considered a Stern-Gerlach experiment where a spin ½ particle enters an inhomogeneous magnetic field which separates the particle into two distinct paths that the spin up state takes versus the spin down state. Bohm argues that this separation cannot be considered to be a bona fide measurement because the two paths could be further recombined into a single path. If the separation into two paths was a sufficient condition for measurement, then the coherence in the particle would be destroyed by an inhomogeneous magnet and the recombination into a single path would be non-coherent. It has been demonstrated experimentally that such recombination does maintain coherence. Because of this and other counter-examples, Bohm required the second stage of irreversibility.

Let us examine how well the additional Orioles-Mompart condition addresses the physical conditions under which irreversibility occurs. Suppose an experiment were done that separated the electron via an inhomogeneous magnetic field and for which the paths were never recombined. Then one could apply the additional Orioles-Mompart condition and conclude that such a device is a bona fide measurement device. In the case that the electron is separated and then recombined, the additional Orioles-Mompart condition implies the initial separation is not a measurement, but in the second case when the electron is separated and not recombined, the initial separation is a measurement. Hence such a condition classifies irreversibility based on what happens in the future.

The additional Orioles-Mompart requirement is rather constraining and renders a lack of unique predictive power to explain when a particular device is a bona fide measurement device based on the current and past state of the device. For example, instead of considering the electron in an inhomogeneous magnetic field, we consider the interaction of a particle with a mesoscopic device. Suppose a state results that can be written in the form in Equation (4.19). If after the initial interaction one does not allow the empty waves to overlap, then Orioles-Mompart would declare the device to be a "good" apparatus. On the other hand, suppose that after the interaction with a device, further interaction is made to occur. In such a case, Orioles-Mompart would declare this not to be a good apparatus. Hence the ability of Bohm's theory with the additional Orioles-Mompart requirement to predict whether or not an apparatus is good, is not based on the history of the state nor the current state of the device but must include the future evolution. Hence a device in one situation would be classified a good apparatus and in another would not be, depending on whether or not the empty waves further interact in the future. However, this is rather contrary to what one might expect based on causal theories. That is under a causal theory whether an apparatus is a measurement device or is not should be based on the past state and the present state; what will happen in the future would not be expected to have bearing on this issue.

Bohm's theory with the added Orioles-Mompart condition is non-unique in the manner that it classifies good measurement devices based on full knowledge of the current and past data, which is typically demanded of any physical theory. The additional Orioles-Mompart condition does not provide predictability to explain in the sense of our requirement R1.1, "What physical situations constitute the divide of irreversibility?" Essentially, we know that it is somewhat more complex than a single electron going through an inhomogeneous magnetic field and perhaps on the order of a macroscopic system consisting of $\approx 10^{23}$ particles. There is no precise criterion in Bohm's theory that provides a prediction of the onset of irreversibility.

Bohm's theory may resolve the philosophers' measurement problem as stated by Lewis and others, and this was indeed an early version of the arguments between Schrödinger and Bohr. However, Bohm's theory does not resolve the measurement problem that grew out of Schrödinger's cat paper in 1935, and was largely the major obstacle to Einstein's incompleteness argument regarding the orthodox quantum theory (as will be further examined in Chapter 5) in the sense of his query, "Is the moon there when nobody looks?" Bohm's theory can be considered one out of the many theories that makes predictions to explain precisely under what conditions measurement occurs. In the case of Bohm's theory and the Chapter 3 UMDT, the CHSH sum is predicted to be $2\sqrt{2}$ and entanglement is predicted to persist. However, such quantum states are incompatible with the existence of a particular outcome, which demands a product state representation.

Master Equations for Deterministic Evolution

A quantum master equation of the deterministic Lindblad form is a deterministic equation for which a system interacts with a larger system such as a thermodynamic bath for which the overall evolution is often taken to be unitary. Another type of master equation is a stochastic master equation for which the overall evolution is generally non-unitary. Stochastic Schrödinger equations will be considered later in this chapter for which the average density matrix obeys a Lindblad equation.

Consider a Hamiltonian consisting of a system and an environment of the form

$$H = H_S \otimes I + I \otimes H_E + H_{S,E} \tag{4.20}$$

where H_S is the self-Hamiltonian of the system, H_E is the self-Hamiltonian of the interacting environment and $H_{S,E}$ is the interaction Hamiltonian between system and environment. We denote $\mathfrak{M}(\mathcal{H})$ as the set of density matrices in a Hilbert space $\mathcal{H}$, the system Hilbert space as $\mathcal{H}_S$ with density matrix $\rho_S \in \mathfrak{M}(\mathcal{H}_S)$ and the system-environment Hilbert space by $\mathcal{H}_{S,E}$ and system-environment density matrix $\rho_{S,E} \in \mathfrak{M}(\mathcal{H}_S \otimes \mathcal{H}_E)$. When the overall evolution of the system and interacting environment is unitary the evolution of the system density matrix ρ_S is generally non-unitary and can be computed by tracing out the environmental degrees of freedom from the joint density matrix $\rho_{S,E}$ as follows:

$$\rho_S = \mathrm{Tr}_E\left[\rho_{S,E}\right], \tag{4.21}$$

where Tr_E denotes the operation of tracing the environment from the joint density matrix $\rho_{S,E}$. It can be proven that in the case where the system is coupled to a Markovian environment, the trace operation results in the following form [157]:

$$\frac{d\rho_S}{dt} = -\frac{i}{\hbar}[H_S, \rho_S] + \gamma \sum_j \left(L_j \rho_S L_j^\dagger - \left(\frac{1}{2} \rho_S L_j^\dagger L_j + L_j^\dagger L_j \rho_S \right) \right). \tag{4.22}$$

This is the most general form of Markovian master equation describing evolution that is trace-preserving and completely positive. Note that the first term on the right-hand side of Equation (4.22) is the Schrödinger equation for ρ_S in the form of a density matrix, which would be the unitary evolution of the system in absence of system-environmental coupling. The second term of the right-hand side of Equation (4.22) specifies non-unitary evolution with respect to H_S in a manner that maintains the unitary evolution of $\rho_{S,E}$ with respect to the overall Hamiltonian of Equation (4.20). As the deterministic Lindblad master equation is non-unitary, it makes sense to examine whether or not such equations are useful toward resolving the measurement problem.

In this respect, let us examine what the deterministic Lindblad equation would predict regarding the UMDT of Chapter 3. Consider as before an apparatus that has a latency time τ_l for registration and that the uncertainty of the time of emission of the photon is Δt_e. It was assumed in Chapter 3 that one can separate the two devices in Figure 3.4, via an arbitrary large distance d so that $d > 2c(\Delta t_e + \tau_l)$. In this case, one only has to include the particles of the environment that are within a time-like window, or those that are within a distance of $c(\Delta t_e + \tau_l)$. If one were to model the particles via a deterministic Lindblad equation, there can be no interaction between the two separated devices. Interaction can only occur between the B mode of the photon and Device *1* and as well between the C mode of the photon and Device *2*. This is precisely the setup of the UMDT of Chapter 3 for which it has already been proven that unitary evolution that includes the local environment within the time-like window $d < c(\Delta t_e + \tau_l)$ will always produce the CHSH sum of $2\sqrt{2}$. Granted, one could restrict the application of the UMDT test of Chapter 3 to the system only. In this case the system is in a diagonal state as found in Equation (4.3) reproduced below for convenience,

$$\begin{pmatrix} |c_1|^2 & \cdots & 0 \\ \vdots & \ddots & \vdots \\ 0 & \cdots & |c_N|^2 \end{pmatrix}.$$

The UMDT test of Chapter 3 if applied only to the system would yield the CHSH sum less than the value *2* since, even unitarily, there would be no direct entanglement between the two system subspaces without including the Devices and their local environments. However, one cannot conclude from a UMDT test on the system alone that the system is in a particular or pure state for which one of the N possibilities has indeed occurred. The Chapter 3 UMDT test on the system alone will always produce a value less than or equal to 2, whether or not interaction is unitary or occurs via the measurement postulate. Hence, applying the test to only the system provides no experimental discriminating power to address the hypothesis test in Chapter 3. In order to have discriminating power, one must include the system, Device *1* and its local environment, for which the CHSH sum of $2\sqrt{2}$ implies that entanglement exists, and the system is *necessarily* in the mixed state of Equation (4.3). That is, the system is not a pure state that would be indicative that a particular outcome had occurred.

Deterministic Lindblad equations are a useful tool to compute the reduced density matrix of the system. If one assumes that the initial state of the devices and surrounding environment are in a pure state, it is known that unitary evolution maintains the purity of the state, i.e., the overall composite state remains pure under unitary evolution. In this case the finding of the system being in a mixed state such as in Equation (4.3) indicates, at least theoretically, that the system is further entangled with other degrees of freedom. That is, a bipartite system that has an overall pure state with subsystems that are in a mixed state indicates (at least theoretically) that the system is further entangled with the environment. As a deterministic Lindblad equation would lead to a mixed state description of the system similar to Equation (4.3) the experienced quantum information theorist would confirm our dictum that:

To the extent there is entanglement, there is no measurement.

Superdeterminism

Superdeterminism is a concept in which everything is pre-determined and that free will is an illusion. In the case that Nature is superdetermined, an experimenter's choice of settings in performing the CHSH test cannot be done independently of the preparation of the two spin ½ particles. Bell assumes in his derivation of his inequality that the settings of the CHSH test can be chosen independently of any hidden variables that might be used to predict the outcome of the experiment.

Superdeterminism is a natural consequence of Unitarianism. If one believes that everything obeys a universal wave function that evolves according to Schrödinger's equation, then in some sense everything is gear-like or mechanical.

Suppose that a gear (Gear 1) is thrown toward another spinning gear (Gear 2) such that the two gears can either mesh perfectly or can be misaligned and cause grinding. If it is found that the two gears always mesh perfectly and are never misaligned, then it must have been that when the gear was thrown that it was already time-synchronized with the second gear. Suppose that unitary evolution is found to have a similar property. Then the Unitarian might conclude that the question of hidden variables is irrelevant. Free will (in the sense of non-unitary physics) is an illusion to the Unitarian, and the preparation of the photons could be such that the results are already compatible with the selection of the angles in the CHSH test, for which the experimenter has no real say. Let us consider if unitary physics provides a back action when Gear 1 is not aligned with Gear 2. Consider a simple energy-conserving interaction Hamiltonian between two two-level systems. One can ask if the initial condition on the phase of one of the particles can in any manner change the time for the second two-level system to become excited. It turns out that the initial phase of the first particle is irrelevant and the time to become excited is independent of the initial phase. Hence in the sense of the two classical gears, there is no grinding to be expected in unitary theory. It does appear perfectly reasonable if one is an unwavering Unitarian to conclude that the gears of unitary evolution will *always* be found to be meshed: in the plus x direction and in the -x direction, in the +y direction and in the -y direction, in the +z direction and in the -z direction, and also forward in time +t and backward in time -t.

Remarkably, there is an important characterization in the mathematics of functions that appears to lend some support to the physical theory of superdeterminism [158]. A function is defined as an analytic function when it can be represented by a convergent power series. One might ask what should this have to do with physics? An analytic function has the property that one can determine the entire function from the function in any small interval. That is, if a wave function is analytic, then one can determine the wave function 100 miles away simply by analyzing it in any small interval. And many important wave functions that are eigenstates of the Hamiltonian that have been derived in quantum mechanics are analytic wave functions. For example, the theoretical solutions of the electron orbits that define the hydrogen atom are analytic, and these analytic functions have been so successfully verified in experiments that one might even consider the hypothesis that such electron orbits are truly analytic. These properties of unitary evolution appear to leave open the possibility that superdeterminism is indeed relevant to certain issues in the foundations of physics. There have been a number of arguments that are counter to superdeterminism. Bell states [159, p. 101]

> *A respectable class of theories, including contemporary quantum theories as it is practiced, have 'free' 'external' variables in addition to those internal to and conditioned by the theory. These variables are typically external to and conditioned by the theory. These variables are typically external fields or sources. They are invoked to represent experimental conditions. They also provide a point of leverage for 'free willed experimenters', if reference to such hypothetical metaphysical entities is permitted. I am inclined to pay particular attention to theories of this kind.*

Furthermore, [159, p. 103]

> *A theory may appear in which such conspiracies inevitably occur, and these conspiracies may then seem more digestible than the non-*

localities of other theories. When that theory is announced I will not refuse to listen, either on methodological or other grounds. But I will not myself try to make such a theory.

Bell is rather negative in terms of endorsing superdeterminism, although he does not close the door entirely on superdeterminism. Additionally, Zeilinger has criticized superdeterminism on the basis of the scientific process which allows one to vary parameters of an experiment [160, p. 266]:

This fundamental assumption is essential to doing science. If this were not true, then, I suggest, it would make no sense at all to ask nature questions in an experiment, since then nature could determine what our questions are, and that could guide our questions such that we arrive at a false picture of nature.

It is interesting that the arguments by Bell and Zeilinger are not scientific proofs that conclusively show that superdeterminism is false; rather their arguments are based largely on the existence of free will to modify external fields or parameters as allowed in regards to the established scientific method. Still such arguments are heuristically compelling to many and very few have seriously considered superdeterministic theories.

A proposal based on superdeterminism has however been put forward recently by 't Hooft [161] [162]. Don't be surprised to see more superdeterministic theories put forward by devoted Unitarians.

Transactional Interpretation

The transactional interpretation (TI) was put forward by J. Cramer in 1986 [163]. Before describing the transaction interpretation, we first examine how Cramer defines the measurement problem. In the paper [164], Cramer states that it is not the case that a clever experimentalist could go into the laboratory and determine which interpretation is correct by testing experimental predictions. Moreover, that interpretations are transformable or equivalent [158]:

... Prof. von Weizsacker and Dr. Görnitz have argued that to the extent that the Copenhagen interpretation, the transactional interpretation, and other interpretations are both self-consistent and also consistent with the quantum mechanics formalism, one can deduce a "dictionary" or set of interpretational transformations which can render one interpretation in the terms or "language" of another. This demonstrates a kind of equivalence principle for interpretations

Cramer also believes that interpretations cannot be tested experimentally [158]:

> *My discussion of quantum mechanics interpretations stressed the point that the interpretation of a mathematical formalism cannot be tested experimentally and must be judged on other grounds.*

For which Cramer confirms the above statement does include the TI, see for example [165]. It should by now be clear to the reader that Cramer believes the measurement problem is defined via the philosophers' measurement problem. However, there most definitely is a physical measurement problem that we have defined for which experimental tests are needed to resolve the problem. Cramer desires to reproduce the orthodox formalism via an interpretation and not to augment the formalism as we claim is required to resolve the measurement problem. Cramer's understanding of the measurement problem appears similar to that of Lewis's at least in the sense that no augmentation of the formalism is required for its resolution.

Due to Cramer's conceptual attitude toward the measurement problem, one cannot expect that the transactional interpretation will meet the Requirements R1.1-R1.4 for a Category 1 theory or R2.1-R2.3 for a Category 2 theory. And this is correct, for it will be shortly demonstrated that TI is just another *a posteriori* theory which provides an interpretation of how a measurement occurred, *given* that a measurement has occurred. This common *a posteriori* theme has now been repeated several times when interpretations are proposed by Unitarians that simply define the measurement problem in a manner similar to the philosopher's definition.

The TI proposes that there is a nonlocal transaction that occurs when any quantum measurement occurs. The mechanics of the transaction are likened to a *handshake* that occurs via an exchange of advanced and retarded wave solutions. Such solutions were proposed in the *absorber theory* of Wheeler and Feynman. The TI utilizes concepts from the absorber theory of Wheeler and Feynman. Consider a particle that travels from an emitter to an absorber. There are three stages that describe a quantum event in TI [164]. First, the emitter sends out an *offer* wave Ψ to the absorber. In the second stage, the absorber sends out a *confirmation* wave in response to the offer wave, a confirmation wave that is the complex conjugate of the offer wave, Ψ^*. The emitter's offer wave is sent backward in time so that at the moment that the particle might be emitted, there exists an overlap $\Psi \Psi^*$ between the offer wave and confirmation wave. At this point, this overlap represents a probability that is used to determine if the offer wave is accepted. The offer is either accepted or rejected in the third stage of TI. Once accepted, energy is transferred from the emitter to the absorber. If not accepted, the process is begun again. The transaction when completed constitutes a quantum event.

In terms of TI, it is explained in [163] that Born's probability law is a statement for which the probability of particular transaction is proportional to the magnitude of the echo corresponding to the transaction that is received by the emitter. It does appear that TI has succeeded in reproducing Born's rule, *given* that a measurement has occurred. Hence TI appears to be a valid solution to the philosophers' definition of the measurement problem. However, the physical measurement problem demands that one fulfill several requirements, among these the conditions under which a measurement occurs.

In this regard, it might be asked whether or not photon absorption or emission is a sufficient condition for measurement. Consider the photon cavity QED experiments of Haroche [121, p. 281] for which $\pi/2$ microwave pulses are used to produce superpositions of atom and photon states. If photon absorption were truly a measurement in such cases, it would not be possible to conduct a second $\pi/2$ pulse and always end up in the same initial state. Additionally, the successful quantum computing experiments in NMR and other areas where $\pi/2$ pulses are successfully utilized strongly indicates that such superpositions are already occurring. That is, the atom-photon goes into an entangled state, for example $(|0\rangle_{\text{atom}}\,|1\rangle_{\text{field}} + |1\rangle_{\text{atom}}\,|0\rangle_{\text{field}})/\sqrt{2}$, which is not indicative of a measurement i.e., the factual occurrence $|0\rangle_{\text{atom}}\,|1\rangle_{\text{field}}$ or non-occurrence $|1\rangle_{\text{atom}}\,|0\rangle_{\text{field}}$ of a photon emission. Moreover the fact that superposition eigenstates, which were predicted by the Jaynes-Cummings model [166] in light-matter interaction, have been experimentally confirmed spectroscopically in experimentation with microwave cavities [167] [168], optical cavities [169] and also more recently in circuit QED [170]. Additionally, time dynamics have been shown to be in agreement with collapse and revivals as predicted by the unitary Jaynes-Cummings rotating wave model [171] were confirmed experimentally in 1987 [172]. Taken together, such work provides rather strong evidence that neither photon absorption nor emission alone can be considered a sufficient condition for resolving the measurement problem. Moreover, the experiments by Haroche [173], in which superpositions of $|0\rangle_{\text{atom}}\,|1\rangle_{\text{field}}$ and $|1\rangle_{\text{atom}}\,|0\rangle_{\text{field}}$ have been produced and confirmed, put the proverbial final nail in the coffin regarding either photon absorption or emission as a sufficient condition for measurement, which is often considered by many as a measurement event and used in proposed solutions to the measurement problem.

Let us examine what TI provides when examining the Chapter 3 UMDT. Since photon absorption does not appear to be a sufficient condition for a measurement event, consider two atoms in place of the devices in Figure 3.4. Evidence from the Haroche entanglement experiments in [173] can be directly applied to each of the two devices and would indicate that the photon absorption is unitary in this case. Hence the Chapter 3 UMDT test would be expected to yield the value $2\sqrt{2}$. On the other hand, if TI claims that a bona fide measurement can occur such that the energy of the photon has been transferred to exactly one of the two atoms with the final state of the two atoms in a product state, then this can be falsified by performing the Chapter 3 UMDT experiment. If such an experiment were performed and the proponents of TI claimed that this is only because the process is unitary and there is no need to use TI, then it is clear that TI is only being invoked as an *a posteriori* theory, with no predictive power to explain the conditions under which a device functions as a measurement device. What is missing from TI is precisely what is needed to provide a solution to the measurement problem as we define it. L. Marchildon states [174] regarding TI and the measurement problem:

> *The question now is: What distinguishes a reversible entanglement process, which gives rise to no confirmation wave and no transaction, from an irreversible one, which does give rise to a confirmation wave and, possibly, to a transaction. Of course, this question has no answer within the strict Hilbert space formalism of quantum mechanics, where all entanglement is in principle reversible. The upshot is that a transaction finds no room within the limits of that formalism. Just like the notion of a classical apparatus in the Copenhagen interpretation, or the one of wave function collapse in von Neumann's theory of measurement, the notion of a transaction must be added to the minimal quantum-mechanical formalism. In particular, the transactional interpretation cannot be considered complete unless the conditions for the possible occurrence of a transaction are spelled out in detail.*

Marchildon here has expressed, in a manner that is in-agreement with the overall theme of this book, as to why further conditions and requirements are needed in order for the TI to be considered as a potential solution to the measurement problem.

Other Interpretations

There are a large number of other interpretations that have been put forward to explain measurement. While it is beyond the scope of this book to present every interpretation, it is instructive to examine several more.

Humpty Dumpty

For example, consider the Humpty Dumpty proposal in [175] [176] [177]. In this proposal the authors' state [177]

> *More generally put, we support the view that the loss of coherence in measurements on quantum systems can always be traced to correlations between the (relevant) degrees of freedom of the measuring apparatus and the system being observed. The correlations are built up in the course of the measurement, and their temporal evolution is correctly described by quantum mechanics. In particular, one need not resort to invoking the notions of "state reduction" or the "collapse of the wave function" as dei ex machina, whose dynamical properties are allegedly outside the framework of quantum mechanics.*

The authors analyze spin coherence in which a particle in an eigenstate of spin x is separated into two paths via a Stern-Gerlach apparatus oriented in the z direction and then recombined. Afterwards, a measurement is made in the x direction. They add a which-way detector using micro-masers in the upper path of the device which is disturbed if the spin takes the upper path. The authors show that the statistics of the output of the Stern-Gerlach apparatus can be found if they trace out the effects of the micro-masers. In the case that the masers are initialized with a coherent state, they predict that there is no effect on the coherence in the final measurement. When the maser is initialized to a number state, there is a complete loss of coherence.

The results of the Humpty Dumpty analysis are certainly correct. In the case that the initial state of the maser is a coherent state, there is no effect on coherence, so consider the case that initial state of the micro-maser is a number state. The analysis in [177] in terms of the effect of the final spin measurement can be seen to be the same whether or not a projective measurement is made on the micro-maser by projecting into a number state. Hence the experiment proposed in [177] is not capable of distinguishing entanglement from measurement.

One can ask what predictions would be found if a similar setup as proposed in [177] were used in the Chapter 3 UMDT. In the Chapter 3 UMDT one must compare Schrödinger predicted entanglement versus the product state demanded under measurement. One is not allowed to trace-out the which-way micro-maser in the computation of entanglement if one were to use the same setup as [177] in each of the two detectors. If the micro-maser interaction with a particle constitutes a bona fide measurement device in terms of projecting the particle, then the Chapter 3 UMDT would show entanglement to break down. On the other hand, if the micro-maser interaction is unitary, the entanglement would survive and be seen in an actual experiment. This would be an interesting experiment to conduct, but we believe that at microwave and optical frequencies, experiments have already established that there exist interactions that are unitary and hence entanglement would be predicted to be seen in such an experiment. Therefore, the micro-masers considered in [177] are not guaranteed to be bona fide measurement devices, and the analysis found in [177] simply confirms that there exist atom-field interactions in cavity QED that are unitary.

One might be tempted to conclude that because there exist micro-maser unitary interactions, that all particle evolution must be unitary. Such conclusions are certainly reasonable in an inductive framework, but when a deductive framework is called for, such an approach can be expected to lead one to ruin, as will be more thoroughly discussed in Chapter 6.

Macroscopic Interaction

In [178] Allahverdyan, Balian, and Nieuwenhuizen propose a scheme that is a variation on Rosenfeld's solution. Their approach has similar shortcomings as Rosenfeld's solution. The authors have a means of providing a measurement basis *given* a measurement occurs. Hence, they have a solution to the philosopher's measurement problem. However neither in [178] nor in their proposal using the Curie–Weiss model in [179], have the authors' met the important requirement of providing credible necessary conditions under which a particular outcome or reduction occurs. That is, it is known that there exist various decoherence and relaxation processes that are experimentally reversible and unitary. Hence neither decoherence nor relaxation is a sufficient condition for measurement. In order to show significant progress on the physical measurement problem, the authors would need to further justify theoretically why a particular specific and sufficiently detailed model is either a necessary and/or sufficient condition for which a particular outcome or reduction occurs, and then experimentally confirm that the unitary predicted entanglement is eliminated for the configurations that they propose. These might be considered to be tasks of tall order. Perhaps, however minimal scientific requirements required for credible acceptance are not simply waived because one decides to engage a difficult problem.

Quantum-Bayesian or QBism

In the Quantum-Bayesian or QBism approach, the wave function is updated according to the manner in which the phenomenon creates a particular experience within an agent. It is claimed that there is no nonlocality in QBism and that the "notorious collapse of the wave-function" is nothing but the updating of an agent's state assignment on the basis of her experience [180].

In QBism, there is a separation between object and observer, observers now being agents that can have individual experiences. QBism further directs the observer, referred to as Alice in [180], to treat "all external systems on the same footing, whether they be atoms, enormous molecules, macroscopic crystals, beam splitters, Stern-Gerlach magnets, or even agents other than the observer." QBism applies Schrödinger's equation to all such external systems.

Now, suppose that the observer Alice has built the Chapter 3 UMDT test and is able to utilize various devices with a range of complexity in the devices. Initially she sees the value $2\sqrt{2}$ in her experiments. But one day, Alice found the CHSH sum to be $\sqrt{2}$. She performs the test again and again. She sees if others can reproduce the results of the experiment. After the results of the experiment were reproduced by several groups, she concludes that there really are particular configurations of systems that function as measurement devices. She now is obligated to NOT treat all external systems on the same footing, whether they are atoms, enormous molecules, macroscopic crystals, beam splitters, Stern-Gerlach magnets, or even agents other than the observer. She discards QBism as yet another interpretation (YAI) that did not give the correct results to an actual experiment. She concludes that the von Neumann formalism is incomplete. And it reminds her of what was being said in some book on the Quantum Measurement Problem that she had read several decades ago. A book she recalls that she never had really fully understood. She finds the book and re-reads it and now it suddenly makes perfect sense to her! She looks up the authors and finds they are in their 90s. She doesn't contact them because she figures they probably by now have gone senile or are in some old-age home. And she is correct on both counts. She agrees that the use of inductive methodology is not an appropriate research methodology to be using as the problem lies outside the problem set that induction is designed to address. And she takes their advice and does undertake a serious *deductive* investigation of the measurement problem. She begins to investigate both theoretically and experimentally the physical reasons and conditions under which measurement occurs.

Properties and their Relation to Measurement

Now that we have examined many YAIs of quantum mechanics, of which none seriously addresses the measurement problem as we have defined it, we will turn to the examination of theories that do propose a physical rationale for the occurrence of measurement. Before this is done, there are several properties of measurement that are common to any new theory that attempts to go beyond standard quantum mechanics. When the measurement problem has been fully resolved, it is conceivable that there will be a substantial number of new properties that are related to a measurement process. However, at present there are at least the following properties that are often associated with measurement: quantum jumps, wave function reduction, nondeterminism, irreversibility and entropy, loss of coherence, amplification, localization, particle absorption, and the Zeno effect.

When an interaction-free or negative measurement occurs, a particle is not absorbed and the measurement device does not register a detection. As has been discussed, in the von Neumann theory of quantum mechanics, the measurement postulate is invoked even in the case of IFM. On the other hand, when a measurement device does register a detection, it is generally due to the presence of a particle that the device is designed to detect. In order to discern negative measurement from the case when a device registers a detection, the latter will be referred to as *positive measurement.*

Quantum Jumps

Quantum jumps are a property that are often associated with measurement. Quantum jumps occur when a particle such as a photon is absorbed in a measurement. Quantum jumps are generally modeled by a physical discontinuity and within the von Neumann theory, such jumps are occurring nondeterministically or statistically. Planck's discovery in 1900 of the quantum of action was met with an interpretation by Bohr and others of the need for a non-causal or nondeterministic process to explain various phenomenon such as black-body radiation. Many researchers seem to be under the impression that quantum jumps were proposed only after Schrödinger's equation was put-forward. That is, quantum jumps were an outgrowth of the statistical Born rule and subsequent Copenhagen interpretation. This is historically inaccurate: the concept that there is a discontinuous action inherent in quantum theory was thought to be a property of the quantum of action and related to the problem of wave-particle duality well before the discovery of Schrödinger's equation. For example, Born refers to energy jumps in 1927 as always having been regarded as the basic pillar [181]

> *... the fact of "energy jumps," which has always been regarded as the basic pillar of quantum theory, as well as the most egregious*

contradiction to classical mechanics.

As further discussed in Chapter 5, Bohr had also attributed such discontinuous actions to the existence of a noncausal or nondeterministic process, well before the discovery of Schrödinger's equation. When Schrödinger originally proposed his equation, he was invited by Bohr to his home for further discussions. According to the account by Heisenberg in [182, p. 75], Bohr made several points regarding the quantum of action in regards to Schrödinger's equation:

> *Bohr: But just take the case of thermodynamic equilibrium between the atom and the radiation field—remember, for instance, the Einsteinian derivation of Planck's radiation law. This derivation demands that the energy of the atom should assume discrete values and change discontinuously from time to time; discrete values for the frequencies cannot help us here. You can't seriously be trying to cast doubt on the whole basis of quantum theory!*
>
> *Schrödinger: There is no reason why the application of thermodynamics to the theory of material waves should not yield a satisfactory explanation of Planck's formula as well—an explanation that will admittedly look somewhat different from all previous ones.*
>
> *Bohr: No, there is no hope of that at all. We have known what Planck's formula means for the past twenty-five years. And, quite apart from that, we can see the inconstancies, the sudden jumps in atomic phenomena quite directly, for instance when we watch sudden flashes of light on a scintillation screen or the sudden rush of an electron through a cloud chamber. You cannot simply ignore these observations and behave as if they did not exist at all.*
>
> *Schrödinger: If all this damn jumping were really here to stay, I should be sorry I ever got involved in quantum theory.*

The dynamics of quantum jumps traces back to Bohr's (1913) transitions between stationary states in his model of the atom and Einstein's (1917) theory of stimulated and spontaneous emission between Bohr's stationary states. In a sense, Einstein's theory was an open quantum system that allowed stochastic jumps to lead to a thermal equilibrium state consistent with Planck's radiation law. After the development of quantum mechanics, quantum jumps were interpreted in terms of state reduction resulting from measurement. Quantum jumps were first directly observed in the 1980s in laser-cooled ions in radio-frequency traps [183]. Since the 1990s, quantum jumps have been described in terms of stochastic evolution equations [184] [185] in which the atom is weakly coupled to photodetectors, and the atom jumps due to entanglement between atom and detectors, which are represented by measurement operators. From this point of view, there would be no jump in the absence of the measurement. The detection scheme conditions the response of the atom so that direct detection results in quantum jumps similar to Einstein and Bohr (though not necessarily between eigenstates) whereas heterodyne detection gives rise to quantum state diffusion [186]. Although such models can be used to consistently analyze experiments, it is not known whether or not quantum jumps are actually detector-dependent. However, Wiseman and Gambetta have proposed an experimental test for whether quantum jumps are objective phenomena or else depend on the measurement device [187].

We have in Chapter 3 presented a technique that is used to show why the measurement problem is an actual problem versus only a problem that requires an interpretation to solve. The method presented in Chapter 3 requires one to test the existence of entanglement. As there are a number of properties of measurement, one might consider whether the detection of such a property might be used to confirm the existence of a measurement process versus unitary evolution. The ability of each of these properties of measurement is now examined with respect to a means to discern between measurement and unitary evolution.

Discerning Quantum Jumps

Suppose that quantum jumps do indeed occur within a measurement process. Are such jumps unique to measurement or are they also expected under unitary evolution? To simplify matters, let us consider a quantum jump that is deterministic. In the development of the quantum theory of radiation by Dirac [188], a photon undergoes annihilation and creation in the interaction of light and matter. Consider a two-level atom with states $|g\rangle$ and $|e\rangle$ interacting with a photon of energy E_P that is in the Fock state $|1\rangle$. If the initial state of the photon-atom is $|1\rangle|g\rangle$, then this will evolve under unitary evolution to a superposition of $|1\rangle|g\rangle$ and $|0\rangle|e\rangle$. In the Dirac model suppose that the state $|1\rangle|g\rangle$ evolves to $\sqrt{\alpha}|1\rangle|g\rangle + \sqrt{1-\alpha}|0\rangle|e\rangle$. Then the reduction of the amplitude of the state $|1\rangle|g\rangle$ from unity amplitude to $\sqrt{\alpha}$ coincides with the rise of the population of the state $|0\rangle|e\rangle$. Hence in some sense part of the amplitude of the state $|1\rangle|g\rangle$ is being converted to the state $|0\rangle|e\rangle$. But this Dirac process requires the annihilation of the photon state $|1\rangle$ and the excitation of the atom from the state $|g\rangle$ to state $|e\rangle$. One can see that even in the unitary process, there is a sense of a quantum jump occurring in that the photon energy is not being slowly and continuously reduced from energy E_P to energy 0, but rather the new term that emerges in the Dirac theory is $|1\rangle|g\rangle$ which represents a jump from $|0\rangle|e\rangle$. Hence jumps, of and by themselves, seem to be present in both unitary theory and measurement.

Let us suppose that an energy conserving Jaynes-Cummings Hamiltonian in quantum optics (or similarly in spin manipulation experiments such as NMR) is considered in which a π pulse is applied. This will entirely transfer the photon-atom state $|1\rangle|g\rangle$ to the state $|0\rangle|e\rangle$. Such a state transfer is indistinguishable from a quantum jump in which the state jumped from $|1\rangle|g\rangle$ to $|0\rangle|e\rangle$. If unitary evolution occurs the states at time zero and time π are given respectively by the two elements of the set $\{|1\rangle|g\rangle, |0\rangle|e\rangle\}$. Now suppose that measurements are made at time zero and time π of the EM-field and atom. Such measurements will also reveal $\{|1\rangle|g\rangle, |0\rangle|e\rangle\}$. Hence unless one can demonstrate that a deterministic quantum jump under measurement occurs faster or slower than a unitary π pulse or via some other distinguishing characteristic, then a quantum jump of and by itself does not appear to be unique to measurement. That is, something more is needed for a deterministic quantum jump to distinguish it from unitary evolution.

Distinguishing unitary evolution from measurement via a Chapter 3 UMDT may be possible if a quantum jump is represented as *steering*. In this way, the state of a distant entangled system exhibits a jump similar to that of an atom entangled with an electromagnetic field. As discussed in Chapter 5, Schrödinger introduced the phenomenon of steering in his 1936 paper in which he analyzed the Einstein-Podolsky-Rosen thought-experiment and demonstrated a nonlocal effect on a distant system so that by using entanglement, "a sophisticated experimenter can … produce a non-vanishing probability of steering the system into any state he chooses." The UMDT test of Chapter 3 used the Bell nonlocality or violation of the CHSH form of the Bell inequality. However, Bell nonlocality is a stronger concept than steering so that Bell-nonlocal states are a subset of steerable states. Similarly, steerability is a stronger concept than entanglement so that steerable states are a subset of entangled states [189]. Using steering, measurements by Alice could bring about a quantum jump in Bob's distant entangled system. However, Bell nonlocality and entanglement are both concepts that are symmetric between Alice and Bob, whereas steering is inherently asymmetric. There are situations where Alice can steer Bob's state but Bob cannot steer Alice's state, referred to as *one-way steering*, very much like the historical view of a quantum jump. Despite the differences between Bell nonlocality and steering, the phenomenon of steering can also be characterized by an inequality analogous to that of the CHSH inequality used in the Chapter 3 UMDT test. Cavalcanti et al. [190] have formulated an inequality that is necessary and sufficient for steering, which takes the form:

$$\sqrt{\langle (A+\acute{A})B\rangle^2 + \langle (A+\acute{A})\acute{B}\rangle^2} + \sqrt{\langle (A-\acute{A})B\rangle^2 + \langle (A-\acute{A})\acute{B}\rangle^2} \leq 2$$

where $\{A, \acute{A}\}$ and $\{B, \acute{B}\}$ represent the two dichotomic measurements of Alice and Bob. When the CHSH inequality is violated, the steering equality is also violated. When Alice and Bob share a maximally entangled state, the maximum violation of the inequality becomes $2\sqrt{2}$, just as for the CHSH inequality, although for steering this is independent of the relative angle between Alice's and Bob's measurements. The phenomenon of steering and the steering inequality have been experimentally demonstrated [191]. The steering inequality can then be used to devise a UMDT test for a steering model similar to that of the Bell-nonlocality test in Chapter 3. This would similarly distinguish unitary evolution from measurement for this situation of a steering model for a quantum jump.

Consider also the possibility the relationship between physical discontinuities and measurement. In the experiment of Steinberg, Kwiat, and Chiao [192], it was found that quantum wave functions can be reshaped so that the group velocity is higher than the speed of light c. However, information in the form a discontinuity or "front" must propagate at the front velocity which is limited by c [193]. An excellent discussion of the various velocities and their relationship to the work by Sommerfeld and Brillioun can be found in [194]. As it is widely accepted that information within a photon propagates unitarily, the discontinuity would be expected to propagate no faster than the speed of light and unitarily. The existence of a discontinuity within a wave function has not been proven to be a sufficient condition for a measurement.

Wave Function Reduction

Measurements are generally of one of two types. In one type, within the process of measurement, the particle that is being measured is annihilated entirely and completely absorbed within the measurement device. The second type of measurement leaves the system particle intact but can serve to reduce or affect the wave packet. Quantum non-demolition measurement is a form of the second type of measurement in which it is attempted to minimally reduce or affect the wave-packet in the process of obtaining information regarding the particle. In this section, we will restrict attention to the latter process whereby the system particle remains intact in the process of interaction with the measurement device.

When an individual electron is emitted, it exists with energies that are quantized and propagates as a wave and exhibits path interference. In [7], Heisenberg examined the problem for which a position measurement is made on an electron via observing a photon of wavelength λ. Starting with an electron in a superposition of position states, Heisenberg states [7]:

> *... each position determination reduces the wave packet again to its original dimension λ.*

In the Copenhagen interpretation, a wave function that is initially in a superposition of the observable's eigenstates will generally change non-locally and discontinuously via a projection into the subspace of eigenstates corresponding to an eigenvalue of the observable when a measurement occurs. Such a change of the wave function is termed wave function reduction or wave function collapse.

An important possibility to consider occurs when a measurement device does not detect the presence of a particle. In this case, the result is negative measurement [195] or interaction-free measurement (IFM) [106]. Because the IFM subspace is not complete, IFM is considered a type of measurement that also involves a projection and will have an associated wave packet modification similar to reduction. The wave function in IFM is projected into the subspace that is orthogonal to the linear space spanned by the eigenstates corresponding to the device detecting the presence of the particle.

Discerning Wave Function Reduction

In terms of whether or not wave function reduction by itself can be justified to be a property of measurement versus unitary evolution is not immediately obvious. The original justification of the reduction in the paper by Heisenberg [7] uses the concept of the Heisenberg uncertainty principles. That is, if one reduces the position uncertainty of an electron through measurement, then its momentum uncertainty is increased. However, unitary evolution of the quantum wave function also respects the Heisenberg uncertainty principle. That is, suppose that through a unitary interaction with a device the position uncertainty as measured by the standard deviation of the electron's wave function was found to decrease, then one would expect the electron's momentum, which is related to the Fourier transform of the positional wave function, to similarly increase.

On the other hand, the Chapter 3 UMDT demonstrates that it is possible in principle based on the current quantum formalism, to distinguish unitary evolution from measurement. The Chapter 3 UMDT was based on a multiple degree-of-freedom reduction and not on a single degree-of-freedom reduction.

Consider a single degree of freedom of a particle that has undergone external orthogonalization such that under unitary evolution the state of system plus apparatus is given by $\sum_r c_r \mid \phi_r\rangle \otimes |\Psi_r\rangle$. The system is in the mixed state given by Equation (4.3) reproduced below for convenience:

$$\begin{pmatrix} |c_1|^2 & \cdots & 0 \\ \vdots & \ddots & \vdots \\ 0 & \cdots & |c_N|^2 \end{pmatrix}.$$

Suppose the final state of the particle under measurement occurs definitely as $|\phi_r\rangle$ for some r. One might argue that the state being definitely in some $|\phi_r\rangle$ is different than the mixed state of Equation (4.3). However, suppose an experiment prepares a set of K states by one of two methods. In Method 1, Alice prepares $|\phi_r\rangle$ using a random number generator that corresponds to probability $|c_r|^2$. This is repeated K times and the states deposited in a box. In Method 2, Alice prepares the mixed state of Equation (4.3) experimentally for each particle, and deposits K such states in a box. If the box is subsequently given to a second experimenter Bob, and he is asked to experimentally discriminate which of the two methods Alice used to prepare the states. Bob, however, finds that it is impossible based on current quantum theory to determine which method Alice used. Hence for a single degree of freedom, it does not appear that a measurement interpretation for which the outcomes definitely have occurred, can be experimentally falsified from a mixed state prediction. This has been referred to as the ignorance interpretation by Hughes [196]. A mixture of pure states that can be interpreted via the ignorance interpretation for which the state is in a pure state with some probability is termed a proper mixture. Mixtures that cannot be interpreted via the ignorance interpretation are termed improper [197, p. 58] by d'Espagnet. Generally, a subsystem that is part of a larger entangled state will result in an improper mixture because the state cannot be properly interpreted to exist as any particular pure state of which we are ignorant, and so a probability distribution is assigned over a set of pure states to represent the mixed state. If the latter ignorance interpretation were true, a hidden variable model for the local subsystems would exist and entangled states would not violate the Bell inequality. However, as such entangled states do violate the Bell inequality, it cannot be said that the mixed state of a subsystem of an entangled state actually exists as a pure state but that we are simply ignorant of it and thereby need to treat the matter probabilistically. On the contrary, no local probabilistic treatment suffices to explain the violation of Bell's inequality, and hence the ignorance interpretation of proper mixture absolutely fails and an improper mixture must be used to interpret the subsystems of entangled states.

On the other hand, the Chapter 3 UMDT demonstrates that it is possible to experimentally discriminate unitary evolution from measurement. However, the UMDT requires two measurements on separate degrees of freedom. The cases of measurement and unitary evolution now become experimentally distinguishable because the difference between the two cases can be manifested in a correlation difference between the measurements rather than a proper mixed state interpretation of the state. In this case the mixed state prediction of unitary evolution is an improper mixture. Wave function reduction, which is predicted to break the unitarily predicted entanglement under measurement, can result in an experimentally different result when multiple degrees of freedom are considered.

Nondeterminism

One might theoretically consider a quantum jump via a deterministic and causal physical discontinuity, but such a jump would not need a statistical interpretation. Once the process itself is said to be governed by nondeterministic or noncausal physics, then a multitude of possible results must present themselves. Suppose on the contrary that only a single result presented itself—in such a case the physics would be deterministic and non-statistical and one might expect the physics to be fully governed by Schrödinger's equation.

One often sees reference to quantum theory as being strictly the application of Schrödinger's equation. However, even before the discovery of Schrödinger's equation, quantum theory was alive and well as the quantum of action was discovered twenty-five years previously. For example, in 1925, prior to the discovery of Schrödinger's equation, Bohr wrote [198]:

> *Planck demonstrated, retaining Boltzmann's account of the second law of thermodynamics, that the laws of heat radiation demand an element of discontinuity in the description of atomic processes quite foreign to the classical theories.*

This is similar to the argument that Bohr used in his debate [182, p. 75] with Schrödinger that his equation could not be universally valid.

Bohr also states [198]:

> *From these results it seems to follow that, in the general problem of quantum theory, one is faced not with a modification of the mechanical and electrodynamical theories describable in terms of the usual physical concepts, but with an essential failure of the pictures in space and time on which the description of natural phenomena has hitherto been based.*

Such statements by Bohr can be found to be repeated over-and-over in many of his works, throughout his lifetime. Until his death, Bohr appears to be unwavering in this respect.

The fact that Schrödinger's equation on its own fails to predict many phenomena and requires augmentation by Born's statistical rule was not seen as a coincidence by Niels Bohr. On the contrary, Born's rule was seen as a principle part of quantum theory for which both a nondeterministic as well as a deterministic description are two complementary methodologies that are both necessary for describing phenomenon in a full and complete manner.

After Born's rule and Schrödinger's equation became part of the Copenhagen Interpretation, von Neumann formalized the mathematics of quantum mechanics largely according to the Copenhagen Interpretation. There are two processes in von Neumann theory, Process 1 is the measurement process and Process 2 is the unitary Schrödinger process.

Von Neumann considers the measurement process a primary process. Von Neumann [13, p. 349] contrasts the measurement process with Schrödinger's equation in a manner similar to Bohr's statements:

> *Another type of intervention in material systems, in contrast to the discontinuity, non-causal, and instantaneously acting experiments or measurements, is given by the time-dependent Schrödinger differential equation.*

Discerning Nondeterminism

Nondeterminism that occurs within a single system is subject to a probability interpretation. However, a single system that is predicted on average to evolve to a proper mixture under unitary evolution (for example in the process of external orthogonalization) cannot be distinguished from the case where the individual measurements occurred, each with the corresponding Born probability. On the other hand, when a quantum wave function impinges on more than a single detector, such as in the case when there are two devices in the Chapter 3 UMDT, it becomes possible to discriminate the cases of measurement from unitary evolution. In this case, the mixture that results must be considered an improper mixture under measurement. The nondeterminism that results in the process of wave function reduction does appear to provide a means to distinguish unitary evolution from measurement when more than a single degree of freedom is present so that an improper mixture has resulted.

Irreversibility and Entropy

The concepts of irreversibility and entropy were presented in Boltzmann's theory, however, whether or not such irreversibility was a consequence of a large number of reversible interactions that are for all practical purposes not distinguishable from the existence of actual elementary non-reversible operations, was not known at the time of Boltzmann. Von Neumann in [13, p. 365] examined the reversibility of the two processes and states:

> *The simplest process would be to refer to the time dependent Schrödinger differential equation, i.e., our process 2. The process is also reversible.*

In regards to the measurement process, Process 1 [13, p. 380]:

> *... each measurement on a state is irreversible, unless the eigenvalue of the measured quantity has a sharp value, in which case the measurement does not change the state at all. As we see, the non-causal behavior is thus unambiguously related to a certain concomitant thermodynamical phenomena.*

Later [13, p. 388]

> *Therefore, if only interventions 1., 2., are taken into consideration, then each process 1., which effects a change at all, is irreversible.*

Again, the property of irreversibility as present in the formalism of quantum mechanics is understood by Bohr and in the mathematics of von Neumann's Process 1. Irreversibility is consistent with Bohr's viewpoint of the non-causal or statistical nature of observation.

Von Neumann also considered in [13] the theory of entropy which is related to the theory of information change. Von Neumann introduced a measure of entropy (often referred to as the von Neumann entropy) and showed that under unitary evolution such entropy remains unchanged. Von Neumann also showed, under projective measurement, that the von Neumann entropy of the system increases except when the system is already in an eigenstate of the measurement observable.

On the other hand, we have already made mention of two types of measurement: a measurement whereby a device detects the presence of a particle such as a photon, and another type termed interaction-free measurement (IFM) whereby the detector does not undergo a change that indicates the presence of the particle.

In the case of IFM it cannot be said that measurement is a necessary condition for irreversibility. In the paper [199] by Elitzur and Dolev, it is shown that IFM's can sometimes be reversed. The interested reader should read the original paper, as a detailed exposé of their work is beyond the scope of this book; a synopsis will be given here. The authors consider a photon that is in a superposition of polarization and consider using detectors to create a partial measurement on the polarization superposition state. They show when the detector does not detect and partly disrupts the state, it is sometimes possible to completely reverse the effect and re-establish the original superposition. The authors state that in the case of partial measurement that in comparison to the case of positive measurement, the case of negative or IFM can sometimes be reversed.

Discerning Irreversibility and Entropy

Irreversibility and entropy increase are predicted to accompany certain acts of measurement as shown by von Neumann. And as reversibility is intrinsically linked to unitary evolution, the question arises as to whether or not irreversibility and entropy change are properties that can be used to distinguish measurement from unitary evolution. As has been discussed, irreversibility does not appear to be a sufficient condition for all cases of measurement, due to the possibility of reversing certain interaction-free measurements.

Now, supposing that one can demonstrate that a system has truly undergone an irreversible time evolution, one can ask if this is a sufficient condition for measurement. This appears to be a reasonable hypothesis at this time. A Category 1 theory that includes irreversible evolution under certain conditions would require augmenting the orthodox theory. Such augmentation could very well also include an entropy change that is associated with an informational change.

A Category 2 theory of measurement has already been proposed as one for which the very conditions under which measurement occur prevent to some extent the experimental verification of the theory. Consider the possibility of a Category 2 theory for which there exist certain configurations of particles for which unitary evolution can never be reversed. Furthermore, that all measurement configurations necessarily belong to such configurations. Then a logical possibility is that it is no longer possible to experimentally falsify an improper mixture that resulted from unitary evolution, in which case the improper mixture might be interpreted as a proper mixture. If such a theory were proven correct, then the property of irreversibility could still logically follow from certain configurations of unitary evolution. However, in a Category 2 theory it does not appear that von Neumann entropy of a system plus apparatus can change.

Amplification

Many associate quantum measurements with an act of amplification. For example, if one intends to measure a single photon that is registered in the classical domain in a macroscopic manner, amplification is often used. Devices that might be hypothesized to be bona fide measurement devices include, for example, a photographic film or a photo-avalanche detector, both of which utilize methods of amplification when a single photon impinges on the device. Hence it is a reasonable question as to whether or not amplification is a necessary and sufficient condition for measurement.

Discerning Amplification

A fundamental tool in the experimental area of quantum information is the use of a non-linear quantum optical phenomenon known as spontaneous parametric down conversion (SPDC). This is achieved by using a crystal to split a single photon from a pump beam into two photons commonly referred to as the signal and idler in a manner that conserves energy and momentum. SPDC can be considered a simple form of amplification in which a single photon is input and two photons are output. Such down conversion has been demonstrated to be in agreement with the unitary prediction in numerous experiments. Many experiments akin to amplification have been conducted. For example, in the paper [200] two photons were produced by parametric down conversion and one of these photons impinged on a metal plate. The photon that was absorbed by the plate created a number of surface plasmons that propagated to the other side of the plate at which time they recombined to release a photon. It was found that it was possible that the photon that was re-emitted was found to exhibit a Bell correlation and therefore entanglement with the original photon. This shows that such conversion to a number of plasmons and subsequent re-emission was not enough to measure the photon.

Due to the numerous experiments that have been conducted using SPDC and the verification of the unitary prediction which violates Bell's inequalities, it can be considered to be a scientifically validated conclusion that there exist amplification mechanisms that are not measurement devices. Hence amplification does not appear to be a sufficient condition for measurement. Whether or not amplification is a necessary condition for measurement is unknown at this time.

Localization

In the case of light, a single photon that has a wave packet that propagates in a vacuum retains the shape of the wave packet or can be said to propagate in a dispersion-free medium. This is because the energy and momentum of a photon are directly proportional and hence the momentum wave function change that occurs in unitary evolution of time proportional to t_0 is multiplication by e^{ikt_0}. Since the momentum and position wave functions are Fourier transforms of each other, the effect of the Schrödinger unitary propagation of a photon wave function is simply that the entire wave function translates or propagates with a phase proportional to t_0, as can be seen by applying the shifting property of the Fourier transform.

A property of unitary evolution found by de Broglie is that not only light but also matter can be associated with a wave function. However, unlike light, the relationship between the energy and the momentum of matter is generally not linear. Hence the wave function of isolated matter particles generally disperses or spreads over potentially wide areas of space when one computes the wave function at later times via Schrödinger's equation.

A characteristic of Einstein relativity, in classical physics, and generally if you simply look around your surroundings, is that objects appear to have very well-defined positions. If unitary evolution were strictly correct, then matter wave functions will generally disperse under unitary evolution. For example, consider the well-known experiment that demonstrated the diffraction of fullerene molecules [26]. This has now been extended to rather large molecules in [26]. Suppose that it is found that larger systems could be diffracted as well, or even a single cell. If such cells always had wave functions that were spread across miles of space, the world might be appear rather chaotic. On the other hand, if one has a microscope and sees a cell, then there is no doubt that the cell is localized. Hence it seems very reasonable to consider whether or not system localization is a necessary and/or sufficient condition for measurement.

Discerning Localization

It has not been proven, at least to the authors' knowledge, that the condition of localization is a necessary and sufficient condition for measurement. While it does appear that macroscopic objects are localized, this does not logically imply that a macroscopic object could not be produced in a laboratory that is not localized.

Consider an atom that interacts unitarily with a photon in a cavity. Suppose the initial state of photon-atom is $|1\rangle|0\rangle$. After waiting for a time related to a π pulse, one will find that the atom has absorbed the photon and the photon is not found in the cavity but is localized in the atom. Hence localization of a photon that was previously external to the atom has occurred and has been demonstrated by only unitary evolution. However, one could include such unitary mechanisms in the Chapter 3 UMDT for each device, and entanglement of the two devices, or a CHSH sum of $2\sqrt{2}$, would be predicted under such unitary devices. One might argue that it cannot be established which particular device unitarily absorbed the photon in the UMDT due to the predicted entanglement, but nonetheless the wave that existed outside the two devices does cease to exist after time even in unitary evolution (time given by π pulse, on-resonance).

Suppose that a measurement has occurred, and we know which of the two devices has absorbed the photon. Now we also know which device absorbed the photon, but is the photon localized within that detector any better than it was during unitary evolution in time π? Perhaps or perhaps not. If not, then the major difference between unitary evolution and measurement would seem to be that in measurement the entanglement is removed and we know which atom absorbed the photon, but unitarily after time π, we also know that the wave of the photon has ceased to exist in all space outside the matter. Localization of a photon therefore does not appear to be a sufficient condition for measurement.

Let us now consider whether or not localization is a necessary condition for measurement. In the case of interaction-free measurement or negative measurement, the wave-function is projected outside a device that does not detect the particle. If the wave-function is spread outside the detector, it will remain spread and hence full localization cannot be said to have occurred in IFM. If a positive detection occurs, it is generally the case that the particle is then found localized to the detector area. Hence for positive detection, it does appear to be a reasonable hypothesis that localization is a necessary condition for measurement.

Loss of Coherence

A property often associated with measurement is some or all loss of coherence of the particle due to the process of observation. For example, consider an experiment with an electron that is put through an interferometer based on its spin. Assume the electron is initially in an equal superposition of up and down spin, i.e. $|\psi\rangle = (|\uparrow\rangle + |\downarrow\rangle)/\sqrt{2}$, and the interferometer is set up so that spin-up takes the upper path of the interferometer, and spin-down takes the lower path. The upper and lower paths of the electron are recombined so that the electron exists in one of two Ports A and B. It is assumed, given the initial equal superposition state, that the interferometer is configured so that when the state $(|\uparrow\rangle + |\downarrow\rangle)/\sqrt{2}$ is input into the interferometer, the electron will always exit in Port A and when $(|\uparrow\rangle - |\downarrow\rangle)/\sqrt{2}$ is input, the electron will always exit in Port B. Now suppose that in the lower path there exists an interaction with a measurement device that can give information that the electron has passed through that path. If the device provides a readout that the electron has passed through the lower path, then the electron is projected into the state $|\downarrow\rangle$ and the electron experiences a total loss of coherence.

If the electron passes through the upper path then the device will not register the electron in the lower path. Hence, we know that when the device does not register the electron in the lower path, then the state of the electron via the measurement postulate is projected into the state $|\uparrow\rangle$, and again experiences a total loss of coherence. If a partial measurement were made by an imperfect device or via a weak measurement that does not reliably indicate the path, then one will generally find that the electron coherence is reduced but not eliminated entirely. Hence loss of particle coherence appears to be a property that is associated with measurement.

Discerning Loss of Coherence

Let us now ask whether or not the loss of particle coherence is either a necessary or sufficient condition for discerning measurement from unitary evolution. Consider that the device is composed of an atomic sample that is affected by the electron passing through either strongly (which will be assumed can be determined via a follow-up experiment on the atomic sample) or very little, depending on the device's setting. Furthermore, we assume that the electron interaction with the atomic sample is described via Schrödinger's equation. The initial state of the sample is assumed to be $|\psi_1\rangle$ which remains the same if the electron does not pass through the sample, and changes to $|\psi_2\rangle$ if the electron does pass through the sample. We assume that $\langle\psi_1|\psi_2\rangle = 1-\beta$ where $0 \leq \beta \leq 1$ is the setting of the device. When $\beta = 0$ the sample is not affected at all by the electron passing through and when $\beta = 1$ the final state of the device is orthogonal to the initial state and hence which-way information of the electron can be perfectly obtained.

Suppose the atomic sample is set to $\beta = 1$ such that the electron strongly alters the quantum state of the sample upon passing through it. Then, after the electron passes through the interferometer, it is possible to probe the sample and determine which-way information. In such a case, the final state of electron-sample is

$$(|\uparrow\rangle \otimes |\psi_1\rangle + |\downarrow\rangle \otimes |\psi_2\rangle)/\sqrt{2}.$$

In this case the density matrix of the electron can be computed by tracing the atomic sample for which one finds that the state of the electron is a completely mixed state. Hence there has been a complete loss of coherence of the electron, and yet the entire treatment has been via Schrödinger unitary interaction. On the other hand, as the setting of the atomic sample is decreased $\beta < 1$, one finds that coherence is restored as a function of β. Hence both the loss of coherence and the restoration of coherence if there were partial measurement of the sample can be seen to be attributable to a unitary process. This indicates that measurement is not a necessary condition for loss of first-order coherence of the particle.

On the other hand, let us examine in the context of coherence, why the Chapter 3 experiment is successfully able to discriminate unitary evolution from measurement within the von Neumann formalism of quantum mechanics. It appears that when the entire device plus system is considered, the second-order coherence attributable to entanglement is predicted unitarily while under measurement the coherence is not predicted. Hence measurement is sufficient but not necessary for loss of coherence. As entanglement is a multi-degree of freedom effect, this indicates that, similar to the properties of wave function reduction and nondeterminism, entanglement does provide a means to discriminate unitary evolution from measurement when an improper mixture is predicted.

The breakdown of unitarily predicted entanglement is therefore a sufficient condition for measurement. This leads again to our dictum:

To the extent that there is entanglement, there is no measurement.

Particle Absorption

Positive measurement generally is often associated with the absorption of a particle into the detector. On the other hand, there are measurement techniques such as non-demolition measurement (see Chapter 7) in which information is gained regarding a particle state for which the particle is not demolished or absorbed into the detector. However, it is conceivable that other particles are absorbed into the detector in non-demolition measurement. Hence it appears to be reasonable to consider whether or not absorption is a necessary and/or sufficient condition for measurement.

Discerning Particle Absorption

In the discussion of the Transactional Interpretation, it was concluded that photon absorption is not a sufficient condition for measurement, due to the experimental evidence in cavity QED experiments for which such interaction is confirmed to be unitary. In the unitary case, there exists an evolution time when a photon interacting with an atom in a cavity will be 100% in the state for which the atom has absorbed the photon, assuming the interaction is on-resonance.

On the other hand, one might ask if absorption is a necessary condition for measurement. Suppose that it is found that measurement devices exist that emit a particle in the process of measurement. Such a case has been considered in [201, p. 612] via stimulated emission. If such devices were found to exist, then absorption of the incoming photon would not be a necessary condition for its measurement. Particle absorption is not a sufficient condition and may or may not be a necessary condition.

The Zeno Effect

The Zeno effect [202] is often considered to be a property of measurement. Consider the phrase, "A watched pot never boils." This statement implies that act of watching or observing the pot through the measurement process, prevents it from boiling.

The Zeno effect is predicted to occur within the von Neumann framework when the same measurement is repeated. Consider a system that is initially in an eigenstate (Eigenstate 1) of the measurement observable and evolves in time via Schrödinger's equation slowly into a superposition of Eigenstate 1 and Eigenstate 2 of the measurement observable. If a measurement is made and the result that represents Eigenstate 1 is obtained, one is obliged to apply the projection postulate to the quantum system which resets the system back to its initial state of Eigenstate 1. If the measurements are made quickly in succession with time between measurements equal to τ_r then the effect of the measurements, as seen in [202], is to delay the transition to Eigenstate 2 as τ_r is decreased.

The Zeno effect has been demonstrated in a number of experiments. In the experiment reported in [203] the quantum Zeno effect was first demonstrated using a three-level atomic system consisting of approximately 5000 Beryllium ions. The authors demonstrated a Zeno effect when the ions were subject to repeated measurement. There have been a number of experimental demonstrations of the quantum Zeno effect.

Discerning the Zeno Effect

Initially, upon publication in 1990 of [203], it was thought by many that the experiment was proof that a measurement effect was causing reduction or collapse of the wave function. Later, analysis [204] [205] showed that the use of strong pulsed lasers applied to the ions was sufficient to cause an effective loss of coherence that could also be attributable to a unitary theory. Later, it was thought that a single system was needed to show that the Zeno effect was a measurement effect. In an analysis of the Zeno effect in the paper [206], the authors recommended the use of a single system to determine if fluorescence still occurs. It is stated in [207]

> *... if one wishes to argue that the term "Zeno effect" should be*

reserved exclusively for environment interactions of a projective measurement type, then the motto is that one person's Zeno effect is another's decohering random precession, when taken at the statistical level. The distinction between these effects can only be made by considering individual systems.

Balzer et al. later experimentally demonstrated the existence of the Zeno effect using a single system in the paper [208]. Still, a question that remained is whether or not such a demonstration is clear proof that the Zeno effect is due to measurement and not unitary evolution.

Consider the following potential method to discriminate measurement from unitary evolution in the Zeno effect. The basis of the method is to utilize different dephasing characteristics predicted under unitary interaction versus measurement. This is required at small times when the system initially couples to what is believed to cause measurement. As one can expect the latency effect of the strong optical laser to be rather fast, this would need to be done over short times. In [209] it is claimed that it is very difficult to induce and to detect experimentally the difference between unitary evolution and reduction. The authors expect that in order to accomplish such a discrimination, no interactions with the environment can take place.

These statements underscore the many subtleties that exist regarding what is required to resolve the measurement problem. Consider in Kwiat [210], using a design that appears to incorporate unitary elements, i.e., beam splitters, etc., an effect was demonstrated that is similar to the Zeno effect. The theory of interaction-free measurement is utilized in the paper because an object that can absorb the photon is either inserted within a path of an interferometer or is not inserted. If the object is inserted, the presence of the object can be completely discerned without the object absorbing the photon. The question that remains is whether or not a unitary interaction versus a bona fide measurement will change in any manner the results that were obtained in the experiment. In Exercise 4.12 a unitary interaction is considered in place of an object that blocks the device. It can be shown by working through Exercise 4.12, that there is no difference between the prediction of unitary evolution and interaction-free measurement, so as long as the interaction with the device is sufficiently short that dephasing cannot be distinguished from measurement during the latency time.

In neither the Itano [203] nor the Kwiat [210] experiment does it appear that one can conclude that the Zeno effect is either a necessary or a sufficient condition for the occurrence of measurement phenomenon. In [211], this issue is further discussed and it is concluded that any mechanism that inhibits the unitary evolution (whether unitary or non-unitary) is sufficient to exhibit a Zeno effect. Hence a Zeno effect does not appear to be sufficient for measurement, and whether or not some Zeno effect is necessary for measurement is not known at present.

Summary

Several properties that might be attributed to measurement have been described and examined with respect to whether or not they can be used to distinguish measurement from unitary evolution. As has been seen, what one might have initially thought to be a strictly measurement effect may have a perfectly reasonable explanation in the unitary domain.

What has been found is that it is not a simple matter to discern unitary evolution from measurement in a scientifically valid method, and hence tackling the measurement problem both theoretically and experimentally clearly requires a serious effort. The many subtleties of the measurement problem have led to a substantial number of statements regarding the measurement problem that are either outright wrong, or for which we would consider to be scientific hypotheses at this point, yet have often been published as a matter of fact by many peer-reviewed journals.

Many statements by researchers are contrary to what is required to be established via scientific methodology and have no place in any scientific journal. There appear to be two extremes of such reporting. On the one hand, there are many who claim that all particles evolve unitarily. Such a claim is contrary to what has been presented in Chapter 3 and is similar to the spreading of political propaganda. Strong statements that are not scientifically validated only spread confusion, and publication of such statements has no part in the scientific process and we suggest should be rejected out of hand. On the other hand, if one is investigating the measurement problem and believes that one has a contribution to make, then one has a responsibility not to go overboard as well. That is, a signature of measurement may very well be a signature of unitary evolution.

In either case, due to the subtleties and experimental difficulties surrounding this problem, one should be extra careful when investigating and reporting on results regarding the measurement problem.

Physical Measurement Theories

As we have seen, none of the interpretations of quantum mechanics in the literature provide a resolution to the measurement problem as we define it. At this point, we will consider physical theories that attempt to resolve the measurement problem. Physical theories provide criteria that address the requirements R1.1-R1.4 in the case of a Category 1 theory or R2.1-R2.3 in the case of a Category 2 theory. Each theory will be described, classified as either Category 1 or 2 and will be evaluated in terms of how well it currently meets the respective requirements that have been set forth to resolve the measurement problem.

Physical theories can also be categorized as to whether or not a measurement device can only detect certain stimuli that are above a single quantum threshold. If the theory predicts that a measurement can only reliably occur after a given threshold is reached that is above a single quantum, then such a theory is called a *threshold measurement theory*.

Consciousness

Many of the pioneers of quantum mechanics believed that consciousness and/or free will are related to quantum mechanics, among them Bohr, Heisenberg, von Neumann, Renninger, Wigner, London, and Bauer. A number of contemporary physicists continue this line of inquiry, including Penrose and Stapp. There are those who also believe consciousness is a necessary condition for measurement. An argument is provided by London and Bauer [212, p. 217]. The authors consider the coupling of a system to a device that is then seen by a conscious observer. The system is desired to be measured for which the outcome of the system $F = f_k$ corresponds to the system being measured into its eigenstate $u_k(x)$. The readout $G = g_k$ is the apparatus eigenvalue corresponding to the apparatus eigenstate $v_k(y)$. The interaction between system and apparatus is such that when the system is definitely in eigenstate $u_k(x)$, the apparatus will readout the eigenvalue g_k corresponding to the apparatus eigenstate $v_k(y)$. The authors show that the interaction of the system and apparatus will cause the system to evolve to a mixed state in the basis of the apparatus eigenstates, in a similar manner as we have already shown via the methodology of external orthogonalization. From there, the authors describe the role of the conscious observer:

> *We note the essential role played by the consciousness of the observer in this transition from the mixture to the pure case. Without his effective intervention, one would never obtain a new ψ function. In order to see this point clearly, let us consider the ensemble of three systems, (object x) + (apparatus y) + (observer z), as a combined and unique system. We will describe it by a global wave function...*

$$\Psi(x, y, z) = \sum_k \psi_k \, u_k(x) v_k(y) w_k(z)$$

> *where the w_k represent the different states of the observer.... The observer has a completely different impression. For him it is only the object x and the apparatus y that belong to the external world, to what he calls "objectivity." By contrast he has with himself relations of a very special character. He possesses a characteristic and quite familiar faculty which we can call the "faculty of introspection." He can keep track from moment to moment of his own state. He attributes to himself the right to create his own objectivity—that is, to cut the chain of statistical correlations summarized in $\sum_k \psi_k \, u_k(x) v_k(y) w_k(z)$ by declaring, "I am in the state w_k" .... It is only the consciousness of an "I" who can separate himself from the former function $\Psi(x, y, z)$ and, by virtue of his observation, set up a new objectivity in attributing to the object henceforward a new function $\psi(x) = u_k(x)$.*

By claiming that a conscious system can cut the chain by declaring "I see $G = g_k$," London and Bauer are claiming that a conscious system that becomes aware of a phenomenon and apparatus that was hitherto in the superposition $\sum_k \psi_k u_k(x)v_k(y)$ is a sufficient condition that will cut the state from the unitarily predicted state of $\sum_k \psi_k u_k(x)v_k(y)w_k(z)$ to the product state $u_k(x)v_k(y)w_k(z)$.

London and Bauer state that,

> *a measurement is achieved only when the position of the pointer has been observed.*

London and Bauer are also claiming that consciousness is a necessary condition for measurement. One might contrast this to Bohr [213] [214]:

> *... whether we have to do with marks on a photographic plate or with direct sensations the possibility of some kind of remembrance is of course the necessary condition for making any use of observational results. It appears to me that the permanency of such results is the very essence of the ordinary causal space-time description.*

> *Indeed, from our point of view, the feeling of the freedom of the will must be considered as a trait peculiar to conscious life, the material parallel of which must be sought in organic functions, which permit neither a causal mechanical description nor a physical investigation sufficiently thorough-going for a well-defined application of the statistical laws of atomic mechanics.*

From this statement it appears that Bohr does not consider a conscious system to be a necessary condition for measurement but rather "some kind" of remembrance related to "permanency" is a necessary condition. In other writing, Bohr refers to the additional conditions of "amplification" and "irreversible functioning" to represent permanent marks [215, p. 73]. However, by referring to direct sensations it appears that Bohr considered that conscious sensation is a sufficient condition for measurement. This is confirmed in a letter that Bohr wrote to Pauli in 1953 [213]:

> *I put the main emphasis on the fact that so-called conscious experiences involve something that can be remembered and hence must correspond to special situations in the organism which have precisely the permanent character required as a basis for observation.*

The properties of irreversibility, permanency, direct sensations, remembrance, suitable amplification, are all properties that Bohr believes are related in some manner to measurement. Although such properties are important, as has already been presented, none of these conditions as of yet provides a resolution to the measurement problem in a manner to fully satisfy the Requirements R1.1-R1.4 for a Category 1 theory and R2.1-R2.3 for a Category 2 theory.

Heisenberg [216, p. 54] in explaining the Copenhagen Interpretation states:

> *... and we may say that the transition from the 'possible' to the 'actual' takes place as soon as the interaction of the object with the measuring device, and thereby with the rest of the world, has come into play; it is not connected with the act of registration of the result by the mind of the observer. The discontinuous change in the probability function, however, takes place with the act of registration, because it is the discontinuous change of our knowledge in the instant of registration that has its image in the discontinuous change of the probability function.*

Heisenberg attributes the discontinuous change in the observer's mind with the discontinuous change of the probability function. On the other hand, Heisenberg attributes the transition from possible to actual with the interaction of system and device. Heisenberg attributes the transition from possible to actual with the effect of unitary decoherence due to the uncontrollable interaction of system and device.

However, Heisenberg appears, similar to London and Bauer, to be attributing the breaking of the unitary interaction which results in loss of entanglement and discontinuous change in the probability function only at the instant of registration in the observer's mind. The attitude of Heisenberg is explained by statements by Renninger [106]:

> *In a personal correspondence Prof. Heisenberg kindly informed me about his opinion with respect to the investigations explained in the present work, and allowed me kindly to summarize his ideas as follows: It is incorrect to think, that the Copenhagen interpretation of quantum theory claims that the principal unavoidability of the influence of a measurement on the object system is related to a proper measurement "process," and that possibly, becoming aware of it later on, has the effect of reducing the wave function "retroactively." However, it is not possible at all to objectify a "measurement process" in the sense of Re in all possible cases. Simply and solely it is the circumstance of becoming aware of the measurement result that is objectifiable and responsible for state reduction, which, hence, is relegated to the "cut" between the object system and the apparatus system. However, what is meant by the unavoidable disturbance of the physical process by the measurement is already the possibility of a measurement, i.e., the existence of the*

measurement apparatus. It is precisely this existence, that brings about a partly undetermined interaction between the measurement apparatus and the object system, and that after carrying out the experiment leads to the uncertainty relation. On the other hand, the act of the registration, that leads to the reduction, is in fact not a physical, but in a sense a mathematical process. Of course, with the erratic change of our knowledge there corresponds also an erratic change of the mathematical representation of our knowledge.

Renninger also is in agreement with Heisenberg's viewpoint:

If Mr. Heisenberg's described view were the common opinion, then my considerations were indeed unnecessary, because basically they are the same, as follows from the last three sentences. At any case, it seems to me that there exists a proper measurement process, and that becoming aware of the result reduces the state (e.g., a – present or absent – to be developed mark on a recorder), at a time instant that is determined more or less exactly by the measurement proper

Wigner also believed that the measurement problem and consciousness are related. In the paper [217] the well-known Wigner's friend argument is presented. Wigner has a friend that observes a quantum object (such as a photon) that is initially in a superposition of positions and either sees a flash, indicating that the friend has observed the quantum object, or does not see a flash. The object has two states ψ_1 and ψ_2. The observer sees the object and Wigner asks his friend what was seen. If the observer sees the object in state ψ_1 (ψ_2), the observer evolves to the final state χ_1 (χ_2) and the joint state of object and observer evolves to a product state $\psi_1 \otimes \chi_1$ ($\psi_2 \otimes \chi_2$). The observer's final state χ_1 is representative of the case when the observer responds to Wigner's question that he has seen the flash, whereas when the observer responds not to have seen the flash, the observers final state is χ_2. Now the object is initially prepared in a linear superposition of $\alpha\psi_1 + \beta\psi_2$ and the experiment is repeated. If one assumes that the quantum state obeys Schrödinger's equation, then the final state of object and observer is $\alpha\psi_1 \otimes \chi_1 + \beta\psi_2 \otimes \chi_2$. Wigner then asks the observer whether or not a flash was seen. According to the measurement postulate, the observer will respond with a probability $|\alpha|^2$ that he did and $|\beta|^2$ that he did not. Furthermore, according to the measurement postulate, the object-observer state changes to $\psi_1 \otimes \chi_1$ when the observer responds that he has seen a flash or to $\psi_2 \otimes \chi_2$ if the observer responds that he has not seen a flash. However, suppose that Wigner then asks his friend what he saw just before Wigner had asked him whether or not a flash was seen. Wigner's friend is found to always answer that he already told him that he saw the flash. Wigner concludes that the state of the wave function right before Wigner asked his friend the question was already either $\psi_1 \otimes \chi_1$ or $\psi_2 \otimes \chi_2$ and not $\alpha\psi_1 \otimes \chi_1 + \beta\psi_2 \otimes \chi_2$. Hence the unitary predicted state is contradicted by such an experiment.

On the other hand, suppose that Wigner's friend is replaced by a single atom initially in a ground state which, in the process of interaction with light, could change state to an excited level. Then Wigner contends that the final state of light-atom is $\alpha\psi_1 \otimes \chi_1 + \beta\psi_2 \otimes \chi_2$. Moreover, Wigner expresses the fact that the differences between the state $\alpha\psi_1 \otimes \chi_1 + \beta\psi_2 \otimes \chi_2$ and the state of either $\psi_1 \otimes \chi_1$ or $\psi_2 \otimes \chi_2$ include observable consequences. The fact that there are observable consequences is an important observation by Wigner and is a major theme in Chapter 3 and throughout this book. But suppose now the atom is replaced by a conscious being. Wigner claims that the state $\psi_1 \otimes \chi_1 + \beta\psi_2 \otimes \chi_2$ is not realistic because it would imply that Wigner's friend was in a state of, "suspended animation" before he answered Wigner's question. Wigner concludes [217, p. 263]:

> *It follows that the being with a consciousness must have a different role in quantum mechanics than the inanimate measuring device: the atom considered above. In particular, the quantum mechanical equations of motion cannot be linear if the preceding argument is accepted.*

Wigner is making the argument that it is not necessary for Wigner to ask his friend what he saw in order to conclude that the state of the wave function is either $\psi_1 \otimes \chi_1$ or $\psi_2 \otimes \chi_2$. It is only necessary to know that Wigner's friend had already observed the system. That is, even before Wigner's friend responds to Wigner's question of whether he saw a flash, the state of Wigner's friend must have been either $\psi_1 \otimes \chi_1$ or $\psi_2 \otimes \chi_2$. If Wigner knows that a conscious system has interacted with a quantum system in a manner which would be in the unitary predicted state $\alpha\psi_1 \otimes \chi_1 + \beta\psi_2 \otimes \chi_2$, then Wigner is claiming the actual state of the system plus conscious system is actually $\psi_1 \otimes \chi_1$ or $\psi_2 \otimes \chi_2$. Wigner is arguing is that such interaction of a quantum wave function by a conscious observer is a sufficient condition for measurement. In regards to whether or not a conscious observer is a necessary condition, Wigner states: [217, p. 256]:

> *In an article which will appear soon G. Ludwig discusses the theory of measurements and arrives at the conclusion that quantum mechanical theory cannot have unlimited validity. This conclusion is in agreement with the point of view here represented. However, Ludwig believes that quantum mechanics is valid only in the limiting case of microscopic systems, whereas the view here represented assumes it to be valid for all inanimate objects. At present, there is no clear evidence that quantum mechanics becomes increasingly inaccurate as the size of the system increases, and the dividing line between microscopic and macroscopic systems is surely not very*

sharp. Thus, the human eye can perceive as few as three quanta, and the properties of macroscopic crystals are grossly affected by a single dislocation. For these reasons, the present writer prefers the point of view represented in the text even though he does not wish to deny the possibility that Ludwig's more narrow limitation of quantum mechanics may be justified ultimately.

Philosophical Reflections and Syntheses, Remarks on the Mind-Body Question, 1995, p. 256 , E. P. Wigner

Ludwig believed that unitary evolution is valid only for microscopic systems and Wigner for all inanimate objects, which for Wigner includes both microscopic systems and a subset of macroscopic objects. Hence for Wigner it appears that consciousness is a subset of only macroscopic objects. However, Wigner does not reject Ludwig's argument--as he states such a narrow limitation may be justified ultimately. A logical possibility in Ludwig's argument is that there exists a macroscopic object that is not conscious but nonetheless is a measurement device. Hence it appears that Wigner believed that consciousness is a sufficient condition for measurement and would have agreed that from a scientific basis, consciousness has not yet been established as a necessary condition. d'Espagnet in [218] reached a similar conclusion:

The doctrine that the world is made up of objects whose existence is independent of human consciousness turns out to be in conflict with quantum mechanics and with facts established by experiment.
Scientific American, Vol. 241, 158-181, 1979 excerpt reprinted with permission of Scientific American.

Several models have been put forward for which consciousness and/or free will are related to quantum phenomenon. These include Eccles [219], Stapp [220] [221] and Penrose-Hameroff [222]. Penrose-Hameroff propose [222] that gravitational reduction results in consciousness via microtubular mediated orchestrated gravitation objective reduction. Such orchestrated reduction is a self-collapse mechanism for which the neurons are synchronized in a manner to mediate gravitational collapse (to be discussed further) resulting in consciousness:

Consciousness depends on biologically 'orchestrated' coherent quantum processes in collections of microtubules within brain neurons, that these quantum processes correlate with, and regulate, neuronal synaptic and membrane activity, and that the continuous Schrödinger evolution of each such process terminates in accordance with the specific Diósi–Penrose (DP) scheme of 'objective reduction' ('OR') of the quantum state.

Physics of Life Reviews, Consciousness in the universe: A review of the 'Orch OR' theory, R. Penrose and S. Hameroff, Vol. 11, p. 39 is licensed under CC BY NC ND

It is further stated in [222] regarding orchestrated objective reduction:

If, however, a quantum superposition is (1) 'orchestrated', i.e.,

> *adequately organized, imbued with cognitive information, and capable of integration and computation, and (2) isolated from non-orchestrated, random environment long enough for the superposition to evolve by the U formalism to reach time τ by $\tau \approx \hbar/E_G$, then Orch OR will occur and this, according to the scheme, will result in a moment of consciousness. Thus if the suggested non-computable effects of this OR proposal are to be laid bare, where DP is being adopted and made use of in biological evolution, and ultimately orchestrated for moments of actual consciousness, we indeed need significant isolation from the environment.*

The general idea behind the Penrose-Hameroff proposal, whereby consciousness is related to quantum measurement, appears to be a reasonable concept. The specifics of the Penrose-Hameroff proposal, whereby microtubules mediate objective gravitational reduction, appears to be at the level of conjecture/hypothesis at this time, as there is little evidence that gravitational forces could be sufficiently isolated and distinguished from other stronger forces that will exist at room temperature. This does not mean that such hypotheses are not a step in the right direction. On the contrary, hypotheses are necessary to be put forward as regards the scientific method. However as will be discussed further in Chapters 6 and 7, society has established bulwarks from which defenders can be expected to be put forth that defend the status quo and attack. This is unfortunate yet has been the reality of the situation going back to the time of Aristotle.

Sufficiency of Consciousness

It appears that if one puts one's emotions aside on the matter and examines this issue in only a logical deductive manner, that one only has to consider whether or not one phenomenon implies another phenomenon. In this case the two phenomena under question are consciousness and measurement. From the arguments made by Bohr and Wigner, it does appear that the ability to become aware of a phenomenon is at least a sound scientific hypothesis of being a sufficient condition for measurement. For example, let us suppose that the human eye can distinguish the case of darkness versus the stimulus of N optical photons reliably. That is, if one generates N optical photons and directs them toward the eye, one sees a flash that can be reliably distinguished from no photons impinging on the eye. There will be some reaction time or latency time for an observer to state unequivocally that he became aware or not aware of the phenomenon, call this time T_L.

Now consider that we replace the beam splitter and single photon state in the UMDT of Figure 3.4 by the state $(|0\rangle_B|N\rangle_C + |N\rangle_B|0\rangle_C)/\sqrt{2}$ that is an entangled state of N photons and the vacuum. Furthermore, both Devices 1 and 2 are removed and replaced with two conscious observers that can distinguish the photons. In fact, such a state could be produced and the experiment performed with the development of a reliable optical controlled-not gate in quantum computing. The question is what would happen if this experiment were performed. If the entire chain of sight through the human mind is Schrödinger unitary, then both conscious observers would have to exist in a superposition of seeing and not seeing the photon.

Suppose that a quantum optics group actually performed this experiment and found that both individuals report the experience that either they have seen the light, or not seen the light in time T_L. This would indicate that the entire chain through the human mind is not Schrödinger unitary. One could argue that the two individuals initially exist as an entangled superposition of quantum states for a short time, and this is only resolved when a measurement mechanism, other than the individual's consciousness, causes the collapse. But then, one would expect that the individual should experience conscious outages when the quantum state is in a superposition, and individual awareness could only occur when some other measurement mechanism causes the collapse. This might be possible if an individual experiences time passage as a stroboscope effect.

Consider that a conscious system measures a system whereby 1) the direct interaction of the system with a conscious system or 2) the initial interaction with the system is unitary and later an external measurement mechanism acts on a conscious system in a superposition. It appears to be inconsequential whether 1) occurred or 2) occurred in order to establish the condition of sufficiency of consciousness to account for measurement. That is, suppose a conscious system becomes aware of an external phenomenon because either 1) the direct interaction caused measurement or 2) an external mechanism acted on a superposition of consciousness. In both cases a conscious observer ultimately becomes aware of the phenomenon; hence all such accounts indicate consciousness is a sufficient condition for the measurement of such phenomenon.

Is consciousness a necessary and sufficient condition for measurement? Maybe and maybe not. We don't know at this time if conscious systems becoming aware of a system is the only reason that measurement of the system can occur. Consciousness as a complete explanation of the measurement problem at this time only partially fulfills Requirement R1.1, and substantially more work would be needed in order to determine how it meets R1.2, R1.3, and R1.4.

The Contrapositive

Consider the sufficiency of consciousness to measurement as believed by Bohr and Wigner expressed by the following statement:

> Given that a system of particles A becomes conscious of a phenomenon B, then the interaction between A and B is a measurement process.

The contrapositive statement is:

> If the interaction between A and B is not a measurement process, then A cannot become conscious of the phenomenon B.

Let us now consider the restriction that evolution takes only two forms: 1) Schrödinger unitary evolution that occurs in non-measurement phenomenon and 2) non-Schrödinger unitary evolution that occurs in measurement phenomenon. Then such a restriction would imply the statement:

> If *A* is a device and *B* is a stimulus to the device, and the interaction is via Schrödinger evolution, then *A* will not become conscious of *B*.

Under such conditions neither *A* nor *B* can become aware of the other via their unitary interaction. Hence suppose that an individual atom is verified experimentally to couple unitarily in its interaction with a given stimulus such as a photon. Then in the interaction with the stimulus, the atom would have to be said to not become aware of the stimulus during its interaction, including when the stimulus (if it is a photon) is completely absorbed by the atom.

Under the restriction considered, this would imply that there exist configurations of matter that cannot become self-aware of a particular stimulus, and only some proper subset of all matter configurations can be considered to be systems that can become self-aware of a particular stimulus.

Threshold vs. Non-Threshold Theory

Consider a conscious system that interacts with an external phenomenon. It is stated in [223, p. 113]:

> *Visual studies conducted by Hecht et al. and by others indicate than an individual sensory receptor in the human eye responds when it absorbs a single quantum of light. In animal species such as the silk moth, an individual olfactory receptor responds when it absorbs a single molecule (Schneider, (1974)... De Vries and Stuiver (1961)) deduced from measurements of human absolute sensitivity to the compound mercaptan (skunk odor) that a single human olfactory cell probably responds to the presence of one molecule, certainly requires no more than eight molecules. The nose, like the eye and the ear, performs at its best about as well as laws of the physical world permit.*
>
>

Generally, organs such as the eye and nose provide an amplification of the input stimulus. It is not known whether or not the amplification in the eye and/or nose is itself a measurement or whether such amplification is unitary in which case the output from such organs would not be subject to measurement until further in the chain. And has already been discussed, amplification is not a sufficient condition for measurement and it is not known currently whether or not amplification is a necessary condition for measurement. Hence it is not known at this time whether or not consciousness, as a theory for which measurement occurs, is a threshold or non-threshold theory.

Mind-Body Dualism

The dualistic demand for two substances has been called for by Descartes regarding the philosophy of mind-body. In Descartes' theory, the substance that is responsible for the creation of being and soul is different than non-animate substance.

The Copenhagen interpretation does advocate two complementary descriptions of matter, one of which is deterministic for non-interacting systems, and another nondeterministic description which Bohr would say is forced upon us due to uncontrollable and unknowable interactions in the process of macroscopic amplification, which constitutes measurement. But the two complementary descriptions in the Copenhagen interpretation were meant to both apply to systems of protons and electrons and other particles, only in different circumstances. There are not two different categories of substances that are responsible for measurement versus unitary evolution in the Copenhagen interpretation.

That consciousness and volition might be related to measurement has indeed appeared to some extent in investigations by Bohr, Wigner, von Neumann, London, Penrose, Stapp, and others. Nevertheless, it has not been shown nor claimed by any of these authors, to the authors' knowledge, that the existence of consciousness and/or volition demands a theory with two distinct types of matter.

Ghirardi, Rimini, and Weber (GRW)

There are a number of theories that address the measurement problem through spatial localization. Transposing the measurement problem completely to the problem of spatially induced measurement localization is relevant if spatial localization is a property that is related to the resolution of the entanglement predicted under unitary evolution versus the product state that occurs under measurement. There is an implicit assumption being made by purveyors of such theories, that this is the case.

Localization may or may not be a necessary condition for positive measurement. However, as macroscopic objects are generally seen as localized in classical physics, it seems reasonable to examine mechanisms and theories that would localize a wave function when positive measurement occurs.

With this in mind, consider theories that address the localization of objects. A theory proposed in 1986 by Ghirardi, Rimini, and Weber (GRW) put forward a model for spontaneous localization theories that utilize a stochastic methodology [224]. Such theories are a valid attempt to solve the measurement problem and are not simply a unitary interpretation that also produces Born's rule *a posteriori* given that a measurement has occurred. The original mathematical description in the paper [224] has since been considerably mathematically simplified. Bell [225] indicated that the original GRW proposal could also be viewed from the effect of the wave function as being transformed to another wave function by the multiplication of a Gaussian function. This was subsequently incorporated into the paper [226].

Consider N particles with wave function

$$\psi(x^{(1)}, x^{(2)}, \cdots, x^{(N)}, t). \tag{4.23}$$

The position of the i th particle is given by $x^{(i)}$. Over a given time interval τ, the wave function either evolves according to Schrödinger's equation or undergoes a stochastic process referred to as a hitting process in [224]. For a single particle, the hitting process is described by a Poisson process with average number of events per time interval τ described by the parameter λ_h. For N particles the average number of events per time interval τ is $N\lambda_h$. Suppose a hit occurs to the ith particle at time $t = t_h$. Let $\hat{\psi}(x^{(1)}, x^{(2)}, \cdots, x^{(i)}, \cdots, x^{(N)}, t_h)$ be the wave function had there not been a hit, i.e. the wave function that is found from Schrödinger's equation. The function ϕ is defined by

$$\phi(x^{(1)}, x^{(2)}, \cdots, x^{(i)}, \cdots, x^{(N)}, t_h, z) \equiv h(x^{(i)}, z)\, \hat{\psi}(x^{(1)}, x^{(2)}, \cdots, x^{(i)}, \cdots, x^{(N)}, t_h)$$

whereby $h(x^{(i)}, z)$ is a positive function parameterized by z that multiplies the wave function. It is assumed that z is a real random variable with probability distribution function given by

$$p(z) = \frac{r(z)}{\int r(z)dz}$$

where

$$r(z) \equiv \int \cdots \int |\phi(x^{(1)}, x^{(2)}, \cdots, x^{(i)}, \cdots, x^{(N)}, t_h, z)|^2 dx^{(1)} \cdots dx^{(N)}.$$

Let the random variable z be chosen stochastically with result $z = z_h$. The new wave function at time t_h is given by

$$\psi(x^{(1)}, x^{(2)}, \cdots, x^{(i)}, \cdots, x^{(N)}, t_h) = \frac{\phi(x^{(1)}, x^{(2)}, \cdots, x^{(i)}, \cdots, x^{(N)}, t_h, z_h)}{\sqrt{r(z_h)}}.$$

In the case of GRW, it is assumed that $h(x^{(i)}, z)$ is a Gaussian probability function:

$$h(x^{(i)}, z) = f(x^{(i)}|z) = \frac{1}{\sqrt{2\pi\sigma_h{}^2}} e^{-\frac{1}{2\sigma_h{}^2}(x^{(i)}-z)^2}.$$

For simplicity, it will be assumed that

$$h(x^{(i)}, z) = \sqrt{f(x^{(i)}|z)},$$

for which $\int r(z)dz$ reduces to unity. Note that $p(z)$ will generally be peaked with ψ which is related to where the particle would be found in position space under Schrödinger's equation. Hence by choosing a particular $z = z_h$, GRW theory will aid in localizing the wave function and decrease the effect of the superposition terms predicted under unitary evolution. Also note that as the number of events per time interval τ is $N\lambda$, the probability of a hit will increase with the number of particles that forms a device and for which all are predicted within Schrödinger's equation to go into superposition. For example, consider a macroscopic device that has a physical pointer in position space that changes its pointer from an initial Position 1 to a final Position 2 upon interacting with a particle. If the pointer is on the order of $N{\sim}10^{23}$ particles, then all these particles will either be found in Position 1 or Position 2 in the conventional theory. Hence GRW theory for large N rapidly helps in localizing, with the exception of the tails of the Gaussian distribution, which form a positional superposition.

GRW as a POVM

It was shown in [227] that there exists a POVM representation for a range of GRW-like dynamical reduction models. The proof is based on the Riesz representation theorem. Here we express GRW as a POVM based on the development in [228] from which can be derived several features of GRW. A single particle is assumed for simplicity; however, the theory can be extended to multiple particles.

We introduce here the use of continuous kets $|\Psi(x,t)\rangle$ which are a function of the wave functions of Equation (4.23). That is, we define the ket

$$|\Psi(x,t)\rangle \equiv \int \psi(x,t_h)|x\rangle dx$$

where $\psi(x,t_h)$ is a complex wave function of the form of Equation (4.23) with $x \in (-\infty,\infty)$, $\int |\psi(x,t_h)|^2 dx=1$. The kets $|x\rangle$ represent continuous position eigen-kets and satisfy $\langle x|x'\rangle = \delta(x-x')$ where $\delta(x-x')$ is the Dirac delta function. Similar to above, $|\widehat{\Psi}(x,t)\rangle$ denotes the continuous ket assuming no hits have occurred at time t_h.

Consider the measurement of $|\Psi(x,t)\rangle$ by positive operators of the form

$$M_i = \int_{-\infty}^{\infty} \sqrt{f(x|z_i)}\,|x\rangle\langle x|dx$$

and each M_i $(i = 1, \dots, N_p)$ is associated with a value of z_i which offsets the center of the Gaussian. Note that

$$\int M_i{}' M_i dz = I.$$

The i th experimental outcome associated with the POVM at time $t = t_h$ is the projection of the state onto M_i, i.e.

$$|\Psi(x,t_h)\rangle = \frac{M_i|\widehat{\Psi}(x,t_h)\rangle}{||M_i|\widehat{\Psi}(x,t_h)\rangle||}. \tag{4.24}$$

$$|\Psi(x,t_h)\rangle = \frac{\int_{-\infty}^{\infty}\phi(x,t_h,z_i)\,|x\rangle\langle x|dx}{\sqrt{r(z_i)}}.$$

The probability of the i th outcome, denoted p_i, is given by

$$\begin{aligned} p_i &= \mathrm{Tr}[M_i^{\prime M_i}|\widehat{\Psi}(x,t_h)\rangle\langle\widehat{\Psi}(x,t_h)|] \\ &= \int_{-\infty}^{\infty}|\phi(x,t_h,z_i)\,|^2 dx \\ &= r(z_i). \end{aligned}$$

We see that the POVM formulation is identical to the GRW formulation. Note that the i th outcome of the POVM defines the position of the Gaussian z_i. In terms of what outcome occurs by a measurement in the POVM formulation of GRW, it is the particular position z_i that is chosen from the Gaussian that is used to project the state.

The projections of the state are accomplished via the operators M_i that are diagonal in the position basis. To further understand the effect of GRW, consider simplifying from the positional infinite dimensional Hilbert space consisting of the real numbers, to an approximation over a finite interval $x \in [-A, A]$ whereby the positions occur on a finite partition $(x_1 < x_2 < \cdots < x_N)$ that is equally spaced $|x_{j+1} - x_j| = \Delta, j = 1, \cdots N-1$. One can convert the integral representations of the operators to summations and the operators M_i can be written in a matrix representation that is strictly a diagonal matrix of the form

$$M_i = \Delta\begin{pmatrix} \sqrt{f(x_1|z_i)} & 0 & 0 & 0 \\ 0 & \sqrt{f(x_2|z_i)} & 0 & 0 \\ 0 & 0 & \ddots & 0 \\ 0 & 0 & 0 & \sqrt{f(x_N|z_i)} \end{pmatrix}.$$

Note that the matrices M_i are strictly full-rank positive diagonal matrices and invertible. The expected value of z (defined as <z>) is now shown to be the expected value of $|\widehat{\Psi}(x,t)\rangle$ [228]:

$$< z > = \int_{-\infty}^{\infty} z\,p(z)dz$$

$$= \int_{-\infty}^{\infty} z\, r(z) dz$$

$$= \int_{-\infty}^{\infty}\int_{-\infty}^{\infty} z\ |\phi(x, t_h, z_i)\,|^2 dz\, dx$$

$$= \int_{-\infty}^{\infty} |\hat{\psi}(x, t_h)|^2 \int_{-\infty}^{\infty} z f(x|z) dz dx$$

$$= \int_{-\infty}^{\infty} |\hat{\psi}(x, t_h)|^2 x\, dx$$

$$= < x >.$$

This shows that the value of *z* that is expected is the value of *x* that is expected. This result relates the statistics of *z* to statistics of the wave function.

Born Rule

On any single trial, GRW has been proven to be invertible, but is clearly not distortionless. However, it is now proven that in terms of the average distribution of the square of the wave function, the GRW operation is in-fact an identity operator. This is also related to the common interpretation of the Born rule [6, p. 152] for which a measurement occurs with probability related to $|\hat{\psi}(x, t_h)|^2$, where $\hat{\psi}$ is the unitarily evolved wave function of the system at time t_h, i.e. given that no hits have yet occurred. If one considers $\psi(x, t_h)$ to be the wave function after GRW is applied, which represents the outcome or pointer of a measurement device, then in terms of the common interpretation of Born's rule one would desire that the average of $|\psi(x, t_h)|^2$, which will be denoted as $\overline{|\psi(x, t_h)|^2}$ is given by $|\hat{\psi}(x, t_h)|^2$.

We define $f(|\psi(x)\rangle | z = z_h)$ as the probability of finding the new state ψ at x, that is after a GRW Gaussian centered at $z = z_h$ is multiplied by the state:

$$f(|\psi(x)\rangle | z = z_h) = |\psi(x, t_h | z = z_h)|^2$$

$$= \frac{|\phi(x, z_h)|^2}{r(z_h)}.$$

Weighting this by the probability of choosing z_h and integrating over z_h it is seen that $f(\,|\psi(x)\rangle) = \overline{|\psi(x, t_h)|^2}$ that is, the overall average probability distribution function of $|\psi(x, t_h)|^2$ after GRW is applied. Now,

$$f(|\psi(x)\rangle) = \int f\big(\,|\psi(x)\rangle | z = z_h\big) f(z = z_h) dz_h$$

$$= \int \frac{|\phi(x, z_h)|^2}{r(z_h)} \frac{r(z_h)}{\int r(z) dz} dz_h$$

$$= \int |\phi(x, z_h)|^2 dz_h$$

$$= \int_{-\infty}^{\infty} f(x|dz_h)|\hat{\psi}(x,t_h)|^2\, dz_h.$$

$$= |\hat{\psi}(x,t_h)|^2 \int_{-\infty}^{\infty} f(x|z_h)\, dz_h$$

which gives when $f(x|z_h)$ has a Gaussian form

$$f\big(\,|\varphi_P\rangle(x)\big) = |\hat{\psi}(x,t_h)|^2.$$

This shows that in terms of the average distribution of the wave function, GRW is an identity operation in terms of the square of the wave function.

Such an identity relationship has been generalized to POVM operators in [229]. These authors consider a POVM with positive Hermitian operators $\{A_i\}$ with the property $\sum A_i = I$ where $i \in \Omega$ represents an outcome of the experiment and utilize the framework of *quantum instruments*. The POVM is implemented via a particular instrument that maps a given input state to an unnormalized output state given by $\Im_i(\varrho)$ such that $||\Im_i(\varrho)|| = \mathrm{Tr}[\varrho A_i]$. An example of an instrument that implements a POVM $\{A_i\}$ is the Lüder instrument define via the relation $\Im_i(\varrho) = A_i^{1/2}\varrho A_i^{1/2}$. There is generally not a unique instrument for a given POVM $\{A_i\}$. The density matrix averaged over all $i \in \Omega$ is given by

$$\Im_\Omega(\varrho) = \sum_i \Im_i(\varrho).$$

A first kind measurement is defined in [229]. An instrument $\Im_i$ that implements a POVM $\{A_i\}$ is a first kind measurement if

$$\mathrm{Tr}[\Im_\Omega(\varrho)A_i] = \mathrm{Tr}[\varrho A_i], \forall \varrho, i \in \Omega.$$

for which the measurement statistics are identical for both the original state ϱ and the state after measurement $\Im_\Omega(\varrho)$. Note that such a measurement may very well disturb the state on any particular outcome but does not disturb the statistics. There exists an adjoint map of $\Im_\Omega(\varrho)$ that maps bounded operators to bounded operators, denoted $\Im_\Omega^*(B)$ such that the probabilities are maintained $\mathrm{Tr}[\Im_\Omega(\varrho)A_i] = \mathrm{Tr}[\varrho \Im_\Omega^*(A_i)]$. A first kind measurement is also equivalent to the condition that the A_i be fixed points of the adjoint map, i.e.,

$$\mathrm{Tr}[\Im_\Omega^*(A_i)] = A_i, \forall i \in \Omega.$$

Criticisms of GRW

There have been a number of criticisms of the GRW theory. Energy and momentum conservation are violated. There is a problem when one attempts to absolutely localize matter. If one does not absolutely localize matter, there are other problems due to the inherent tails of the matter that must exist. These issues are now discussed in more detail.

Violation of Energy and Momentum Conservation Laws

There is a substantial problem in attempting to localize a pure matter wave function. Consider an electron with an initially pure wave function that has some spread Δx. The energy of (a stationary) electron is related to the spread Δx. If Δx were extremely small, the energy of the electron would be typically higher than, for example, a stationary electron that is widely spread. Hence providing a mechanism that affects the pure state of a matter wave function could easily become problematic in terms of conservation of energy and momentum.

Generally, the energy of a matter wave function changes when multiplied by a Gaussian function as in Equation (4.24). As shown in [230] energy can increase without bound and diverge with time. The authors propose a modification for which the energy increases asymptotically to a finite value rather than an infinite value. However, even with this modification, the model continues to violate energy conservation. Energy (and momentum) conservation is in our view a major problem with GRW that has not been resolved to-date. However, not all see this as a major problem. Violations of energy conservation that might be observed as a function of parameters of GRW has been further considered in [231].

In [232], Pearle proposes that there is a white noise background that can be incorporated into the Hamiltonian of the particles. The energy that is used to increase the particles energy in the process of reduction comes from a decrease in the noise background. If the white noise background is from external particles with mass, then if the system was sufficiently isolated in a vacuum chamber and experiments performed at sufficiently low temperature, then one would expect such noise sources could be reduced or eliminated. If the white noise background is from external radiation, then one could employ a Faraday cage to reduce these effects.

On the other hand, it has been suggested in [233] that there could be a fluctuating noise component in the gravitational metric which cannot be screened by a Faraday cage. More work is needed to confirm the claim of the existence of a physical white noise background that exists to balance energy in GRW collapse. A step in this direction has been made when considering the validation of the continuous version of GRW known as continuous spontaneous localization (CSL), which will be introduced shortly. The CSL parameters required to form a latent image are estimated in [234]. This work indicates that the original parameters of the noise coupling that were expected by GRW were too small, and a larger value is required. In the paper [235], high accuracy measurements of cantilever thermal fluctuations revealed a nonthermal force noise of unknown origin. This excess noise is stated to be compatible with the CSL noise predicted by Adler. With improving experimentation, this issue is expected to be resolved.

No-Tail Energy Conservation Problems

If one were to localize a particle in the classical sense, then the particle should be absolutely localized. If one attempts to localize an electron wave function to a single point, the wave function is fully localized but the energy of the electron is infinite as the momentum uncertainty will broaden to levels that are experimentally known not to occur in the measurement process. McQueen states regarding such localization [236],

> *Collapse in the idealized theory reduces all but one of the coefficients to zero, while the mod-square of the chosen coefficient goes to one. But this collapse function is unphysical. This is due to position/momentum incompatibility.*
>
> Studies in History and Philosophy of Science Part B: Studies in History and Philosophy of Modern Physics, Vol. 49, K. J. McQueen Four tails problems for dynamical collapse theories, P. 10, Copyright 2017 with permission from Elsevier.

McQueen claims that in order to avoid energy increases associated with absolute localization of matter wave functions, GRW employed a Gaussian wave function to create a hit which has infinite tails, but such tails fall-off rapidly:

> *The relationship between energy and momentum then yields drastic post-collapse violations of energy conservation: ones that we know by experiment do not occur. So GRW formulated the collapse function as a Gaussian. The collapse then raises the mod-square of the chosen coefficient - the collapse centre - close to one while reducing the mod-square of all other coefficients close to zero but never actually to zero. GRW carefully chose the probability per unit time for spontaneous collapse and the width of the bell curve of the Gaussian ... so that the energy conservation violations are consistent with known experiments.*

Causality and Hegerfeldt's Theorem

No absolute localization of a positive energy particle can occur without developing tails due to Hegerfeldt's theorem [237]. Hegerfeldt proved that any positive energy particle that is absolutely localized will immediately upon unitary evolution develop instantaneous tails. From Hegerfeldt's theorem, one cannot absolutely localize a positive energy particle such as an electron without violating causality.

Although approximate solutions are known [238], tails are generally a requirement of any positive energy solution. The fastest rate that the tail of a photon that is not absolutely localized can fall has been investigated by [239] and found to be exponential. The exponential falloff of a photon wave function can be of the order $e^{-f(r)}$where $f(r)$ increases with r slower than linearly. Note that Gaussian tails fall faster than exponentially. It was further shown in [240] that the tails of free particles such as electrons that satisfy the Dirac equation fall faster than exponentially but cannot be strictly positive energy electrons and require a superposition of an electron and positron for representation. A similar result appears in [241], which appears to strictly rule out Gaussian tails. GRW would seem to necessitate the production of positrons if it is applicable to electrons.

Tail problems

As discussed, when faced with either unphysical energy when absolute localization is employed, or finite energy when a tail is employed, McQueen states that GRW chose to have tails. Additionally, tails are needed to avoid Hegerfeldt causality violations associated with absolute localization.

The use of a wave function with an infinite tail may eliminate the large energy associated with absolute localization; however, the issue now becomes whether or not the use of wave functions with tails have problems of its own. Consider the argument put forward by Albert and Loewers in [242]. The authors consider a modification of Schrödinger's cat in the following manner. Consider a cat initially in the state $|R\rangle$ that observes an electron's spin in a given basis. If the electron spin is spin-up $|\uparrow\rangle$, unitary evolution is designed to allow the cat to live and evolve to state $|\text{Alive}\rangle$ which will be assumed to be in Position x_1 but if the electron is spin-down $|\downarrow\rangle$, the unitary evolution kills the cat which evolves to state $|\text{Dead}\rangle$ at Position $x_2 \neq x_1$. Now if the overall evolution were truly Schrödinger unitary, then if the electron is started initially in the superposition state $(|\uparrow\rangle + |\downarrow\rangle)/\sqrt{2}$, the overall state $|\psi\rangle$ of cat and electron would evolve to:

$$|\psi\rangle = (|\text{Alive}\rangle \otimes |\uparrow\rangle + |\text{Dead}\rangle \otimes |\downarrow\rangle)/\sqrt{2}.$$

Now suppose GRW is applied to $|\psi\rangle$ which consists of a superposition of a cat at locations x_1 and x_2 by means of choosing a Gaussian that is centered at z_h. Let us assume that z_h is approximately x_1. Then the state above will reduce to

$$|\psi\rangle = (\sqrt{1-a}\,|\text{Alive}\rangle \otimes |\uparrow\rangle + \sqrt{a}\,|\text{Dead}\rangle \otimes |\downarrow\rangle)/\sqrt{2}.$$

However, since measurement in GRW occurs in the preferred basis of position and GRW utilizes a Gaussian distribution which has infinite tails, it means that the cat which is in the state $|\text{Dead}\rangle$ will always have some amount $\sqrt{1-a}$ in which the tail continues to populate the state $|\text{Dead}\rangle$, i.e., while a can be nearly zero, a must remain strictly greater than zero. Albert and Loewer state in [242]

> *Note that while most of GRW (alive)'s amplitude is indeed ... concentrated in the 'alive' region of the state space, it also has non-zero tails which extend into the 'dead' region. And so it follows from the eigenstate-eigenvalue rule that the cat is as a matter of fact not determinately alive (or dead), when GRW (alive) obtains, after all. And so the GRW theory, as we have stated it above, patently fails to solve Schrödinger's paradox.*
>
>

GRW appear to have skirted some (but not all) of the issues of absolute localization in the prior sections by means of introducing infinite tails. However, now they have created another problem for which the cat is always in a superposition for which the measurement problem does not appear to be resolved.

Complete Reversibility

GRW has been shown to be invertible, hence completely reversible. Consider a photon that is in a superposition of two locations with a given phase. The complete absorption of the photon by either detector destroys the photon coherence in the von Neumann theory. One would therefore expect that there should be an element of irreversibility to the theory that resolves the measurement problem. Yet complete reversibility is in principle possible in the GRW POVM given the POVM outcome. This doesn't prove that the theory is incorrect, but invertibility does not appear to be a strong point of GRW theory.

Validating GRW

Localization theories have the potential to address the requirements that have been required of a Category 1 Theory, R1.1-R1.4. In addressing R1.1, the theory predicts specific physical conditions or configurations under which non-Schrödinger evolution occurs versus Schrödinger evolution. GRW theory employs the two parameters λ_h and σ_h which specify the frequency of the GRW stochastic hits as well as the localization that occurs when the wave function is multiplied by a Gaussian of standard deviation σ_h. Hence when there are no hits, GRW predicts unitary evolution and when there are hits, the theory predicts non-Schrödinger evolution. Hence the theory partially meets requirement R1.1 as the parameters λ_h and σ_h are unknown.

In addressing R1.2, experiments can be used to verify that requirement R1.1 is met or violated. Firstly, one can determine what values of parameters λ_h and σ_h can be considered to be eliminated by results of known experiments. In the paper [243], the authors examine a number of experiments in relation to these parameters. The authors conclude that λ_h should be greater than 10^{-17} s^{-1} (termed the conventional value) and assume σ_h is on the order of 10^{-5} cm. Testing this value would require an experiment that can diffract molecules 1000000 times larger than fullerenes. Assuming that the interaction of a particle with photographic film constitutes a bona fide measurement, this would require that λ_h should be at least 10^8 times larger than the conventional value. Testing this value would require an experiment that can diffract molecules 100 times larger than fullerenes. Other possible experiments include cantilevers. Due to the lack of experimental evidence, R1.2 can only be considered to be very partially fulfilled.

Requirement R1.3 is addressed, as GRW provides a nondeterministic prescription for the average change of state of the system, given that a GRW hit occurs. It is specified in [226] that the measurements are in a preferred basis which is that of localized states. And as well, GRW theory does provide a mechanism for the change of the state to a particular state via the nondeterministic Brownian motion. Hence

GRW largely meets requirement R1.3. In terms of R1.4, the mechanics of both GRW and CSL reduce the state to a new state that agrees with the statistics associated with the von Neumann measurement postulate. However, the detailed mechanics of this change predicted either in GRW or in CSL have not been confirmed or seen experimentally, and as such have not yet met R1.4.

Stochastic Differential Equations

GRW theory as originally proposed has discrete time jumps, for which the wave function evolution is discontinuous. Pearle in 1973 suggested the use of stochastic differential equations to account for the reduction process [244]. With the use of a stochastic differential equation, changes do not occur discontinuously in a single jump, but can be made to change in a continuous manner. The heuristic idea of stochastic differential equations is to increase the rate of occurrence of the jumps while simultaneously decreasing the effect that the stochastic process has on the change on the state for any given occurrence. For example, in the case of GRW one might consider increasing the Poisson parameter λ_h so that in the limit, events are expected to occur continuously, while simultaneously decreasing the effect of any given hit. In this manner, a stochastic process can arise that is expected to be continuous in the manner that the state-vector changes, but still leads to the stochastic changes that are in agreement with the Born rule.

Stochastic differential equations often utilize continuous Brownian motion in their description. A stochastic process $\{B_t, t > 0\}$ is called a real Brownian motion if the following hold [245]:

1. The initial value B_0 is a real value (for a real Brownian motion process).
2. The process has independent increments, i.e. for all times $0 \leq t_1 \leq \cdots \leq t_N$ the increments $B_{t_N} - B_{t_{N-1}}, B_{t_{N-1}} - B_{t_{N-2}}, \cdots, B_{t_2} - B_{t_1}$ are independent random variables
3. For all t the increments $B_{t+\sigma^2} - B_t$ are normally distributed random variables $N(0,\sigma)$, i.e. with mean 0 and standard deviation σ.
4. Almost surely, the function $t \rightarrow B_t$ is continuous.

A standard Brownian motion $\{B_t, t > 0\}$ has initial value $B_0 = 0$. Brownian motion can be extended to a vector form of N dimensions. By considering a two-dimensional real Brownian process, a standard complex Brownian motion can be defined in which both real and imaginary components are normally distributed. We will use the convention in [246] for which a complex motion is defined as

$$B_t = \frac{1}{\sqrt{2}}\left(B_t^{(1)} + iB_t^{(2)}\right)$$

where $B_t^{(1,R)}, B_t^{(2,R)}$ are independent real standard Brownian motions. This definition has the advantage that B_t continues to have variance parameter σ^2. The Itô rule for this case is [247] $dB_t' dB_t = dt, dB_t dB_t = 0$.

The development of stochastic differential equations was mathematically formalized by Itô and in a different form by Stratonovich. These two forms are based on different definitions and the use of one form over another has various advantages depending on the situation [185]. The theory of stochastic differential equations developed by Itô was extended in the seminar paper by Hudson and Parthasarathy [248], which also relates to the work of [249] and [250].

We utilize the Itô form which can be written as an equation that has an added stochastic term of the form:

$$dx_t = \mu(x_t, t)dt + \sigma(x_t, t)dB_t, \tag{4.25}$$

where B_t is a complex Brownian stochastic process.

Denoting the change in x_t as Δx over a small interval of time Δt, it is seen that the distribution of Δx is given a normally distributed random variable with mean $\mu(x,t)\Delta t$ and variance $\sigma(x,t)^2\Delta t$. Moreover, as Δx can be written independently of the x, one sees that the process is a Markov process. The term $\mu(x_t, t)$ is referred to as the drift coefficient and $\sigma(x_t, t)$ as the diffusion coefficient. While B_t is a complex Brownian motion process, the stochastic process x_t is often referred to as a diffusion process.

In order to apply the Itô form to the evolution of a wave function $|\psi_t\rangle$ consider associating $|\psi_t\rangle$ with x above and restricting initially the evolution to the linear stochastic differential equation

$$d|\psi_t\rangle = C|\psi_t\rangle dt + A|\psi_t\rangle dB_t \tag{4.26}$$

where C , A are linear operators on $|\psi_t\rangle$. This form can be extended to a summation over multiple A_i, but for now we assume a single A_i. It is shown in [244] that unless further restrictions are put in place, $|\psi_t\rangle$ will generally have a norm that changes in time. It is found that the norm can be written as:

$$\begin{aligned}\langle\psi_t|\psi_t\rangle = \langle\psi_0|\psi_0\rangle + \int_0^t \langle\psi_s|(C + C^\dagger + A^\dagger A)\psi_s\rangle ds + \\ \int_0^t \langle\psi_s|(A + A^\dagger)\psi_s\rangle dB_s.\end{aligned} \tag{4.27}$$

If one further restricts the above so that the second term is zero via

$$C + C^\dagger = -A^\dagger A$$

then $\langle\psi_t|\psi_t\rangle = \langle\psi_0|\psi_0\rangle + \int_0^t \langle\psi_s|(A + A^\dagger)\psi_s\rangle dB_s$. Now if we set the anti-Hermitian part of C to $-\frac{i}{\hbar}H$, Equation (4.26) becomes

$$d|\psi_t\rangle = -\frac{i}{\hbar}H|\psi_t\rangle dt - \frac{1}{2}A^\dagger A|\psi_t\rangle dt + A|\psi_t\rangle dB_t. \tag{4.28}$$

The above equation is a linear equation in the states $|\psi_t\rangle$. In order to further restrict $|\psi_t\rangle$ so that it has constant unity norm, non-linear terms are introduced. Let

$$|\widehat{\psi}_t\rangle = \frac{|\psi_t\rangle}{\| |\psi_t\rangle \|}.$$

It is shown in [244] that this restriction results in the Itô equation

$$d|\widehat{\psi}_t\rangle = -\frac{i}{\hbar}H|\widehat{\psi}_t\rangle dt + \left(-\frac{1}{2}(A^\dagger - \overline{R})|\widehat{\psi}_t\rangle + \frac{1}{2}(A - \overline{R})\overline{R}|\widehat{\psi}_t\rangle\right) dt + (A - \overline{R})dB_t|\widehat{\psi}_t\rangle \tag{4.29}$$

where $\overline{R} \equiv \frac{1}{2}\langle\widehat{\psi}_t|A + A^\dagger|\widehat{\psi}_t\rangle$. Note that once the restriction is made to the normalized form $|\widehat{\psi}_t\rangle$, the resulting Equation (4.29) is now nonlinear in the states $|\widehat{\psi}_t\rangle$. Assuming self-adjoint operators $A = A^\dagger$ the above can be written as

$$d|\widehat{\psi}_t\rangle = -\frac{i}{\hbar}H|\widehat{\psi}_t\rangle dt - \frac{1}{2}(A - \bar{A})^2|\widehat{\psi}_t\rangle dt + (A - \bar{A})dB_t|\widehat{\psi}_t\rangle \tag{4.30}$$

where $\bar{A} \equiv \langle\widehat{\psi}_t|A|\widehat{\psi}_t\rangle$. This can be rewritten in the form

$$d|\widehat{\psi}_t\rangle = -\frac{i}{\hbar}H|\widehat{\psi}_t\rangle dt + \left(\bar{A}A - \frac{1}{2}A^2 - \frac{1}{2}\bar{A}^2\right)|\widehat{\psi}_t\rangle dt + (A - \bar{A})dB_t|\widehat{\psi}_t\rangle. \tag{4.31}$$

This can be further extended to a summation over multiple A_i which results in

$$d|\widehat{\psi}_t\rangle = -\frac{i}{\hbar}H|\widehat{\psi}_t\rangle dt + \sum_i \left(\bar{A}_i A_i - \frac{1}{2}A_i^2 - \frac{1}{2}\bar{A}_i^2\right)|\widehat{\psi}_t\rangle dt + \Sigma_i (A_i - \bar{A}_i)dB_t|\widehat{\psi}_t\rangle. \tag{4.32}$$

State Reduction

A result established by Gisin in [251] is that for a single adjoint operator A, the evolution of $|\widehat{\psi}_t\rangle$ converges (in the sense that the mean square deviation approaches zero) asymptotically to an eigenstate of A. In [226] conditions were found for the case of multiple operators A_i that the evolution of $|\widehat{\psi}_t\rangle$ moves into subspaces that are associated with the operators A_i in the stochastic differential equations. A similar result was established in [252] showing that the dispersion entropy, which is a measure of localization of $|\widehat{\psi}_t\rangle$ to one of several subspaces, generally decreases in time in a manner for which $|\widehat{\psi}_t\rangle$ converges to evolve into one of the subspaces.

A condition on the subspaces that appears to be met in both papers is described.

Let us assume that the operators A_i in Equation (4.32) are Hermitian, and for simplicity assume that A_i can be represented by orthogonal projection operators P_i ($P_i^2 = P_i$, not necessarily rank 1, $P_i P_j = 0$ for all $i \neq j$). Let $\Re(P)$ denote the range space of an operator P. Note that if $x \in \Re(P)$, $P_i x = x$, and P_i is the identity map for all $x \in \Re(P)$.

An initial state $|\hat{\psi}_t\rangle$ is found to evolve to $\Re(P_i)$ for some P_i. However, the particular outcome i occurs stochastically. That is, under any given trial for which the stochastic differential equation is evaluated for all time, the evolution will converge into $\Re(P_i)$ for some particular P_i. This shows that such stochastic differential equations with Brownian motion converge to a single subspace, which is a prediction of standard von Neumann measurement theory. Asymptotic spectral stability has been proven in [253].

Furthermore, the probability of the state asymptotically belonging to one of the subspaces, i.e., $|\hat{\psi}_t\rangle \in \Re(P_i), t \to \infty$, is given by $\langle\hat{\psi}_0|P_i|\hat{\psi}_0\rangle$. Hence the stochastic differential equations with Brownian motion reproduce the quantum prediction of Born's rule which is also a prediction of standard von Neumann measurement theory. Quantum diffusion equations developed by Gisin and Percival were shown to be related to the work of Hudson and Parthasarathy in [254].

Although stochastic differential equations generally converge into eigenstates of observables, such equations often do not conserve energy or momentum. If the conditional expected value of the random variable representing the dynamic energy $H_t \equiv \mathrm{Tr}(\rho(t)H)$ has the martingale property $\mathrm{E}(H_t|\{H_\tau : \tau \leq s\}) = H_s$ for $s \leq t$ then energy will be conserved on average. Consider setting $A = H$ in Equation (4.30)

$$d|\hat{\psi}_t\rangle = -\frac{i}{\hbar}Hd|\hat{\psi}_t\rangle\,dt - \frac{1}{2}(H - \bar{H})^2|\hat{\psi}_t\rangle dt + (H - \bar{H})dB_t|\hat{\psi}_t\rangle. \tag{4.33}$$

Equation (4.33) has the desired martingale property, and stochastic equations of this form are defined in [255] [256] as energy-driven stochastic reduction models. As such, one might consider such a model for a closed system collapse reduction model. However, on-average conservation is not the same as strict energy conservation on every trial and, as will be further discussed in Chapter 5, was at the heart of the Bohr-Kramers-Slater debate. Additionally, it has been argued by [257] that energy-driven models cannot lead to spatially localized states in commonly occurring cases. Whether or not energy is strictly conserved or conserved on-average is an important question that will be further discussed in Chapters 5, 7, and 8.

Master Equations for Nondeterministic Evolution

Master equation approaches have been developed that provide an equation for the density matrix of a reduced system that interacts with a larger system such as the environment [258]. In 1976 both Gorini, Kossakowski, and Sudarshan (GKS) for the case of discrete systems [249] and Lindblad [250] for the case of continuous systems, developed the most general Markovian master equation that is trace-reserving and completely positive. The Lindblad form for the reduced system dynamics has been shown to be equivalent to tracing out the unitary interaction of a system with a Markovian environment. For example, it has been shown in [157] using the Jaynes-Cummings model as the interaction between a system and oscillator bath, that a Lindblad form results for the evolution of the density matrix ρ of the system. A general Lindblad form for the evolution of a reduced density matrix is given as

$$\frac{d\rho}{dt} = -\frac{i}{\hbar}[H,\rho] + \sum_i \left(L_i \rho L_i{}^{\dagger} - \frac{1}{2} L_i{}^{\dagger} L_i \rho - \frac{1}{2} \rho L_i{}^{\dagger} L_i \right) \tag{4.34}$$

where L_i are operators that represent the interaction of the system and environment.

When starting in a pure state, the density matrix can evolve deterministically to a mixed state in a master equation. Although master equations such as the Lindblad form are deterministic equations in the evolution of the density matrix, one might nevertheless consider the use of stochastic (nondeterministic) differential equations to model the individual stochastic outcomes such that, on average, the stochastic states average exactly to the deterministic mixed state predicted in a master equation. In such a case when the individual trials are stochastic, the average system evolution may be described by a nondeterministic master equation.

Gisin and Percival in [259] have shown that the stochastic differential equation

$$\begin{aligned} d|\widehat{\psi}_t\rangle &= -\frac{i}{\hbar} H|\widehat{\psi}_t\rangle dt + \sum_i \left(\bar{L}_i L_i - \frac{1}{2} L_i{}^{\dagger} L_i - \frac{1}{2} \bar{L}_i{}^{\dagger} \bar{L}_i \right) |\widehat{\psi}_t\rangle dt \\ &\quad + \sum_i (L_i - \bar{L}_i) dB_t |\widehat{\psi}_t\rangle \end{aligned} \tag{4.35}$$

has the mean density matrix given by

$$\rho = \overline{|\widehat{\psi}_t\rangle\langle\widehat{\psi}_t|}$$

where ρ satisfies the Lindblad form of Equation (4.34). This shows that one can effectively compute or simulate the Lindblad reduced density matrix of a system that interacts unitarily with an environment via the use of stochastic differential equations. One can also derive a Lindblad form for energy-driven reduction models [255].

Furthermore, note that the equation of the density matrix ρ in Equation (4.34) is linear in ρ. Hence the general evolution of such equations is non-linear in the wave function and linear in the density operator. We will see later in Chapter 7 that such nondeterministic evolution is within a class of time-evolution that does not suffer from problems of signaling.

The fact that stochastic equations of the form of Equation (4.35) have the property of quantum state convergence to one of the operator's range space in agreement with von Neumann theory, and furthermore that the evolution is non-linear in the wave function yet linear in the density operator, gives impetus to consider the possibility that stochastic equations of the form of Equation (4.35) govern the theory of measurement. This possibility will be further examined in Chapter 7. One might also consider as well that the ensemble average in terms of a density matrix corresponds exactly to the Lindblad equation. However, this has positive and negative consequences: it is positive as it appears to correspond to real-world effects in open-system theory. It is negative in the sense, as has been already discussed previously, that such master equations also follow from a unitary deterministic theory that does not resolve the measurement problem as a Category 1 theory. Hence to experimentally distinguish such a Category 1 theory from a purely unitary theory would require isolating or minimizing a system due to decoherence to rule out the possibility that a non-unitary effect is actually a unitary effect in a higher Hilbert space.

Continuous Spontaneous Localization (CSL)

The discrete time localization theory proposed in the original paper [224] was later extended in the paper [226], in which the time of the jumps is decreased to zero via the use of a stochastic differential equation. It has been proven that under certain conditions the stochastic Brownian motion is almost surely continuous. This means that, unlike the original GRW proposal, CSL appears to move in a continuous manner. However, such processes are also generally almost surely non-differentiable [245, p. 3]. In CSL a locally averaged density operator is defined as

$$N(\boldsymbol{x}) = \sum_{s} g(\boldsymbol{y}-\boldsymbol{x})a^{\dagger}(\boldsymbol{y},s)a(\boldsymbol{y},s)d\boldsymbol{y}$$

where $a^{\dagger}(\boldsymbol{y},s)$ is a particle creation operator at point $\boldsymbol{y}$ and spin s and $a(\boldsymbol{y},s)$ the respective destruction operator, $g(\boldsymbol{x})$ is a spherically symmetric positive real function that integrates to unity and is assumed to be a zero-mean Gaussian distribution with standard deviation σ. An advantage of [226] is that reduction preserves the symmetry properties of identical particles, whereas the original proposal did not maintain symmetry. Equation (4.28) can be extended to a summation over operators A_i,

$$d|\psi_t\rangle = -\frac{i}{\hbar}H|\psi_t\rangle dt - \frac{1}{2}\sum_i A_i{}^{\dagger}A_i|\psi_t\rangle dt + \sum_i A_i\,|\psi_t\rangle dB_t.$$

Associating the operators A_i with the operator $N(\boldsymbol{x})$ and the index i with the space point $\boldsymbol{x}$ and converting the summations to integrals in the infinite dimensional case, it is found [244] that the equivalent Itô equation is given by

$$d|\psi_t\rangle = -\frac{i}{\hbar}H|\psi_t\rangle dt + \int N(\boldsymbol{x})|\psi_t\rangle\, dB(\boldsymbol{x})d\boldsymbol{x}\, dt - \frac{1}{2}\int N^2(\boldsymbol{x})\,|\psi_t\rangle d\boldsymbol{x}dt$$

where $\overline{dB(\boldsymbol{x})} = 0, \overline{dB(\boldsymbol{x})dB(\boldsymbol{y})} = \delta(\boldsymbol{x}-\boldsymbol{y})dt.$

Let us suppose that the general form of the solution to the measurement problem to be of the form described by a Schrödinger equation plus an additional term that allows variation from the purely Schrödinger evolution via wave function coupling to a noise term. In such a case, it has been proven in [260, pp. 165-175] that, up to the choice of operator to which the noise is coupled, the CSL theory is a unique form that yields the Born rule as a deduced consequence.

Criticisms of CSL

CSL theory has remedied some problems that the original GRW suffered. CSL preserves the symmetry properties of the particles. However, CSL suffers from many similar problems as the original GRW. The original GRW was parameterized by the two parameters λ_h and σ_h. In CSL there is no reason to consider λ_h as the localization process is continuous, rather only the parameter of the standard deviation of the Gaussian $g(\boldsymbol{x})$ is relevant.

Additionally, GRW is in principle reversible and is also a time-symmetric collapse theory [261]. On the other hand, as we have already discussed, measurement is sufficient but not necessary for loss of coherence within the von Neumann theory. That is, the absorption of a photon by one of many possible detectors is, within the von Neumann theory, generally an irreversible action that destroys the phase coherence of the photon and is not generally time-reversible. Such time-reversible theories are not consistent with this fundamental theoretical tenet. Within von Neumann theory, it is theoretically impossible to reverse a photon that has been positively measured by a specific detector. There have been experiments that have reversed unitary evolution, but this is hardly different than throwing a ball up in a gravitational field and having it come down. Nobody has yet reversed a photon that has been positively measured in the measurement process, and to expect otherwise is due the improper application of inductive logic to a problem that demands a deductive approach, which will be further discussed in Chapter 6.

The remaining criticisms of GRW appear to go over into CSL such as lack of energy and momentum conservation as well as the issues with the tail of the Gaussian.

Mass Threshold Theory

Consider a pure state of a matter wave function that enters an interferometer. For example, consider a molecule that interacts with a beam splitter mechanism that causes the molecule to separate into two paths. If the general state of the molecule was not preserved, it is possible that one could find the molecule dissociated into various atoms in the two arms when measurement occurs that localizes the molecule to one or the other arm. However, this is not the case—such molecules are found to be entirely in one or the other arm when measured.

Interference experiments are being performed with larger and larger systems. In 1999 an experiment showed that the fullerene molecule C_{60} exhibited interference [23]. In 2011 organic molecules of sizes up to 6 nm composed of 430 atoms were shown to exhibit interference [262]. Also demonstrated was the resilience of such molecules to resist decoherence. In 2013, [26] demonstrated high-contrast quantum fringe patterns with molecules composed of 810 atoms consisting of approximately 5000 protons, 5000 neutrons, and 5000 electrons, with a large number of possible internal energy levels. In 2010 [263] [264] and more recently 2016 [265], the possibility of creating superpositions of living organisms has been suggested using quantum optomechanics. In [265] it was suggested that the range of viruses of 10^{-17}grams to bacterium on the order of 10^{-12}grams could be experimentally probed. In [266] a measure has been proposed for quantifying the macroscopic size of a quantum superposition. The authors also review technologies that have enabled larger and larger superpositions to be experimentally probed resulting in a plethora of advances on the macroscopic scale between 1980 and 2013.

With such advancements, one might expect that there is no inherent limitation on the size of a spatial superposition. Although no inherent limitation has yet been seen experimentally, there is already an answer to this question.

Consider a mesoscopic object of mass m that can take two paths through a double-slit of size d as shown in Figure 4.2. Such a device was considered by Feynman in 1957 [267] when the mass m is sufficiently large that the gravitational field can be used to discern which-way information and by Bohr [268] if an object with a sufficiently large charge can be used to discern which-way information via the Coulomb field.

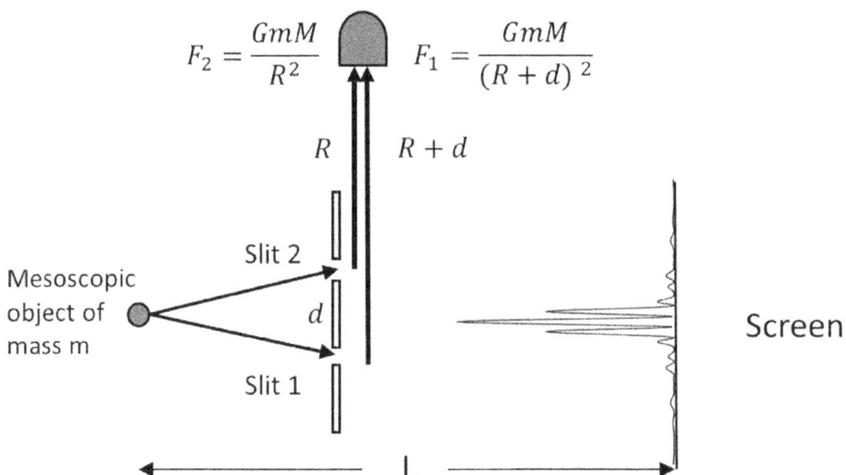

Figure 4.2: Double-slit experiment with gravitational detectors, below the detection threshold, double-slit interference pattern is retained.

A device of mass M is located at a distance R from Slit 2 and at a distance $R+d$ from Slit 1. The force on the device if the object were to pass through Slit 1 is given by $F_1 = GmM/(R+d)^2$ whereas the force on the device if the object were to pass through Slit 2 is given by $F_2 = GmM/R^2$. Suppose that no device is able to make a measurement such that the force can be inferred to an accuracy smaller than $F_2 - F_1$.

In such a case it is possible for the particle to exhibit interference. We will denote this case as *below threshold*, and the interference pattern for the below threshold case is shown in Figure 4.2. Suppose that a device exists that is able to make a measurement such that the force can be inferred to an accuracy smaller than $F_2 - F_1$. In such a case, in-principle which-way information could be obtained and first-order interference is eliminated. Note that it is not required that one actually obtains distinguishing which-way information via an actual measurement—it is only required that one could obtain such which-way information. Such cases will be referred to as *above threshold*. This shows that once the mass of the object m (or possibly mass in-combination with other parameters) is such that the gravitational field of the object can be determined in principle by a remote detector that is capable of resolving whether or not the object passed through Slit 1 versus Slit 2, then interference through the interferometer will be lost.

An analysis has been given by Feynman [267] for the case where the particle is detected based on its mass. If a measurement occurs in a cubic volume l^3 in a time given by l/c, the dimensionless gravitational potential (which is normalized by dividing the non-dimensionless gravitational potential by c^2) will remain uncertain to [269, p. 1192] [270]

$$\begin{aligned}\Delta g &= \sqrt{\frac{\hbar G}{c^3 l^2}} \\ &= \frac{l_p}{l}.\end{aligned} \tag{4.36}$$

where $l_p = \sqrt{\hbar G/c^3}$ is the Planck length. In order to detect the mass of an object, the gravitational potential $g(x)$ must be greater than Δg within the detection volume l^3. The dimensionless gravitational potential produced by an object of mass m at a distance x from the object is given by

$$g(x) = \frac{Gm}{xc^2}. \tag{4.37}$$

Assuming the object is placed outside the detection volume, if the minimum gravitational potential due to the object that is within the detection volume is greater than uncertainty in the gravitational field Δg, then the object should be detectable. If the object is placed on the boundary of the detection volume, then the minimum gravitational potential due to the object would be (for a two-dimensional detector) at a distance l. In such a case, the gravitational potential is always greater than $g(l)$ in the detection region. As an uncertainty in the gravitational potential is equivalent to an uncertainty in the mass, one finds in detectable regions

$$\Delta m \geq m_p \tag{4.38}$$

where $m_p = \sqrt{\hbar c/G}$ is the Planck mass. Hence it appears from this argument that the detectable mass in time l/c should be greater than the Planck mass m_p.

Note that as one places the object farther from the detection volume, the minimum gravitational potential will decrease, hence the maximum (with regard to distance from the detection volume) over the minimum (with regard to within the detection volume) occurs when the object is on the boundary. The Planck mass m_p is approximately 0.02 mg, which is quite large in comparison with the mass of most typical atomic particles (e.g.10^{19} times the proton mass). It would appear that if a mass has an uncertainty that is less than the Planck mass, that such a mass would have a gravitational potential that necessarily falls below Δg somewhere within the detection volume and may not be detectable. It appears a reasonable hypothesis that a single Planck mass is on the threshold of being externally reliably detectable by its gravitational field. Hence it is no surprise that atomic particles are observed to often exist in coherent first-order superpositions of position. Also note that as the reduced Compton wavelength of a mass m is $\lambda_c^{(r)} = \hbar/(mc)$, then $m = m_p$, it is seen that $\lambda_c^{(r)} = l_p$ or approximately 1.6e-35 m which is quite small compared with typical atomic particles. Hence a Planck mass has a rather large mass in comparison with the typical mass of atomic particles and if such a mass were to be confined to a Planck length, would have an inordinately high density.

Although this experiment has not yet been accomplished, related experiments to test for collapse once the mass m is sufficiently large have been proposed by [271] [272] [273] as well as others to determine if quantum superposition features of the gravitational field can be revealed. Although experimental methods are improving, the testing of gravitational superpositions has not yet been achieved due to the relatively weak coupling in comparison to electromagnetic coupling. Many have been touting the benefits of quantum optomechanics for the purpose of probing quantum macroscopic superpositions [274] [275].

A black hole is predicted to form if a mass to be confined to a sphere of radius smaller than the Schwarzschild radius given by $r_s = 2GM/c^2$. The Schwarzschild radius of a Planck mass is $r_s = 2l_p$. A limitation of a particle mass m is (approximately) that its size is greater than $\lambda_c^{(r)}$. Hence it appears that a Planck mass that is confined to its reduced Compton radius is not only at the limit of external gravitational detectability but also at or nearby the threshold of black hole formation.

A bound similar to Equation (4.38) has also been derived based on a derivation of the uncertainty of the measurement of an electric field that was developed by Bohr and Rosenfeld [276]. Bohr and Rosenfeld utilized a test charge within the detection volume that would be subject to a change in momentum depending on the electric field. Their relationship was extended to the gravitational field by Peres and Rosen [277] which yields the result for the gravitational potential

$$\Delta g \geq \left(\frac{l_p}{l}\right)^2.$$

Furthermore, [278] propose a lower bound given by,

$$\Delta g \geq \left(\frac{l_p}{l}\right)^{\frac{2}{3}}.$$

Note that as $\Delta g \leq 1$, the largest lower bound is given by [278] followed by the bound proposed by Feynman and lastly the bound of Peres and Rosen based on the work of Bohr and Rosenfeld. It is shown in [278] that using their improved bound, the minimal mass for which the measurement time is less than the expected decoherence time is also the Planck mass. Hence by many accounts, the Planck mass is a reasonable hypothesis for establishing the classical-quantum gravitational boundary.

Baym and Ozawa [268] have analyzed a modern gravity detector that is sensitive to changes in the gravitational field. The detector consists of two mirrors that form a laser interferometer that are separated by a distance $S + \eta(t)$. When the gravitational potential $\phi(x,t)$ is zero $\eta(t) = 0$, but when $\phi(x,t)$ changes, there is a force exerted that changes the distance between the mirrors. It is assumed that $\eta(0) = d\eta/dt|_{t=0} = 0$ and at the time of the measurement t_f the potential has changed to a value $\phi(x,t_f)$ and is constant during the measurement. Under these conditions, the equation for $\eta(t)$ is given by [268] :

$$\eta(t) = -\frac{d^2}{dt^2}\phi(x_0,t)S\frac{1-\cos wt}{w^2} \tag{4.39}$$

where x_0 is the midpoint between the mirrors. The accuracy of the device that is required in order to discriminate whether the object passed through Slit 2 versus Slit 1 is given by

$$\Delta\frac{d^2}{dt^2}\phi(x_0,t) = 2Gm\left(\frac{1}{R^3} - \frac{1}{(R+d)^3}\right)$$

$$\sim \frac{Gmd}{R^4}.$$

Using the relation $(1 - \cos wt) \leq (wt)^2/2$, it is seen that

$$\Delta\eta(t) \leq \frac{GmdST^2}{R^4}.$$

It is assumed $R > cT$ where T is the time-of-flight of the object so that the detector is sufficiently distant from the apparatus that there is no back-reaction from the measurement and hence,

$$\frac{GmdST^2}{R^4} < \frac{GmdS}{R^2c^2}.$$

Assuming $S \ll R$

$$\Delta\eta(t) < \frac{Gmd}{Rc^2}.$$

Squaring both sides and multiplying the top and bottom of the left-hand-side by $\hbar$ gives

$$\frac{\hbar\Delta\eta^2(t)}{\hbar G^2} < \frac{m^2}{c^4}\left(\frac{d}{R}\right)^2$$

and substituting the Planck length $l_P = \sqrt{G\hbar/c^3}$ results in

$$\frac{\hbar c\Delta\eta^2(t)}{{l_p}^2 G} < m^2\left(\frac{d}{R}\right)^2.$$

There are several compelling reasons presented in [279] for believing that $\Delta\eta(t)$ must exceed the Planck length. Assuming this is the case,

$$\frac{\hbar c}{G} < m^2\left(\frac{d}{R}\right)^2.$$

Substituting the Planck mass $m_p = \sqrt{\hbar c/G}$

$$m > m_p\left(\frac{R}{d}\right). \tag{4.40}$$

We define $m_{th} \equiv m_p R/d$. The Planck mass $m_p = \sqrt{\hbar c/G}$ is approximately 2 x 10^{-5}g and for R/d on the order $1/m_p$ gives a mass on the order of 1 gram. The fringe separation seen on a screen that is a distance L from the particle is given by [268]

$$d_f \approx \frac{L}{d}\frac{\hbar}{mv}.$$

Assuming the velocity $v \approx L/T$ and $T \approx R/c$ gives

$$d_f \approx \frac{\hbar R}{cdm}.$$

Substituting into Equation (4.40) gives

$$d_f \approx \frac{\hbar R}{cdm} < \frac{\hbar}{cm_p}.$$

Noting that for $\hbar/(cm_p) = l_P$ it is found that an object with mass $m > m_{th}$ that satisfies Equation (4.40) also has an associated fringe separation that satisfies (subject to the approximations made in [268])

$$d_f < l_P.$$

This indicates that whereupon the mass is great enough that it can be detected by an external gravimeter, the fringe separation that could possibly be used to detect first order coherence of the object becomes less than the Planck length and is undetectable.

Gravitationally Induced Collapse

Penrose [280] has put forward a mass threshold theory of gravitationally induced collapse termed objective reduction. Penrose considers an object of mass m that unitarily evolves to a superposition of two locations f_1 and f_2 that are separated by a given distance $|f_2 - f_1| = d_p$. Although such a superposition is stationary in standard quantum mechanics, Penrose proposes that such a superposition is ultimately not stationary due to the existence of two different gravitational fields that define two different space-time structures. Penrose proposes that the existence of two different space-time structures is untenable and hence such a superposition will decay to one of the two locations f_1 or f_2. Penrose proposes that the characteristic decay time is given by

$$\tau \cong \frac{\hbar}{E_G} \tag{4.41}$$

where E_G is the gravitational self-energy of the difference of the two locations, which is equivalent to the energy required to separate two instances of the object initially at a common point and then moved to the two locations f_1 and f_2. The same time scale has been proposed by Diosi [281] and it is claimed by Diosi [282] that his and Penrose's scheme yield similar results, although Diosi's theory is based on a dynamical master equation whereas Penrose assumes a single characteristic time for collapse. Now, if such a collapse mechanism exists and is verified by loss of first-order coherence to occur on a time scale on the order of $\hbar/E_G$, *and* this is different than predicted by the proposed non-local unitary quantum theory, then the necessity of adding an additional collapse mechanism would be warranted. On the other hand, if the unitarily predicted quantum state itself, when including a gravitational field that becomes externally detectable, leads to a similar finding of loss of first-order coherence, then more work would be needed to ascertain whether or not a collapse mechanism is necessary to explain the phenomenon observed in an experiment. That is, a unitary explanation might not be ruled out, depending on the experiment and the parameters involved.

A review of other theories of collapse, including gravitational collapse is given in Bassi et al. [283]. Various models of gravitationally induced decoherence are contrasted in Bera et al. [284]. Diosi [285] expects that the natural width a_0 of the wave packet of any macroscopic object is

$$a_0 \approx \frac{\hbar^2}{Gm^3}.$$

This estimate also coincides with work of Károlyházy in [286].

Note that as has been previously discussed, such collapse has also been proposed to be associated with the formation of consciousness in the Penrose-Hameroff orchestrated reduction model. In this model, the objective reduction in gravitational collapse is orchestrated or mediated within microtubules which also results in consciousness.

Quantum Mechanics with Fields

There must be inherent limitations of a spatial superposition, with all other parameters held constant. This is because there exist devices such as gravimeters that can measure differences in gravitation from mass variation. There is no doubt that a particle will eventually lose first-order coherence once $m > m_{th}$.

Typically, when one computes a first-order interference pattern in quantum mechanics, one evolves the local particle state that moves through the interferometer and ignores the gravitational field. Baym and Ozawa [268] re-examined Bohr's analysis of two-slit interference with charged particles and the implications for quantization of the electromagnetic field as well as the gravitational counterpart to such an experiment using massive particles. The Baym-Ozawa work does not explain how to compute the first-order interference pattern from standard quantum mechanics—only that the interference fringes are not in principle measurable when $m > m_{th}$. However, when the gravitational field can be detected and the information can discern which-path information, this indicates that standard local non-relativistic quantum mechanics requires modification. Feynman stated at a 1957 conference at Chapel Hill [267]

> *One can conclude that either gravity must be quantized because a logical difficulty would arise if one did the experiment with a mass of order* $10e-4$ *grams, or else that quantum mechanics fails with masses as big as 10e−5 grams.*

We show here why this is and examine the possibilities in modifying the standard local non-relativistic Schrödinger unitary evolution to a generalized non-local non-relativistic Schrödinger unitary evolution. Let us consider a point (x_s, t) that is located a vertical distance d_s from the peak of the pattern on the screen in Figure 4.2. The state at x_s after the two slits can be examined by considering the phase coherence of the vector $|\psi\rangle = [\psi_1(x_s, t)\ \psi_2(x_s, t)]^T$, where ψ_1 represents the wave function that travels only through Slit 1 and ψ_2 through Slit 2, T denotes the transpose operation. In the von Neumann theory, these vectors actually exist and change in time. However, from the Baym-Ozawa result they are not observable when $m > m_{th}$ due to limitations of measurement devices which are imposed largely by uncertainty relationships. Now $|\psi\rangle = [\psi_1(x_s,t)\ \psi_2(x_s,t)]^T$ will generally have some phase relationship, for example

$$|\psi\rangle = [\sqrt{a}\, e^{-iwt}\ \sqrt{1-a} e^{-iwt+\theta}]^T.$$

Computing the corresponding density matrix gives

$$\rho(x_s) = \begin{pmatrix} a & a(1-a)e^{-\theta} \\ a(1-a)e^{\theta} & 1-a \end{pmatrix}.$$

When an object's mass exceeds m_{th}, the Baym-Ozawa result shows that the fringe separation is less than the Planck length, and there can be entire oscillations of fringes within a single Planck length. Since the fringes are oscillating, the density matrix is a function of x, due to alternating constructive interference and destructive interference. In such a case, if a physical detector integrates over a size greater than l_P, then two-slit interference would be expected to be lost. What we are left with as seen by a detector of size greater than l_P is essentially

$$\rho(x) = \begin{pmatrix} 1/2 & 0 \\ 0 & 1/2 \end{pmatrix}. \tag{4.42}$$

Remarkably a detector in this case can only see an effective mixed state of the system. Yet Schrödinger's equation tells us that a pure state should exist for an object in isolation, unless the object cannot be considered to be in isolation. And this provides an explanation as to precisely what has happened. When an object's mass exceeds m_{th}, the object cannot be treated as if it is in isolation. The gravitational field of the object now allows in-principle discrimination of the which-way path information that the object took.

The state of the problem (without assuming collapse) is suddenly enlarged when the mass exceeds m_{th}. Whereas the object could previously be considered to be an isolated system described by a pure state that exhibits first-order interference when $m < m_{th}$, the object can no longer be considered an isolated system when $m > m_{th}$; additional degrees of freedom now need to be appended to the system for its full description. In this case, the detectable gravitational field needs to be considered.

Let us denote the state of the gravitational field at the detector in Figure 4.2 as $|\phi_1\rangle$ if the object travels through Slit 1 and as $|\phi_2\rangle$ if the object travels through Slit 2. Then the overall state of the particle plus its gravitational field is $|\psi\rangle = (|1\rangle|\phi_1\rangle + |2\rangle|\phi_2\rangle)/\sqrt{2}$ and one finds that the unitary prediction of the state of the particle is entangled with its gravitational field. In such a case when one traces the system from the entangled state it is found that $\rho(x)$ is mixed in agreement with Equation (4.42).

This shows that an object with mass in quantum mechanics cannot simply be considered to be in the Hilbert space spanned by the states $|1\rangle$ and $|2\rangle$ but rather the particle's gravitational potential must be included so that the state of the particle cannot be considered as a local state for the purpose of making calculations. The Hilbert space must be enlarged to include both state of the local particle and its gravitational potentials to the extent that they are externally detectable,

$$|\psi\rangle = (|1\rangle|\phi_1\rangle + |2\rangle|\phi_2\rangle)/\sqrt{2}.$$

Note that if the gravitational potentials are not detectable, then the extent that they are detectable is the same for both $|1\rangle$ and $|2\rangle$ and the state $|\phi_1\rangle = |\phi_2\rangle$. The state $|\phi_1\rangle$ is not the gravitation potential but represents a state of the gravitational potential to be externally detectable and differentiated from $|2\rangle$. When $m > m_{th}$, the states must be included and so long as the uncertainty in the gravitational field produced when the particle is locally in state $|1\rangle$ versus state $|2\rangle$ is less than the measurable uncertainty equivalent to a single Planck mass, then the states are orthogonal, i.e., $|\phi_1\rangle \perp |\phi_2\rangle$. In such a case, any measurements of the local particle states that might reveal first-order local coherence would be impossible as the local particle state is now a mixed state and the interference pattern of a double slit experiment shown in Figure 4.3 results.

In terms of the precise unitary physics one might expect that each term $|1\rangle|\phi_1\rangle$ and $|2\rangle|\phi_2\rangle$ evolve according to Schrödinger's equation. The fields $|\phi_1\rangle$ and $|\phi_2\rangle$ are quantum objects in the sense that they could appear or disappear depending on what measurements are made on the particles $|1\rangle$ and $|2\rangle$. However, so far, there has not been a need to quantize the fields, that is, they are continuous and individual gravitons are not needed to establish this argument yet. Interestingly, when Feynman in 1957 stated the requirement that the gravitational field would need to be quantized, Rosenfeld replied [267],

> *I do not see that you can conclude from your argument that you must quantize the gravitational field. Because in this example at any rate, the quantum distinction here has been produced by other forces than gravitational forces.*

The interaction of quantized and non-quantized systems has been further considered by Peres [287]. Others such as in [288] [289] have considered methods of providing a quantum state representation that includes both matter and gravitational states. Often such quantum models that incorporate gravity are being developed to understand processes such as black-hole formation and the formation of event horizons. Further work along these directions is needed to determine unitary predicted evolution which is required to discern unitary predictions from measurement collapse models.

As shown in Chapter 3, under unitary evolution first order coherence is simply replaced by higher order coherence or entanglement. An issue addressed in Chapter 3 is how to experimentally discriminate entanglement from a product non-entangled state that is predicted under measurement. The problem at hand is somewhat similar in that first-order coherence of the object has now been unitarily replaced by object-gravitational field entanglement coherence.

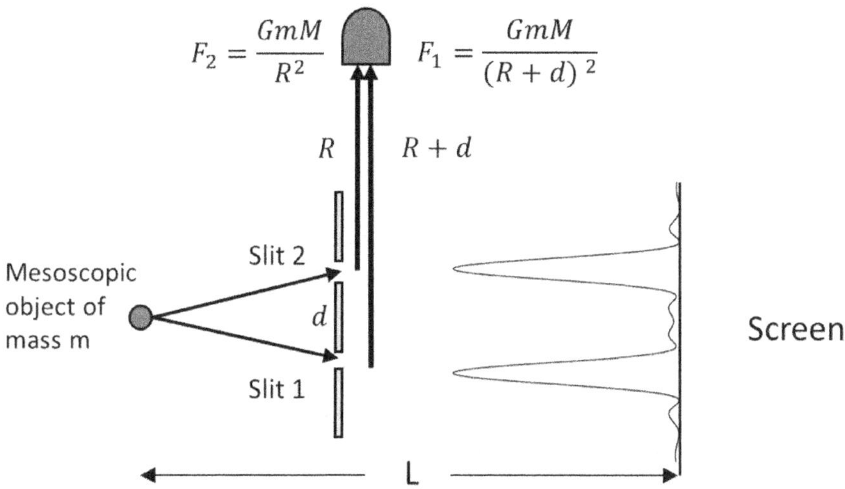

Figure 4.3: Double-slit experiment with gravitational detectors, above detection threshold allowing which-way information, double-slit interference pattern eliminated.

A bona fide measurement requires that the state of the particle and field be either $|1\rangle|\phi_1\rangle$ or $|2\rangle|\phi_2\rangle$, i.e., particle-field are in a product state. That is, the particle travels through only one slit and the field exists as if the particle took the respective path. On the other hand, the unitarily predicted solution for this problem is non-local, and similar to the unitary analysis of detection in Chapter 3 that predicts an entangled state of particle-path and gravitational field. Experiments that investigate the loss of first-order coherence of the particle and not the loss of entanglement required in collapse may not be sufficient to resolve these issues. It is possible that experiments show the eventual loss of first-order coherence, which is predicted under *both* measurement and our proposed non-local extension of unitary quantum mechanics. If one can experimentally quantify entanglement due to gravitational superpositions such as proposed in [273] [290], then there still remains the possibility that measurement cannot be distinguished from our proposed non-local extension of unitary quantum mechanics if the loss of entanglement due to a physical collapse is precisely concurrent with the theoretical external detectability of the difference in the gravitational fields of the two objects. On the other hand, if the loss of entanglement due to a physical collapse occurs below the threshold for which exists theoretical external detectability, then such an experimental endeavor could prove to be successful.

Experiments on superpositions in gravity are nonetheless important, but there is little doubt that first order coherence must eventually be lost as m increases. Whether entanglement is also lost for $m > m_{th}$ is unknown. That is, whether or not the final state is $|\psi\rangle = (|1\rangle|\phi_1\rangle + |2\rangle|\phi_2\rangle)/\sqrt{2}$ versus the state being identically either $|1\rangle|\phi_1\rangle$ or $|2\rangle|\phi_2\rangle$ in the case of measurement, does not appear to be a simple matter to discern.

If it is not plausible to directly discern entanglement from a product state when gravitational fields are involved, a manner in which at least indirect evidence could be obtained that could provide support to a detailed theory is that the theory provides a specific collapse model that is different than predicted under unitary evolution. For example, if the lifetime of a superposition that is proposed under Penrose's collapse proposal is verified and is different than that under unitary evolution, or a theory predicts loss of first-order coherence before unitary evolution makes such a prediction, then such results would have to be taken into account as providing indirect evidence in support of that theory.

Related issues concerning entanglement in terms of gravity as regards to such measurement theories have been examined in [291] for which they claim that gravitational decoherence models given by [292] cannot be correct. If it were to be proven to be impossible to discern entanglement from product state measurement, and no other parameters of a proposed theory are distinguishable from unitary theory, this would render such a proposed theory a Category 2 theory, which will be discussed later.

Charge Threshold Theory

Consider that, instead of using a large mass for which the gravitational field can be detected in a manner that discerns which-way information in Figure 4.2 a charged object with charge $Q = Ze$ is used (it is remarked in [268] that this experiment was developed by Bohr but unpublished). Instead of measuring the gravitational force in Figure 4.2, the Coulomb force is measured. If Z is large enough, the Coulomb force can be detected in a manner that the accuracy is sufficient to discriminate which slit the charged object took. At that point, first-order interference must be lost. A detailed analysis of this experiment was also conducted in the paper [268]. The authors utilized much of the work of Bohr and Rosenfeld for which they proposed a specific apparatus composed of positive charge Q_B and negative charge $-Q_B$ separated by a distance. The negative charge is tethered to the box, while the positive charge is displaced at equilibrium by an electric field. The positive charge can move further in response to an external electric field. The motion can be detected depending on the parameters for which detailed analysis utilizing uncertainty relationships was completed by Bohr and Rosenfeld. It is found that the critical threshold of charge Z_{th} is given by

$$Z_{th} \cong \frac{1}{\sqrt{\alpha}} \frac{cT}{d}.$$

where α is the fine structure constant. Typically, it is found for most typical parameters of d and T, that $Z \gg 1$ which is why single electrons, protons and most molecules exhibit wave properties. However, once the charge is sufficiently large, such first order wave interference will be lost. Upon introducing a measure of distinguishability $\mathcal{D}$ that quantifies how well the Coulomb field of the two slits can be distinguished, it was found that

$$\mathcal{D} = \operatorname{erf}\left(\frac{Z}{2\sqrt{2}Z_{th}}\right),$$

where erf represents the error function. From the distinguishability, it is seen that the threshold $Z = Z_{th}$ is not a hard threshold, that is, although distinguishability decreases as Z decreases, the distinguishability does not reduce to zero except at $Z = 0$.

Now, in the case of the detection of the gravitational field, it was found that first-order interference was lost once the gravitational field could be discerned. Furthermore, the mechanism of first-order interference loss was due to the fringe separation distance falling below the Planck length. According to standard quantum mechanics, there must also be a loss of first-order interference for the case when the object is a Coulomb charge above threshold $Z > Z_{th}$. What then is the mechanism for first-order interference in the case when the object consists of charge?

In [268] it is argued that when the charge changes direction when passing through the slits, there will be radiation emitted by the charge which will disturb the charge's phase. This disturbance will reduce the visibility $\mathcal{V}$ of the interference pattern. A calculation is performed in [268] of the expected visibility $\mathcal{V}$ which indicates that such an explanation is consistent with the loss of first-order interference. Furthermore, the loss of first-order interference is shown to be consistent with the simultaneous increase in $\mathcal{D}$ found above, and for which it is shown $\mathcal{V} + \mathcal{D} \leq 1$ for which a similar general condition on the tradeoff regarding visibility and which-way information had been derived by [293].

In this case of Coulomb charge, like the previous gravitational case, the loss of first-order coherence comes at the expense of the production of a detectable Coulomb charge. In terms of the measurement problem, the unitary prediction of entanglement between slit path and Coulomb field would need to be experimentally distinguished from the measurement prediction of a product state of slit path and Coulomb field. Like the prior case of gravitational field, this does not appear to be a simple matter. On the other hand, experimentally demonstrating the loss of first-order interference at the expense of the production of the detectable Coulomb field would certainly be of interest, but such a demonstration (without further alteration) would not contribute to the resolution of the measurement problem. That is, there is no doubt that there will be loss of first-order coherence upon exceeding some threshold, but it is possible that such a demonstration would not distinguish entanglement from product-state, which is key to resolving the measurement problem. The dependence of the measurement problem on such minutiae, underscores why a thorough understanding of the issues involved is important.

Impossibility of Detecting Coherence

A logical possibility to be considered in regards to the resolution of the measurement problem is that there exists some physical limitation that prevents meeting the requirements of a Category 1 theory. For example, consider the previous mass threshold and charge threshold theories which indicate that double-slit interference is lost once the mass (or charge) becomes large enough that which-way information can be discerned from external measurements of the gravitational (or charge) field. If it were additionally found (this is not known as of yet) that it is impossible to discern the unitarily predicted entangled state from the measurement case of product states, then such theories would be rendered a Category 2 theory.

Note that one might argue that it is impossible to detect coherence because it requires reversibility of a macroscopic object. However, we have taken pains to develop the Chapter 3 UMDT in a manner that does not require the original interaction between photon and particle to be reversed. Hence claims that reversibility is required in order to conduct a UMDT appear to be falsified by the Chapter 3 UMDT.

In a paper by Hartle and Gell-Mann, they add an additional requirement of decoherence to the consistent history theory. However, we have already seen that decoherence is simply an orthogonalization methodology that by itself, is not sufficient to resolve the measurement problem. They state in [294]:

> *When the past is permanent, we may still lose the ability to retrodict the probabilities of alternatives in the past through the impermanence or inaccuracy of present records but not from the failure of those past alternatives to decohere in the face of the projections that describe information we acquire as we advance into the future. Yet we know that such continued decoherence of the past is not guaranteed in general by quantum mechanics. Adjoining future alternatives to a set of histories is a fine graining of that set and in general a fine graining of a decoherent set of histories may no longer decohere. Verifying the continued decoherence of all the past alternatives as we fine-grain our set of histories to deal with the future would in general require significant computation. We would have to check that the branches corresponding to every alternative past that might have happened continue to be orthogonal in the presence of their newly adjoined sets of projections. Yet we adjoin sets of projections onto ranges of quasiclassical operators without making this calculation, secure in the faith that previous alternatives will continue to decohere despite this fine graining. It is this assumption of continued decoherence of the past that permits the focus for future predictions on the one branch corresponding to our particular history and the discarding of all others. In other words, we pointed out above, it is the permanence of the past that permits the "reduction of the state vector."*

That is, the authors make an assumption of continued decoherence, in order to focus future predictions. However, such an assumption may or may not prove to be a sufficient condition for measurement and this issue is at the heart of a proper resolution of the measurement problem. For example, consider a spin particle that enters a Stern-Gerlach apparatus and splits into two paths. Suppose an experiment is performed in which the particle is allowed to split through a Stern-Gerlach apparatus and continues to propagate in a superposition. In such a case, there is continued decoherence of the past. If one concludes that continued decoherence of the past is sufficient for measurement, one is in for a surprise. It is possible within quantum mechanics to perform a different experiment and reverse the splitting and recombine the two paths in a manner that maintains coherence. Therefore, in terms of the Chapter 3 development, there is no measurement that has occurred due to the initial splitting.

One option is to demand or simply assume that it is impossible for reasons of thermodynamics or otherwise to physically reverse decoherence at some level. Omnès, who has written substantially regarding the measurement problem [295], has more recently examined the possibility of necessary randomness in the environment in tandem with decoherence theory [296]. Omnès considers the use of predecoherence of waves that randomly effect the growth of other waves that carry entanglement. The resultant of the wave interactions is allowed to generate random fluctuations in a manner that is consistent with the Born probability rules. However, Omnès admits that the randomness that would be required in the environment is still unexplained.

In cases such as in mass threshold theory, it does appear that first-order decoherence will be lost forever once a particle of substantial mass passes through the two slits in Figure 4.3. In cases other than mass or charge threshold theories, for which the underlying physics is in principle unitarily reversible, no physical rationale has been rigorously established as to why there should be measurement.

Although it a logical possibility, there is no experimental evidence to date nor theoretical rationale for expecting that unitary evolution *below* force thresholds (gravitational, etc.) should not be reversible. At its very core, the channels of reversibility are present in unitary evolution below threshold which can in principle be accessed to reverse any given operation. It would seem that a much more intelligent approach would be not to look toward unitary reversible processes in order to find irreversible processes, but rather into the existence of physical process that are indeed irreversible. These two rather different approaches will be further discussed in Chapter 6.

Philosophers and the Measurement Problem

Philosophers are well-versed in the general theory of philosophical classification. Philosophers have developed methods of classifying ideas and theories into categories, for example, "Absolutism," "Anti-realism," "Anti-reductionism." Philosophers might consider a given theory regarding the quantum measurement problem and classify its ontological and epistemological properties. Such classification is certainly important as regards the advancement of philosophy. However, it seems doubtful that application of such concepts will lead to significant advancement toward the resolution of the measurement problem.

Maudlin in [297] defines three measurement problems: 1) the problem of outcomes, 2) the problem of statistics, and 3) the problem of effects. The problem of outcomes is stated to be due to the inconsistency of the following three statements:

> 1. *The wave function of a system is complete, i.e., the wave-function specifies (directly or indirectly) all of the physical properties of a system.*
>
> 2. *The wave function always evolves in accord with a linear dynamical equation (e.g., the Schrödinger equation).*
>
> 3. *Measurements of, e.g., the spin of an electron always (or at least usually) have determinate outcomes, i.e., at the end of the measurement the measuring device is either in a state which indicates spin up (and not down) or spin down (and not up).*
>
> Topoi, Three measurement problems, Vol. 14, 1995, p. 7, T. Maudlin © Kluwer Academic Publishers 1995. With permission of Springer.

The problem of statistics is related to the problem whereby one starts with a common initial wave function but for which the final state is not necessarily unique and that occurs according to Born statistics. The problem of effect is related to the projective effect that a measurement has on a particle and the predictive power such outcomes provide for follow-up measurements.

One might ask if successful resolution to the measurement problem would be the proposal of any theory that does not suffer from these three problems. For example, suppose that one adds additional non-local forces and hidden initial state variables beyond the wave function such as in Bohm's theory, so that the problem of outcomes, statistics, and effects, all appear to be addressed. Then one could argue that Bohm's theory is a valid resolution to the measurement problem, based solely on such a definition. For example, Maudlin in [154] states regarding Bohm's theory,

> *Bohm's theory solves the measurement problem completely and without remainder.*

Additionally, GRW is acknowledged in [297] as a theory that survives these three problems. But there are many resolutions to the measurement problem based on such a definition.

More recently, Maudlin in [298, p. 281] appears to agree that a stumbling block of the orthodox interpretation is the inability to specify the conditions under which measurements occur. While we agree it is a stumbling block for the orthodox interpretation, it is also related to the requirements that we have set forth to be theoretically addressed and experimentally confirmed by any (Category 1 or 2) proposed solution to the physical measurement problem.

While it does appear that Bohm's theory solves the philosophers' measurement problem completely and without remainder, Bohm's theory has a long way to go before it can be considered to have resolved the physical measurement problem that we have defined. The physical definition of the quantum measurement problem goes beyond the definition in [297] because Schrödinger's prediction of an entangled state is in principle experimentally distinguishable *within the current formalism* from the case of measurement. Because of this problem, we propose that a fully scientifically acceptable solution must provide a compelling theoretical argument from well established principles such as conservation of energy and momentum as well as validating experiments that address the requirements we have laid out for Category 1 or Category 2 type theories.

Philosophers are often guided by "Occam's Razor." This is essentially that the theory with the least assumptions and the simplest answer should be selected. In the philosophical literature, it is claimed in [299] that many-worlds theory is superior to Bohm's theory because many-worlds theory satisfies ontological issues in philosophy better than Bohm's theory. And in [300] the question is addressed as to which quantum mechanical interpretation is the best ontology. Certainly, from the philosophers' definition of the measurement problem, these are reasonable issues to examine. On the other hand, the scientific community should be able to converge toward a solution from our physical definition of the problem, without the need to rely on such philosophical issues. The physical measurement problem is not a problem of interpretation but demands that the current theory be scientifically investigated and augmented, whether or not the solution is of Category 1 or Category 2, and irrespective of categorizations.

The investigation into the measurement problem we argue in Chapter 6 should be conducted via a careful deductive scientific investigation. In a deductive investigation, it is very possible that exceptions, rather than what follows from inductive theory, can emerge as the solution. Many (but not all) philosophers are currently of the opinion that the measurement problem is nothing beyond an interpretational issue of explaining von Neumann's theory. This is incorrect: there is no doubt that the measurement problem is not an interpretational issue from the scientific perspective: the current theory will demand significant augmentation in the case that a successful Category 1 theory emerges and some augmentation in the case that a successful Category 2 theory emerges.

Proof by Pejorative

Let us examine the types of statements that do not belong in a scientific investigation of the measurement problem. Consider the following statements made by S. Goldstein in [301]:

> *"Many physicists pay lip service to the Copenhagen interpretation, and in particular to the notion that quantum mechanics is about*

observation or results of measurement. But hardly anybody truly believes this anymore—and it is hard for me to believe anyone really ever did."

Consider the statement, "hardly anybody truly believes this anymore." This is neither a mathematical statement nor a scientific statement. In order to show this in a scientific manner, one would expect at a minimum a poll of physicists that are queried in a scientific manner, which does not appear in the paper. This would beg the question: why would Bohr, Born (who won the Nobel prize for the statistical interpretation), von Neumann, London, Bauer, Wigner, Heisenberg, Penrose, Renninger, Stapp, and many others ever have believed this? From [301] Goldstein states:

"However, the Bohr-Einstein debate has already been resolved, and in favor of Einstein: What Einstein desired and Bohr deemed impossible—an observer-free formulation of quantum mechanics, in which the process of measurement can be analyzed in terms of more fundamental concepts—does, in fact, exist. Moreover, there are many such formulations, the most promising of which belong to three basic categories or approaches: decoherent histories, spontaneous localization, and pilot-wave theories."

If Einstein won the debate with Bohr, then apparently somebody forgot to tell him. Interestingly Einstein was alive when Bohm's pilot-wave theory was proposed, and in a letter in 1952 Born claimed [302]

Have you noticed that Bohm believes (as de Broglie did, 25 years ago) that he is able to interpret the quantum theory in deterministic terms? That way seems too cheap to me.

The pejoratives "Hardly Anybody," "Difficult to take Seriously," "Obviously," "Clearly," "Of Course," etc., are inappropriate when one should be well aware there are scientists that have been working hard to resolve this problem with a contrary viewpoint. No matter how it is worded, dogma is not a scientific methodology.

Summary

The measurement problem exists because there is a contradiction between the unitary predictions of Schrödinger's equation and the prediction of measurement. The results of Chapter 3 indicate that the problem can be reduced to two-qubit entanglement and can be further investigated both theoretically and experimentally. This leads to the establishment of the incompleteness of Quantum Mechanics in its present form and the need to augment the current theory. Two classes of potential solutions consisting of Category 1 and 2 type theories were proposed as well as reasonable requirements that should be met in order to verify a given theory. Generally, for a Category 1 theory, the conditions under measurement can be determined and experimentally verified. In a Category 2 theory one or more of the requirements of Category 1 are prohibited to be achieved by virtue of the conditions under which a measurement occurs. The main point of this book that is that the measurement problem is a real problem with either a Category 1 or 2 outcome. This is a fact; it is neither conjecture nor hypothesis.

We have discussed why prior proposed solutions are not satisfactory in this day and age of quantum information. This is because entangled states are now routinely experimentally discriminated from product states. We have also shown that necessary and sufficient conditions on measurement are important but may not necessarily resolve the measurement problem.

The method of external orthogonalization has been examined and which applies to numerous proposals that aim at resolving the measurement problem. It is found that unitary external orthogonalization is not a satisfactory solution.

Various popular interpretations have been discussed. It is seen that interpretations are put forward to provide some semblance of understanding to the Copenhagen interpretation but do not provide any significant physics that is required to resolve the measurement problem. We believe that while such interpretations might be useful under certain conditions, the main task of physics is to propose physical measurement theories that can be theoretically and experimentally investigated.

Numerous properties that are oft-associated with measurement were examined. For each property, the necessity and/or sufficiency regarding resolution of the problem was also discussed.

Physical measurement theories that put forward a physics-based resolution to the measurement problem have been discussed. Consciousness based theory, GRW, stochastic differential equations and CSL, as well as threshold theories such as mass threshold and charge threshold have been critically examined.

In terms of a consciousness-based theory explaining measurement, in order to meet the Category 1 requirements, it would be desirable to have a physical theory that explains the conditions under which consciousness occurs. However, there is currently no physical theory as to what actually physically constitutes consciousness. Fundamental issues regarding consciousness and free will were proposed by Aristotle ca 350 BCE and very little progress has been made since then.

GRW and stochastic differential theories have been examined and found that none of these theories can be considered at this time to address the requirements that we have specified in order to be considered as an established theory.

Mass threshold theories and charge threshold theories proposed by Penrose, Diosi, and others have been examined. It is our contention that first order interference will be lost once sufficient mass or charge exists in a superposition. We expect this to occur both in conventional quantum theory as well as in collapse models. In such a case, experimental demonstration that first-order interference is lost cannot be considered to be a resolution of the measurement problem. The task at hand is to discriminate the unitary prediction from the measurement prediction. This is possible if the predictions of a given measurement collapse model are different than the unitary prediction.

Lastly, Category 2 theories have been discussed which have the characteristic that the proposed theory itself imposes that unitary theory cannot be distinguished from measurement. For example, the proposal that unitary decoherence is fundamental and can never be reversed. This is a logical possibility and cannot be dismissed at this time. Category 2 theories would scientifically gain credibility if it was found that there is some inherent physical limitation. The experimental investigation of mesoscopic superpositions is only in its infancy, and experiments in quantum information have only pointed to the continued ability to control systems in larger-and-larger systems without any fundamental Category 2 roadblocks. There is no experimental work to date that the authors are aware of that are proof of the necessity of a Category 2 theory.

It is our contention that none of the theories that we have considered have yet met the requirements that have been specified for either a Category 1 or Category 2 theory.

Appendix 4.A

Rosenfeld considers a partitioning of measuring device states first into a shell of all states of approximate energy E that would correspond to stationary states of the measurement device, for which there are S such states. Any macroscopic change of the device corresponds to quantum transitions within the same shell. Furthermore, a shell is composed of a set of cells. To each cell corresponds multiple states whereby the k th cell $\{\Omega_{ki}\}$ has s_k states where the Ω_{ki} are quantum states of approximate energy E of the device. After a short interaction time τ the device is considered to be non-interacting with the system but in a temporary superposition of device states Ψ_r. The device states Ψ_r are represented by a partitioning of $\{\Omega_{ki}\}$ into non-overlapping sets of cells $\{\Omega_{ki}^{(r)}\}$. After time τ, the device continues to evolve unitarily toward equilibrium in a relaxation time T which is assumed unknown and variable. Rosenfeld assumes that a device that is in in state Ψ_r evolves unitarily from within the cells $\{\Omega_{ki}^{(r)}\}$ toward a final equilibrium state that consists of only a particular single cell e_r consisting of s_{e_r} states $\{\Omega_{e_r i}^{(r)}, i = 1 \ldots, s_{e_r}\}$ for which the device populates all states within the e_r cell such that s_{e_r} is on the same order of magnitude as S. Rosenfeld considers the correlation coefficient

$$C_{k,rr'} = \sum_r \langle \Omega_{ki} | \Psi_r(t) \rangle \langle \Omega_{ki} | \Psi_{r'}(t) \rangle^*$$

for which the summand is related to the product of the device being in cell Ω_{ki} having evolved from $\Psi_r(t)$ and $\Psi_{r'}(t)$. Rosenfeld defines

$$\mathfrak{W} = \lim_{T\to\infty} \int_0^T \frac{1}{T} dt\,,$$

whereby an average is taken over the time it takes for the device to reach equilibrium after the initial interaction occurs that results in the state. It is claimed via ergodicity that $\mathfrak{W}(C_{k,rr'})$ will approach zero for $r \neq r'$ and $\mathfrak{W}(C_{k,rr'}) = s_k/s$ when $r = r'$, for which $s_k = s_{e_r} \approx S$.

Exercises

4.1 Show that the von Neumann entropy of a density matrix remains the same for all unitarily evolving wave functions.

4.2 Suppose that a system is initially in a given state $|\varphi_1\rangle$ and that after a given a measurement occurs, the final state of the system is $|\varphi_2\rangle$:

a. Show that one can always find a unitary operation that maps $|\varphi_1\rangle$ to $|\varphi_2\rangle$.

b. What are the deficiencies of such a methodology when proposed as a solution to the measurement problem?

4.3 Consider a system that is in a superposition of two states. Consider a swap model of unitary evolution such that an environmental qubit state is allowed to be any function of the degree of system superposition a:

a. Show that such a model can be used to construct a unitary model of measurement.

b. What are the deficiencies of such a model in terms of being an accepted scientific theory?

4.4 Consider applying Rosenfeld's methodology to each device of Figure 3.4 Assume the overall initial state of the two devices is pure and the system is initially in a pure state. When the device is in the r th pointer position, the device is found to occupy of a cell that corresponds to multiple states denoted $\{\Omega_{ki}^{(r)}\}$ in Appendix 4.A.

a. Show that if the partial density matrices of the devices are in completely mixed states as found as in Equations (4.8) and (4.9) when the overall state of the devices is a pure state, that the entanglement, as measured by the entropy of entanglement, is non-zero.

b. Conclude that the entanglement is independent of the number of degrees of freedom specified by the multiple device states $\{\Omega_{ki}^{(r)}\}$ and the unitarily predicted entanglement is not eliminated or reduced by Rosenfeld's methodology.

4.5 Show that $\mathcal{P}(|\psi_1\rangle \otimes |\psi_2\rangle = \mathcal{P}(|\psi_1\rangle) \otimes \mathcal{P}(|\psi_2\rangle)$, where $\mathcal{P}(|\psi\rangle) \equiv |\psi\rangle\langle\psi|$.

4.6 Consider three spin measurements each oriented in either the x, y, or z direction at times t_0, t_1, t_2 on a spin ½ particle for which it is found that either the spin is up ($|\uparrow\rangle$) or down ($|\downarrow\rangle$) at each time. Denote $P_{x,|\uparrow\rangle}$, $P_{x,|\downarrow\rangle}$, $P_{y,|\uparrow\rangle}$, $P_{y,|\downarrow\rangle}$, $P_{z,|\uparrow\rangle}$, $P_{z,|\downarrow\rangle}$, as the projectors of the respective x, y, and z orientations that represent the respective spin up and down measurements. Using these spin projections, exhibit a consistent history family for which not all of their component projectors associated at the same time commute. Assume the Hamiltonian is zero.

4.7 Using three spin measurements as defined in Exercise 4.6, find an example of a family of three events for which all of the events are mutually orthogonal i.e. $Y_{\alpha_i}(k)Y_{\alpha_j}(k) = 0, \forall i \neq j$, but the family is not a consistent history family. Assume the Hamiltonian is zero.

4.8 Show that if a history set is composed in a manner that all pairwise histories are mutually orthogonal in the first or last elements, that the history is consistent.

4.9 Show why the partial trace operation is a counter-example to the statement that all positive linear operators are completely positive.

4.10 Consider the Great Smoky Dragon problem by Wheeler and the delayed choice experiment performed by Shih, Scully and Zeilinger (e.g., for discussions of both, see [303]). Consider the issues of causality, nonlocality, nondeterminism, irreversibility, loss of coherence, and the determination of whether or not a given phenomenon is unitary versus a measurement phenomenon:

a. Explain how the Smoky Dragon experiment affects our understanding of each of these issues.

b. Which issues do you expect the experiment to have the most significant bearing and which issues do you expect to have the least bearing.

4.11 Consider the argument made by Hughes called the ignorance interpretation [196]. Hughes argues that during unitary interaction the composite system under the ignorance interpretation will lead to a mixed state that is incompatible with the pure entangled state:

a. Suppose that among the possible outcomes, a single outcome does occur. Is the composite state in such a case a mixed state or pure state?

b. Is Hughes argument a single trial argument or an ensemble argument?

c. How can the argument be modified that would show an experimental difference between the cases of measurement versus unitary evolution on any trial?

4.12 Utilizing results from Kwiat et al. [210], consider substituting a measurement device for the object that is placed in the object's path:

a. Show the statistics of the final detectors in the case of IFM are the same as would correspond to the measurement device not detecting the photon and the object being placed but not absorbed in the apparatus.

b. Now instead of a measurement device that blocks the path of the photon, consider a unitary interaction between the photon with a device. Assume that the response time of the device is sufficiently short that dephasing cannot be distinguished from measurement during the latency time. Show that the statistics at the final detectors are indistinguishable between the case when a measurement device was placed and a unitary interaction occurred, for the case when IFM occurs.

c. Conclude that this experiment does not distinguish between whether or not unitary interaction occurs or bona fide measurement, and the Zeno effect would appear to occur in either situation.

4.13 Show that GRW as a POVM is a first kind measurement [48].

4.14 Consider the paper [200] in which a modified Bell experiment is conducted. A nanostructure that consists of a metal film perforated with a periodic array of subwavelength holes is placed in the path of each of the down converted photons. The nanostructure converts the photon into many surface plasmon waves which involve a macroscopic number of propagating electrons. The plasmon waves continue to tunnel through the holes in the metal until a photon is re-radiated at the other side. The object is to determine if entanglement survives this process. Note that if conversion into a macroscopic number of plasmon waves constitutes a measurement which measures a photon to a particular polarization, then entanglement would be eliminated. However, the authors found that this did not occur, and that entanglement survived the conversion to a macroscopic number of electrons and back. The authors state:

> *Furthermore, a simple estimate shows that SPs are very macroscopic, in the sense that they involve some* 10^{10} *electrons. Our experiment proves that this macroscopic nature does not impede the quantum behaviour of SPs, because they can act as intermediates in transmitting entangled photons to yield the expected fourth-order.*
>
>

Conclude from this experiment that the conversion from light to a macroscopic number of matter particles is not a sufficient condition for measurement.

4.15 Examine the experiment in the paper [203] in 1990 and consider the statement:

> *The number of scattered photons per ion per pulse was therefore about 72, more than enough to cause the collapse of the wave function. We emphasize that it is the number of scattered photons*

which is important, not the number that can be detected by the apparatus.

a. Based on more recent experiments such as reported in 2002 in [200], has 72 photons been rigorously established as a sufficient condition for wave function collapse? (note that in [211] the issue of measurement versus unitary evolution in regards to the Zeno effect has been elaborated further, whereby it is recognized that unitary disruption could also have accounted for the results of the 1990 paper.)

b. If there are 10^{10} electrons that are shown to evolve unitarily in a given situation such as in [200], does this imply that 10^{25} electrons will evolve unitarily in the given situation in [200]? Does this imply that 100 electrons will evolve unitarily in all situations?

CHAPTER 5

Historical Perspective

Each epoch dreams the one to follow.

Jules Michelet 1839

Introduction

Measure for Measure

Measurement is among the most fundamental concepts, tracing back to the question of what exists and what is knowable about the world. Every era searches for the hidden assumptions of the world around us. Underlying the state of the measurement problem a chorus is heard behind the many conflicting voices, beginning with the Greeks of the 5th and 6th centuries BCE and through to the scientific revolutions of the 16th and 17th centuries and coming together as the 20th century witnessed the formulations and interpretations of quantum physics. Democritus and Aristotle, Newton and Maxwell, Einstein and Bohr, Heisenberg and Schrödinger, these among many others are all here today in the minds of working scientists, in a sense still protecting their ideas and saving them from oblivion as they continue to be debated and interpreted just as when they were first being developed and refined. There has been an urgent need in the first decades of the 21st century to finally come to grips with the quantum measurement problem in a way that conceptually and quantitatively describes the many attributes required to understand our physical existence. Scientists significant in the history of these ideas are shown in Figure 5.1. The history of their ideas is an essential part of our journey.

Dividing the World

Physics is at its core a science of measurement, which requires clearly dividing the

Figure 5.1: Scientists Significant in the History of the Measurement Problem.
AIP Emilio Segrè Visual Archives: James Clerk Maxwell; Max Planck; Niels Bohr (Photograph by A. B. Lagrelius and Westphal, W.F. Meggers Gallery of Nobel Laureates Collection); Erwin Schrödinger (Brittle Books Collection); John Bell (Photograph by Kurt Gottfried)

world into an observer, an observed system, and a method of interacting with the system. The ability to successfully characterize a system in this way went hand-in-hand with our understanding of the world around us. What properties can an observer exploit to reliably quantify any aspect of a system of interest? The specifics of what this entails rests on the full range of details of our physical theory of nature. From the beginning, a central system of interest of course was our fundamental platform for observation, the Earth itself. An early example is the Greek mathematician Eratosthenes' (276-195 BCE) measurement of the circumference of the Earth [304, p. 75] based on his observation of the sun's angle of elevation. His physical theory

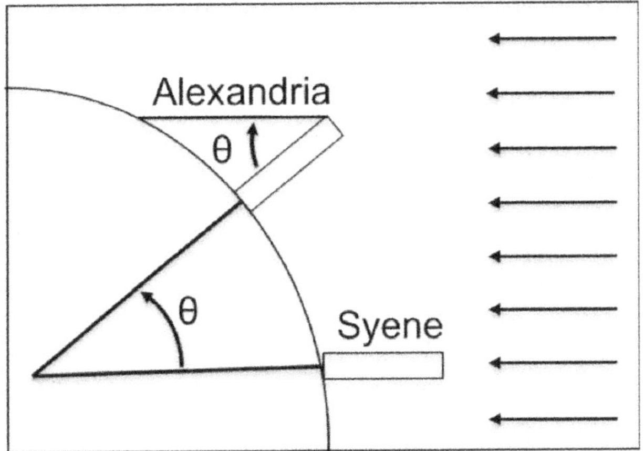

Figure 5.2: Eratosthenes' Measurement of Earth's Circumference.

consisted of the assertion that the sun's rays are parallel since the sun is located at a great distance and the knowledge that the rays fall vertically at noon on the summer solstice at Syene, a city located a known distance southeast of Alexandria, Figure 5.2. At the same date and time, Eratosthenes could observe in Alexandria that sunlight fell at an angle of approximately 1/50 of a circle (≈7.2 degrees) from the vertical, allowing him to find the Earth's circumference using straightforward geometry. This method can remarkably give results to within 1% accuracy, showing the power of even the simplest underlying theory of light-matter interaction if applied with insight and imagination to the act of measurement.

This is the key to the methodology of physics, as we understand it today: an astute combination of empirical measurement and conceptual argument is necessary to deduce the state of the world around us. As discussed in detail in Chapter 6, neither an empirical tabulation of facts nor conceptual argument alone is sufficient to carry out the deductive construction of a consistent picture of the physical universe. Both the empirical and the conceptual are necessary and how to combine and balance these human activities has been argued for centuries as the modern view of science gradually emerged. One of the earliest recorded conceptual arguments for the form of the universe was the Greek philosopher and mathematician Archytas' (428-327 BCE) thought experiment to deduce the infinity of space. Archytas argued that if he extends his staff, the end of the staff will designate a point beyond which the staff can again be extended. This process can then be repeated without limit, implying the infinity of space [305, p. 125], Figure 5.3 [306]. However, the argument was not accepted by Plato (429-347 BCE) or Aristotle (348-322 BCE). Nonetheless, Archytas' argument had great influence and was adapted by the Epicureans and the Stoics. Variations of it are found in versions by John Locke (1632-1704) and Isaac Newton (1642-1727). The application of empirical measurement to this fundamental question would have to wait until the 20th century when both observational methods and theory were sufficient to address cosmology. Modern physics would eventually answer that space could be finite without having an edge, as presupposed by Archytas' argument. However, this was a precedent in the role of thought experiments for the deduction of physics.

Figure 5.3: Archtyas extends his staff and peers under the edge of the firmament discovering the hidden workings of the universe and its cosmic machinery. The end of the staff will designate a point beyond which the staff can again be extended. This process can then be repeated without limit, implying the infinity of space.

Meanwhile, the development of measurement would gradually be extended to all parts of science. Measurement was of interest for both understanding our world and for the utilitarian purpose of exploiting the observed properties of the world. As civilizations have developed since antiquity, it had been essential to establish increasingly accurate and standardized weights and measures for the range of physical quantities needed for architecture, exploration, agriculture, manufacturing, commerce and other aspects of societies. The earliest attempts used arbitrary standards centered on practical measures and were not based on fundamental understanding. This gradually became more sophisticated, as requirements were more exacting and advances in the sciences led to the elements resembling modern metrology utilizing rational systems of units. It was found that the relation of measurement to the underlying physical properties of the world can be subtle and necessarily requires fundamental research in the physical sciences. The discoveries in thermodynamics, electromagnetism, atomic physics and other fundamental areas could be applied to standards of measurement. Measurement can generally be defined as the correlation of numbers with entities [307], usually in terms of ratios of quantities using standards. The result is an extraction of information or a reduction of uncertainty about a quantity using the methods of statistics and sampling. Each particular type of quantity (e.g., length, time, mass, temperature, etc.) has its associated specialized methods developed for the accurate determination of the measurement. This concept of quantity, although foreshadowed by the ancient Greeks, can be traced back to Newton who had defined in the *Principia Mathematica* [308]:

> *Quantity of matter is a measure of matter that arises from its density and bulk conjunctly.*

Quantity of matter is today referred to as mass and played a crucial role in his theory of gravity. Newton concluded in Volume 3 of the *Principia*:

> *Gravity acts on all bodies universally and is proportional to the quantity of matter in each.*

Newton estimated the constant of proportionality, called G, but more than 100 years elapsed before Henry Cavendish (1731-1810) measured it in the laboratory using his innovative method of the torsion balance, in which two small test masses are suspended from a fine wire and gently twisted [309, p. 136]. Gravity is so much weaker compared to other forces and cannot be shielded from outside gravitational influences that measurements of "Big G," as it is often called in metrology circles today, have still not improved much beyond Cavendish's result. The torsion balance was another inspired method of coupling to the world to carry out an act of measurement. Cavendish claimed he was "weighing the world" since the mass of the earth can be obtained once G is known. Attempts by our civilizations to weigh or "measure the world" in all its variety eventually extended from the deepest parts of the cosmos to the most basic constituents of matter making up our world and including that of the investigators themselves.

The observations of our universe now span some 15 billion years in time and roughly 34 orders of magnitude in size ($10^{-12} - 10^{22}$ cm). The investigations that we are privy to necessarily are those which take place from an unexceptional planet orbiting an ordinary star located at the edge of a typical galaxy. The biology of these same investigators traces back to protein molecules containing the same hydrogen produced in the embryonic processes of the universe. The terrestrial world of the investigators may be modest compared to the grand structures observed in the universe at large but it contains the one known instance of consciousness with its capacity to be self-aware: we are aware that we are measuring the world. This becomes significant when considering the peculiar problem of measuring our own thinking activity, as the division of the world into an observer and observed system can then become ambiguous. This issue of identifying where the border between the observer and the observed lies will be seen to play a key role in our story of measurement. This is especially evident in the contrast between classical and quantum physics.

Back-Action from the World

The classical mechanics of Newton was initially found to be consistent with the fundamental characteristic of a successful measurement: from the results of an experiment, it could derive conclusions regarding the results of subsequent experiments. This is due to the property that the states of the observed system before and after the measurement can be regarded as identical. This continued to be the case when classical mechanics was extended to Albert Einstein's (1879-1955) view of space-time coordination of events from his theory of relativity in 1905 to account for the finite speed of light and its invariance for inertial observers. Although the implications were not immediately realized, all of this had already changed in 1900 with the discovery by Max Planck (1858-1947) that the existence of a new elementary quantity called the *quantum of action* designated by h, now called Planck's constant, was needed to understand atomic radiation (*action* is a quantity with dimensions of energy-time). The introduction of the fundamental quantum of action by Planck was the beginning of the end for determinism. It took several decades for this to be generally accepted by the scientific community but eventually led in 1925 to a new corresponding theory, *quantum mechanics*. For Niels Bohr (1885-1962), who was one of the major figures in quantum measurement as well as atomic and nuclear physics, the essence of quantum theory could be expressed in terms of a *quantum postulate*, which [310]

> *attributes to any atomic process an essential discontinuity, or rather individuality, completely foreign to the classical theories and symbolized by Planck's quantum of action.*
>
>

This postulate implies a renunciation of the causal space-time coordination of atomic processes. Classically, quantifiable properties exist as an objective reality and measurement determines the value of these pre-existing quantities with in principle arbitrary accuracy. For quantum mechanics, with intrinsically non-deterministic physics, the situation is very different since the measurement generally causes a change in the state of the system (except for special cases such as *non-demolition measurements* discussed further in Chapter 7), and this change is in principle impossible to predict. Bohr wrote in 1935 [311]:

> *The procedure of measurement has an essential influence on the conditions on which the very definition of the physical quantities in question rests.*
>
>

This property of quantum measurement reflects the *back-action* occurring between measurement device and measured system. In *Measure for Measure*, Shakespeare (1564-1616) took the meaning of his title from Mathew 7:2, "with what measure ye mete, it shall be measured to you again," the title then being aptly suggestive of the back-action between measurement device and system characteristic of quantum measurement.

Individual quantum systems need not have well-defined states but instead may be in correlated arrangements with other quantum systems, "entanglement", where only the entire superposition carries information about the whole. The term *entanglement* was introduced by Erwin Schrödinger (1887-1961) in 1935 [156], though aspects of it had been seen by Einstein and Bohr over the previous twenty years [111]. As will be discussed, Bohr and Einstein in many ways pioneered the deepest inquiries into quantum measurement during the first half of the 20th century. Entanglement can occur between different particles or between two or more properties of the same particle. In particular, according to Schrödinger's equation, a measurement device interacting with a quantum system may become entangled with its observable properties before measurement occurs. Other signatures of the quantum had also been identified in the quest to understand the essential differences between classical and quantum. As demonstrated in Chapter 4, the quantum measurement problem is related to the unitary prediction of *entanglement,* as given by the mantra:

To the extent there is entanglement, there is no measurement.

In 1927, Werner Heisenberg (1901-1976), working at Bohr's institute in Copenhagen, obtained his indeterminacy or "uncertainty" relations when he realized that quantum mechanics must allow for approximate values of position and momentum with trajectories that are not sharply defined [7]. In relation to this, Heisenberg and Bohr also examined the possibility of "disturbances" occurring during quantum measurements, such as Heisenberg's famous example of the γ-ray microscope [9], in which the act of observing would impart uncontrollable momentum kicks to any quantum objects showing the futility of determining its trajectory. However, Bohr later considered this description as misleading and instead emphasized the role of the interaction between measurement device and system, because the interaction required during measurement is uncontrollable. This necessarily affects the state of the closed system, and the results have an apparent random or non-deterministic appearance.

Complementarity in the World

Theories of physics are built up using elementary physical quantities such as space, time, energy, and momentum. Measurement devices utilize the properties of these physical quantities to extract information about the observed system. For example, rigid rulers extended in space might be used to determine the position of an object. However, the same object could be allowed to collide with a movable piece of the measurement device with known mass in order to determine its momentum. An important difference between classical and quantum mechanics is that in classical physics, various measurement tools may be used in combination to supplement one another in the measurement of a state of the same system. Combining the disparate information from separate measurements is what allows a deterministic causal description of a system to be possible in classical physics. However, in quantum physics, the results from different types of measurements cannot always be combined and a quantum physical phenomenon found by observing the same system with different experimental arrangements can be mutually exclusive. The separation between the observer and observed system in realizing measurements of events may be arranged in many ways corresponding to different conditions of observations and type of apparatus determining the particular aspect of the phenomenon we wish to observe. At the quantum level, the deterministic chain of events in classical physics instead becomes lines of similar possibilities, each weighed by an amplitude for probability of occurrence and closed by the irreversible click of a detector [312, p. 124].

In his 1927 lecture at the Volta Congress in Como, Italy [310], Bohr called this logical exclusion of phenomena from different experimental arrangements *complementarity*, a view that generalizes the concept of wave-particle duality:

> *The very nature of the quantum theory thus forces us to regard space-time coordination and the claim of causality, the union of which characterized the classical theories, as complementary but exclusive features of the description, symbolizing the idealization of observation and definition respectively.*
>
>

Bohr's initial attempts at justifying the complementarity picture used the uncertainty relations and arguments in terms of disturbances to the system occurring during measurement; see Plotnitsky [313] for a detailed account. However, depending on the particulars of the experiment, complementarity has more generally appeared to be enforced by a variety of other signatures for quantum behavior that have since been identified within quantum mechanics: entanglement, uncertainty, measurement disturbance, which-way information, visibility, and distinguishability, among others. The role of these various signatures in quantum interference experiments have turned out to be neither completely logically independent nor logical consequences of one another, recently leading to arguments in the literature [314] [315] and showing that there are numerous ways of dissecting complementarity. Bohr's concept of complementarity continues to play a role today with studies of interference involving the most current experimental measurement techniques [316].

The logic of quantum physics can also appear similar to that of introspection. The concept of observation occurs in both physics and psychology with a long history of repercussions. When we describe our thinking activity, it can also be made part of the content of consciousness when attempting self-perception, Figure 5.4. In this case, the

Figure 5.4: Ernst Mach's 1886 self-portrait presenting a "view from the left eye," illustrates issues occurring when attempting self-perception; with the right eye closed, the accompanying cut is presented to the left eye in a frame formed by the ridge of the eyebrow, nose and mustache.

subject and object can even appear to alternate in an infinite regression [317, p. 302]. Bohr's favorite presentation of this situation was Paul Møller's humorous *Tale of a Danish Student*, in which the student becomes dismayed when contemplating his own consciousness [318, p. 48]:

> *"And then I come to think of my thinking about it; again, I think that I think of my thinking about it, and divide myself into an infinitely retreating succession of egos observing each other. I don't know which ego is the real one to stop at, for as soon as I stop at any one of them, it is another ego again that stops at it. My head gets all in a whirl with dizziness, as if I were peering down a bottomless chasm, and the end of my thinking is a horrible headache."*

Bohr compared this arbitrariness of the separation between object and subject in psychology with the arbitrariness in the distinction between system and measurement device in quantum physics [319, p. 24]:

> *The unavoidable influencing by introspection of all psychical experience, that is characterized by the feeling of volition, shows a striking similarity to the conditions responsible for the failure of*

causality in the analysis of atomic phenomena.

As will be discussed further in the section *Free will, Consciousness, and Soul*, for much of Bohr's life he considered not only the analogies of the complementarity of psychological perception with observation of quantum processes but also seriously considered the possible role of consciousness and volition in relation to the quantum of action.

Bohr's complementarity principle was an important historic step in quantum measurement, but only one of several lines of approach to understand measurement after the initial development of quantum mechanics. Within this initial overview of the historical perspective of the measurement problem, complementarity was the natural starting point for this "measure for measure" aspect of observation. More details of complementarity will be discussed in later sections; however. a myriad of other historical details not yet mentioned underlie the array of physical concepts required to arrive at even this level of understanding in the early 1930s. As will be seen, subsequent interest in the deeper aspects of measurement waned with the necessity of understanding nuclear physics, the onset of World War II and the subsequent focus on a range of new developments and applications of quantum theory, from quantum electrodynamics to condensed matter. However, beginning in the 1950s there was a low-level resurgence of interest both in *interpretations* of quantum theory and in new approaches to the question of quantum measurement. These efforts continued to grow in the following decades, particularly after John Bell's (1928-1990) theorems in the 1960s allowing tests to discriminate the underlying workings of quantum theory, of Einstein's unholy choice between *hidden variables* and *spooky action at a distance*. However, the subject of measurement was also transformed in the last twenty years due to advances in experimental techniques, which make it possible to perform repeated measurements, feedback, and partial measurements on single continuously measured quantum systems. An idea of this surge in interest in measurement in the last decades, both theoretical and experimental, can be seen from a sampling of the cast of characters involved [320, p. 339]. Nevertheless, the quantum measurement problem has remained unresolved.

Developments in measurement followed many intertwined paths of concepts and experiments extending from the ancient Greeks and the later development of science as we view it today, up to the present state of the quantum measurement problem. These details of our theory of nature developed over centuries and have involved many diverse concepts. The history of the quantum measurement problem can be viewed broadly as *The Rise and Fall of Classicality* followed by *The Rise of the Measurement Problem*. As will be seen in the following, these developments were fought on several different fronts across overlapping historical times, each battle a dichotomy of conflicting and complementary notions described in sections of this chapter, such as: *Space-Time versus Quanta, Deductive versus Inductive Thought, Atomism versus Continuum, Clockwork versus Free Will,* and *Irreversibility versus Demon.*

The Rise of Classicality

The Clockwork Universe

Centuries before the achievements of Newton and his successors allowed deterministic celestial motions to be predicted and understood in the sense of modern science, the clockwork analogy appeared in works on astronomy, from the late medieval period of Robert Grosseteste (1175-1253) and Nicole Oresme (1325-82) up to Nicholas Copernicus' (1473-1543) epic work of the heliocentric theory, *De revolutionibus.* In the 17th century, the view of a clockwork universe was articulated by René Descartes (1596-1650), Johannes Kepler (1571-1630), and Robert Boyle (1627-1691) in which all evolutions must be cyclic but with orbits initially arranged by a designer. Johannes Kepler (1571-1630) wrote in 1605 that his aim is (Letter to Herwart von Hohenburg, February 10, 1605)

> *to show that the heavenly machine is not a kind of divine, live being, but a kind of clockwork...insofar as nearly all the manifold motions are caused by a most simple, magnetic and material force...*

In Kepler's case, the orbits took the form of a nested arrangement of Platonic solids. The most popular world model in the 17th and 18th centuries was the clockwork universe proposed by Boyle, comparing it to a real clock in Strasbourg Cathedral. The presence of a creator's role in the universe's clockwork continued with Newton, whose General Scholium of the *Principia Mathematica* comments on the designer of this system whose operations it was Newton's honor to give a detailed mathematical description. Newton states in the *Principia*:

> *This most elegant system of the sun, planets, and comets could only proceed from the counsel and dominion of an intelligent and powerful being.*

The prevalent view of an all-embracing deterministic explanation of celestial and terrestrial phenomena based on Newton's achievements eventually emerged, however Newton's writings raised additional issues not associated with what later became accepted as *Newtonian Mechanics.* Though mechanical cause and effect eventually became the business of Newtonian science, the issue of a first cause still remained for Newton himself. In Query 28 of his *Opticks*, Newton maintained [321]:

> *...the main Business of Natural Philosophy is to argue from Phenomena without feigning Hypotheses, and to deduce Causes from Effects, till we come to the very first Cause, which certainly is not mechanical.*

In Newton's *Opticks*, he comments on his two roles for God in the universe, that of creating and sustaining. He argued that only divine intervention could explain why the mutual gravitational attraction of the planets does not destabilize the solar system. In Book III of the *Opticks*, Newton expressed the view that the solar system is unstable and requires intermittent adjustment, to which Leibniz criticized Newton that if God had to intervene in the creation, this would surely demean His craftsmanship [322, p. 147]. By the late 20th century, this issue had been related to deterministically chaotic properties of the solar system with the result that it cannot be predicted beyond about 5 million years [304, p. 245].

Newton had created a theory that produced many features of the solar system such as Kepler's geometrical properties of planetary orbits as a result of the universal law of gravitation. However, Newton noticed that other observed features of the solar system were apparently outside of his system. For example, the planets all rotate in a counterclockwise direction around the sun and the planes of the orbits are nearly coincident. Given such configurations of the planets, their future course could be determined, but why these configurations? This led to considering whether such features were a result of chance or else evidence of design in the world. In Query 31 of the *Opticks*, Newton contrasted the choice versus chance issue regarding motions of planets that orbit the sun in the same direction and nearly identical plane with those of comets whose orbits were at every possible angle to ecliptic plane of the planets:

> *For while Comets move in very excentrick Orbs in all manner of Positions, blind Fate could never make all the Planets move on and the same way in Orbs concentrick, some inconsiderable Irregularities excepted, which may have arisen from the mutuall Action of Comets and Planets upon one another, and which will be apt increase, till this System want a Reformation. Such a wonderful Uniformity in the Planetary System must be allowed the Effect of Choice.*

Although Newton did not make probabilistic arguments in the *Principia*, his attitude in the *Opticks* and in private correspondence was that even if we cannot discover a mechanism for the unexplained regularities, we could be assured that there is such a mechanism [323, p. 273]. After Newton's death, a number of scientists attempted to justify the orientations of the orbits based on probabilistic arguments, including Daniel Bernoulli (1700-1782), Georges Buffon (1707-1788), and Pierre Laplace (1749-1827). Explanations of these issues would eventually require the new concept of angular momentum conservation, which preserves the rotational direction of an initial gas cloud as well as the effects of inelastic collisions that provide a mechanism by which the gas cloud can settle to a state that minimizes mechanical energy while conserving angular momentum. That minimal energy configuration is a flattened disc. As a result of high-resolution astronomy and numerical computations, stars are now believed to form within clouds of gas and dust that collapse under gravity. Over time, the surrounding dust particles stick together, growing into larger rocks, which eventually settle into a thin proto-planetary disk where asteroids, comets, and planets form. Once these planetary bodies acquire enough mass, they dramatically reshape the structure of the original disk, forming rings and gaps as the planets sweep their orbits clear of debris and guide dust and gas into tighter and more confined zones.

This type of understanding can be extended to the scale of galaxies though details from atomic and nuclear physics must be included to fill in all the details and has been made possible only with advances in computer power and computational algorithms. The complexity on this scale now increases, galaxies comprising radiation, normal matter as well as the more recently recognized *dark matter,* which is nonluminous and interacts essentially only via gravitation. The universe is thought to have rapidly expanded during an early *inflationary* era after the Big Bang leading to the growth of tiny fluctuations in the density of matter. Denser regions become gravitationally bound capturing both normal and dark matter. These gases become cooled by the emission of photons resulting from the interactions between electrons, hydrogen, and helium. A galactic disc is thereby formed as the rotating gas contracts and denser regions continue to accumulate matter. Then, occasionally, stars are born as nuclear fusion ignites a clump that is sufficiently dense and massive. After forming a sufficient number of stars, gases become expelled from the galaxies in the form of *superwinds*, leading to a competition between gravitation pulling gases together and violent supernova explosions breaking it apart, enabling galactic formation to be now understood over the entire scale of cosmic time [324] [325].

Laplace's Demon

However, all of this detailed understanding still reaches back to the deterministic Newtonian worldview. By the late 18th century, physical science had become thoroughly divorced from religion and determinism prevailed. In 1814, Laplace characterized the nature of a deterministic universe but without the possibility of free will [326]:

> *We may regard the present state of the universe as the effect of its past and the cause of its future. An intellect which at a certain moment would know all forces that set nature in motion, and all positions of all items of which nature is composed, if this intellect were also vast enough to submit these data to analysis, it would embrace in a single formula the movements of the greatest bodies of the universe and those of the tiniest atom; for such an intellect nothing would be uncertain and the future just like the past would be present before its eyes.*
>
>

Such a hypothetical intelligence that can know all the forces upon it and predict the future has come to be known as *Laplace's Demon* [327]. This was the view taken into the late 18th and early 19th centuries and also involved extending the deterministic Newtonian picture from particles to include a new type of quantity, the concept of *field*. Laplace's *Mécanique Céleste* (1799-1825) included a reformulation of Newtonian mechanics in terms of a field, a quantity in which every point of space is assigned a magnitude and direction representing the acceleration produced by all the surrounding masses. This facilitated computation involving multiple bodies such as the 1846 prediction by John Couch Adams (1819-1892) and Jean-Joseph Leverrier (1811-1877) of the existence of the planet Neptune from irregularities found in the orbit of Uranus. The field concept played a further crucial role in the understanding of electricity, magnetism, and light by Faraday and Maxwell to profound effect, however, still under the influence of Newtonian determinism.

The determinism of Laplace's Demon stems from Newton's second law of motion and the existence and uniqueness theorems for solutions of the corresponding simultaneous differential equations of second order for which initial values of the dependent variables and their first derivatives are known at an initial time. However, exceptions to the uniqueness theorems were known at least since the work of Poisson at the beginning of the 19th century. In the 1870s, Boussinesq searched for exceptions to Laplace's Demon by way of singular solutions of certain classes of differential equations that have multiple solutions for certain initial conditions [328]. Later researchers focused on singular solutions in systems in which the force on a particle fails to satisfy a local Lipschitz continuity condition [329], which is known to be a sufficient condition for uniqueness. A simple example is *Norton's Dome,* the motion of a mass initially balanced at the summit of a symmetrical dome which has multiple solutions for the same initial conditions [330]. However, such exceptions did not make much impact on researchers. The success of deterministic Newtonian descriptions of the universe was so extensive and convincing that conditions guaranteeing existence and uniqueness of solutions were viewed as the essence of a good physical theory. As always, *nothing succeeds like success* [331].

Even if we know that a unique solution exists, predictions often require that we know the initial conditions with a particular accuracy. Laplace's Demon in some cases must be able to make its initial measurements with arbitrary accuracy. This is particularly true for *deterministically chaotic* systems, after Poincaré's discovery in the 1890s of the possibility of instabilities of celestial bodies whose long-term dynamics is exponentially sensitive to the initial conditions. Laplace's Demon was a precursor of many developments that explore whether there are limits on various aspects of knowledge. Even if Laplace's Demon knows the initial conditions to arbitrary accuracy, prediction may even be computationally undecidable or intractable and require computational power beyond that of any deterministic algorithm as was demonstrated later in the 20th century [332] [333]. Such behavior can be exhibited for example by *Turing machines* [334], idealized universal computers that use predefined rules to determine a result from a set of input variables. An intriguing property of Turing machines is the famous *Halting Theorem*: configurations exist for which it cannot be known *a priori* whether the machine will give a result and stop computing. As a consequence, the long-time behavior of Turing machines is completely unpredictable [335]. In a clockwork universe, the state of a system at an initial time completely determines the state at a later time. However, the *state* in classical mechanics becomes identified with the *measurement-outcome*, and these two concepts are very different in quantum theory where complementarity disrupts the identification of measurement and state. That determinism and causality are logically independent notions was noted by Bertrand Russell [336] and the distinction later became important with the emergence of quantum theory. There are counter examples of theories which are causal and nondeterministic as well as examples which are deterministic but non-causal [337]. Quantum physics is a probabilistic theory that is causal but not deterministic. Other limits on knowledge include restrictions on information transmission via the constancy of the speed of light within Einstein's Special Relativity, Heisenberg's uncertainty principle of quantum mechanics, and the restrictions imposed by the existence of quantum entanglement leading to a variety of *no-hidden-variable* theorems that distinguish classical and quantum phenomena. A continuing question throughout the 20th and 21st centuries has been how these various limits relate to each other. As physics expanded its horizons, the clock in any clockwork universe was no longer assembled from merely cogs and gears.

Maxwell's equations had carried the determinism of Newton over into the realm of electromagnetism. However, a decade prior to Poincaré's studies of chaos, Maxwell questioned how free will enters into nature. In 1873, he gave a talk entitled, "Does the progress of Physical Science tend to give any advantage to the opinion of Necessity (or Determinism) over that of Contingency of Events and the Freedom of the Will?," in which he delineated these distinctions [338]:

> *There are other classes of phenomena which are more complicated, and in which cases of instability may occur, the number of such cases increasing, in an exceedingly rapid manner, as the number of variables increases....In all such cases there is one common circumstance,—the system has a quantity of potential energy, which is capable of being transformed into motion, but which cannot begin to be so transformed till the system has reached a certain configuration...For example, the rock loosed by frost and balanced on a singular point of the mountain-side, the little spark which kindles the great forest, the little word which sets the world a fighting, the little scruple which prevents a man from doing his will, the little spore which blights all the potatoes, the little gemmule which makes us philosophers or idiots. Every existence above a certain rank has its singular points.*
>
> *If, therefore, those cultivators of physical science...are led in pursuit of the arcana of science to the study of the singularities and instabilities, rather than the continuities and stabilities of things, the promotion of natural knowledge may tend to remove that prejudice in favour of determinism which seems to arise from assuming that the physical science of the future is a magnified image of that of the past.*

Maxwell emphasizes that there are phenomena which do not easily fit into the familiar deterministic framework and suggests that understanding of the physics of nondeterminism may arise from study of the physics associated with singularities and instabilities. He gave this talk shortly after he had devised his thought experiment for violating the Second Law of Thermodynamics in terms of yet another type of Demon, an entity, perhaps with free will, who could sort fast and slow molecules, which has come to be known as *Maxwell's Demon*. As will be discussed in the section *Irreversibility versus Demon*, this was yet another piece of the puzzle determining the meaning of measurement. That there may be exceptions to determinism was acknowledged occasionally and mostly in private or within small circles, but the applications of science toward the end of the 19th century almost exclusively embraced physical determinism. Any exceptions were in the use of statistical methods which were viewed as representing ignorance of underlying deterministic phenomena. But ideas of nondeterminism did drift in the background, at least in the minds of the more deductive thinkers.

As a student at the ETH Zurich around 1900, Einstein had been profoundly impressed by the ability of mechanics to provide explanations in areas that apparently had nothing to do with mechanics, for instance the mechanical theory of light, the kinetic theory of gases, and the deduction of the laws of thermodynamics from the statistical theory of classical mechanics [339, p. 18]. From the time of Newton until the close of the 19th century, the prevalent view among physicists was that mechanical

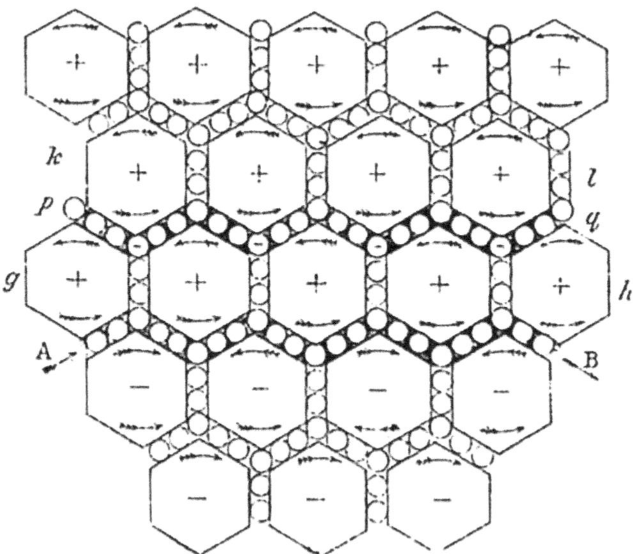

Figure 5.5: James Clerk Maxwell's mechanical model of gears and idler wheels for propagation of light through the luminerferous ether.

concepts ultimately would suffice to explain all physical phenomena. Newtonian mechanics was so widespread and familiar in the 19th century that mechanical analogues of electromagnetic effects in terms of gears, idler wheels, and vortices were commonly devised to better understand and illustrate these phenomena. In 1863, Maxwell actually used a mechanical model as a guide to conclude that the vacuum displacement current must be added to the electromagnetic equations, which was the crucial element allowing him to derive the existence of electromagnetic waves in empty space moving at the speed of light [340, p. 356], Figure 5.5. He used this model to arrive at his Proposition 14, which modifies the equations for electric currents for the effect due to the elasticity of the medium by adding the displacement current $\dot{E}$ to Ampère's Law: $\nabla \times H = 4\pi j + \dot{E}$ [341]. Based on this result, Maxwell concluded that there should be electromagnetic waves: "We can scarcely avoid the inference that light consists in the transverse undulations of the same medium which is the cause of electric and magnetic phenomena." [342] Maxwell emphasized that any mechanical aides should be regarded as scaffolding to be taken down after achieving understanding, leaving the electromagnetic fields as the only reality. Whereas in Newton's formulation of the theory of gravitation the interaction between masses occurred instantly at a distance, we now know that the phenomena of electromagnetism have the subtlety that signals between charges proceeding at the speed of light, and this is implemented by waves of the electromagnetic field. However, those investigating electromagnetism in the late 19th century initially took the view that electrical signals also acted instantly at a distance similar to Newtonian mechanics.

Backstory to Atomism

Newton's ideas concerning the reciprocal actions of the particles of material bodies had inspired many physicists and chemists interested in the structure of matter. In particular, John Dalton was influenced by Newton's treatment of the interparticle relations rather than the particles themselves [343, p. 323]. Newton's theory of matter, called the *nutshell theory* by Joseph Priestley (1733-1804), consisted of hierarchically arranged particles and voids, the smallest of which were called *ultimate particles.* Newton's scheme only had a minor influence on science during the 18th century, but Dalton's borrowing of the interparticle relationships in Newton's scheme led to the first modern atomic theory. Newton's view of matter was influenced by Robert Boyle (1626-1691), incorporating both the indivisible particles of the ancient Greeks and Aristotle's *minima naturalia* which embody properties of the bulk. Dalton likewise would establish a link between the microscopic and macroscopic regimes [344]. Newton had a justification for *Boyle's Law* based on a repulsion of particles that would decrease as the square of the distance between them. Newton regarded his theory as a working hypothesis, but Dalton instead took Newton's corpuscular views of matter straightforwardly and, by never wavering, achieved remarkable success in a short time. If atoms in proximity repel each other, then Dalton took chemical combinations as having as few atoms as possible. Dalton had concluded that "the ultimate particles of all homogeneous bodies are perfectly alike in weight, figure, etc.," which recalls Newton's description that Dalton had transcribed into his notebook [344]:

> *All these things being consider'd, it seems possible to me, that God in the Beginning form'd Matter in solid, massy, hard, impenetrable Particles, of such Sizes and Figures, and with such other Properties, and in such Proportion to Space, as most conduced to the End for which he form'd them; and that these primitive Particles being Solids, are incomparably harder than any porous Bodies compounded of them; even so very hard, as never to wear or break in pieces; no ordinary Power being able to divide what God himself made one in the first Creation...*

During the 18th century, Newton's view of matter had been forced to compete with the theory of Georg Ernst Stahl (1660-1734) in which matter comprises identical particles which are formed by atoms of one and the same basic material and from which molecules are formed. The Newtonian-Stahl ideas were revived during the last decades of the 18th century by the chemical revolution brought about by Lavoisier, giving birth to the molecular theory of Dalton, which was a crucial event in the genesis of modern atomism [343, p. 328]. Dalton's atomic theory first appeared in an 1807 book by Dalton's advocate Thomas Thomson and then in Dalton's *A New System of Chemical Philosophy* in 1808. As early as 1803, Dalton began deducing relative atomic weights. For example, he knew that water was comprised by weight of 87.5 percent oxygen and 12.5 percent hydrogen. Then if he *assumed* that water was a binary compound HO so that one atom of water and one atom of oxygen goes into each water molecule, then he could deduce that every oxygen atom weighs seven times as much as every hydrogen atom; e.g., O=7, if H=1. Or if water instead was H_2O, he could deduce that O=14. H=1 [345]. Dalton's chemical atomic theory gave a basis for giving relative *atomic weights* to chemical elements, where chemically indivisible units combined with each other in integer multiples consistent with the laws of stoichiometry. Dalton could deduce each of his atomic weights using more than one separate line of deduction, assuming a formula for the compound in each approach. Consistency among the results would give confidence in the chemical formulas. Although Dalton regarded these units as the actual constituents of matter, most chemists regarded them as a convenient scheme for developing their work. Dmitri Mendeleev (1834-1907), who developed the Periodic Table for predicting the existence and properties of new chemical elements, considered Dalton's atoms as being [344]

> *indivisible, not in the geometrical abstract sense, but only in a physical and chemical sense...units with which [chemists were] concerned in the investigation of the natural phenomena of matter.*

The influence of Dalton's work can be traced from 1803 to the present, and it enabled chemistry to have the status of the earliest robust area of science that had an impact on both fundamental issues and technology. Dalton's physician, Joseph Ransome, recalled a walk with Dalton (ca. 1820) in which Dalton explained that it had occurred to him that if Newton's ultimate particles were indestructible, then there must be many varieties varying in size and weight to account for so many known chemical elements and compounds. Then he emphasized, illustrating this with a piece of limestone picked up from the path, that since compounds are formed by one-to-one association with their elements, then the principle of multiple proportions must exist [345]:

> *Thou knows it must be so, for* ***no man can split an atom***.
>
>

Deductive thought experiments can have very long-lasting influence.

In the period of approximately 1860-95, comparison of Dalton's atomic hypothesis with spectroscopy experiments was hampered by the complexity of the internal structures revealed by experimental observations [346]. However, the kinetic theory of heat of Maxwell and Boltzmann, which was on an underlying molecular picture, began to allow more precise predictions, for example allowing specific calculated values for the mean free paths and relative diameters of various gas molecules as well as Avogadro's number. Gas laws began to be applied to liquids based on van der Waal's law of corresponding states [346]. Evidence for atoms and molecules continued to accumulate from the advances of both experimental and theoretical developments including cathode rays, radioactivity, the photoelectric effect, electron-photon scattering, leading up to the discovery of quantum mechanics.

However, atomism was still under debate in the late 19th century, when Maxwell would describe the state of atomism in an 1873 lecture [347]:

> *An atom is a body that cannot be cut in two. A molecule is the smallest portion of a particular substance...Do atoms exist or is matter infinitely divisible? The discussion of questions of this kind had been going on ever since man began to reason, and to each of us, as soon as we obtain the use of our faculties, the same old questions arise as fresh as ever. They form as essential a part of the science of the nineteenth century of our own era as that of the fifth century before it.*

In the 19th century, there were conflicting approaches for what is acceptable as a description of nature. Some insisted on eliminating hypothetical elements and only including directly measurable quantities. These included Ernst Mach (1838-1916), Wilhelm Ostwald (1853-1932), Pierre Duhem (1861-1916), and Gustav Kirchhoff (1824-87). Henri Poincaré's view was that the atomic hypothesis was not capable of being either proved or disproved. Alternatives to the discrete atomic approach were continuous formulations of matter such as hydrodynamics, the use of *equivalents* and volume elements, and the use of thermodynamic approaches based on energetics [346]. Others, such as Ludwig Boltzmann (1844-1906) and Heinrich Hertz (1857-94) were able to make progress by using the atomic hypothesis. The effort to discredit atomism was led by Ostwald and Mach. Ostwald had won the 1909 Nobel Prize for his research on catalysis and insisted that it cannot be explained in terms of atomic theory but instead results from transformations of energy. Ostwald's approach, which he first summarized in his 1887 lecture in Leipzig, became known as *energism* and held that energy and not matter was the basis of all natural processes. He was led to this view by studying Gibbs' (1839-1903) approach to statistical mechanics which he concluded did not require any concepts beyond changes in "volume elements" without assuming the existence of atoms [348]. Concepts such as atoms, molecules and ions were convenient fictions without basis. In Ostwald's 1895 lecture in Lübeck, "The Superseding of Scientific Materialism," he had summarized [349]:

> *...everything we know of the world is transmitted to us through our sense organs. In order that a sense organ should function, a transmission of energy between it and the external world is necessary and sufficient. Thus, the only direct knowledge we have of the external world is of its energy conditions. Everything beyond that is subjective addition.*
>
>

Mach also dismissed atomism but criticized energism as being an abstract theory. His objections were partly philosophic ones and also being a physiologist, he emphasized the preeminence of sense data, Figure 5.4. He regarded the concept of *atom* as only a symbol for how sensations are manifested [348]:

> *Let us look at the facts in an unbiased way, without preconception. The world consists of colour, sounds, heats, pressures, spaces, times, etc., which we shall now call neither sensations nor phenomena, because either name already implies a one-sided arbitrary theory. We shall call them **elements.***
>
>

Mach admitted that the atomic hypothesis had served a useful purpose historically in the development of physics, but he insisted that this framework now needed to be dismantled. As Phillip Frank (1884-1966) recalled from his student days in Vienna with Mach in 1907 [346]:

> *Two characteristic beliefs of nineteenth-century science broke down during its last decades...the belief that all phenomena in nature can be reduced to the laws of mechanics, and the belief that science will eventually reveal the "truth" about the universe.*
>
>

In spite of their differences, Mach and Ostwald led a united front against atomism. At the 1897 meeting of the Academy of Sciences in Vienna, Boltzmann defended his kinetic theory of gases based in terms of atoms, when Mach abruptly stated during the discussion, "I don't believe atoms exist." Boltzmann later recalled that "The utterance ran in my mind" [348]. Regarding these statements by Ostwald and Mach, it should be kept in mind that in the 1890s, theoretical physics was still not a fully formed discipline [350]. Whereas what was in the process of becoming *theoretical physics* had previously been synonymous with *natural philosophy* and focused on seeking out the underlying causes of phenomena, it was now also becoming *mathematical physics*. Scientists such as Boltzmann and Planck would now separate theoretical physics from the philosophy and anthropomorphism of Mach [351]. As discussed in the section *The Fall of Classicality*, Planck had earlier been aligned with Mach against atomism. However, in his work leading to the discovery of the quantum, Planck had been drawn to accept Boltzmann's statistical approach to the Second Law, coming to believe that atoms were as real as planets [352, p. 49].

Atomism versus Continuum

Newton had memorably produced a light spectrum (ca. 1670) by directing sunlight through a prism, producing "a confused aggregate of rays induced with all sorts of colors" instead of discrete lines. Twenty-five years after Newton's death, Scottish natural philosopher Thomas Melvill (1726-1753) reported observations of discrete spectra resulting from heated sodium within the salt molecules held in a flame. Joseph von Frauenhofer (1787-1826) observed discrete dark lines in the solar spectrum, later explained by the presence of sodium in the outer layers of the sun. These were the beginnings of experimental optical spectroscopy. As discussed in the section *The Fall of Classicality*, the radiating power of a substance seen in its emission spectrum could be shown to equal its absorbing power as seen in its dark line absorption spectrum, as first quantified by Kirchhoff in 1859. He was able to derive from these results *Kirchhoff's Law*, the crucial property of blackbody radiation that eventually led Planck to discover the quantum of action in 1900. Pais points out that, by the time Bohr entered the picture, Kayser's handbook of spectroscopy already contained 5000 pages, a tremendous backlog of spectral information, which lacked fundamental interpretation [9, p. 141]. Bohr's remarkable success in developing his quantized atomic theory in 1913 began with the case of the hydrogen atom. The hydrogen spectrum was first detected by Anders Ångström (1814-1874) in 1853 and within a few years the frequencies of four of the lines of the hydrogen spectrum were identified and measured.

The view had been generally widespread in the 19th century that continuity was a dominant aspect of the description of the natural world, tracing back at least to Aristotle's definition [353]:

> *The continuous is a species of the contiguous; two things are called continuous when the limits of each, with which they touch and are*

> *kept together, become one and the same.* [Metaphysics XI.12]

However, Aristotle also anticipated discontinuities in nature:

> *Nor, again, can there be anything between that which suffers and that which causes increase; for that which starts the increase does so by becoming attached in such a way that the whole becomes one. Again, the decrease of that which suffers decrease is caused by a part of the thing becoming detached. So both that which causes increase and that which causes decrease must be continuous; and if two things are continuous there can be nothing between them.* [Physics VII.2]
>
> *It is evident, therefore that between the moved and the mover – the first and the last – in reference to the moved there is nothing intermediate.* [Physics VII.2]
>
> *Nor again is there anything intermediate between that which undergoes and that which causes alteration.* [Physics VII.2]
>
> *And this is evident from what happens in respect of sensation; for the same thing never appears sweet to some and bitter to others...* [Metaphysics XI.6]

Continuity here is being used as adjacent in time. It appears that what is being described are actually discontinuous processes since they are extremities and are adjacent. It may be that had Aristotle been alive when quantum mechanics was founded, he might have associated Planck's quantum of action with these discontinuities and concluded that because particles are quantized and divisible, that this would require their entire absorption into the measuring device upon measurement in order to meet the requirement of "becoming attached in such a way that the whole becomes one." This would be akin to accepting wave function collapse.

Some two millennia following the beginnings of atomism with the Greeks, the mid-18th and early 19th centuries saw the first stirrings of quantitative formulations of atoms, Figure 5.6, with Daniel Bernoulli (1700-1782) and James Joule (1818-1889) followed by the full development of the kinetic theory of gases by Rudolph Clausius (1822-1888), James Clerk Maxwell (1831-1879), and Ludwig Boltzmann (1844-1906). At the turn of the 19th century, Antoine Lavoisier (1743-1794) initiated the notion of chemical elements and determined that chemical compounds occur in proportion to their weights. On the basis of an atomic picture, John Dalton (1766-1844) explained that elements are identical and that materials can be viewed as built up from atoms. Guy-Lussac (1778-1850) went further and formulated the law of volumes for reacting gases, which provided the context in which Avogadro (1776-1856) was able to state that all gases contain an equal number of particles at equal volume, temperature and pressure, *Avogadro's Number* $N_A = 6.022 \times 10^{23}$. The 1840's saw the crucial formulation of the law of conservation of energy by Joule, Hermann von Helmholtz (1821-1894) and Robert Mayer (1814-1878). Clausius demonstrated the equivalence of heat and mechanical work, namely the First Law of thermodynamics, which more generally is the principle of conservation of energy. However, he found that thermodynamics required an additional concept: heat cannot of its own accord pass from a colder body to a warmer one, the Second Law of Thermodynamics. An equivalent statement of the Second Law is the impossibility of constructing a perpetual motion machine of the second kind: an apparatus that without

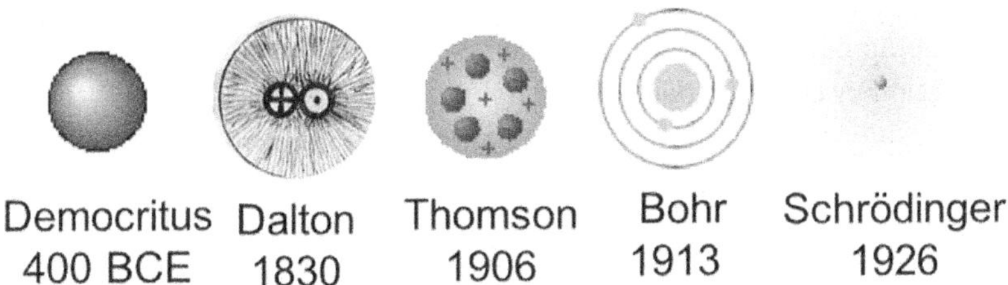

Figure 5.6: Evolution of the conception of the atom.

violating the conservation of energy nevertheless transforms heat into mechanical work with 100% efficiency. For a system composed of atoms, Maxwell's thought experiment involving a demon, "a very observant and neat-fingered being," showed how the Second Law might be violated and later led to insights into the role of information in physical systems. Statistical mechanics began with the work of Maxwell and Boltzmann and gave a statistical interpretation of transport and thermal equilibrium involving the vital relationship between entropy and the probability of the state of a gas. It was developed by J. Willard Gibbs (1839-1903) into a form general enough to address all classical systems. All of these developments were essential precursors to Planck's discovery of the quantum of action in 1900. The theories which dominated quantitative physics were all based on the continuum world picture: Newton's equations of classical mechanics, Maxwell's equations of electromagnetism, and Einstein's special relativity. The dramatic transition to the discontinuities of quantum theory began in physics and eventually spread to chemistry and biology.

Atomism Prevails

In the 19th century, there had been a crucial distinction between *chemical* and *physical* atomism, as described by August Kekulé (1829-96), the founder of the theory of chemical structure [354]:

> *I regard the assumption of atoms, not only as advisable, but as absolutely necessary in chemistry, I will go even further, and declare my belief that chemical atoms exist, provided the term be understood to denote those particles of matter which undergo not further division in chemical metamorphoses. Should the progress of science lead to a*

theory of the constitution of chemical atoms—it would make but little alteration in chemistry itself. The chemical atom will always remain the chemical unit...whether matter be atomic or not, thus much is certain that, granting it to be atomic, it would appear as it now does.

The concept of the chemical atom had become indispensable regardless of whether matter was actually composed of atoms or not. However, for physicists the atoms in atomic theory were real in the sense that they gave meaning to phenomena that were inhomogeneous on an atomic scale. Lord Rayleigh's (1842-1919) explanation for why the sky is blue in terms of scattering by air molecules depended on the wavelengths of visible light being comparable to the average spacing between molecules. The value of Avogadro's number could be determined using many different methods, each sensitive to the atomic scale in a different way, but with each method consistently giving a similar result. In 1870, Lord Kelvin (1825-1907) described several different physical relations, all of which had been used to determine comparable values for Avogadro's number: the relation between mean-free path and viscosity of a gas; the relation between capillary energy and heat of vaporization; and relation between electrical contact energy and chemical heat [349]. Although Ostwald had been correct in claiming that the Gibbs' ensemble formulation of statistical mechanics does not explicitly invoke atoms, the Gibbs' grand canonical ensemble can be used to calculate the value of the number fluctuations in a fluid and compared to the observed value from opalescence measurements to determine a value for Avogadro's number that is consistent with other methods. By the last decade of the 19th century, atomic theory was consistent with known physics and chemistry except for heat and electromagnetism. As will be discussed in the section *Is Irreversibility Intrinsic?*, this was accomplished by Boltzmann's formulation of statistical mechanics which was explicitly based on the scattering of molecules and was also consistent with Gibbs' theory. This allowed all of macroscopic thermodynamics to be based on an underlying atomic theory. Although the identification of heat with the kinetic motion of molecules had been suggested as early as the 17th century by Francis Bacon (1561-1626) and Robert Hooke (1635-1703), Boltzmann's theory could explicitly relate the temperature to the average energy of the molecular degrees of freedom using Maxwell's law of equipartition of energy.

Despite the apparent successes in the application of Boltzmann's theories, Planck would still be led to say in 1895 [355]:

It was impossible to prevail against the authority of men like Ostwald and Mach.

Three years later, Boltzmann would write pessimistically [356]:

It would be a tragedy for science if the theory of gases should fall into oblivion because of the currently fashionable hostile attitude toward it.

Probabilistic Thinking, Thermodynamics and the Interaction of the History and Philosophy of Science. Synthese Library, Boltzmann's Conception of Theory Construction: The Promotion of Pluralism, Provisionalism, and Pragmatic Realism, Vol. 146, 1980, p. 175, E.N. Hiebert, © Springer Science+Business Media Dordrecht 1980. With permission of Springer.

However, at the beginning of the 20th century, the reality of atoms became undeniable. Ostwald maintained his position on energism until the work of Jean Baptiste Perrin (1870-1942), who had long believed that atoms were real, determined Avogadro's number from observations of Brownian motion in colloids. After this, Ostwald could no longer reject discontinuity of matter. In 1909, Oswald wrote [357, p. 249]

I am now convinced that we have recently come into possession of experimental proof of the discrete or grainy nature of matter, for which the atomic hypothesis had vainly sought for centuries, even millennia.

Republished with permission of University of Chicago Press, from Chapter Seven, The Experimental Basis from which to Test Hypotheses: Brownian Motion, in book: Error and the growth of experimental knowledge, D.G. Mayo, © 1996 University of Chicago Press – Books.

Brownian motion had been known since 1827 and had been interpreted in terms of kinetic theory by 1879. Perrin's experimental confirmation of Einstein's 1905 theory of Brownian motion was carried out in 1910. Sadly, this was four years after Boltzmann's death by suicide. Mach also apparently changed his mind, as physicist Stefan Meyer recalled in 1950: Mach saw the apparatus developed by Erstel, Geitel, and Crookes that allowed the flashes created by single α-particles to become visible on a screen, whereupon Mach exclaimed [348], "Now I believe in atoms." However, in spite of this, Mach's writing continued to imply denial, as in this 1910 reply to a criticism by Planck [348]:

If belief in the reality of atoms is so crucial, then I renounce the physical way of thinking. I will not be a professional physicist, and I hand back my scientific reputation. In short, thank you so much for the community of believers, but for me freedom of thought comes first.

Journal for General Philosophy of Science, Saving Mach's Views on Atoms, Vol. 41, 2010, p. 1, M. Bächtold, © Springer Science+Business Media B.V. 2010. With permission of Springer.

Lise Meitner (1878-1968) was a student of Boltzmann with a dissertation on heat conduction in an inhomogeneous body directly related to kinetic theory. She attended his lectures from 1902 until his death in 1906. Meitner later recalled the difficulties he encountered by the resistance to kinetic theory [358]:

Boltzmann had no inhibitions whatever about showing his enthusiasm while he spoke, and this naturally carried his listeners along. He was also very fond of introducing remarks of an entirely personal character into his lectures—I particularly remember how in

> *describing the kinetic theory of gases, he told us how much difficulty and opposition he had encountered because he had been convinced of the real existence of atoms, and how he had been attacked from the philosophical side, without always understanding what the philosophers had against him.*
>
> Looking Back, L. Meitner, Bulletin of the Atomic Scientists, November 1, 1964, Taylor & Francis Ltd, reprinted by permission of the publisher.

The loss of Boltzmann led Meitner to move to Berlin where she became the first woman that Max Planck allowed to attend his lectures. The following year she became Planck's assistant and began the collaboration with chemist Otto Hahn that twenty-two years later would involve her in the discovery of nuclear fission. Meitner and

Figure 5.7 Photograph of the 1st Solvay Conference in 1911 Seated (L-R): W. Nernst, M. Brillouin, E. Solvay, H. Lorentz, E. Warburg, I. Perrin, W. Wien, M. Curie, and H. Poincaré. Standing (L-R): R. Goldschmidt, M. Planck, H. Rubens, A. Sommerfeld, F. Lindemann, M. de Broglie, M. Knudsen, F. Hasenöhrl, G. Hostelet, E. Herzen, J.H. Jeans, E. Rutherford, H. Kamerlingh Onnes, A. Einstein, and P. Langevin.

Hahn led the group that first discovered nuclear fission of uranium. Meitner and her nephew Otto Frisch were the first to correctly interpret the results as being nuclear fission. Remarkably, Meitner's research had spanned the eras from the denials of atomism up to the splitting of the atom. For the majority of the scientific community, the atomic debates ended at the 1st Solvay Conference in 1911, Figure 5.7, which brought together several generations of prominent scientists to focus on the details of the history and successes of atomism, and the new developments of radiation theory and quanta [346]. Atoms and molecules were no longer hypothetical but existential.

Einstein's Space-Time

Einstein took the first step in the quantization of light in his miracle year, the *annus mirabilis* of 1905, while a patent clerk at the Patent Office in Bern Switzerland. The four papers published in 1905 had all addressed important problems of the time with an impact in physics that reverberates to today. Here are the titles and the physical dominoes that fell as Einstein marched through the year:

1. Light quanta hypothesis (March)
   On a Heuristic Point of View Concerning the Production and Transformation of Light

2. Brownian motion (May)
   On the Motion of Small Particles Suspended in Liquids at Rest Required by the Molecular-Kinetic Theory of Heat

3. Special relativity (June)
   On the Electrodynamics of Moving Bodies

4. $E = mc^2$ (September)
   Does the Inertia of a Body Depend upon its Energy Content?

These include the origins of both space-time and quanta: the view of deterministic causal space-time coordination of dynamics emerging from relativity and the view of wave-particle duality when particles are endowed with the quantum of action. However, Einstein viewed the light quanta paper as the truly revolutionary one, although not necessarily the path he viewed toward the future. Einstein had proposed that under certain circumstances monochromatic light with frequency ν behaves as if it consists of light quanta or particle-like objects with energy $E = h\nu$, now called *photons*. This included understanding the important photoelectric effect, the emission of electrons from a metal surface irradiated by light with frequency ν above the threshold given by the work function of the metal, which led to Einstein's Nobel Prize in physics in 1921. At that point, the Nobel committee appears to have been in a state of indecision regarding Einstein's work on space-time; Einstein was awarded the Nobel prize only for the photoelectric effect and not for the special theory of relativity.

Prior to developing his theory of gravitation, Einstein's special relativity of 1905 had incorporated the implications that the speed of light c is invariant for all inertial observers in constant motion. This led to the four-dimensional space-time picture, Figure 5.8, in which events are constrained by *light cones* pointing into the past and future along which light can travel, which confine the time-like trajectories of any particles with mass and put restrictions on communication with other particles outside the light-cone due to the finite speed of light. The four-dimensional space-time view was actually put forward by Hermann Minkowski (1864-1909), Einstein's former professor, as the proper setting for Einstein's special relativity, although he did not live long enough to witness its dominance within 20th century physics. An important contribution by Minkowski is the concept of *proper time* τ comprising both space and time, which is the same for all inertial observers. And likewise, the *proper mass* m (also the *rest mass*) relates to the energy and momentum through the equations

$$(c\tau)^2 = (ct)^2 - r^2 \tag{5.1}$$

$$(mc^2)^2 = E^2 - (pc)^2 \,. \tag{5.2}$$

Minkowski had summarized this as [359]:

> *Henceforth space by itself, and time by itself, are doomed to fade away into mere shadows, and only a kind of union of the two will preserve an independent reality.*
>
> Reprinted by permission of Dover Publications.

That is, significant quantities in terms of an independent reality are not individually space or time or energy or momentum, but rather proper time and proper mass.

The Fall of Classicality

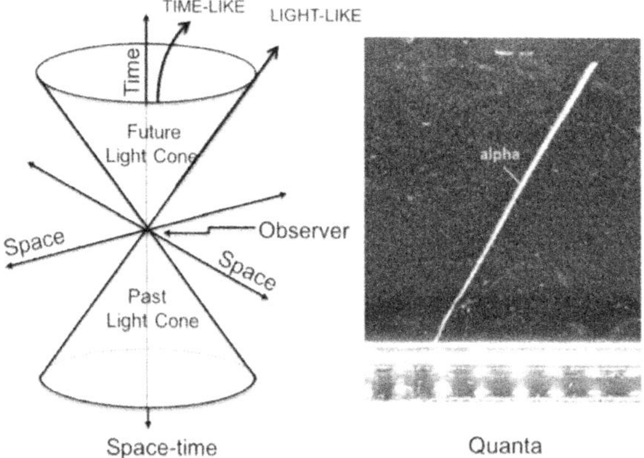

Figure 5.8: Space-time: Constraints on dynamics within space-time as required by relativity. Quanta: Linear tracks in space-time within a cloud chamber originating from spherical wave-functions of alpha particles, an early enigma for measurement.
Cloud chamber image: Julien Simon, Cloudylabs

The introduction of the fundamental quantum of action by Planck in 1900 was the beginning of the end for determinism, although it took several decades for this to be generally accepted by the scientific community. More and more anomalies were discovered which challenged mechanistic explanations: blackbody radiation, radioactivity, X-rays, and specific heats of solids and gases.

The era of quantum theory commenced on December 14, 1900, when, at a meeting of the German Physical Society in Berlin, Max Planck presented his derivation of the blackbody radiation formula, using a statistical mechanical method that had the unprecedented feature that energy exchange between radiation and radiating body was discontinuous with quanta of energy transfer given by $E = h\nu$, which is:

$$B(\nu, T) = \frac{8\pi\nu^2}{c^3}\frac{h\nu}{\exp\left(\frac{h\nu}{kT}\right) - 1}. \tag{5.3}$$

Here k is the Boltzmann constant that gives a measure of the amount of energy (i.e., heat) corresponding to the random thermal motions of a particle, c is the speed of light, and h is a new constant, now called Planck's constant or Planck's *quantum of action*. Equation (5.3) describes the density of radiant energy at frequency ν given off in thermal equilibrium at temperature T by a *blackbody*, an ideal perfect absorber. This exact expression represents the maximal radiation that a body can emit at thermal equilibrium, whatever the composition or structure and it is governed by the quantum of action.

As has always been known, a heated body glows. The corresponding problem of the heat radiation formula is to understand the dependence of the emitted light on the frequency and temperature, as well as the nature and shape of the body. By the 1860s, it had become accepted from the work of Michael Faraday (1791-1867) and James Clerk Maxwell (1831-1879) that light was electromagnetic radiation and thermal radiation represents a conversion of thermal energy into electromagnetic energy, including visible light. The radiation law should then be understood in terms of the thermodynamics of electromagnetic processes. Quantum theory was thus born at the crossroads of the two sciences of Thermodynamics and Electromagnetism.

In 1859, Gustav Kirchhoff (1824-1887) discovered that for a body in thermal equilibrium with radiation, the ratio of the emission and absorption is independent of the particulars of the body, depending only on frequency and temperature. This is due to the Second Law of Thermodynamics and violation would imply the impossibility of a perpetual mobile of the second kind, i.e., the impossibility of spontaneous conversion of thermal energy into mechanical work. Kirchhoff proposed studying this fundamental result by way of perfect absorbers called *blackbodies*, which can be approximated in the laboratory by an oven with a small hole. It took 40 years of efforts by many physicists to measure the explicit spectrum for blackbody radiation, but by the 1890s, experimental measurements of the spectral distribution had improved rapidly, stimulated by the need for improved temperature measurements and the absolute temperature scale.

As seen in Figure 5.9, at any temperature T there is a preferred frequency ν of emitted light and higher frequency emission is suppressed. In contrast to this, the radiation formula based on classical theory by Lord Rayleigh (1842-1919) and James Jeans (1877-1946) predicted a divergence as ν^3 at high frequencies (later dubbed the *ultraviolet catastrophe*). This is due to the classical equipartition theorem, which predicts that all degrees of freedom will have an average energy of $kT/2$ and therefore most of the energy will be at higher frequencies where most of the modes are. This is contrary to the observations, which are consistent with Planck's expression in Figure

5.9. This means there must be a mechanism preventing energy from going into the high frequency modes and it must be governed by a physical constant with dimensions mixing energy and time so as to relate temperature to frequency. This is precisely the role of Planck's constant h which has the dimensions of energy-time or *action*:

$$h = 6.626\, x\, 10^{-34} \text{ Joule-second.} \tag{5.4}$$

A reduced Planck's constant $\hbar \equiv h/2\pi$ (pronounced "h-bar") is also often used in expressions to eliminate the frequent appearance of factors of π. Planck also used the newly expanded set of fundamental constants k, c, G and h to propose a now-famous system of natural units, saying they [360]

> *necessarily retain their significance for all time and all cultures, even alien and non-human ones...*
>
>

Today these units have found their way into the area of quantum gravity where the *Planck length* $l_p = \sqrt{\hbar G/c^3}$, *Planck time* $t_p = \sqrt{\hbar G/c^5}$, and *Planck mass* $m_p = \sqrt{\hbar c/G}$ naturally play a role.

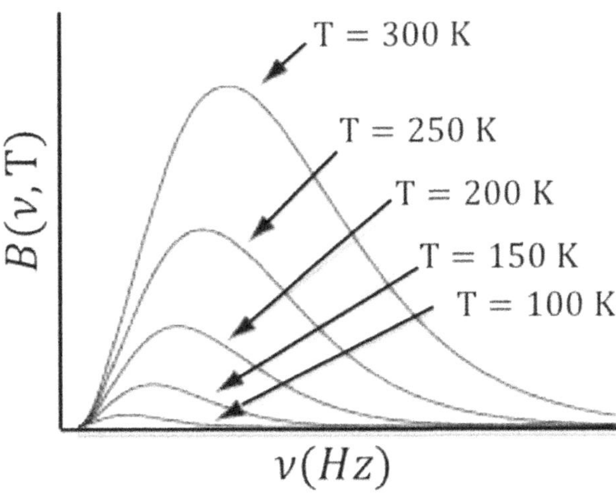

Figure 5.9: Planck's Blackbody Radiation Law.

There were controversies over the explanation of irreversibility and the second law of thermodynamics between reversible classical mechanics and Boltzmann's statistical-mechanical theory. Although the statistical nature of radioactive decay of atoms was recognized since its discovery, there were attempts for over a decade to explain it from a deterministic and mechanical basis [361] [362]. And for some, Bohr's model of the atom still did not fully rule out the possibility of an underlying causal explanation of radioactivity. That radioactivity was truly nondeterministic was largely accepted only with the more general nondeterminism of quantum mechanics after 1925. The first quantum mechanical explanation of α-particle radioactivity and derivation of the decay law didn't appear until 1928 [363] [364], with a decisive difference in interpretation as concluded by Condon and Gurney:

> *We have had to consider the disintegration as due to the extraordinary conjunction of scores of independent events in the orbital motions of nuclear particles. Now, however, we throw the whole responsibility on to the laws of quantum mechanics, recognizing that the behavior of particles everywhere is equally governed by probability.*
>
>

This had followed the 1926 assertion by Max Born (1882-1970) that, although the quantum state evolves deterministically, an individual measurement occurs irreducibly at random and is nondeterministic. Not only was the cause of the measurement event not known, cause did not play a role. Einstein had written the oft-quoted letter to Born of December 4 from that year, objecting that God "is not playing at dice" [365, pp. 90-91]:

> *I, at any rate, am convinced He is not playing at dice.*
>
>

Einstein had previously introduced probabilities into quantum dynamics in three papers of 1916-17 [366] [367] [368] on the emission and absorption of the radiation of atoms. In doing so, he had found an alternate way of obtaining Planck's blackbody radiation law and this was also related to the quantum jumps of Bohr's atomic model. Einstein postulated the existence of transition probabilities for spontaneous and induced emission and absorption between the discrete energy levels of an atomic system. He mentioned that his scheme for spontaneous emission was essentially identical to Rutherford's 1900 description of radioactive decay [9, p. 191]. However, he was disturbed about the causality of the model in that he could not predict the direction in which a light-quantum moves after spontaneous emission, saying: "The weakness of the theory lies on the one hand in the fact this it does not get us any closer to making the connection with the wave theory; on the other, that it leaves the duration and direction of the elementary processes to chance" [369]. Rutherford's decay was within his planetary picture of the atom with electrons orbiting a massive nucleus. As discussed in the section *Deductive Thought versus Inductive Thought,* in order for Bohr to deductively construct the first theory of atoms and molecules in 1913 that addressed their structure in terms of the configurations of electrons, he was led to incorporate Planck's quantum of action. Rutherford had responded to Bohr [370]:

> *How does an electron decide which frequency it is going to vibrate at when it passes from one stationary state to the other? It seems to me*

> *that you have to assume that the electron knows beforehand where it is going to stop.*
>
> Compendium of Quantum Physics, Quantum Jumps, p. 599, 2009, K. Hentschel, © Springer-Verlag Berlin Heidelberg 2009. With permission of Springer.

Rutherford's instincts led him to notice these symptoms of nondeterminism, describing the "grave difficulty" in this case almost as an act of free will, an aspect of the model that Bohr would eventually answer with his correspondence principle.

Max Planck, the discoverer of the quantum, never completely came to terms with it and retreated into the 'God's-Eye-View' of his religious beliefs. On the eve of his death, Bohr commented during a final interview, Figure 5.10 [371]:

> *Planck was religious...he said that a God-like eye could certainly know what was the energy and the momentum. And that was very difficult you see. And then I said to him when we came back from it... You have spoken about such an eye: but it is not a question of what an eye can see, it is a question of what you mean by knowing.*
>
> Reprinted with permission by AIP Oral History Interviews: Interview of Niels Bohr, October 31, 1962.

Figure 5.10: Bohr's Last Blackboard, from the eve before his death, illustrating (top) a complex function as an analogy to free will and (bottom) Einstein's photon box which Bohr had shown was consistent with quantum mechanics.
Used by permission of the AIP Emilio Segrè Visual Archives.

Planck's God-like-eye is not unlike Laplace's Demon. Bohr actually had criticized Heisenberg's initial argument of the gamma-ray microscope thought-experiment for the uncertainty principle along similar lines [372, p. 88]. Bohr objected that Heisenberg began by assuming that the electron actually has a definite position and

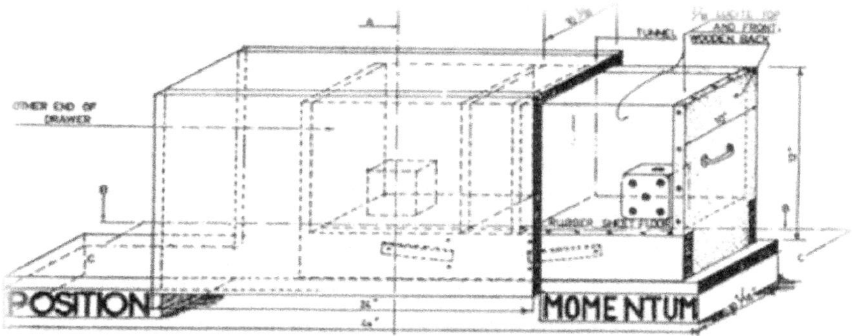

Figure 5.11 Problematic mechanical model of complementarity similar to the 1939 New York World Fair exhibit by University of Copenhagen, with two dice representing complementary features of nature and its values on the die faces. Drawer allows only one die to be shown at a time, however sliding drawer to reveal the complementary die activates flipper to throw the die and randomize its value. Actual quantum complementarity in nature would not allow the wooden tunnel to be removed revealing Planck's God's-Eye-View of both features.

momentum but that the act of observing one of these disturbs the other. Heisenberg had assumed a fixed underlying existence, essentially permitting a "God's-Eye-View" [373], Figure 5.11. A famous "Addition in Proof" in Heisenberg's 1927 indeterminacy paper acknowledges that Bohr, "has brought to my attention that I have overlooked essential points". For Bohr, what we can know determines the conceptual framework of our description and, in particular, our observations are under conditions in which we are also part of the world. Bohr later summarized his view of Laplace's Demon [372, p. 89]:

> *Quite apart from the fact that such [an external intelligence] would not be able to communicate with us and apart from the problem of how such an observer would be able to keep account without interfering with the course of the phenomena (very problematic), we must maintain that by science—especially as the development of physics has taught us—we mean the possibility of collecting human observations and our possibilities of ordering them.*
>
>

Nondeterminism had been argued by Heisenberg in 1927 to follow from the uncertainty relations, preventing some observables from having definite values with outcomes being indeterminate before a measurement. The generalized versions of Heisenberg's uncertainty relation [374] [375] specify how precisely non-commuting observables X and Y can be measured in a quantum state in terms of the variances

ΔX^2 and ΔY^2:

$$\Delta X^2 \Delta Y^2 \geq \frac{1}{4} |\mathrm{Tr}\rho[X,Y]|^2 \,,$$

where $\Delta X^2 = \mathrm{Tr}\rho X^2 - (\mathrm{Tr}\rho \mathrm{X})^2$ is the variance. Non-vanishing ΔX can represent a randomness due to measurement from non-commuting operators or can also be due to classical noise as well. Although either ΔX and ΔY are allowed to vanish, they cannot vanish simultaneously for a given state ρ unless they commute, $[X,Y] = 0$. It is not straightforward to determine ΔY in the case that ΔX vanishes but more recent developments of uncertainty relations in the form of sums in terms of entropy and variances have been able to address these issues [376]. The entropic approach has generalized the uncertainty paradigm to allow further understanding of joint measurability of non-commuting observables, quantum steering and situations where the measured system is entangled with its environment [377] [378]. The full story of randomness in quantum mechanics can be seen only in the presence of nonlocal quantum correlations. We will see that the local uncertainty of Heisenberg has since become subsumed by the global nondeterminism of Bell.

Irreversibility versus Demon

Imagining an intelligence intervening into the Laws of Nature has historically been a useful way of delving into fundamental principles, as we had previously seen with the determinism of Laplace's Demon. In 1867, Maxwell wrote in a letter to P.G. Tait concerning two compartments *A* and *B* connected by a diaphragm: "Now conceive of a finite being who knows the paths and velocities of all the molecules by simple

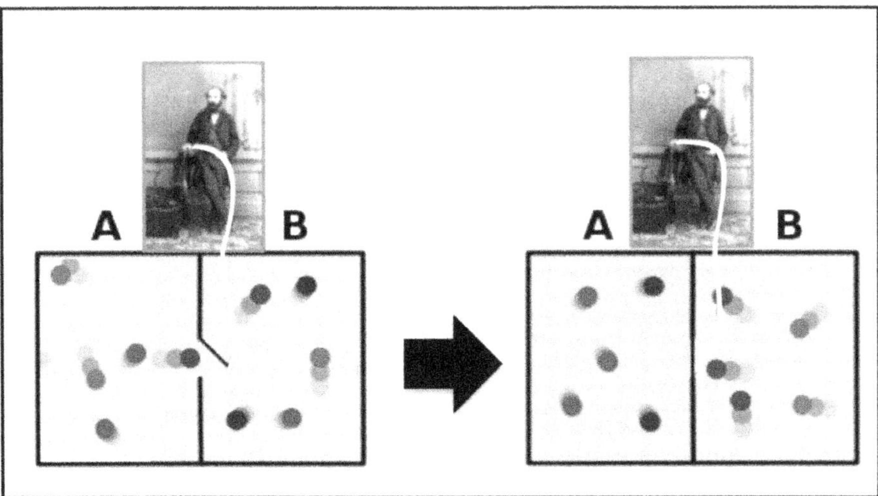

Figure 5.12: Maxwell's Demon. Maxwell re-enacting his thought experiment of "a being whose faculties are so sharpened that he can follow every molecule in its course," allowing the temperature of B to be elevated above A without the expenditure of work, violating the second law of thermodynamics.

inspection, but who can do no work except open and close [the] hole in the diaphragm by means of a slide without mass" [379, pp. 213-215]. His task would be to allow molecules to pass from *A* to *B* if they have greater than average speed in *A*, and from *B* to *A* if they have less than the average speed in *B*. The result of these tactics is that molecules in *B* become more energetic than they were originally and those in *A* less energetic, resulting in heat flowing the wrong way. The being has arranged for entropy to decrease and thus violate the Second Law of Thermodynamics. In his important 1874 Nature paper on energy dissipation, William Thomson (Lord Kelvin 1824-1907) dubbed this selective influence a *demon*, "an intelligent being endowed with free will, and fine enough tactile and perceptive organization to give him the faculty of observing and influencing individual molecules of matter" [380]. This was in the original Greek sense of *daemon* as an intelligent deity and not the malevolent scarlet imp often pictured. The demon as intelligent guide is no doubt an alter-ego for Maxwell himself, Figure 5.12.

The three laws of thermodynamics have been debated since they were formulated in the late 19th and early 20th centuries. The First Law states that energy cannot be created or destroyed; the Second Law that the amount of disorder or entropy in an isolated system can never decrease; and the Third Law that it is impossible to cool an object to absolute zero in a finite number of steps. In 1847, James Joule (1818-1889) stated the principle of conservation of energy (and thereby the First Law), concluding that "Experiment has shown that whenever living force [kinetic energy] is apparently destroyed or absorbed, heat is produced," [381] and further he gave the amount of heat equivalent to the converted kinetic energy. Rudolph Clausius (1822-1888) formulated the Second Law in 1850 based on empirical evidence that there is a favored direction for spontaneous processes. He quantified the notion of irreversibility of macroscopic processes by defining the thermodynamic entropy *S*, which can only increase for isolated systems. Decrease of entropy required energy exchange with the environment. Lord Kelvin framed the Second Law equivalently in terms of the work produced by a cyclic engine. Walther Nernst (1864-1941), during the years 1906-12, formulated the Third Law. The Second Law implies that heat flows from hot to cold, devices of perpetual motion of the second kind cannot exist and that the production of disorder is irreversible although this appears to conflict with the reversible laws of mechanics at the microscopic level. Understanding the issues involved with the classical mechanical demon took another 115 years, and much progress has been made on Maxwell's demon in the context of quantum mechanics [382]. Maxwell's demon by construction did not change the energy but its "selective influence" did involve observation or measurement. A final understanding of the quantum demon must wait for a complete theory of measurement so that it is known with confidence which assumptions made in various formulations of the demon are valid.

The Second Law characterizes the difference between reversible processes which do not change the entropy and irreversible ones in which work done is dissipated as heat, resulting in increased entropy of the system. In thermodynamics, entropy is a state variable whose change is defined for a reversible process at temperature T as $dS = \delta Q/T$ where δQ is the heat absorbed, so the task of a potential demon is to somehow decrease the entropy for each cycle of the process. For example, in 1912 Smoluchowski (1872-1917) [383] considered an inanimate object without intention as a possible demon in the form of a spring attached to a door that only opens to one side, so that slow moving particles could be prevented from returning from *A* to *B* by the door's snapping shut. However, Smoluchowski showed that thermal fluctuations prevent the spring door from acting as a demon as they cause the door to execute a Brownian motion so that it occasionally gets kicked in the wrong direction just enough to foil the demon.

However, it was Leo Szilard (1898-1964) in 1929, in the heightened period after the discovery of quantum mechanics, who analyzed an intentional demon in terms of *information*. He analyzed a single-particle in a two-chamber heat engine in which a Maxwell demon could extract work from the closed cycle and once again violate the Second Law [384]. This was the same Szilard who conceived of the nuclear chain reaction in 1933 and in 1939 wrote the confidential letter that the reluctant Einstein signed to convince President Roosevelt to begin building the atomic bomb. Szilard's two-state engine was limited by the Second Law to producing a maximum amount of work equal to $k_B T \ln 2$. The two-state engine essentially constituted a computational bit, thus $k_B T \ln 2$ is also the maximum energy that can be obtained by one bit of information. Here was a first indication that information and energy were related and Szilard's article may have been the first information theory paper. Claude Shannon (1916-2001) later defined an informational entropy that increased with information content establishing the field of information theory in 1948. Szilard concluded that there was an entropic price paid caused by the demon's measurement that balanced the extracted work.

Rolf Landauer (1927-1999) showed in 1961 that the entropy change in Szilard's model actually came in the step that erases the information previously acquired during the measurements [385]. The dissipation of heat into the environment due to the process of erasing compensates the entropy decrease caused by Maxwell's demon. That the erasure of one bit of information is necessarily accompanied by the release of at least $k_B T \ln 2$ of heat is called *Landauer's erasure principle*. For this to be valid, the erasure process should not depend on the initial state of the system. Arguments by Brillouin and Gabor had previously claimed that the Second Law was not violated because the demon's measurement resulted in an energetic penalty [386] [387]. However, Charles Bennett demonstrated that general-purpose computation can be performed in a logically and thermodynamically reversible manner, and in 1982 he proposed that Maxwell's demon was prevented from breaking the Second Law due to the thermodynamic cost of erasing the information by Landauer's principle and not due to acquiring it during information processing [388]. This erasure is necessary after a full cycle of information processing and energy production because the demon's memory must be reset to allow the next iteration. Landauer's principle guarantees that the erasure dissipates more energy than the demon produces during each cycle, in agreement with the Second Law. Landauer erasure has been experimentally verified in classical systems such as trapped colloidal particles and RC-circuits [389] [390] [391]. Experiments have also demonstrated that a vanishing amount of heat is dissipated for reversible systems [392]. Therefore, any attempts by classical mechanical demons to intervene in irreversible processes are vanquished by Landauer's principle.

Is Irreversibility Intrinsic?

Einstein was convinced that thermodynamics was the only universal physical theory that will never be overthrown because he believed thermodynamics stands on general principles that are independent of the underlying details [339, p. 33]. However, a microscopic understanding of thermodynamics relies on the framework of the statistical mechanics of atoms and molecules. The term *irreversibility* was introduced into physics primarily in the context of the Second Law of Thermodynamics and determining whether the law was exact or an approximation became an overriding issue. In yet another letter from Maxwell to Tait, he explained that the point of his demon argument had been to demonstrate that the Second Law "has only a statistical certainty" [393] [394]. Ludwig Boltzmann (1844-1906) developed a statistical approach to explain the time-asymmetric and irreversible non-equilibrium behavior of macroscopic systems and insisted that such processes were determined by which states were the most probable to occur. Although Boltzmann's view eventually prevailed, the issue of entropy and irreversibility were highly contentious subjects in the second half of the 19th century, leading up to the discovery of the quantum [395] [396].

Einstein explained Brownian motion using Boltzmann's ideas and in Planck's discovery of the quantum, he was forced in desperation to accept Boltzmann's methods in order to explain the black-body radiation spectrum when nothing else would work. Objections to Boltzmann's ideas, in terms of the reversibility paradox in 1876 by Josef Loschmidt (1821-95) and by the recurrence paradox in 1896 by Ernst Zermelo (1871-1953), were part of the conflicts at the end of the 19th century over the roles of atoms and energy in scientific explanations of matter. In addition to these specific objections was the general but strenuous opposition by the *energeticists*, such as Wilhelm Ostwald, and Pierre Duhem, who tried to understand all physical processes merely with the help of concepts involving only pure energy, as well as by Ernst Mach who rejected atomic theory.

In a lecture at the 1944 centennial of Boltzmann's birth, the great mathematical physicist Arnold Sommerfeld (1868-1951), who made important contributions to the early formulations of quantum mechanics and whose students were Heisenberg and Pauli, described a debate at a meeting in Lübeck in 1895 between Ostwald, who denied the existence of atoms, and Boltzmann [357, p. 46] [397]:

> *The champion for energetics was Helm; behind him stood Ostwald, and behind both of them the philosophy of Ernst Mach (who was not present in person). The opponent was Boltzmann, seconded by Felix Klein. The battle between Boltzmann and Ostwald was much like a duel of a bull and a supple bullfighter. However, this time the bull defeated the toreador in spite of all his agility. The arguments of Boltzmann struck home. We young mathematicians were all on*

> *Boltzmann's side; it was at once obvious to us that it was impossible that from a single energy equation could follow the equations of motion of even one mass point, to say nothing of those for a system of an arbitrary number of degrees of freedom.*
>
> Life and Personality of Ludwig Boltzmann, Acta Physica Austriaca, Suppl. X, The Boltzmann Equation, Vol. 10, 1973, p. 17, D. Flamm, © Springer-Verlag 1973. With Permission of Springer.

This was an example of Planck's lament that "A scientific truth does not triumph by convincing its opponents and making them see the light, but rather because its opponents eventually die and a new generation grows up that is familiar with it." The opposition to Boltzmann by the older generation persisted despite the lack of new adherents.

However, Boltzmann's ideas were relevant to Maxwell's demon. Without a demon present, the gas of particles in Maxwell's box will eventually become uniformly distributed across both compartments *A* and *B*. This is irreversible in the sense that, in our experience, the particles do not find themselves all grouped in compartment *A* at a later time. Josef Loschmidt (1821-1895) objected that this irreversibility cannot be explained using only the laws of mechanics since these laws are reversible in time. Loschmidt demonstrated that any solution for the motion of the particles remains a solution when the velocities of all the particles are reversed at a particular time: $\bar{v}(t) \longrightarrow -\bar{v}(t)$, so that all processes will proceed in reverse. The particles could all very well end up in *A* at a later time if governed only by the reversible laws of mechanics. Some contemporary commentators would note with approval Boltzmann's supposed retort to Loschmidt: "Go ahead and reverse them!" However, Boltzmann's actual argument was that the distribution of particles depends not only on the laws of mechanics but also on the initial states of the particles. Loschmidt's theorem becomes a concern only for those initial states which lead to a very non-uniform distribution at a certain time but it doesn't prove that there are not more initial conditions that lead to a uniform distribution at the same time. Loschmidt's reversed swarm of particles was not impossible, just quite improbable [357, p. 98]. In fact, Boltzmann could demonstrate that there are vastly more uniform than non-uniform distributions so that there are a correspondingly much greater number of initial states which lead to them.

Niels Bohr had a strong interest in statistical mechanics beginning with his thesis work and with his 1913 quantum model of the atom. From Bohr's 1912 lecture notes is his hand-drawn sketch of an isolated mountain and a succinct description of what constitutes entropy that illustrates Boltzmann's argument in a very simple way: in terms of a man who finds himself wandering in a flat land, so that it may be said that in a year, it would also be expected to find him there, because it would be a strange coincidence if, on his way, he should just hit on the one sharply peaked mountain [398, p. 320].

Thirty years earlier, both Maxwell and Thomson (later Lord Kelvin) had also agreed with Bohr's view of Boltzmann's arguments in that they accepted an underlying reversibility for classical systems but they acknowledged that there are physical processes that were effectively irreversible. Maxwell gave an example of the consequences of listening only to Loschmidt [399]:

> Now *one thing in which the materialist (fortified with dynamical knowledge) believes is that if every motion great & small were accurately reversed, and the world left to itself again, everything would happen backwards: the fresh water would collect out of the sea and run up the rivers and finally fly up to the clouds in drops which would extract heat from the air and evaporate and afterwards in condensing would shoot out rays of light to the sun and so on. Of course, all living things would regrede from the grave to the cradle and we should have a memory of the future but not of the past.*
> The Scientific Letters and Papers of James Clerk Maxwell, Volume II, P.M. Harman (Ed), Cambridge University Press, Reissue Edition, 1995.

Boltzmann also responded to Loschmidt by working out his statistical interpretation of the Second Law in detail, which appeared in his 1877 paper, "On the relation between the second law of the mechanical theory of heat and the probability calculus with respect to the theorems on thermal equilibrium." For each microscopic state of a macroscopic system, Boltzmann allocates an entropy, S_B, which accurately characterizes isolated systems and increases to describe the approach to equilibrium where its value then agrees with Clausius' entropy S. Boltzmann associates each macroscopic state M with an entropy that also represents every microscopic state m in the phase-space volume Γ_M associated with M:

$$S_B(M(m)) \propto \log \Gamma_{M(m)}.$$

Max Planck introduced the notation for the entropy that now appears on Boltzmann's tombstone:

$$S_B = k \log W,$$

where k is Boltzmann's constant and W is the number of microstates by which the macroscopic state M can be realized. For non-equilibrium systems, this identifies the increase in entropy with increases in the volume of the occupied phase space region. Macroscopic systems have a very large number of degrees of freedom and Boltzmann's approach captures the *typicality* of the actually observed irreversible behavior. Rare events, such as the spontaneous separation of gas particles into side A, would occur only over times longer than the age of the universe. This feature of Boltzmann's statistical theory also answered Zermelo's objection based on Henri Poincaré's (1854-1912) *recurrence theorem*. Poincaré had shown that any bounded system will almost certainly return to a state that is arbitrarily close to its initial state. However, Boltzmann easily showed that the recurrence times for any macroscopic system would be so enormous as to be irrelevant.

Boltzmann's approach brought probability into science at a time when determinism was dominant. The entropy can be expected to increase with time if the system has been initially prepared in a low-entropy state, which can be easily arranged by [400]

> *"experimentalists who are themselves in low-entropy states...born in such states and maintained there by eating low-entropy foods, which in turn are produced by plants using low-entropy radiation coming from the Sun, and so on."*
>
>

This line of reasoning leads unavoidably to the question of the initial state of the universe having a very small Boltzmann entropy. Richard Feynman notes that understanding irreversibility requires adding to the physical laws the hypothesis that in the past the universe was more ordered than it is today [401]. Roger Penrose argues that the initial state just after the Big Bang has an approximately uniform energy density with low entropy [402, p. 702].

Boltzmann generalized Maxwell's approach for the kinetic theory of gases to non-equilibrium processes, deriving what is now called the *Boltzmann equation*, describing the approach from non-equilibrium to equilibrium in terms of collisions of the particles. Based on this, he attempted a derivation of irreversibility in thermodynamics based on the underlying classical mechanics of colliding particles that he considered equivalent to the law of entropy increase. This was formulated in terms of a function H identified with negative entropy for which

$$\frac{dH}{dt} \leq 0$$

where the equality holds if and only if thermal equilibrium is reached. This is Boltzmann's *H-theorem*. Since classical mechanics is reversible in time, Boltzmann had to have made an assumption in his derivation that breaks time symmetry but identifying it precisely turned out to be subtle. It was finally found during a debate in the pages of Nature during the 1890s [396]. The assumption (later called the assumption of *molecular chaos* or *Stosszahlensatz*) was that the velocities of colliding particles are uncorrelated and independent of position. Therefore, an assumption of disorder had been made. The status of Boltzmann's molecular chaos assumption was later reconstructed by Paul and Tatiana Ehrenfest in 1912 [403], which led to renewed interest in the related but difficult concept of the ergodic theorem equating statistical averages and time averages [395]. Observations of physical quantities during experiments last a finite duration which may be extremely long on the microscopic level, allowing the microstate to experience many collisions, so that time averages are empirically significant. The ergodic theorem was revisited and clarified by von Neumann and Birkhoff in the 1930s after the development of quantum statistical mechanics.

Demon versus Photon

Following Planck's opening address on *Radiation Theory and Quanta* at the 1st Solvay Conference on Physics in 1911 (the "witches Sabbath in Brussels" as Einstein described it), the indefatigable Marie Curie (1867-1934) noted that the emission of energy was instantaneous in Planck's theory and this implied that Maxwell's equations would not hold even in vacuum. She further asked how the emission of energy in Planck's theory could be interrupted and suggested a parallel to Maxwell's Demon [404, p. 39]:

> *It is quite probable that such a mechanism will not exist at our [human] level, but rather be comparable to Maxwell's demon. It would allow one to obtain the deviations based on the laws of radiation as envisaged by statistics just as Maxwell's Demon allows one to obtain the deviations following from the consequences of Carnot's principle.*
>
> The Solvay Conferences on Physics, Aspects of the Development of Physics Since 1911, 1975, J. Mehra, © D. Reidel Publishing Company, Dordrecht, Holland. With permission of Springer.

Both comments were intuitively perceptive. The Photoelectric Effect semi-classically does not conserve energy. At short times, the energy that has fallen on the photo-detector cannot exceed the work function. However, a discrete photon is annihilated and circumvents the energy problem [405, pp. 20-22]. Gregory Wentzel (1898-1978) presented his semi-classical theory of the photoelectric effect at the 1927 Solvay Conference where Einstein would have heard him. At that time, Einstein was confronting the apparently conflicting issues of energy conservation, nonlocality and a photon that does not split at a beam splitter.

Backstory to Wave-Particle Duality

In Query 29 of his *Opticks*, Newton had famously asked:

> *Are not the Rays of Light very small Bodies emitted from shining substances?*

Thus, Newton conjectured his corpuscular view of light and was able to account quantitatively for the optical phenomena known in the 17th century, including refraction of light. It was not until the 19th century that the observation of diffraction showed conclusively the inadequacy of Newton's picture. Meanwhile, Christiaan Huygens (1629-1695) in the 1690s proposed that light "spreads, as sound does, by spherical surfaces and waves: for I call them waves from their resemblance to those which are seen to be found in water when a stone is thrown into it." [406] To support the transmission of such waves, Huygens proposed the existence of "ethereal matter," the thorny concept of an ether going back in some form to the Greeks and persisting up to applications of the electromagnetism of Maxwell and finally found to be unnecessary by Einstein's development of special relativity in 1905. However, Huygens' waves were actually longitudinal as with the sound waves of his inspiration, and not the transverse waves that we now know from Maxwell's electromagnetism. And they were considered in terms of impulses in a medium and not as periodic waves in the modern sense because methods for treating periodicity had not yet been formulated [309, p. 343]. Another property of light familiar to Newton and Huygens was interference though then only incompletely understood. Thomas Young (1773-1829) made a decisive advance by explaining the physics behind constructive and destructive interference of overlapping waves followed by Augustin Fresnel (1788-1827) who formulated the mathematics of interference as well as experiments to verify the ideas. This included the double-slit arrangement as previously discussed in Chapter 1. An important confirmation of Fresnel's theory was obtained by François Arago (1786-1853) who experimentally observed a bright spot at the center of the geometrical shadow of a circular obstruction. Formed by constructive wave interference, this phenomenon was surprising to many at the time and became known as "the spot of Arago." By the late 19th century, the success of Maxwell's approach had merged optics and electromagnetism, with light now being recognized as a type of electromagnetic wave. And so it was that light was firmly in the category of *wave* when the quantum of action was discovered in 1900.

However, Newton was somewhat of a moving target as usual. In order for Newton to be able to describe diffraction, he knew that the motion of the light corpuscles would have to be affected at a distance by the diffracting body. An example was Francisco Grimaldi's (1618-1663) observation of the fringes of light diffracted from an edge in 1665. Regarding this situation, Newton also famously said in another query:

> *Are not the Rays of Light in passing by the edges and sides of Bodies, bent several times backwards and forwards, with a motion like that of an Eel? And do not the three Fringes of colored Light above-mentioned arise from such bendings?*

Thus, within Newton's picture, a force necessarily reaches out from the surface of the body. For Newton's theory to be successful in all applications, including diffraction, it would actually have had to be something of a corpuscular-wave! However, rather than being ahead of his time, anticipating a type of *eel-corpuscle duality*, we can see in hindsight that Newton's approach to optics did not stand a chance in 1704 when the *Opticks* was published. However, Newtonian mechanics from the *Principia* was enormously successful and changed the way science was done. Why was the *Principia* so successful and the *Opticks* much less so?

As discussed in the section *Deductive versus Inductive Thought*, the striking aspect of Newton in the *Principia* was his deductive approach in that he would "feign no hypotheses" in contrast to other scientists of the era. What sets Newton apart from Leibnitz and others is that he does not insist that nature act in certain ways just because he doesn't like it. Nature did conform accurately to his universal gravity in which a force between masses reaches out instantaneously over arbitrary distances. Leibnitz would object to Newton's theory and simply declare that "matter cannot act where it is not." But that was his loss. Newton did not understand how "matter cannot act where it is not" either. In a private letter in 1692 to a confidant, Newton stated [407, p. 52]:

> *That gravity should be innate, inherent & essential to matter so that one body may act upon another at a distance through a vacuum without the mediation of anything else by & through which their action or force may be conveyed from one to another is to me so great an absurdity that I believe no man who has in philosophical matters any competent faulty of thinking can ever fall into it.*
>
> Newton's Third Law and Universal Gravity, in Newton's Scientific and Philosophical Legacy, 1988, p. 25, J.B. Cohen,

Newton repeatedly did search for some explanation of how universal gravity might act. He sought to describe universal gravity in terms of other phenomena: aether particles, electrical effluvia, an aether with variable density, etc. [407, p. 40]. None of these approaches could produce a force varying inversely as the square of the distance and that acts mutually between masses. But he did not allow this to affect or prejudice the science that did. One historical study has argued that Newton did not follow his deductive approach in the *Opticks* but instead it was Huygens who more closely followed Newton's deductive method from the *Principia* in his own more successful development of optics [408]. Perhaps the paradigm of deductive versus inductive thinking also explains this aspect of Newton's research in optics.

The impact of Newton's *Principia* was so overwhelming that his universal theory of gravity, with its force instantaneously acting at distance, became the template for performing successful science. In particular, it was also applied to the laws of electricity and magnetism in the century to follow. The formulation of electrodynamics by André-Marie Ampère (1775-1836) was in terms of an inverse-square force acting instantaneously at a distance [340, p. 348], and this approach was adopted by others into the next century including Gauss, Weber, Lorentz, Lienard and Wiechert. However, the mostly self-taught but prodigious Michael Faraday introduced the concept of lines of force traversing space between conductors and made this tangible in his demonstrations using sprinklings of iron filings. This led to the view of an electromagnetic field occupying the space in the regions between current carrying conductors. The field became an independent entity that could fill space itself, even in a vacuum. Faraday's field concept was taken over by Maxwell leading eventually to his electromagnetic field equations. Ampère's action at a distance formulation would give identical results as long as the current changes were not too rapid. The concept of electromagnetic fields ultimately would be needed as experiments became more refined. However, at the time of Faraday's investigations electromagnetic waves were unknown and they were first detected by Heinrich Hertz (1857-1894) in 1888, twenty years after Faraday died. This gave concrete evidence for the field concept which would become central to much of the physics of the 20th and 21st centuries. One implication of the presence of electromagnetic fields that was later discovered was that they would require a finite time to propagate. However, Faraday actually anticipated this and he wrote a sealed letter on March 12, 1832 stating his views on the propagation of electromagnetism, intending it to be read after one hundred years. The letter was given to the Secretary of the Royal Society of London where it lay forgotten for the next century until it was finally opened by Sir William Bragg (1862-1942) on June 24, 1937 [409, p. 11]:

> *I am inclined to compare the diffusion of magnetic forces from a magnetic pole to the vibrations upon the surface of disturbed water, or those of air in the phenomenon of sound; i.e. I am inclined to think the vibratory theory will apply to these phenomena as it does to sound, and most probably to light. By analogy, I think it may possibly apply to the phenomenon of induction of electricity of tension also. These views I wish to work out experimentally; but as much of my time is engaged in the duties of my office, and as the experiments will therefore be prolonged, and may in their course be subject to the observation of others, I wish, by depositing this paper in the care of the Royal Society, to take possession as it were of a certain date; and so have right, if they are confirmed by experiment, to claim credit for the views at that date; at which time as far as I know, no one is conscious of or can claim them but myself.*
>
> G.R.M. Garratt, The Early History of Radio: From Faraday to Marconi, The Institute of Engineering and Technology, London 2006, Reproduced by permission of the Institution of Engineering & Technology.

Planck's Fortunate Guess

In response to improved experimental measurements, Planck had initially guessed Equation (5.3) in October 1900 using thermodynamic entropy arguments to interpolate two expressions: first, a heuristic formula of Rayleigh's that accurately fit experiments at low frequencies. In this regime, it is equivalent to the improved expression of Rayleigh-Jeans which was derived later, given by:

$$B^{RJ}(\nu, \mathrm{T}) = \frac{8\pi}{c^3}\nu^2 \mathrm{kT}, \tag{5.5}$$

and secondly, a formula proposed by Wilhelm Wien (1864-1928) that accurately fit experiments at high frequencies:

$$B^{W}(\nu, \mathrm{T}) = \frac{8\pi a}{c^3}\nu^3 \exp(-b\nu \mathrm{T}). \tag{5.6}$$

Here the universal constants a and b are of course precursors of combinations of k and h. By introducing the new constant h, Planck used an entropy argument to successfully find an interpolation between the two laws yielding Equation (5.3). He described his expression as a "fortunate guess," but it proved to be in perfect agreement with all the experiments to date and would continue to agree with continually refined future experiments. Today this includes the thermal radiation of the cosmic microwave background left over from the Big Bang of cosmology. It has been said of Planck's interpolation that [410, p. 18]:

> *Never in the history of physics was there such an inconspicuous mathematical interpolation with such far-reaching physical and philosophical consequences.*
>
>

By way of his "vehement obstinacy" in obtaining this law, Planck could be said to have reached a "conflagration of clear sight" [411]. Ineluctable modality of the quantum. The auxiliary variable Planck had needed for his interpolation was called "h" after the German word *hilfsgrösse* [412] which translates to "auxiliary variable," humble nomenclature for a quantity with grand implications. Such is how history is made. The fortunate guess was later shown to be exact within quantum mechanics.

However, at this point there was not yet any quantization in Planck's work despite the appearance of h. Although not appreciated at the time, there was, however, foreshadowing of the quantum wave-particle duality at this stage. Planck had followed closely the works of Rudolf Clausius (1822-1888), one of the founders of thermodynamics who had formulated the Second Law of Thermodynamics in terms of the irreversible increase of entropy. The entropy S was a new quantity introduced by Clausius as a measure of how much thermal energy is unavailable for conversion into mechanical work. Planck had taken the second law to be absolute and that entropy would increase without exception and he was skeptical of the atomic hypothesis that all matter is composed of indivisible particles. In contrast, the pioneers of statistical mechanics viewed the world in term of atoms: Maxwell, Ludwig Boltzmann (1844-1906), and Josiah Willard Gibbs (1839-1903). Planck had spent the previous several years in attempts to understand the radiation problem from first principles without success and had reached the point of taking the desperate steps of using Boltzmann's statistical mechanical methods.

Since Kirchhoff had shown that thermal radiation was independent of the particulars of the radiating body, Planck chose the simplest model of a set of electrically charged oscillators with energy U (one for each frequency) filling a cavity with reflecting walls. As summarized in Appendix 5.A, this led him to guess an interpolation expression, $\alpha U + \beta U^2$, which exactly echoes the first identification of wave-particle duality later found by Einstein in 1909. Einstein had applied his previously derived formula for the fluctuation in energy to Planck's law and found:

$$\Delta E^2 = h\nu < E > + \frac{c^3}{8\pi\nu^2} < E >^2. \tag{5.7}$$

This is given per unit volume and frequency range and $< E > = U$ is the average oscillator energy. The partition into the linear and quadratic terms in Equation (5.7) just matches that for Planck's fortunate guess for his interpolation and explains why

Planck's resulting radiation law Equation (5.3) exactly matches that later found from quantum mechanics.

Einstein immediately recognized the linear term in Equation (5.7), corresponding to the Wien limit, as having a particle origin corresponding to the fluctuations of a gas of independent particles. The quadratic term, corresponding to Rayleigh-Jeans, could be identified with fluctuations of the superposition of random standing waves in a small cavity. What had emerged was apparently a type of "wave-particle" duality but it was perplexing to Einstein in how to think of this in terms of his previous 1905 proposal of light-quanta, now called *photons*, as independent localized particles. In a 1909 letter to Lorentz, Einstein said [413]:

> *I am not at all of the opinion that one should think of light as composed of quanta which are independent of each other, and localized in relatively small regions. For the explanation of the Wien limit of the radiation formula this would indeed be the easiest way. But even the splitting of a light ray at the surface of a refracting medium completely forbids this approach. A light ray divides, but a light quantum indeed cannot divide without an alteration of frequency.*

As will be seen, this moment is the onset of Einstein's continuing difficulties in facing the issues of wave-particle duality and later entanglement, even though Schrödinger did not coin that term until 1935 in relation to the measurement problem. Although not then recognized in full, wave-particle duality goes hand-in-hand with entanglement and the implications of this can already be seen emerging in 1909, with much more to come in the story. One piece of the puzzle was that the independent light-quanta had to have the additional quantum property of *indistinguishability*, which affects the counting of the possible energy states they may occupy. And this property implies that they are actually not quite independent but in a subtle way, the details of which would have to wait for the discovery of Bose-Einstein statistics in 1924. However, this is what allowed the Wien limit at low temperature to be handled in terms of "classical" independent particles while the explicit appearance of h counterintuitively appears at high temperatures in the Rayleigh-Jeans limit.

Therefore, we see in hindsight that Planck's fortunate guess already contained the seeds of wave-particle duality. But Planck did not yet have a deep understanding of the physics that lie behind the radiation law that had emerged from his guess. With the new interpolation expression Equation (5.3) now in hand, he then spent eight strenuous weeks between October and the December meeting of the German Physical Society searching for a theoretical understanding of it. In his derivation, Planck reluctantly drew upon Austrian physicist Ludwig Boltzmann's 1877 result for the increase of entropy utilizing a relation between entropy S and probability W, summarized by the equation $S = k \log W$. Although Planck had previously wanted to avoid the statistical conception of irreversibility advocated by Boltzmann, he reinterpreted the probability in Boltzmann's relation as a measure of the elementary disorder depending on the internal structure of his cavity oscillators. Both Planck's and Boltzmann's investigations were of the evaluation of the maximum entropy, the most probable equilibrium states of systems with energies in random distributions. In order to calculate the probability of a state with an energy shared among many oscillators of the same frequency, it was essential that Planck treat the energy as being composed of finite magnitude cells of size E in phase space. In Boltzmann's derivations, the size of the cells, representing molecules with different velocities, were arbitrary and could be taken to zero, while Planck could arrive at a derivation of Equation (5.3) only with finite cells of blackbody cavity oscillators with different frequencies, proportional to multiples of $E = h\nu$ (see Appendix 5.A).

Planck described his plight in taking the step of discrete energies in a letter to R.W. Wood thirty years later, describing it as "an act of desperation" [414]:

> *"I knew then that the problem is of fundamental significance for physics; I knew the formula that reproduces the energy distribution in the normal spectrum; a theoretical interpretation* ***had*** *to be found at any cost, no matter how high."*
>
> Reproduced from M. J. Klein, Thermodynamics and Quanta in Planck's Work, Physics Today 19 (11), 23 (1966) by permission of the American Institute of Physics.

Max Planck's discovery of the quantum in 1900, which combined relentless trial and error on Planck's part using the newest available ideas from statistical mechanics along with his awareness of the results emerging from the precision black-body developed at the Physikalisch-Technische Reichsanstalt nearby after he had moved to Berlin [415],

> *a thing of beauty of porcelain and glass, a heavy insulated cylinder with interior reflecting walls…communicated with the world via a hole from which radiation could escape for measurement.*
>
> Max Planck's compromises on the way to and from the Absolute. In: Quantum Mechanics at the Crossroads, New Perspectives from History, Philosophy and Physics, 2007, p. 21, J.L. Heilbron, © 2016 Springer-Verlag Berlin. With permission of Springer.

As seen in the section *The Quantum Triumvirate,* Planck's quantum of action led to the rise of nondeterminism and the insights for how the triumvirate of nondeterminism, entanglement and nonlocality shape quantum theory. From the beginning, Planck also understood its relation to atomism and their spectral lines [416]:

> *… the dimensions of the quantum are not energy but "action" (energy times time) so that its complete explanation will come not from considerations of a state but from considerations of a process. In other words, we are dealing with atomism not in space but in time since processes that we used to consider as steady in time really show*

> *temporal discontinuities... I was interested that this natural constant remains invariant according to the relativity principle when transferring from a resting to a moving frame of reference although almost all other quantities like space, time, and energy change. It was just this fact that led me to a closer investigation of the relativity principle. I'm fully convinced that the problem of spectral lines is intimately tied to the question of the nature of the quantum of action.*
>
> Max Planck's compromises on the way to and from the Absolute. In: Quantum Mechanics at the Crossroads, New Perspectives from History, Philosophy and Physics, 2007, p. 21, L. Heilbron,

Over the next decade, scientists gradually came to terms with the implications of Planck's result. There was little immediate interest up to 1905 when Einstein took the next step of introducing the light-quanta. As discussed in the section *Atomism Prevails,* he was also resisted by those who opposed the atomic viewpoint: Ernst Mach (1838-1916), Wilhelm Ostwald (1853-1932), and Pierre Duhem (1861-1916) who had also battled against Boltzmann. Some historians have argued whether or not Planck actually accepted the discontinuous aspect of his derivation both from the historical record of Planck's work leading up to the derivation and his subsequent statements seemingly hedging on the issue [417] [418]. The truth may lie somewhere in between since in his Nobel Prize address in 1920, Planck credits Einstein with taking the decisive step:

> *The failure of all attempts.... soon left no doubt: either the quantum of action was a fictitious quantity, or the derivation of the radiation law rested on a truly physical thought...Experiments have decided in favor of the second alternative. But science does not owe the prompt and indubitable character of this decision to tests of the law of the energy distribution of thermal radiation, and even less to my special derivation of this law; it owes that to the unceasing progress of the researchers who have put the quantum of action to the service of their investigations. A. Einstein made the first breakthrough in this domain.*

Planck was clearly the discoverer of the quantum although subsequent researchers uncovered the physical meaning of it, and foremost was Einstein. Einstein's influence stands at the crossroads of space-time and quanta. Shown in Figure 5.8 are trajectories in space-time with the constraints imposed by relativity as well as apparently similar trajectories originating from quanta, the spherical quantum wave functions of alpha particles. Understanding the relation between space-time and quanta occupied the first decades of the 20th century.

Bohr's Correspondence Principle

Bohr's theory was deliberately incomplete so that he could find principles that would allow a systematic search for a "rational generalization" of classical electrodynamics. A major conceptual tool that characterized Bohr's atomic work was the *correspondence principle*, his celebrated asymptotic consistency requirement between quantum theory and classical physics, that took various forms over the years and was often misunderstood by even his closest collaborators [312, p. 114], but guided his research and dominated quantum theory until the emergence of quantum mechanics in 1925-26 [419, p. 81]. It should be emphasized that the correspondence principle is not the elementary recognition that the results of classical and quantum physics converge in the limit $h \rightarrow 0$. Bohr said emphatically that this [420]

> *is not the correspondence argument. The requirement that the quantum theory should go over to the classical description for low modes of frequency is not at all a principle. It is an obvious requirement for the theory.*
>
> The Wave-Particle Dilemma. In: Mehra J. (eds) The Physicist's Conception of Nature, 1973, p. 251, L. Rosenfeld, © 1973 D. Reidel Publishing Company. With permission of Springer.

Instead, the correspondence principle is deeper and aims at the heart of the differences between quantum and classical physics. Even more than the previous fundamental work of Planck and Einstein on the quantum, Bohr's atomic work marked a decisive break with classical physics. This was primarily because from the second postulate of Bohr's theory there can be no relation between the light frequency (or color of the light) and the period of the electron, and thus the theory differed in principle from the classical picture. This disturbed many physicists at the time since experiments had confirmed the classical relation between the period of the waves and the electric currents. However, Bohr was able to show that if one considers increasingly larger orbits characterized by quantum number n, two neighboring orbits will have periods that approach a common value. And from the frequency of the light that is emitted in a jump between two such orbits according to Bohr's quantum postulate, one finds that it approaches the frequency of classical theory with this period. Therefore, even though the ***physics*** of the emission of light in quantum theory is quite different from that in the classical theory, the results approached approximate classicality with increasing n. Bohr's theory in a precise way contained within itself the results of classical theory and is a generalization of classicality in the sense of this correspondence principle.

Bohr first applied the correspondence arguments when he argued that as energy intervals $\Delta E = h\Delta\nu$ between stationary states become arbitrarily small the emitted frequencies would coincide with the frequencies expected from classical theory and he used this correspondence in 1913 to derive the energy levels of thc hydrogen atom and subsequently in his analysis of the Stark effect. It could also function as a selection principle, as Bohr explained [421]:

> *Among the processes that are conceivable and that according to the quantum theory might occur in the atom we shall reject those whose occurrence cannot be regarded as consistent with a correspondence of the required nature.*
>
> Niels Bohr, The Theory of Spectra and Atomic Constitution, Three Essays, Cambridge University Press, 1924.

Bohr's most careful statement of the correspondence principle appeared in a 1921 paper [422]. However, Bohr insisted that the correspondence principle was quantum mechanical and that its physical meaning did not depend directly on the validity of classical mechanics for the motion of an atomic system in stationary states. Bohr's statements following Heisenberg's development in 1925 of matrix mechanics, the first consistent theory of quantum mechanics, including his later complementarity interpretation of the wave-particle duality, continued to be influenced by the correspondence between the quantum and classical regimes. In Bohr's famous Como lecture of 1927 in which he first introduced the concept of complementarity, he talked of [310]

> *a far-reaching correspondence between the consequence of the classical theory and those of the quantum theory.*
>
>

The correspondence principle influenced his interpretation of quantum mechanics and was a key element for his formulation of complementarity. Bohr felt that the principle had been fully incorporated into Heisenberg's matrix mechanics in that [198]

> *the whole apparatus of the quantum mechanics can be regarded as a precise formulation of the tendencies embodied in the correspondence principle.*
>
>

Bohr's friend and colleague Georg von Hevesy (1885-1966) recounted Einstein's reaction to this aspect of Bohr's theory in a 1913 letter to Bohr [423, p. 113]:

> *I told him then that it is established now with certainty that the Pickering-Fowler spectrum belongs to Helium. When he heard this he was extremely astonished and told me: "Then the frequency of the light does not depend at all on the frequency of the electron? ... And this is an* ***enormous achievement****. The theory of Bohr must then be right."*
>
> Andrew Whitaker, Einstein, Bohr and the Quantum Dilemma: From Quantum Theory to Quantum Information, Cambridge University Press 2006.

Einstein later summarized Bohr's achievements in atomic theory [339]:

> *That this insecure and contradictory foundation [of quantum physics in those years] was sufficient to enable a man of Bohr's unique instinct and tact to discover the major laws of the spectral lines and of the electron shells of the atoms together with their significance for chemistry appeared to me like a miracle—and appears to me a miracle even today.*
>
> Albert Einstein, Autobiographical Notes, In: Albert Einstein: Philosopher-Scientist, Paul Arthur Schillp (Editor), Cambridge University Press 1949.

Born's Statistical Interpretation

Shortly after quantum theory was developed with the matrix mechanics of Heisenberg in 1925 and the wave mechanics of Schrödinger in 1926, it was often not clear how to describe even the simplest experiments and how to distinguish space-time and quanta in these situations. As seen in Figure 5.8, the tracks left by quanta within a cloud chamber, a device filled with supersaturated vapor which creates an ionized trail in response to a charged particle, do not seem so different than the time-like trajectories of any massive particle within a light-cone of space-time. This was discussed at the 1927 Solvay Conference, which was dominated by discussion of the recently discovered formulations of quantum mechanics. Max Born (1882-1970), who had proposed his rule for the statistical interpretation of quantum mechanics, raised that very question as prompted by a concern of Einstein's [4, pp. 160, 437]:

> *Mr. Einstein has considered the following problem: A radioactive sample emits α-particles in all directions; these are made visible by the method of the Wilson cloud [chamber]. Now, if one associates a spherical wave with each emission process, how can one understand that the track of each α-particle appears as a (very nearly) straight line? In other words: how can the corpuscular character of the phenomenon be reconciled here with the representation by waves?*
>
> G. Bacciagaluppi and A. Valentini, Quantum Theory at the Crossroads: Reconsidering the 1927 Solvay Conference, Cambridge University Press, 2013.

Born goes on to say

> *The description of the emission by a spherical wave is valid only for as long as one does not observe ionization.*
>
> G. Bacciagaluppi and A. Valentini, Quantum Theory at the Crossroads: Reconsidering the 1927 Solvay Conference, Cambridge University Press, 2013

In other words, Born appeals to observation or measurement as the crucial factor and he follows this by referring to both Heisenberg's "reduction of the wave-packet" as well as an attempted unitary description of measurement by Pauli.

In Heisenberg's uncertainty paper, he also acknowledges in a footnote, "The statistical interpretation of the de Broglie waves was first formulated by Einstein.", and "It was mathematically analyzed in a fundamental paper of M. Born and used for the interpretation of collision phenomena." [7] Heisenberg cites Einstein's second paper from 1925, *Quantum Theory of the Monatomic Ideal Gas*, the first application to such a gas of what now is called *Bose-Einstein statistics*. In this paper, Einstein again applies his fluctuation techniques to a volume of molecules in the gas and arrives at two terms of a wave-particle duality, one corresponding to a particle fluctuation and the other due to interference of waves, similar to his 1909 result which had exhibited a special case of wave-particle duality for photons as discussed in the section *The Fall of Classicality*. Einstein's citations of de Broglie's 1924 thesis, sent to him by his adviser Paul Langevin, brought wide notice to de Broglie's work. In Section 9 of this paper, Einstein remarks on the diffraction of a stream of gas molecules by a slit (similar to de Broglie): "Large diffraction angles occur if λ is of the same order of magnitude as σ or larger. Thus, in addition to deflections due to collision in accord with mechanics, mechanically inexplicable deflections of the molecules also will occur with frequency comparable to the former, which will diminish the mean free path." These conclusions are very close to Born's statistical interpretation.

Born's paper was the first to contain the probability concept within the full theory of quantum mechanics developed independently by Heisenberg in 1925 and Schrödinger in 1926. Born writes: "One obtains the answer to the question, not *what is the state after the collision* but *how probable is given the effect of the collision*." Max Born was awarded the Nobel Prize for his fundamental research in quantum mechanics, especially for his statistical interpretation of the wave-function, but not until 1954, long after his 1926 contribution, suggesting the continuing resistance to the view that nondeterminism is a fundamental property.

Einstein's Quandary

A conflict of views on the nature of quantum mechanics emerged in the first half of the 20th century, with researchers such as Bohr, Born, Heisenberg, Jordan, and Pauli embracing nondeterminism, each with subtle differences in their positions, while Einstein, Planck, Schrödinger, and de Broglie were inclined towards determinism though for differing reasons [320, p. 18]. De Broglie tried to maintain determinism within a picture of waves piloting particles of well-defined position. Schrödinger despaired of the quantum jumps of particles but an acceptable wave picture seemed to him out of grasp after he had uncovered the inevitability of entanglement in 1935 [424] [425] [426]. Einstein put forward a series of *gedanken experiments* to first test the consistency of quantum theory and later its incompleteness while taking the long-term view that the field approach of his gravitational theory held the seeds to also include the quantum. However, not being able to find a way, his thoughts remained conflicted as he grasped at other possibilities. In a 1935 letter to Paul Langevin, Einstein wrote [427]

> *In any case one does not have the right today to maintain that the foundation must consist of a field theory in the sense of Maxwell. The other possibility, however, leads in my opinion to a renunciation of the time-space continuum and to a purely algebraic physics. Logically this is quite possible... Such a theory doesn't have to be based upon the probability concept...*
>
> John Stachel, The Other Einstein: Einstein Contra Field, Science in Context 6, 275-290 Cambridge University Press (1993).

However, the situation remained murky. The issue was clarified further in a 1929 paper [428] by Nevil Mott (1905-1996), a future Nobel laureate, for work in magnetic and disordered systems. Mott used quantum mechanics to analyze the probabilities of exciting atoms at successive positions in a cloud chamber and found that the probability is negligible unless they lie on a linear track. Heisenberg did a similar analysis the same year in his Chicago Lectures [429, p. 66]. From these analyses, there is a set of potential linear tracks pointing in different directions, but with the set of potential tracks still maintaining the spherical symmetry of the incoming spherical wave of the α-particle. The issue then arises as to how one particular linear track is picked out of the infinitely many potential linear tracks, Figure 5.8. This is where the measurement problem is seen—during a measurement which is consistent with only one of the lines of ionization probabilities. This issue clearly has interesting similarities with the issues of *potentiality* and *actuality* that were considered by Aristotle as discussed in the section *Backstory to Deductive Thought.*

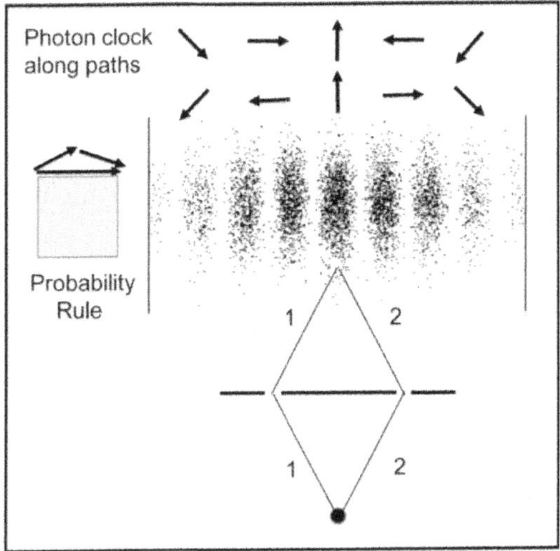

Figure 5.13: Photons interfering as particles via Feynman's sum-over path picture, with clock hands rotating along each path at the photon frequency rate and interference given by a head-to-tail probability rule.

The logical possibilities and pitfalls of following the paths of quanta were explored in Chapter 1, and it was seen that there is always a price to pay for any attempted tampering with wave-particle duality. The formulations of quantum mechanics by Heisenberg, Schrödinger, and Born, give a correct description of wave-particle duality, with knowledge that either the system evolves like a delocalized unitary wave evolution or as a localized particle via Born's rule when measurement has occurred. In addition to the Heisenberg and Schrödinger formations of quantum mechanics, a third way was found in 1948 by Richard Feynman (1918-1988) in terms of a sum over all possible paths of a particle, the summation represented mathematically by a path integral. In contrast to the wave picture of Schrödinger, Feynman's picture has localized particles moving along the paths but the particles are also able to endow the paths with the ability to interfere with each other to produce the quantum properties. At first glance, this may appear as another suspicious tampering of wave-particle duality, but Feynman could show that his scheme is equivalent to the Heisenberg and Schrödinger formulations. For photons moving along paths, Feynman's method can be visualized in a remarkably simple way [430] to unveil what properties localized light quanta need to exhibit correct wave-particle duality. Feynman explained that the machinery of his sum-over paths method implies that the photon essentially carries along a "clock" as it moves along each path, which can be thought of conceptually as an arrow rotating at the rate of the photon's frequency as shown schematically in Figure 5.13 for two photon paths in a double-slit experiment. The clock arrow will point in different directions as the photon moves along a path and the directions will generally be different for different paths. However, the paths can interfere with each other as they have a square probability rule for the sum over paths: the clock arrows are added head to tail and the resultant arrow is squared to give the probability of finding the photon at that location. As seen in Figure 5.13, the arrows add at interference maxima and cancel at minima. This method has been successfully applied to experiments in photon optics and in explaining the limit of how classical lenses and gratings function.

After Feynman had developed the sum-over-paths method, Feynman's professor, John Wheeler (1911-2008), thought this intuitive depiction might sway Einstein's view of quantum mechanics [431]:

> *"Visiting Einstein one day, I could not resist telling him about Feynman's new way to express quantum theory. "Feynman has found a beautiful picture to understand the probability amplitude for a dynamical system to go from one specified configuration at one time to another specified configuration at a later time. He treats on a footing of absolute equality every conceivable history that leads from the initial state to the final one, no matter how crazy the motion in between. The contribution of these histories differs not at all in amplitude, only in phase...This prescription reproduces all of standard quantum theory...Doesn't this marvelous discovery make you willing to accept quantum theory, Professor Einstein?" He replied in a serious voice, "I still cannot believe that God plays dice. But maybe," he smiled, "I have earned the right to make my mistakes.""*
>
>

As seen in all three of the quantum formulations, a picture of localized quanta must also possess subtle additional features in order to exhibit correct wave-particle duality. The history of this begins with Einstein. Planck's discovery of the quantum of action led Einstein to take a further step beyond Planck and consider light-quanta in his miracle year of 1905. Even before Planck did so himself, Einstein had realized that Planck's radiation law meant that a new kind of physics was called for.

Einstein said it was [339]

> *as if the ground had been pulled out from under one, with no firm foundations of physics to be seen anywhere, upon which one could have been built.*
>
> Albert Einstein, Autobiographical Notes in Albert Einstein: Philosopher-Scientist, Paul Arthur Schillp (Ed), Cambridge University Press 1949, pp. 2-49.

However, in 1905 Einstein treated the light-quanta as independent particles with no wave-particle duality yet in sight. Einstein had previously developed expertise in calculating fluctuations in ensembles of molecules, such as in the Brownian motion paper [432], and these were the tools he had at hand for the theory of light quantization. In doing so, he remarkably was able to infer from the macroscopic properties of heat radiation their corresponding microstructure in terms of a fluctuating quantity. Unlike the molecules in his previous fluctuation work where the number of molecules was fixed, most processes of heat radiation are not fixed. However, Einstein focused on the volume fluctuations, essentially the only process of heat radiation that does not alter the number of quanta, and thus could utilize his previous methods to show that the energy E of the heat radiation can be identified with N independent spatially localized quanta of magnitude $h\nu$ [433] (see Appendix 5.A).

In response to Arnold Sommerfeld's (1868-1951) inquiry in 1912 about the status of Einstein's work on the quantum, Einstein wrote [434, pp. 1-2]:

> *Your kind note only adds more to my predicament. But I assure you that with respect to my quantum [theory] I have nothing new to say that should arouse interest...I am now exclusively occupied with the problem of gravitation and hope...to overcome all the difficulties. One thing is certain however, that never in life have I been quite so tormented...Against this problem [of gravitation], the original problem of relativity is child's play.*
>
> Einstein, Hilbert, and the Theory of Gravitation: Historical Origins of General Relativity Theory, 1974, p.1, J. Mehra, © 1974 D. Reidel Publishing Company. With permission of Springer.

Already in 1907, he published an article, *The Principle of Relativity and Its Consequences*, in which he suggested the equivalence between a gravitational field and an accelerated frame of reference, the *Equivalence Principle* that would become the basis for his theory of gravitation, *General Relativity*. This focus on relativity reflected his deepest views on how a physical theory should be approached and would influence his views concerning the meaning of quantum theory for decades to come. In his *Autobiographical Notes* of 1949, Einstein put his views on quantum theory into perspective [339]:

> *It is my opinion that the contemporary quantum theory by means of certain definitely laid down basic principles, which on the whole have been taken over from classical mechanics, constitutes an optimum formulation of the connections. I believe, however, that this theory*

offers no useful point of departure for future development. This is the point at which my expectation departs most widely from that of contemporary physicists...At this point it is my experiences with the theory of gravitation which determine my expectations. These equations give, from my point of view, more warrant for the expectation to assert something ***precise*** *than all other equations of physics...I have learned something else from the theory of gravitation: No ever so inclusive collection of empirical facts can ever lead to the setting up of such complicated equations...Equations of such complexity can be found only through the discovery of a logically simple mathematical condition which determines the equations completely.*

Albert Einstein, Autobiographical Notes in Albert Einstein: Philosopher-Scientist, Paul Arthur Schillp (Ed), Cambridge University Press 1949, pp. 2-49.

Einstein's theory of relativity comprised causal deterministic trajectories within the structure of space-time that extended the classical world-view of Newton. From Bohr's perspective, the first important extension to the depiction of our reality occurred with the physics of Galileo and Newton followed by the field theories of Faraday and Maxwell [435, p. 118]. The introduction of relativity closed this line of development. In contrast, quantum phenomena must refer to observations obtained under circumstances that take into account the details of the experimental apparatus. A question naturally emerges as to how precisely does *space-time* differ from *quanta*?

Einstein's Ghost Field

Einstein's 1905 paper signaled that a foundational result of 19th century physics, the wave character of light, must be reexamined. By 1908 or early 1909 he realized that independent light quanta are not the whole story of radiation by deriving Equation (5.7) from the Planck radiation law, simply viewed as an empirical relation, to see what lurks below the surface. Since both terms are necessary to reproduce the Planck law, this implied that the localized quanta must somehow interfere with one another. In the previously mentioned letter to Lorentz expressing his doubts about the notion of independent light-quanta, he went on to say further [3]:

I conceive of the light quanta as a point that is surrounded by a greatly extended vector field, that somehow diminishes with distance. Whether or not several light quanta are present with mutually overlapping fields one must imagine a simple superposition of the vector fields, that I cannot say. In any case, one must have equations of motion for the singular points in addition to differential equations for the vector field.

Nicht Sein Kann was Nicht Sein Darf, or the Prehistory of EPR, 1909-1935: Einstein's Early Worries about the Quantum Mechanics of Composite Systems, In: Sixty-two Years of Uncertainty, 1990, p. 61, D. Howard,

This was the beginning of Einstein's idea of using guiding or *ghost fields* to somehow deal with the unavoidable but mysterious influence at a distance between light quanta. As early as 1920, Einstein had written to Born:

> *That business about causality causes me a lot of trouble, too. Can the quantum absorption and emission of light ever be understood in the sense of the complete causality requirement, or would a statistical residue remain?*
>
> Max Born, The Born-Einstein Letters, MacMillan and Co. Ltd. 1971, reproduced with permission of Palgrave Macmillan.

However, Einstein did speculate about a probabilistic interpretation of Maxwell's equations to link the wave and particle concepts, using his *ghost-field* concept, a precursor to Born's 1926 probabilistic interpretation of the wave function of quantum mechanics. Although discussed with a number of physicists, he never published a paper regarding this concept [436, p. 382]. In Einstein's ghost-field, a wave of interference radiation of vanishingly small amplitude and which carries no energy prepares the way for the radiation of energy in the form of light quanta. This consists of indivisible quanta of magnitude $h\nu$, which follow the path prescribed by the interference radiation. In Born's extended paper on collision processes in 1926, he states [4, p. 163] [181]:

> *In this, I start from a remark by Einstein on the relationship between the wave field and light quanta; he said, for instance, that the waves are there only to show the corpuscular light quanta the way, and in this sense he talked of a 'ghost field'. This determines the probability for a light quantum, the carrier of energy and momentum, to take a particular path; the field itself, however, possesses no energy and no momentum.*
>
> G. Bacciagaluppi and A. Valentini, Quantum Theory at the Crossroads: Reconsidering the 1927 Solvay Conference, Cambridge University Press, 2013.

Einstein continued to think along the lines of the ghost-field idea in the 1920s, though again without publishing a detailed theory. In 1925, one year before Born's collisions papers, Einstein gave a colloquium in Berlin where he discussed the idea that every particle was accompanied by a ghost or guiding field [437, p. 441] [3, p. 72]. According to Eugene Wigner (1902-1995), who was present at the colloquium, Einstein realized that it conserves energy and momentum only on average and violates conservation principles for individual trials. When a light quantum and an electron collide, both would follow a guiding field. However, guiding fields give only the probabilities of the directions in which the light quantum and the electron will proceed and so they follow their directions independently, giving only statistical energy and momentum conservation. This difficulty could be overcome by the introduction of a guiding field in configuration space as was done in a theory by Louis de Broglie (1892-1981), and later by David Bohm (1917-1992), that determined probabilities for all particles collectively. In Einstein's approach, each particle had its own guiding field and so entanglement was precluded [4, pp. 200-201]. This led to a sort of complementarity between exact conservation laws and entanglement. For Einstein, exact conservation was sacred but he could not tolerate entanglement. This predicament meant Einstein would eventually go his way and the gap between him and the physicists developing quantum theory continued to widen in the following decades. For Einstein, "indeterminacy was but a symptom; entanglement was the underlying disease" [111, p. 67]. In addition to indeterminacy and entanglement, a third essential ingredient in understanding quantum theory turned out to be nonlocality. Einstein struggled for the rest of his life to fit together this triumvirate of puzzle pieces into his view of the universe.

Einstein's ghost field work was one of several related efforts that were never published. He had made many other attempts to understand the quantum. In May 1927, Einstein had read a paper before the Prussian Academy of Sciences in Berlin, entitled "Does Schrödinger's Wave Mechanics Determine the Motion of a System Completely or Only in the Sense of Statistics?" [438]. This was Einstein's attempt at a deterministic hidden-variable completion of quantum mechanics with the aim of showing that Schrödinger's wave mechanics completely determines the motion of a system. Einstein was able to use the methods of non-Euclidean geometry familiar to him from his development of his theory of gravitation, General Relativity, to specify n unique initial conditions at each point in the n-dimensional configuration space that completely determine the system's dynamics. These n degrees of freedom would in effect serve as hidden variables leading to a deterministic version of Schrödinger's equation. However, after submitting the paper to the journal of the Prussian Academy, Einstein noticed the possibility of solutions in which non-local correlations exist between subsystems which was not satisfactory for a physical system. He was not allowed to have it both ways. Therefore, he requested to the editor that the paper be withdrawn before publication. Yet again, the appearance of nonlocality had blocked Einstein's way.

These issues would smolder in the background for Einstein as quantum theory developed further, including Bohr's atomic models beginning in 1913. Although primarily working on his theory of gravitation, Einstein still managed to make important contributions to quantum theory. In dealing with a gas of particles in thermal equilibrium with radiation, Einstein applied statistical considerations to Bohr's stationary states and succeeded in arriving at a simple deduction of Planck's radiation law but only if the transition is given by Bohr's atomic transition hypothesis $E_m - E_n = h\nu$. In 1916, Einstein's published arguments derived the "A and B coefficients" regarding rates of spontaneous and stimulated emission. These arguments set down the principles that decades later would be utilized in the operation of the laser, which relies on stimulated emission. Einstein's work on spontaneous emission sets the scale for all radiative transitions and manifests the fundamental interaction of matter with the vacuum. However, he was concerned that his theory could not predict the direction in which a light-quantum moves after spontaneous emission thereby violating causality. A further major innovation of the 1917 paper was that the momentum as well as the energy of light quanta were included so that the excitation of an electron is accompanied by momentum transfer. This was a crucial step allowing the proper coupling of light-quanta to matter.

Bohr-Einstein Debates Begin

Bohr had significant reservations regarding Einstein's 1905 concept of the light-quanta for the next twenty years as it did not fit into his understanding of radiation. For example, in Bohr's 1922 Nobel address he stated:

> *In spite of its heuristic value…the hypothesis of light-quanta, which is quite irreconcilable with so-called interference phenomena, is not able to throw light on the nature of radiation.*

Bohr joked with Heisenberg [439, p. 6]:

> *Even if Einstein sends me a cable announcing the proof of the light quantum, the message cannot reach me because it has to be propagated by electromagnetic waves.*

As will be discussed, Bohr backed himself into a corner, weighing conceptual arguments against the available experiment evidence. Over the years, he allowed himself to consider several consistent possibilities to address the quantum, including on-average violation of energy conservation. However, in 1925 he finally ran out of alternative arguments to explain electron-radiation scattering phenomena when the development of new coincidence-counting techniques permitted definitive experiments to be performed confirming the existence of light-quanta. This was one more step in the evolution of Bohr's thinking. Just as Einstein had been struggling behind the scenes to understand questions of wave-particle duality that were ultimately related to entanglement, Bohr had gone on his own journey to come to terms with the quantum issues. Even before he had accepted light-quanta and had to confront wave-particle duality, he had been thinking about issues that were also ultimately related to coherence and entanglement. For example, he was perplexed by experiments of Ramsauer from 1920 showing the counterintuitive result that the mean free path of electrons passing through noble gases can increase without bound so that the atoms of the gas become virtually invisible to the electrons [3].

Bohr and Einstein first met in 1919 during a visit by Bohr to Berlin and then the following year when Einstein stopped in Copenhagen. This is where the bond between them really materialized, including losing track of time on an apocryphal streetcar ride, arguing over quanta [440, p. 56]:

> *Bohr: "Einstein, if you should be really able to prove that light is particles more or less in the commonly accepted sense…do you really believe that one could imagine seeing a law passed which would*

make it illegal to use diffraction gratings?"

Einstein: "Or conversely, if you could prove that light is exclusively wavelike, do you then think that you could get the police so stop the use of photocells?"

They immediately developed respect for each other although their subsequent meetings and correspondence were rare but important. But this mutual respect would set the stage for the next three decades for both of them struggling over the essence of quantum theory and each of them coming to terms with a problem that was ultimately about the role of entanglement in the measurement problem.

Bohr was not the only one to initially reject Einstein's light-quanta. This included Planck ("That he may sometimes have missed the target in his speculation, as for example, in his hypothesis of light quanta, cannot really be held against him") and Millikan who later confirmed Einstein's prediction of the photoelectric effect ("The semi-corpuscular theory by which Einstein arrived at this equation seems at present to be wholly untenable").

BKS Showdown over Quanta

By the 1930s, Bohr, had finally accepted the idea of light-quanta. In his earlier work on atomic structure, Bohr had discussed emission and absorption of light in atomic transitions, without examining the mechanisms for these processes nor the nature of light itself. Quantum phenomena had initially cast doubts on classical theory but, apparently these had only to do with interactions between matter and radiation, not with radiation itself. However, there was also the question of conservation of energy and momentum. In 1923, Arthur Compton (1892-1962) observed a change in frequency when light is scattered by electrons, thereafter called the *Compton effect*. The results could be explained by assuming the light beam behaves as light-quanta and that energy and momentum are conserved. However, Bohr found a way of using Einstein's approach without also using the light-quantum hypothesis by reinterpreting the principles of energy and momentum conservation as *statistical* principles. The conservation laws had the status of a fundamental law since the mid-19th century but would not be tested experimentally at the level of individual microscopic processes such as atomic transitions or collisions of electrons until 1925. Thus, it was in 1924 that Bohr, Kramers and Slater published a provocative description of the interaction of matter and electromagnetic interaction, historically known as the BKS paper that combined quantum transitions and electromagnetic waves with energy and momentum being conserved only on average. BKS tried to take [441] the duality between particle and wave-models as the starting point for the interpretation of quantum theory. The waves would play the role of a probability field, even though this forces energy conservation for individual processes to be abandoned. BKS correctly noted that the

Compton experiments had actually only demonstrated conserved energy-momentum averaged over many individual processes.

Energy and momentum are conserved exactly within the formalism of Schrödinger's equation; however, Schrödinger's equation allows entanglement. For strict validity of the conservation laws, Bohr expected a breakdown of the space-time description that is incompatible with the properties of mechanical models [441]. In 1925, Walther Bothe (1891-1957) and Hans Geiger (1882-1945) developed counter-coincidence techniques to verify that in the Compton effect the secondary photon and electron are produced simultaneously as required by causality, disproving the BKS theory. Secondly, Compton and Alfred Simon studied the Compton effect in cloud chamber experiments, enabling them to demonstrate that energy-momentum was conserved in individual events. Soon after the results became known, Bohr wrote [442, p. 126]:

> *It seems ...that there is nothing else to do than to give our revolutionary efforts as honorable a funeral as possible.*
>
> S. Petruccioli, Atoms, Metaphors and Paradoxes, Niels Bohr and the construction of a new physics, Cambridge University Press 1993.

However, BKS had also led to deeper discussions and renewed attention to the difficulties in the foundations of quantum theory and subsequently influenced Heisenberg, Kramers and Born to explore mathematics that strongly inspired the development of Heisenberg's matrix mechanics in 1925.

This immediate shift in point of view illustrates the character of Bohr's deductive approach to research when he readily accepted the experimental results disproving BKS. However, Bohr returned to the possibility of violation of energy and momentum conservation again in the crisis of *β-decay* due to experiments carried out in 1929 [9, p. 364]. With the deductive approach, it's important to emphasize that all possibilities logically remain under consideration even if they had failed in other situations. In α-decay, the α particles emitted by any given species of atoms all have the same energy and the final two particles emerge with equal and opposite momenta. The result is that a particular α-particle momentum corresponds to a particular energy. In contrast, the β-spectrum is continuous so that it is either a two-body process which violates energy and momentum conservation or else β-decay is not a two-body process and additional unobserved particles are involved that account for the conservation laws. In this case, Bohr once again proposed non-conservation as a possibility, but the additional particle interpretation ultimately turned out to be correct as suggested by Pauli and modeled by Fermi. This led to the important discovery of the neutrino though it took until 1956 to finally detect it due the weakness of the interaction of the neutrino with matter. Although Bohr's preference failed once again, it must be emphasized that it is important to keep all consistent options under consideration.

When Bohr finally accepted Einstein's light-quanta in 1925, the developments in quantum theory were in full tilt. This included de Broglie's proposal for matter waves in 1924, extending wave-particle duality to matter. However, a special case of matter wave-particle duality had appeared in 1909 when Einstein derived a version of

Equation (5.7) for matter when he studied the *Gibbs paradox* [3]. Pauli formulated the exclusion principle in 1925 and the mathematical theory of spin in 1927. In 1927, Dirac developed the relativistic equation for the electron which incorporated Pauli's theory of spin and predicted the existence of anti-matter. That indistinguishable particles have two types of counting statistics for allowed energy states was developed by Bose and Einstein in 1925 for one class now called *bosons* (which includes light-quanta or photons) and by Fermi and Dirac in 1926 for the second class, now called *fermions*.

Heisenberg discovered the uncertainty relations of Equation (1.2) in 1927. Although the derivation of the uncertainty between position and momentum can be extended to any conjugate pair of Hermitian operators, Pauli noticed that the semi-boundedness required for the stability of any Hamiltonian implies that there is no time operator in quantum mechanics to allow the uncertainty relations to be extended to energy and time. Although other methods were later developed to justify time-energy uncertainty relations, Bohr gave a very simple argument in his 1927 Como paper where he had also introduced complementarity [9, p. 312]. For a particular wave-packet with finite extension Δx, he considered the time interval Δt during which the bulk of the wave-packet passed a particular point and the frequency interval $\Delta\nu$ where the relevant frequencies lie along with the corresponding interval for inverse wave length $\Delta(1/\lambda)$. Then from the known theory of the resolving power of optical instruments, he could write:

$$\Delta t \Delta\nu \geq 1,\ \Delta x \Delta(\tfrac{1}{\lambda}) \geq 1\,. \tag{5.8}$$

From the wave-particle duality of Einstein or de Broglie for either photons or matter, we have:

$$E = h\nu,\ p = h/\lambda\,. \tag{5.9}$$

Combining Equations (5.8) and (5.9) immediately gives the uncertainty relations, Equations (1.2) for both position-momentum and energy-time. So, Bohr's acceptance of light-quanta could be used in stimulating ways.

Pauli also visited Bohr's Copenhagen Institute in June 1927 and joined in mediating the disparate views of Bohr and Heisenberg on the uncertainty relations. From this time on, Bohr, Heisenberg and Pauli shared a somewhat common set of views, sometimes known as the *Copenhagen interpretation* of quantum mechanics. Heisenberg's uncertainty paper and Bohr's complementarity paper were the two basic pillars, and it was further summarized in Pauli's 1933 *Handbuch der Physik* article [443]. The interpretation included the uncertainty principle, wave-particle duality for photons, electrons, and other particles, Born's probabilistic interpretation of the wave function, the correspondence between eigenvalues and measured values, and the correspondence principle. However, the Copenhagen interpretation was not necessarily coincident with Bohr's epistemology which he continually refined although it was not always widely understood or appreciated.

The new Copenhagen interpretation also created the climate for a consensus on the issue of quanta for both light and matter. The next stage in understanding involved delving into the issue of entanglement, which is the central issue for the measurement problem. As discussed, both Einstein and Bohr had been conceptually dealing with entanglement in various forms decades before Schrödinger coined the term in 1935. For the next two decades, this issue mainly involved debates between Einstein and Bohr along with other important supporting characters, taking place at Solvay Conferences, during private discussions, in correspondence, and several important journal articles. Following Einstein's death in 1955 and Bohr's in 1962, the issue of entanglement would then become re-energized after the formulation of the Bell inequalities in the 1960s and their supporting experiments. This would continue up to the present day with the development of quantum information and new experimental techniques to manipulate entangled states of photons and atoms.

Measure and Meaning

The single-photon measurements described in Chapter 2 are carried out in a quantum optics laboratory equipped with components arranged on an optical table, such as photonic sources, polarizing beam splitters, mirrors, phase shifters, and finally photodiodes or other devices to collect the photons at the edges of the set-up. How is

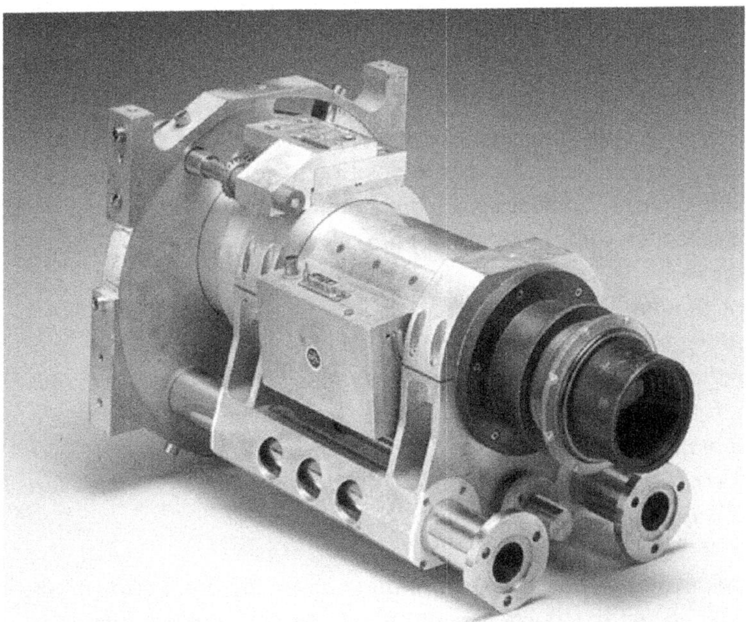

Figure 5.14: Single-photon detector, similar to that used in Hubble Space Telescope. Constructed by the Department of Physics and Astronomy, University College.

it known which, if any, of these components have induced a measurement event to occur? And which components instead operate in a unitary manner? Certain of these devices may have been designated as a *detector*, perhaps as seen on the label of a carton shipped from a manufacturer, or by a colleague in our laboratory who had designed the device, Figure 5.14. If the intensity of the source is sufficiently low, the detector is found to record pulses representing the arrival of a photon separated by intervals in time during which nothing happens. Our experience has always been that, on each particular run, a definite outcome always occurs: the detector either clicks or it does not click. There would never be a half response, either an entire photon arrives or else nothing happens. And two detectors never respond simultaneously (unless the source had emitted two photons within the resolving time of the detector; a further decrease of the source intensity will decrease the chance of such coincidences). Since the energy of a single photon is typically very small ($\approx 10^{-19}$Joule), a device is often designed to register the photon's appearance by utilizing the photon to produce a current pulse by means of a high voltage avalanche process [444]. *On-off* or *click-detectors* produce well-defined "clicks" when excited by any finite number of photons, which can only deliver information about the absence or presence of photons [445]. However, could the clicking of such detectors also constitute a quantum measurement problem event which requires distinguishing unitary and non-unitary processes based on the physics of quantum entanglement as developed and discussed in Chapters 3 and 4? Although they are not designed to reveal such a distinction, detectors do function using many of the elements that we can now appreciate as being involved in the precise formulation of the quantum measurement problem.

Whole Photon or Nothing

Paul Dirac was one of the pioneers of quantum theory and his textbook of 1930, *The Principles of Quantum Mechanics*, an exposition in concise and measured prose and which became the early standard work on quantum mechanics, stated a much-repeated principle of detection early in its pages [446, pp. 8-9]:

> *The result of such a determination must be either the whole photon or nothing at all. Thus, the photon must change suddenly from being partly in one beam and partly in the other to being entirely in one of the beams. So long as the photon is partly in one beam and partly in the other, interference can occur when the two beams are superposed, but this possibility disappears when the photon is forced entirely into one of the beams by an observation.*
>
> *...the wave function gives information about the probability of one photon being in a particular place and not the probable number of photons in that place. The importance of the distinction can be made clear in the following way. Suppose we have a beam of light consisting of a large number of photons split up into two components of equal intensity. On the assumption that the intensity of a beam is connected with the probable number of photons in it, we should have*

half the total number of photons going into each component. If the two components are now made to interfere, we should require a photon in one component to be able to interfere with one in the other. Sometimes these two photons would have to annihilate one another and other times they would have to produce four photons. This would contradict the conservation of energy.

The new theory, which connects the wave function with probabilities for one photon, gets over the difficulty by making each photon go partly into each of the two components. Each photon then interferes only with itself. Interference between two different photons never occurs.

The Principles of Quantum Mechanics by P.A.M. Dirac (1958), By permission of Oxford University Press.

The feature that "each photon then interferes only with itself" is essential to the orthodox interpretation of quantum mechanics and to understanding quantum measurement. It essentially means that self-interference can occur only if two components of the wave packet can overlap, which requires that the optical path length not exceed the coherence length. Note that Dirac's statement applies only to situations where interference is revealed by measuring single photons and not to experiments where correlations between different photons are detected. The latter corresponds to the case of destructive interference of photons originating from separate, independent sources [447] [23].

At the time that Dirac made these statements, experiments had not yet confirmed all of these claims regarding "the whole photon or nothing at all" and "each photon then interferes only with itself." If a photon enters a beam splitter, as in Figure 5.15 (b), with the resulting "click" of either detector D1 or D2, is the decision to either reflect or transmit made at the beam splitter or at the detectors? If a beam splitter is indeed a unitary device, as discussed in Chapter 2, shouldn't the stochastic clicking occur at the detectors and not at the beam splitter? And what are the implications of these possibilities? The status of these questions was raised as early as the landmark 5th Solvay Conference in 1927, at which Dirac was the youngest participant. Among the twenty-nine attendees were such luminaries as Einstein, Bohr, Planck, Lorentz, Marie Curie, Schrödinger, Pauli, Heisenberg, Compton, de Broglie, and Born. During the General Discussion, Einstein proposed a thought-experiment of particles impinging on a small slit in a screen being diffracted onto a hemispherical detector (such as a photographic film), Figure 5.15 (a).

Einstein then discussed two interpretations of this scenario [4, pp. 400-401]:

Interpretation I. The de Broglie-Schrödinger waves do not correspond to a single electron, but to a cloud of electrons extended in space. The theory does not give any information about the individual processes, but only about the ensemble of an infinity of elementary processes.

> *Interpretation II. The theory has the pretension to be a complete theory of individual processes. Each particle directed towards the screen, as far as can be determined by its position and speed, is described by a packet of de Broglie-Schrödinger waves of short wavelength and small angular width. This wave packet is diffracted and, after diffraction, partly reaches the film D in a state of resolution.*
>
> G. Bacciagaluppi and A. Valentini, Quantum Theory at the Crossroads: Reconsidering the 1927 Solvay Conference, Cambridge University Press, 2013.

He argued that by Interpretation I, $|\psi|^2$ represents the probability of a particular particle being at a given point of the detector whereas with Interpretation II, $|\psi|^2$ gives

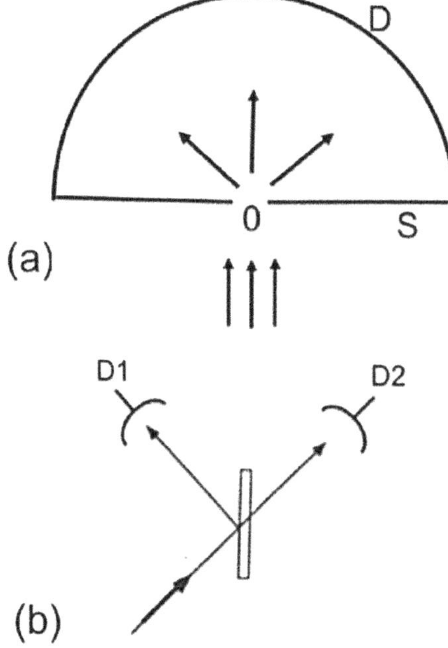

Figure 5.15: (a) Einstein's 1927 thought experiment from the 5th Solvay Conference: photons are diffracted through hole in screen S to detecting hemisphere D; (b) Simplified version with beam splitter and two detectors.

the probability that at a given instant the same particle is present at a given point. He says,

> *Here, the theory refers to an individual process and claims to describe everything that is governed by laws.*
>
> G. Bacciagaluppi and A. Valentini, Quantum Theory at the Crossroads: Reconsidering the 1927 Solvay Conference, Cambridge University Press, 2013.

In other words, Interpretation I does not specify the individual trajectories of the particles and so is incomplete, but with interpretation II, the theory claims to be complete. Einstein also objects that II would allow the same elementary process to produce an action at two or more places across the detector, resulting in

> *an entirely peculiar mechanism of action at a distance, which prevents the wave continuously from producing an action in two places on the screen.*
>
> G. Bacciagaluppi and A. Valentini, Quantum Theory at the Crossroads: Reconsidering the 1927 Solvay Conference, Cambridge University Press, 2013.

Such action at a distance would be viewed by Einstein as conflicting with his theory of relativity as he would also later object in his 1935 Einstein-Podolsky-Rosen (EPR) paper. Einstein is also recognizing here the "whole photon or nothing at all" aspect required by quantum mechanics and later emphasized by Dirac in his 1930 book. These points were also made by Heisenberg in his 1929 University of Chicago lectures [429, p. 39] where he discusses Einstein's 1927 example in terms of a photon impinging on a semi-transparent mirror or *beam splitter*, Figure 5.15 (b).

> *...one other idealized experiment (due to Einstein) may be considered. We imagine a photon which is represented by a wave packet built up out of Maxwell waves. It will thus have a certain spatial extension and also a certain range of frequency. By reflection at a semi-transparent mirror, it is possible to decompose it into two parts, a reflected and a transmitted packet. There is then a definite probability for finding the photon either in one part or in the other part of the divided wave packet. After a sufficient time, the two parts will be separated by any distance desired; now if an experiment yields the result that the photon is, say, in the reflected part of the packet, then the probability of finding the photon in the other part of the packet immediately becomes zero. The experiment at the position of the reflected packet thus exerts a kind of action (reduction of the wave packet) at the distant point occupied by the transmitted packet, and one sees that this action is propagated with a velocity greater than that of light.*
>
> Reprinted by permission of Dover Publications.

However, Heisenberg finds the nonlocality requirement of "reduction of the wave packet" due to the "whole photon or nothing" property does not conflict with relativity, claiming

> *it is also obvious that this kind of action can never be utilized for the transmission of signals so that it is not in conflict with the postulates of the theory of relativity.*
>
> Reprinted by permission of Dover Publications.

This may be the earliest statement of the *peaceful coexistence* of quantum theory and relativity, to use Abner Shimony's persistent phrase [448].

Exact Conservation with Whole Photon or Nothing

As discussed in the section *Einstein's Ghost Field*, since the early 1920s Einstein had been investigating the formulation of quantum mechanics using "ghost fields," guiding fields that give probabilities of the directions in which particles will proceed, and so they follow their directions independently, giving only average energy and momentum conservation. At the same time, de Broglie was developing his own version of a hidden-variable theory, which utilized the entire configuration space and presented it at the 1927 Solvay Conference [4]. Einstein's familiarity with these issues prompted him to comment on the action at a distance feature of his Figure 5.15(a) thought experiment,

> *It seems to me that this difficulty cannot be overcome unless the description of the process in terms of the Schrödinger wave is supplemented by some detailed specification of the localization of the particle during its propagation. I think M. de Broglie is right in searching in this direction. If one works only with Schrödinger waves the Interpretation II of $|\psi|^2$, I think, contradicts the postulate of relativity.*
>
> G. Bacciagaluppi and A. Valentini, Quantum Theory at the Crossroads: Reconsidering the 1927 Solvay Conference, Cambridge University Press, 2013.

With his by now hands-on knowledge of conservation laws and nonlocality, Einstein also emphasized in the Solvay Discussion the implications of his Interpretation II regarding conservation,

> *It is only by virtue of II that the theory contains the consequence that the conservation laws are valid for elementary processes; it is only from II that the theory can derive the result of the experiment of Geiger and Bothe, and can explain the fact that in the Wilson [cloud] chamber the droplets stemming from an α-particle are situated on very nearly contiguous lines.*
>
> G. Bacciagaluppi and A. Valentini, Quantum Theory at the Crossroads: Reconsidering the 1927 Solvay Conference, Cambridge University Press, 2013.

From this statement would emerge the beginnings of a struggle over the next fifty years to ascertain theoretically and experimentally the nature and interrelations of nondeterminism, entanglement, exact conservation laws and the corpuscular aspects of photons. These would be the ingredients needed to characterize the quantum measurement problem.

Recall from the discussion in the section *BKS Showdown over Quanta* that the Bohr-Kramers-Slater (BKS) theory for electron-radiation scattering had predicted energy conservation only on average but this was refuted by the experiments of

Compton-Simon in 1925 demonstrating exact energy and momentum conservation and Bothe-Geiger in 1926 also demonstrating coincidence between the produced photons and electrons. This led to the acceptance of Einstein's light quanta with its "whole photon or nothing" property though Einstein still struggled to find an acceptable way to understand quantum theory. Schrödinger had been skeptical of the reality of discontinuities in quantum theory since he had complained to Bohr about "all this damned quantum jumping" during his visit to Copenhagen in 1926. In his series of Wave Mechanics papers in 1926, he introduced characteristic frequencies as the basic properties of interacting systems and explained dynamics as a resonance phenomenon that maintains space-time continuity. His attitude was that [449],

> *It is hardly necessary to emphasize how much more congenial it would be to imagine that at a quantum transition the energy changes over from one vibration to another, than to think of a jumping electron.*
>
> Erwin Schrödinger, In: Ernst Mach's Vienna 1895-1930, Boston Studies in the Philosophy of Science, Vol. 218, 2001, p. 85, H.W. de Regt, © 2001 Springer Science+Business Media B.V. With permission of Springer.

Schrödinger had also been quite sympathetic to BKS with its adherence to continuity and did not object to violation of exact conservation laws for individual events. His fourth Wave Mechanics paper also developed a treatment of radiation scattering with continuous waves which conserved energy and momentum only on average [450]. However, Schrödinger retreated from publicly discussing his original interpretation following the pressure from Bohr and the results of the photon-electron scattering experiments that clinched the validity of exact conservation laws for every individual interaction [451]. However, he returned to exploring his earlier views in the 1930s after his discovery of quantum entanglement and also possibly encouraged by new experiments of Shankland [452] that apparently conflicted with the earlier Compton-Simon and Bothe-Geiger results. This provoked a flurry of letters in *Nature* by Dirac, Peierls, and others discussing the startling implications if this turn of events was valid. This included a letter by Bohr and a report by Jacobsen from Bohr's Copenhagen Institute who reported his new scattering measurements that did support exact conservation laws [453] [454]. However, there were still lingering doubts in some circles which were finally put to rest in 1950 with experiments by Cross and Ramsey [455] and Hofstadter and McIntyre [456] verifying exact conservation and coincidence, employing modern electronic counting methods. In contrast, the Bothe-Geiger experiments of 1926 were laborious and time-consuming, employing two opposing needle counters and X-rays passing between them. One of the counters responded only to photons and the other only to recoil electrons. The deflections of the counters were recorded on silver bromide film that moved at high speed between the counters to record the timing of the particles from the collisions. The recording of coincidences of the joint impact of a photon and recoiling electron would indicate the conservation of energy at the atomic level. It took Bothe and Geiger almost a year to develop, dry, and analyze the results from the three kilometers of 1.5 cm wide film it took to accomplish this in order to obtain the statistically convincing results that the photon and electron counter were coincident to within 10^{-4} sec. These intense efforts might even be viewed as a prefiguring of the Bell-type correlation experiments. On Einstein's recommendation, Walter Bothe shared the 1954 Nobel Prize with Max Born, Hans Geiger having died in 1945 [457] [458].

Still harboring doubts, Schrödinger persuaded Ádám, Jánnosy, and Varga [459] [460] to carry out Heisenberg's version of Einstein's thought-experiment to test whether light at the half-silvered mirror would always respond as a particle or a wave, Figure 5.15 (b) [461]. However, the result was effectively a null experiment due to insufficient detector efficiency at that time. It was not until 1974 that Clauser finally demonstrated the "whole photon or nothing" property that photons do not split at a half-silvered mirror and that true single photons are not measured at both detectors simultaneously [462]. Only one detector clicks. This experiment used a light source based on atomic cascades in mercury atoms to produce heralded single photons and two joint intensity measurements performed at the outputs of the beam splitter. If the photon is indivisible, detectors 1 and 2 are never triggered simultaneously so that there is an anti-correlation effect, $<\hat{n}_1\hat{n}_2>=0$, i.e., *photon antibunching*. Clauser's experiments on the "whole photon or nothing" photon antibunching also confirmed the exact point-wise conservation of energy in QED, further confirming this aspect of the Compton-Simon and Bothe-Geiger experiments. A calcium atomic cascade source was used by Grangier et al. to also observe antibunching as well as wave interference effects in a Mach-Zender interferometer configuration [72]. Einstein's concern about action at a distance was finally addressed by Guerreiro et al. with the experimental observation of single-photon anti-bunching in which the detectors are separated by space-like distances, confirming that in each round of the experiment only one detector clicks [463].

The Rise of the Measurement Problem

The Characteristic Trait

By 1935, Schrödinger had begun to return to his earlier views which had argued against discontinuous collapse. As a result, he wrote a series of three papers in 1935-36 [156] [119] [464] that appeared just months after the EPR paper and first began introducing the concept and properties of entanglement. The initial paper, *The Present Status of Quantum Mechanics*, also introduced within a discussion of the measurement problem the hypothetical dilemma of the living and dead cat, *Schrödinger's Cat* (discussed in Chapters 1 and 2), with the consequence that there was apparently nothing to prevent microscopic oddities from appearing in our macroscopic world [119]. In these papers Schrödinger coined the term *verschränkung*, now ubiquitously translated into English as *entanglement*, that he described as *the characteristic trait* of quantum mechanics in the paper *Discussion of Probability Relations between Separated Systems* [156],

> *When two systems of which we know the states by their respective*

representatives, enter into temporary physical interaction due to known forces between them, and when after a time of mutual influence the systems separate again, then they can no longer be described the same way as before, viz., by endowing each of them with a representative state of its own. I would not call that one but the characteristic trait of quantum mechanics, the one that enforces its entire departure from classical lines of thought. By the interaction the two representatives (or ψ-functions) have become entangled.

E. Schrödinger, Discussion of Probability between Separated Systems Proceedings of the Cambridge Physical Society 31 (4), 555-563 (1935).

Schrödinger's papers were motivated partly by his correspondence with Einstein and by the 1935 Einstein-Podolsky-Rosen (EPR) paper, which had argued that quantum mechanics is incomplete. By this time, Einstein was at the Institute for Advanced Study (IAS) in Princeton, New Jersey, having fled to America from the turmoil in Germany [465]. Einstein's assistant, Nathan Rosen, had recently written a paper presenting the first reliable calculation of the structure of the hydrogen molecule. This turned out to require wave-functions which could not be written as products of wave-functions for each of the two electrons in the molecule. These were entangled states. At the traditional 3 o'clock IAS tea, Rosen described their properties, and Einstein immediately saw the implications for the interpretation of quantum mechanics. They were joined by Boris Podolsky who later also proposed writing an article. However, the subsequent events quickly became exasperating for Einstein. The *New York Times* headline of May 4, 1935 read, "Einstein Attacks Quantum Theory" after Podolsky leaked the development to the press. Einstein was quite irritated at Podolsky for doing this and avoided him thereafter, telling the *Times* that the information "was given to you without my authority." Podolsky had also written the article and Einstein was not pleased with the result, writing to Schrödinger "For reasons of language, this was written by Podolsky after several discussions. Still, it did not come out as well as I had originally wanted; rather, the essential thing was, so to speak, smothered by the formalism [*gelehrsamkeit*]." [465] Behind the scenes, Einstein often clarified his thoughts in letters which sometimes give more insight than the journal articles. In an August 8, 1935, letter to Schrödinger, Einstein attempts a simpler incompleteness argument by considering a macroscopic example to avoid having to assume locality as was done in the EPR paper [466, p. 78]:

The system is a substance in chemically unstable equilibrium, perhaps a charge of gunpowder that, by means of intrinsic forces, can spontaneously combust, and where the average life span of the whole setup is a year. In principle, this can quite easily be represented quantum-mechanically. In the beginning the psi-function characterizes a reasonably well-defined macroscopic state. But, according to your equation, after the course of a year this is no longer the case. Rather, the psi-function then describes a sort of blend of not-yet and already-exploded systems. Through no art of interpretation can this psi-function be turned into an adequate description of a real state of affairs; in reality, there is no intermediary between exploded and not-exploded.

Schrödinger's letter in reply of September 19, 1935, is the first appearance of his thought-experiment of the living and dead cat, telling Einstein that he had constructed an example very similar to his exploding powder keg, thus suggesting the origin of *Schrödinger's Cat*. However, Einstein was pleased with this extension to a cat, as in this 1950 letter to Schrödinger, though he sometimes conflated Schrödinger's use of cyanide with his use of gunpowder for the demise of the cat,

You are the only contemporary physicist, besides Laue, who sees that one cannot get around the assumption of reality—if only one is honest...They sometimes believe that the quantum theory provides a description of reality, and even a complete description; this interpretation is, however, refuted, most elegantly by your system of radioactive atom + Geiger counter + amplifier + charge of gun powder + cat in box, in which the ψ-function of the system contains the cat both alive and blown to bits.

Einstein had also remarked that de Broglie had "lifted a corner of the great veil" [467] and it was de Broglie's dissertation that had led Schrödinger to wave mechanics. With the presence of entanglement, Schrödinger could emphasize the wave-function viewpoint and instead view measurement as concerning both elements of the entangled system and detector. In this way, there would be no need to refer to particles before an observation. The act of measurement would then be responsible for the collapse into a discrete value [451]. In his 1935-36 entanglement papers, Schrödinger examined a composite system in Hilbert space $H_1 \otimes H_2$ describing systems S_1 and S_2 respectively. He showed that any state vector can be written as the essentially unique biorthogonal decomposition,

$$\psi(x,y) = \sum_k a_k g_k(x) f_k(y)$$

with complete sets of unit vectors $\{g_k\}$ and $\{f_k\}$, which comprise eigenvectors with eigenvalues λ_k^g and λ_k^f of some observables of the systems S_1 and S_2. Measurements in the y-system with eigenfunctions $f_k(y)$ with different eigenvalues occur with probability $p_k = |a_k|^2$, in which case it follows that the x-system must be assigned the wave function $g_k(x)$ in all cases. Schrödinger thus found the troubling result that instantaneous changes in the state of system S_1 result from measurement on the spatially distant system S_2. Therefore, the Copenhagen interpretation of measurement is nonlocal due to entanglement. This includes the special case of states considered by EPR in which all the $|a_k|^2$ are equal so that every observable of one system is determined by an observable of the other one. The EPR argument used a special perfectly correlated state of particles with equal and opposite momenta that we now recognize as entangled,

$$|\psi(x)\rangle_{EPR} = \int dq\, |q - x\rangle\, |q\rangle = \int dp\, e^{ipx} |-p\rangle\, |p\rangle$$

Bohm's version of EPR used two spin-1/2 particles or qubits in the superposition state [468],

$$|\psi_{AB}\rangle_{EPR} = (|0\rangle_A |0\rangle_B + |1\rangle_A |1\rangle_B)/\sqrt{2}\,.$$

EPR had argued that this superposition together with the quantum measurement postulate leads to situations in apparent contradiction to the causality principle of relativity since a measurement by Alice in the $\{|0\rangle, |1\rangle\}$ basis will lead to Bob's spatially separated state to collapse into the state $|0\rangle$ or $|1\rangle$ corresponding to the measurement outcome of Alice. Causality would appear to be violated as if under the influence of a *spooky action at a distance.* In Bohr's reply to EPR, he invoked his concept of complementarity to explain that this state of affairs is not problematic [469]:

> *There is no question of a mechanical disturbance of the system under investigation [in the EPR situation] ... there is essentially the question of an influence on the very conditions which define the possible types of predictions regarding the future behavior of the system.*
>
> N. Bohr, Physical Review 48, 696 (1935), Copyright (1935) by the American Physical Society.

Thus, Alice and Bob end up with a perfectly correlated classical bit after the experiment with no transfer of information. The consequences of the possibility of entanglement becoming degraded at large separations had been considered by Schrödinger and also by Wendell Furry [470] [471] so that entangled states spontaneously localize to mixed product states. Schrödinger had considered this as a way to avoid the "paradox," whereas Furry agreed with Bohr and used the example to show that any such modified version of quantum mechanics must immediately be viewed as wrong. During the period 1960-70's when Clauser was developing his photon anti-bunching experiments, a group led by Edwin Jaynes (1922-1998) had proposed to the quantum optics community a *neoclassical theory* (NCT) of light-matter interactions and challenged them to disprove it with evidence [461]. NCT assumed that an atom's wave-function is governed by Schrödinger's equation while the radiation fields are described by the classical Maxwell's equations. This might be viewed as a modern version of Schrödinger's classical field description of Compton scattering, although NCT was more sophisticated and included effects such as radiation reaction. In a sense, it may even hark back to Planck, with classical fields and quantized energy exchange. Jaynes et al. were able to predict an impressive range of phenomena using NCT including absorption of radiation, spontaneous emission of radiation, the Lamb shift and the black-body radiation spectrum. However, Clauser pointed out that when applied to EPR correlations, NCT would exhibit an example of the Schrödinger-Furry hypothesis with vanishing entanglement at large separations. The Schrödinger-Furry hypothesis was known to be false by that time so Jaynes conceded that NCT had been refuted. Of course, it would also be refuted by the full power of the Bell inequality experiments.

In his 1936 paper, Schrödinger had furthermore demonstrated that unitary transformations, $f_k(y) = \sum_i \alpha_{ik}\, h_i(y)$, can be found for the eigenstates of the y-system so that the state $\psi(x, y)$ can be shown to take the form

$$\psi(x,y) = \sum_i \sqrt{w_i}\left(\sum_k \frac{\alpha_{ik}\sqrt{p_k}}{\sqrt{w_i}} g_k(x)\right) h_i(y)$$

where $w_i = \sum_l p_l |\alpha_{il}|^2$. As a consequence, an appropriate observable corresponding to the eigenfunction $h_i(y)$ can be found to be measured on system S_2 so that one can transform the state of S_2 into *any* state of $S_1$with probability $|w_i|$. Such a process is called *steering,* which has received increasing recent attention from a quantum information viewpoint. In some sense, steering is an even spookier action, giving Alice the ability to affect Bob's state through her choice of measurement basis. EPR is merely a particular case of steering. As Schrödinger put it [464],

> *It is rather discomforting that the theory should allow a system to be steered or piloted into one or the other type of state at the experimenter's mercy in spite of his having no access to it.*
>
> E. Schrödinger, Probability relations between separated systems, Proceedings of the Cambridge Physical Society 32, 446 (1936).

In the hierarchy of nonlocality, it is now known that steerable states are a subset of entangled states and a superset of states that can exhibit a Bell nonlocality [189]. Quantum steering gives a unified description of entanglement for pure bipartite systems with applications in quantum teleportation, entanglement monotones, quantum cryptography and Bell inequalities.

One month after the EPR paper was published, another paper appeared by Einstein and Rosen (ER), called *The Particle Problem in General Relativity* [472]. Dissatisfied with quantum mechanics, Einstein was interested in the extent to which the field theory method of his theory of gravity, General Relativity, could be used to account for atomic phenomena. Einstein sought an approach based on field theory in such a way as to avoid singularities so that the masses of particles are not concentrated at a point. Einstein thought this would ultimately be the real meaning of Heisenberg's uncertainty principle. In the ER paper, the question is asked:

> *Is an atomistic theory of matter and electricity conceivable which,*

while excluding singularities in the field, makes use of no other field variables than those of the gravitational field and those of the electromagnetic field in the sense of Maxwell?

A. Einstein and N. Rosen, Physical Review 48, 73 (1935), Copyright (1935) by the American Physical Society.

The ER paper presented a solution of General Relativity in the form of what is now called an *Einstein-Rosen bridge*, a space-time construction that makes a smooth tube-like connection or bridge between two distinct pieces of space-time. The ER-bridge is now often also called a *wormhole*, as coined by John Wheeler. Einstein was not ultimately successful in finding manifestations of quantum mechanics from field theory. However, there have been recent attempts to relate the ER solution with actual EPR correlations and thus bring entanglement into the picture with gravitational field theory solutions. General relativity contains solutions in which two distant black holes are connected through the interior by ER-bridges. Although the black holes might be spatially distant, General Relativity shows that the ER-bridge cannot be used to transmit information between them. For the purposes of resolving issues in black-hole physics, Maldacena and Susskind postulated that the ER-bridges between two black holes are always created by EPR-correlations between the microstates of the two black holes, calling the proposed relation *ER=EPR* [473]. Namely, quantum entangled particles in an EPR state are connected by an ER-bridge. Just as EPR states cannot be used to signal acausally, neither can ER-bridges because they are not traversable. An ER=EPR relation suggests there would also be a quantum mechanical version of a classical ER-bridge that supports quantum nonlocality. If valid, something like ER=EPR might finally achieve a relation via entanglement between the atomistic and gravity domains sought by Einstein in 1935, although almost certainly not in a way that Einstein would have approved.

Howard summarizes the argument by Einstein against the completeness of quantum mechanics [3] which had not been presented as clearly in the 1935 Einstein-Podolsky-Rosen (EPR) paper [474],

A complete theory assigns one and only one theoretical state to each real state of a physical system. But in EPR-type experiments involving spatio-temporally separated by previously interacting systems, A and B, quantum mechanics assigns different theoretical states, different "psi-functions," to one and the same real state of A, say, depending upon the kind of measurement we choose to carry out on B. Hence quantum mechanics is incomplete.

A. Nicht Sein Kann was Nicht Sein Darf, or the Prehistory of EPR, 1909-1935: Einstein's Early Worries about the Quantum Mechanics of Composite Systems, In: Sixty-two Years of Uncertainty, 1990, p. 64, D. Howard, © 1990 Plenum Press, With Permission of Springer.

The first entangled state generated within quantum mechanics may have been from Born's 1926 paper *On the Quantum Mechanics of Collisions*, in which he introduced the probabilistic interpretation of quantum mechanics [6]. Born carried out a calculation of electron-atomic scattering where the system is initially in a product,

$$\psi_{n,\tau}^{0}(q_k, z) = \psi_n^0(q_k)\sin(2\pi z/\lambda).$$

The electron is in a plane wave momentum eigenstate incident on the atom and the atom is in an energy eigenstate. Thus, there is initially no entanglement between electron and atom. After the interaction, the state after the scattering was calculated using a perturbation method with the result clearly in the form of an entangled state in terms of momentum eigenstates scattered in different directions and excited atomic states,

$$\psi_{n,\tau}^{1}(x, y, z; q_k) = \sum_m \iint d\omega\, \Phi_{n,m}(\alpha, \beta, \gamma) \sin k_{n,m}(\alpha x + \beta y + \gamma z + \delta)\psi_m^0(q_k).$$

Born then makes this conclusion about the result,

> *If one translates this result into terms of particles, only one interpretion is possible. $\Phi_{n,m}(\alpha, \beta, \gamma)$ gives the probability* for the electron, arriving from the z-direction, to be thrown out into the direction designated by the angles α, β, γ, with the phase change δ... Schrödinger's quantum mechanics therefore gives quite a definite answer to the question of the effect of the collision; but there is no question of any causal description. One gets no answer to the question, "what is the state after the collision," but only to the question, "how probable is a specified outcome of the collision."*

The asterix refers to the famous footnote saying that the probability is actually given by the square of $\Phi_{n,m}$ (though it should actually have said "absolute square"). Born is led to hypothesize indetermism in quantum mechanics from the terms of an entangled state decomposition, though how this entanglement would depend on the basis of expansion was of course not yet appreciated in 1926.

Following Born's pioneering work, wave-mechanical solutions for collisions involving entangled states were presented by C. G. Darwin in the paper "A collision problem in wave mechanics," [475] and building on Darwin's techniques, N.F. Mott explained the tracks of α-particles in cloud chambers in the paper "The wave mechanics of α-ray tracks" [428] as discussed in the section *Einstein's Quandary*. According to Darwin,

> *Before the first collision, (the wave function) can be represented as the product of a spherical wave for the α particle, by a set of more or less stationary waves for the atoms... [The] first collision changes this product into a function in which the two types of coordinates are inextricably mixed.*
>
>

The calculations of Darwin and Mott clearly involved entangled states of the entire system. Heisenberg also analyzed the cloud chamber problem in his 1929 University of Chicago lectures but took a more flexible approach for how to divide the compound structure into system and the observational device. For Heisenberg, the arbitrariness in switching from a corpuscular representation to a wave representation whenever convenient reflected the freedom in choice of the *cut* (*Schnitt*), as Heisenberg called it, between system and device [426].

Collisions could evidently produce complicated entangled states. However, a more systematic understanding of entanglement would benefit from the experimental realization of particular entangled states, especially in the form of EPR correlations. Production of the first EPR states would have to wait for the ability to control photon polarization states. The methods for doing this were analyzed in the decade following the EPR paper, during the period of 1947-1949 [476]. Dirac's ideas on particle-pair

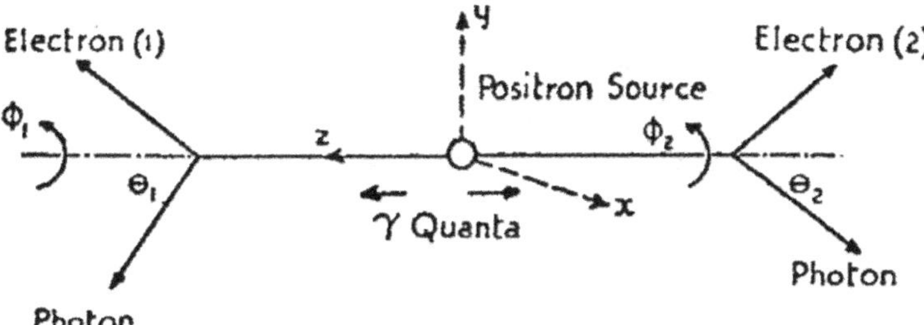

Figure 5.16: Proposal from 1947 of the method which led to the first realization of polarization entangled photons, i.e. EPR correlations [479].

production in 1930 [477] were discussed by John Wheeler in 1946 [478] and the detailed quantum physics of the process and a rendition of the experiment by Pryce and Ward appeared in 1947 [479] with all the elements required for quantum entanglement, Figure 5.16.

The theory was independently derived by Snyder et al. in 1948 [480]. Two groups published experimental efforts by 1948 [481] [482] and more definitive results were reported by Wu and Shaknov in 1950 [483]. This was the same Chien-Shiung Wu (1912-1997) who would become prominent for her 1956 experimental confirmation of the violation of parity conservation in the weak nuclear interaction. The entangled states expected from these works were equivalent to the two-qubit version of the EPR state of the type discussed previously but using instead the states of horizontal and vertical polarization discussed in Chapter 2, $|\psi_{AB}\rangle_{EPR} = (|H\rangle_A|V\rangle_B + |V\rangle_A|H\rangle_B)/\sqrt{2}$. However, none of these works were presented in the context of EPR or referenced the EPR paper. It was an article by Bohm and Aharonov [484] seven years later that discussed EPR in reference to using an annihilation process to produce polarization of two quanta propagating in opposite directions from the Wu-Shaknov paper. Bohm and Aharonov would say that the results of EPR could be tested by polarization properties of pairs of photons. They pointed out that the magnitude of the Wu-Shaknov result agrees with quantum mechanics and disagrees with the Schrödinger-Furry hypothesis. The Bohm-Aharonov paper led subsequent researchers to use optical polarization based measurements to test EPR correlations [83] [85]. All of these experiments verified the expectations of quantum mechanics for EPR entangled states. Thus it took around two decades before the EPR scenario began to be realized. Although EPR originally proposed their state via momentum conservation and the first experimental evidence was obtained via photon polarization, EPR states are now commonplace in quantum optics and quantum information. Entanglement distribution has been demonstrated over more than 1200 kilometers [485] with no evidence of the Schrödinger-Furry effect. EPR states today appear via experimental processes such as spontaneous parametric down conversion, spontaneous emission, Raman scattering, ionization, etc., all verifying the predictions of quantum mechanics [486], Figure 5.17.

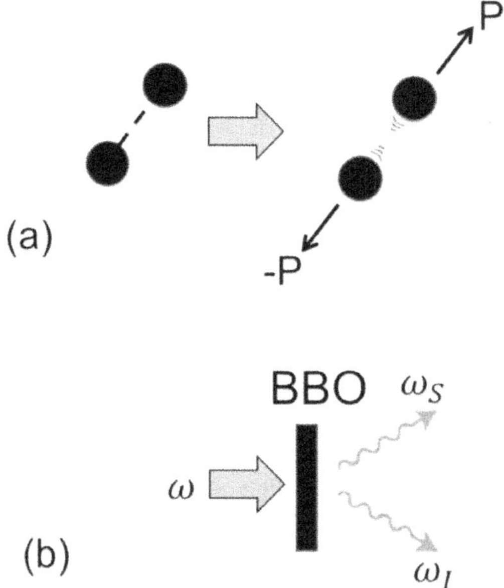

Figure 5.17 EPR state formation (a) Particle momentum conservation (b) Photon pairs from spontaneous parametric down-conversion [486].

Johnny Goes to Göttingen

At the time of the EPR and Schrödinger papers in 1935-1936, the renowned mathematician and polymath John von Neumann was also at the Institute for Advanced Study (IAS) along with Einstein. In addition to his many fundamental contributions to mathematics, mathematical physics and computer science, von Neumann had also published *The Mathematical Foundations of Quantum Mechanics* [13] in 1932, a tour de force which had formulated quantum mechanics with mathematical rigor using the theory of Hermitian operators and Hilbert spaces as well as introducing the *Measurement Postulate* as a means of implementing the non-unitary aspect of the measurement process. The existence of a single unpublished letter dated

April 11, 1936, from von Neumann to Schrödinger gives evidence that Einstein actually had contact with von Neumann regarding quantum mechanics and shared with him some of his correspondence from Schrödinger [487, pp. 211-213]. The letter to Schrödinger begins,

> *Einstein has kindly shown me your letter as well as a copy of the Pr. Cambr. Phil. Soc. manuscript.*
>
>

This refers to the second of the "Probability Relations between Separated Systems" papers [464] in which Schrödinger had introduced steering. Von Neumann would certainly have readily understood the details of Schrödinger's developments regarding the entangled states since he indicated that he was glad that Schrödinger had liked the mixing method. Schrödinger had evidently followed von Neumann's treatment of mixed states from his 1932 book. The mathematical theory of quantum mechanics in von Neumann's book also allows multi-particle entangled states to be analyzed and characterized in terms of their individual subsystems, though von Neumann did not categorize the properties of entanglement as Schrödinger had done in his 1935-1936 papers.

However, von Neumann does not appear to perceive the resulting quantum nonlocality demonstrated by Schrödinger as anything new or problematic. He objects to Schrödinger's concerns about the instantaneous correlations since such correlations can also exist classically,

> *I cannot accept your 4. completely. I think that the difficulties you hint at are "pseudo-problems." The "action at a distance" in the case under consideration says only that even if there is no dynamical interaction between two systems (e.g., because they are far removed from each other), the systems can display statistical correlations. This is not at all specific for quantum mechanics, it happens classically as well.*
>
>

Von Neumann then described a simple classical example of two sealed boxes which both contain either one white ball each and or one black ball each but one does not know which color the balls are. The boxes are then spatially separated with one remaining on Earth and the second taken to Sirius. Of course, on opening the box on Earth, one immediately knows the state of the box on Sirius. To make the point, von Neumann uses the risqué joke: The moment something happened to his wife in Paris, the colonel was cuckolded in Madagascar. This type of correlation is now often called Reichenbach's *Principle of Common Cause* [488], that whenever two events are correlated, either one is a direct cause of the other, or they share a common cause. However, this principle is not straightforward when the correlated events A and B are space-like separated and in conjunction with relativistic causality. Of course, this was the situation later addressed by Bell [159] which considered a common cause in the past of A and B, denoted by hidden variables λ. This line of reasoning leads to the Bell inequalities which are violated by quantum theory and by experiments on systems with EPR correlations. Hence, von Neumann's explanation is incorrect. The principle of common cause plays a subtle but crucial role in the derivation of Bell's inequalities. Attempts at retaining some form of common cause in the face of quantum correlations result in a range of difficulties for any theory [191]. However, this was not known at the time of the letter, thirty years earlier.

Von Neumann's book had included his well-known theorem which demonstrated that hidden variables cannot correspond to operators in Hilbert space. In the proof, he assumes that a linear operator on a Hilbert space is associated with each measurement apparatus and that measurement results are given by eigenvalues of this operator. On this basis, he could show that, within the structure of quantum mechanics, there cannot exist any additional or hidden variables that could be used to distinguish statistical differences between particles in the same pure state. In this sense, quantum mechanics is a complete theory and von Neumann concluded that [13, p. 327],

> *The only formal theory existing at the present time which orders and summarizes our experiences in this area in a half-way satisfactory manner, i.e., quantum mechanics, is in compelling logical contradiction with causality. Of course, it would be an exaggeration to maintain that causality has thereby been done away with: quantum mechanics has, in its present form, several serious lacunae, and it may be even that it is false, although this latter possibility is highly unlikely, in the face of its startling capacity in the qualitative explanation of general problems and in the quantitative calculation of special ones.*

He was well-aware that his proof did not rule out hidden variables in general but that the link between physical quantities and operators would then have to be severed in that case. In his book [489, p. 89], Abner Shimony relates a story by Peter Bergmann, one of Einstein's assistants at the IAS, that Einstein was aware of von Neumann's analysis and once opened von Neumann's book to the page where the proof is given and pointed to the linearity assumption underlying the proof, asking "Why should we believe in that?" In 1952, David Bohm (1917-1992) had resurrected a thirty-year old approach of de Broglie's, though without being aware of it, and took advantage of the fact that von Neumann's theorem concerned hidden-variables theories of only a specific kind to develop contextual models with hidden variables not obeying that same linearity assumption but which could reproduce the predictions of quantum mechanics. It is sometimes stated that von Neumann's linearity assumption was unnecessary, therefore, his impossibility proof is mistaken; e.g., Bell's statement that, "There is nothing to it. It's *not* just flawed, it's *silly*!" [490]. However, this misconstrues von Neumann's carefully worded statement of the intent of his proof, i.e., that hidden variables cannot correspond to operators in Hilbert space [491]. In his book, von Neumann had concluded that the Hilbert space formalism is well supported but a formal proof cannot exist that no more complete theory than quantum mechanics will ever be possible. Thus, it's not surprising that Bohm had said in a letter to Pauli [492],

> *It appears that von Neumann has agreed that my interpretation is logically consistent and leads to all results of the usual interpretation. (This I am told by some people). Also, he came to a talk of mine and did not raise any objections.*

It's interesting then to consider in hindsight what von Neumann may have been thinking in his 1936 letter to Schrödinger concerning the white and black balls on Sirius.

Following Bohr's presentation at the *New Theories in Physics* conference in Warsaw in 1938, von Neumann gave an unscheduled presentation following Bohr, which included a succinct summary of his hidden variables impossibility proof as well as a discussion of quantum logic [493]. Afterwards, Bohr pointed out that the very simple experimental cases which he had alluded to in his own presentation showed, in more elementary form, the same essential points as those which appeared in von Neumann's mathematical analysis [493, p. 38]. Thus, Bohr was publicly comfortable with von Neumann's approach and that it was consistent with his own complementarity views. However, Bohr privately had impatience with formalism, and he had been seen shaking his head skeptically regarding von Neumann's proof [494, p. 263].

The first mention in print of the collapse of the state vector in quantum mechanics occurs in von Neumann's 1927 paper, *The Probability-theoretical Construction of Quantum Mechanics* [495] [496],

> *A system left to itself (not disturbed by any measurements) has a completely causal time evolution [governed by the Schrödinger equation]. In the confrontation with experiments, however, the statistical character is unavoidable: for every experiment, there is a state adapted to it in which the result is uniquely determined (the experiment in fact produces such states if they were not there before); however, for every state there are "non-adapted" measurements, the execution of which demolishes that state and produces adapted states according to stochastic laws.*

We have seen how the interrelated properties of nondeterminism, entanglement, and exact conservation laws are crucial to characterizing the measurement problem. The conservation laws had been experimentally shown to hold exactly for individual processes, beginning with the important Compton-Simon scattering experiment. Born was led to his probabilistic rule to account for measurement by the analysis of scattering processes. Von Neumann also used the Compton-Simon scattering experiments as the basis for a thought-experiment leading to the proposed *Measurement Postulate* in his 1932 book *The Mathematical Foundations of Quantum Mechanics* [13, pp. 212-217]. The Compton-Simon paper is the only experimental reference in the book. The exact conservation laws found by Compton-Simon are used as basis of von Neumann's argument [497], Figure 5.18.

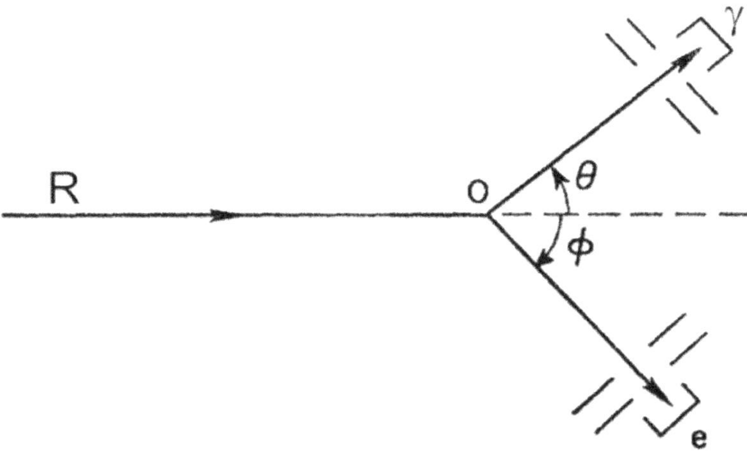

Figure 5.18: von Neumann's motivation for the projection postulate based on Compton-Simon cloud chamber experiments: (1) Central line R of the γ-e recoil collision is unknown but with a distribution of possible directions, conserved due to momentum conservation and (2) recoil of a photon γ and electron e toward detectors. If measurement of the photon γ occurs first, this is sufficient to determine R and therefore e, transitioning from statistical possibilities to a definite state of the electron e [497].

The Compton-Simon experiment involved a monochromatic beam of X-rays passing through a cloud chamber, a vapor-filled vessel. Photographs of the interactions were taken at intervals and a pattern of tracks revealed the result of photons scattering from the orbital electrons of the molecules of the vapor. It was from such photographs that the validity of the conservation laws was deduced. It should be emphasized that the conservation laws also hold for relativistic particles in the relevant reference frame as is the case for these processes. Von Neumann considered a representative collision with the central line R of the collision along with the recoiling photon and electron each of which is measured by a detector, Figure 5.18. The central line is not known before the measurement, but the exact conservation laws enable us to determine R once the path of either recoiling particle is known. Therefore, before the measurement occurs, we can only make statistical statements about R. Suppose the photon reaches its detector first in our reference frame. Then due to exact conservation, the central line R is immediately known and therefore the state of the electron. The electron can be then considered as transitioning from a set of statistical possibilities to a definite state, a discontinuous and instantaneous reduction of the state. The electron momentum after the collision is given by $\acute{p}_e = p_\gamma - \acute{p}_\gamma + p_e$, yielding a transition of the electron to the momentum eigenstate $|\acute{p}_e\rangle$.

As part of his formulation, von Neumann proposed two postulates for evolution of a quantum system.

- Process 1: acausal non-unitary reduction applied for measurement
- Process 2: casual Schrödinger unitary evolution otherwise

Von Neumann's mathematical theory also allowed multi-particle entangled states to be analyzed and characterized in terms of their individual subsystems. As discussed in Chapters 3 and 4, non-unitary and unitary processes can be distinguished as a criterion for defining the measurement problem and entanglement plays a central role. Von Neumann makes the comment [13, pp. 418-420]:

> *Quantum mechanics describes the events which occur in the observed portions of the world, so long as they do not interact with the observing portion, with the aid of [process 2] evolution, but as soon as such an interaction occurs, i.e., a measurement, it requires the application of [process 1] evolution. The dual form is therefore justified.*
>
> *The difference between these two processes $\rho \rightarrow \acute{\rho}$ is a very fundamental one: aside from the different behaviors in regard to the principle of causality they are also different in that the former [Process 2] is (thermodynamically) reversible, while the latter [Process 1] is not.*

A von Neumann measurement is carried out in terms of the spectral decomposition of an observable A so that $A = \sum_i a_i P_i$ with projection operator $P_i = |a_i\rangle\langle a_i|$ for eigenstates $|a_i >$ and the non-degenerate eigenvalue a_i is found as a measurement result. Then the state vector makes a transition from the initial state to the corresponding eigenvector:

$$|\psi >= \sum_k c_k |a_k\rangle \longrightarrow |a_i\rangle.$$

In case of degeneracy, this was later generalized by Lüders in 1951 [498] to the projection

$$|\psi>=\sum_k c_k|a_k\rangle \longrightarrow \frac{P_i|\psi\rangle}{||P_i|\psi\rangle||}$$

with $P_i = \sum_k |a_{ik}\rangle\langle a_{ik}|$ being the projection operator on the subspace spanned by the eigenvectors of A for which $a_{ik} = a_i$. The von Neumann formalism has been generalized by Kraus and to quantum trajectories as discussed in Appendix 5.A.

The road toward developing von Neumann's theory began in the fall of 1926, when the prodigious twenty-two-year-old John von Neumann ("Johnny", younger than Dirac, Heisenberg, and Pauli) arrived at Göttingen University in Germany, a renowned center for mathematics. Its history had included Gauss, Riemann, Dirichlet, and Minkowski and now the driving force was the great mathematician David Hilbert. Max Born was an assistant to Hilbert when von Neumann joined the group. An eminent mathematician of his time, Hilbert also took a great interest in problems of physics and wanted to bring the rigor of mathematics to them and in particular use the axiomatic method, saying "physics is too hard for physicists." As a result, Hilbert gave courses and seminars in physics each year. Heisenberg had previously been Born's assistant and in 1925, the year he had discovered quantum mechanics at Niels Bohr's Copenhagen Institute, he presented his theory at Hilbert's seminar. This led to a burst of activity by Hilbert's group, furthering their interest in quantum mechanics, and initially intending to improve the existing formulations.

However, von Neumann realized that the natural framework for quantum mechanics was given by the abstract theory of separable Hilbert spaces and the associated linear operators. In this formulation, states of the physical system are given by Hilbert space vectors and measurable observables by Hermitian operators. This led to a two-year effort (1927-1929) by von Neumann, analyzing the formal aspects and the interpretation of quantum mechanics and this was developed in an axiomatic way [496]. It also led von Neumann into the mathematics of operator theory in which he achieved prominent results. Important physical quantities such as position and momentum are represented by unbounded Hermitian operators and by 1929 von Neumann had completely solved the problem of extending spectral theory to the unbounded case. In this way, von Neumann completely avoided the methods used by Dirac (including the singular Dirac delta-functions) which he had said "does not meet the demands of mathematical rigor in any way." Among other results, his approach showed the equivalence between matrix mechanics and wave mechanics in a rigorous way. Heisenberg's observables were operators on $\ell^2(N)$ (the space of complex square summable sequences) whereas Schrödinger's wave-functions were unit vectors in $L^2(\mathbb{R}^3)$ (the space of complex square integrable functions), and the inner products provide Born probabilities. A unitary transformation between these Hilbert spaces exhibits the mathematical equivalence between the two versions of quantum mechanics. Von Neumann also introduced the formalism of density operators $\hat{\rho}$ to characterize ensembles and *mixtures,* finding that the expectation value of an operator $\hat{A}$ is given by $\mathrm{Tr}(\hat{\rho}\,\hat{A})$ which holds for both pure and mixed states. The assumptions that went into recovering the Born rule were essentially the same as those for the no-hidden-variables proof [496].

In 1927, von Neumann published a trilogy of papers in the Göttingen Journal of Mathematics in which he developed the rigorous mathematical formulation of quantum mechanics. These became the basis of his 1932 book, *The Mathematical Foundations of Quantum Theory*, which was translated from German into English in 1949. The first three chapters essentially constitute the material from the trilogy, Chapter 4 deals with the completeness of quantum mechanics (i.e., the no-hidden-variables theorem), while Chapters 5 and 6 are devoted to measurement including irreversibility, thermodynamics, and the details of the measurement process. This included defining the entropy of quantum ensembles, addressing the irreversibility of macroscopic measurements due to the reduction of the wave function and the consistency of measurement regarding the interactions between a system and a measuring apparatus.

Von Neumann focuses in Chapter 6 on determining the precise point at which systems switch evolutions between Process 1 and Process 2, saying [13, p. 418]

> *...subjective perception is a new entity relative to the physical environment and is not reducible to the latter. Indeed, subjective perception leads us into the intellectual inner life of the individual, which is extra-observational by its very nature.*

In an example of the measurement of temperature, he gives four options for designating an *observer*: the thermometer, the human taking part in the measurement, the retina of the human, and the brain cells of the human. He concludes that the choice of designating the *observer* is arbitrary and a matter of convenience [13, p. 421],

> *In order to discuss this, let us divide the world into three parts: I, II, III. Let I be the system actually observed, II the measuring instrument, and III the actual observer. It is to be shown that the boundary can just as well be drawn between I and II + III as between I + II and III. (In our example above, in comparison of the first and second cases, I was the system to be observed, II the thermometer, and III the light plus the observer; in the comparison of the second and third cases, I was the system to be observed plus the thermometer, II the light plus the eye of the observer, III the observer, from the retina on; in the comparison of the third and fourth cases, I was everything up to the retina of the observer, II his retina, nerve tracts and brain, III his abstract "ego".)*

Within the framework of his axiomatic exposition of quantum mechanics, von Neumann then encounters the same issue as had the various *Copenhagen* narratives of quantum mechanics from Bohr, Heisenberg and Pauli: that of the dividing line between system and measurement device along with their quantum vs classical characteristics. As previously noted, Bohr, Heisenberg and Pauli shared a somewhat common set of views, sometimes known as the Copenhagen (or "orthodox") interpretation of quantum mechanics, though the persistence of a "Copenhagen viewpoint" may have owed much to Heisenberg's proselytizing for it during the 1950s [499]. More accurately, it could be described as a Copenhagen "spirit" rather than "interpretation" [500].

As emphasized by Zinkernagel [501], an issue from the beginnings of quantum mechanics is understanding the process of acquiring knowledge involving a separation between the observer and the observed system, this *cut* that allows a "detached" position for the observer to read at leisure the records left by the observed system. In a letter to Bohr, Pauli had called [213]

> *the conceptual description of nature in classical physics, which Einstein so emphatically wishes to retain, the ideal of the detached observer.*

Bohr discusses how complementarity unexpectedly embraces the detached observer:

> *Far from indicating a departure from our position as detached observers, the notion of complementarity represents the logical expression for our situation as regards objective description in this field of experience, which has demanded a renewed revision of the foundation for the unambiguous use of our elementary concepts. The recognition that the interaction between the measuring tools and the physical systems under investigation constitute an integral part of quantum phenomena has not only revealed an unsuspected limitation of the mechanical conception of nature characterized by attribution of separate properties to physical systems, but has forced us in the ordering of experience to pay proper attention to the conditions of observation.*

The cut can be made in many ways. Bohr's complementarity meant that the dividing line is determined by the nature of the experimental arrangement. However, Heisenberg and Pauli argued, and demonstrated by calculations, that the cut could be moved freely in the direction of the observer but that it cannot be moved arbitrarily far in the direction of the measured system. Heisenberg had used the existence and flexibility of the cut as the basis for an argument that hidden-variables cannot exist in quantum mechanics [502]. Since the laws of quantum mechanics apply to all systems, including the measuring apparatus, any steps to give a more complete account of the quantum state of a system would prevent the arbitrariness of the cut. Heisenberg appears to relate the cut to the necessity of eliminating what would come to be known as entanglement (i.e., "fine connections") between the system and the measurement device [501]:

> *To get an observation, one must therefore cut out a partial system somewhere from the world, and one must make "statements" or "observations" just about this partial system. There, one destroys the fine connections of phenomena, and the very point where we place the cut between the observed system, on the one hand, and the observer and his apparatus, on the other hand, we have to expect difficulties for our conception.*
>
>

As discussed in Chapter 4, Heisenberg also identifies two subprocesses: 1) where the system state becomes mixed because it interacts with a device and 2) when the state collapses. Subprocess 2) only occurs once an observer becomes conscious of it.

Einstein und Bohr Verschränkten

Einstein and Bohr were the dominant figures in the conceptual foundations of quantum measurement in the early to mid-20th century. They were both exceptionally conceptual and deductive thinkers. The thirty-five-year long series of disagreements over thought-experiments constituting the Bohr-Einstein debates were the sounding chamber for the clarification of the central issues of quantum measurement and to which the various other players reacted and responded. These debates began essentially when they first met in 1920 and carried on at various levels until Einstein's death in 1955. Particularly prominent were the disagreements at the 5th and 6th Solvay conferences in 1927 and 1930, respectively and with the EPR paper in 1935. Einstein initially attempts to undercut quantum mechanics by demonstrating the possibility of intrinsic violations of Heisenberg's uncertainty principle and Bohr finding the often subtle flaws in Einstein's schemes. Einstein eventually lowers his sights with the goal of showing quantum theory to be incomplete rather than incorrect. However, more recently there has been the acknowledgment that both Einstein and Bohr unmistakably appreciated early on that *entanglement* was the principle feature of quantum theory that needed to be confronted, though they may not have always used the term. The evidence for this has been particularly emphasized by Howard [3]. For Bohr, entanglement was the underlying enforcer of complementarity. For Einstein, entanglement was the impediment for his view that real states of spatially separate objects are independent of each other. Their joint recognition of entanglement ultimately overrode other concerns over science and that led to their being themselves intertwined or *entangled – verschränkten* – and in disagreement over quantum theory for over three decades. And as demonstrated and emphasized in this book, entanglement is the key to defining and resolving the measurement problem. Einstein and Bohr were the key players in defining the issue of quantum measurement.

As discussed in the section *Planck's Fortunate Guess*, Einstein had mentioned in a 1909 letter to Lorentz that it would be problematic if his proposed light-quanta behaved as independent particles, since "A light ray divides, but a light quantum indeed cannot divide without an alteration of frequency." For Einstein, this was the source of the infection, the first inkling that nature would not conform to a view where separate parts of space have an independent real existence. The conspiracy involving the ingredients "nondeterminism", "whole photon or nothing," "exact conservation for individual events," and "entanglement" appeared to be built-in features of nature. However, yet another factor entered with Bose statistics in 1924 so that Einstein's light quanta must be bosons, further reinforcing the lack of corpuscular independence for photons but also necessary for Planck's black-body formula. Bose and Fermi statistics thus also enter into Nature's conspiracies and thereby the measurement problem. Wave-particle duality is intrinsic in the proceedings of quantum theory. Bohr's rethinking of an ordinary space-time description was given another jolt four years later with the failure of the 1924 BKS theory after the Compton-Simon and Bothe-Geiger experiments showing exact conservation for individual events. The conspiracies that Nature required to implement the quantum of action into the universe were closing in on both Einstein and Bohr. Over the succeeding three years, with the complete theory of quantum mechanics and in particular Heisenberg's indeterminacy principle, Bohr's thinking moved inexorably toward complementarity.

In the 1927 Solvay conference, Einstein's single slit (i.e. beam splitter) thought-experiment had brought out aspects of nonlocality, incompleteness and energy conservation, Figure 5.15. A more celebrated photon-box thought-experiment occurred at the 1930 Solvay conference. Einstein envisioned a mirrored box that could contain a photon for an arbitrary lapse of time. A precisely timed shutter mechanism allows accurate timing of the photon emission from the box. The box is suspended by a spring so that it can be weighed before and after the photon emission and thus finding the photon's energy by the mass-energy equivalence relation of relativity and apparently violating Heisenberg's time-energy uncertainty principle; i.e., weigh the box to fix the energy and open the box to check the clock. According to Bohr's 1949 retelling of the story [503], after a sleepless night of being stumped, Bohr realized that the weighing of the box requires vertically moving the box in the gravitational field and that Einstein's theory of relativity predicts that this will change the rate that the clock ticks. A calculation showed that this introduced just the uncertainty in the clock's rate to save Heisenberg's indeterminacy. Thus, Einstein was defeated in yet another debate.

However, Einstein explained later to Ehrenfest that he also had another purpose in mind for his photon box [111]. In this variation, the emitted photon is allowed to be reflected back from a mirror at an arbitrarily long space-like separated distance. At the time of the photon's distant reflection, we can make a choice to either weigh the box or check the clock. Einstein argued that this choice can have no effect on the distant photon so it will be identical whenever it returns. However, quantum mechanics would ascribe a different state depending on the choice. Therefore, Einstein had concocted yet another argument for him to claim that quantum mechanics is incomplete. And once again, Einstein had devised an example that intrinsically involves what would come to be known as entanglement. Einstein's own 1949 comments point to Einstein's and Bohr's mutual appreciation of entanglement's role in quantum mechanics [504, p. 681],

> *...Niels Bohr seems to me to come nearest to doing justice to the problem. Translated into my own way of putting it, he argues as follows: if the partial systems A and B form a total system which is described by its Ψ-function* $\Psi(A, B)$*, there is no reason why any mutually independent existence (state of reality) should be ascribed to the partial systems A and B viewed separately, not even if the partial systems are spatially separated from each other at the particular time under consideration. The assertion that, in this latter case, the real situation of B could not be (directly) influenced by a measurement taken on A is, therefore, within the framework of quantum theory, unfounded and (as the paradox shows) unacceptable.*
>
> Albert Einstein, Remarks to the Essays Appearing in this Collective Volume, p.663 in Albert Einstein: Philosopher-Scientist, Schilpp, Paul Arthur (Editor), Cambridge University Press 1949.

Even noninteracting systems that are spatially separated cannot always maintain mutual independence. And this applies when such systems involve an observed subject and a measurement apparatus, leading to inevitable correlations. As Bohr had explained in 1928 [442, p. 78],

> *...in order to make observation possible we permit certain interactions with suitable agencies of measurement, not belonging to the system, an unambiguous definition of the state of the system is naturally no longer possible, and there can be no question of causality in the ordinary sense of the word. The very nature of the quantum theory thus forces us to regard the space-time co-ordination and the claim of causality, the union of which characterizes the classical theories, as complementary but exclusive features of the description...*
>
> S. Petruccioli, Atoms, Metaphors and Paradoxes, Niels Bohr and the construction of a new physics, Cambridge University Press 1993.

Free Will, Consciousness, and Soul

Clockwork versus Free Will

A clock's chiming within the Paradise of Dante's *Divine Comedy* (1320) is the first reference to mechanical timepieces in European literature. Pre-Reformation art had placed much of our world within the otherworldly, and the medieval was a clockwork universe set in movement by divine cogs and machinery under the power of the Prime Mover, a concept tracing back before Aristotle as that "which moves without being moved" [353] [Metaphysics XII.7]. The technology of each era is often taken as a metaphor for understanding the world, Figure 5.19.

Figure 5.19: Fragment of the Antikythera Mechanism (1 BCE, Greece), an early astronomical clockwork device for computing the positions of the sun and moon.

A clock, though created by humans, ticks along predictably, seemingly with determination. But what of the world we see around us? Some aspects of determinism seem to have been expressed by the Greek atomists Leucippus and Democritus as early as the 5th century BCE. The atomists consider a world order forming when some of the infinite mass of randomly churning atoms forms a circular eddy or swirl, interpreted in some readings as initially forming by chance. The atomists' universe within the swirl appears to be deterministic and purposeless, although this view relies on a single fragment of Leucippus [505, p. 185],

> *Nothing happens at random, but everything from reason and by necessity.*

The first part of the fragment ("nothing happens at random, but everything from reason") asserts the Principle of Sufficient Reason, later associated with Leibniz, while the second part ("and by necessity") makes a stronger causal claim that whatever happens has to happen.

Notions along these lines continued with the Stoics in the 3rd century BCE that include the first recorded disputes over free will and determinism. Epicurus, motivated by the need to reconcile the existence of human free will, included the possibility of objective chance by allowing interruptions without cause into the deterministic motion of atoms by allowing *swerves* in their motion which later became known through

Lucretius' *On the Nature of Things* [506]. For example, an atom's falling vertically downward could also include a slight swerve to one side. Such indeterminacies would then make their way up to our macroscopic observable world. The recognition of pure chance also in the initial formation of the swirl by Democritus and the Stoics has been attributed to an ignorance of their causes and not a denial that they have causes [505, p. 187]. Thus, departures from deterministic behavior may have been considered by the Greeks as similar to the origin of classical statistical mechanical properties in the 19th century or the hidden variables sometimes later invoked to account for quantum nondeterminism. Aristotle however rejected the two major concepts of Democritus: atomism and determinism. He asserted that there are both animate and inanimate objects. The animate objects generally behave non-deterministically. Had Aristotle been alive to see the developments in quantum mechanics, he may have argued that Schrödinger's wave equation would not be by itself sufficient to describe the universe as there is no chance inherent in unitary evolution [353],

> *Further, no one could say why a thing once set in motion should stop anywhere; for why should it stop here rather than here? So that a thing will either be at rest or must be moved ad infinitum, unless something more powerful gets in its way.* [Physics IV.8]

> *Chance and what results from chance are appropriate to agents that are capable of good fortune and of action generally. Therefore necessarily chance is in the sphere of actions...Hence what is not capable of action cannot do anything by chance. Thus an inanimate thing or a beast or a child cannot do anything by chance ...The spontaneous on the other hand is found both in the lower beasts and in many inanimate objects.* [Physics II.6]

> *But it is a wrong assumption to suppose universally ... in virtue of the fact that something always is so or always happens so. Thus Democritus reduces the causes that explain nature to the fact that things happened in the past in the same way as they happen now: but he does not think fit to seek for a first principle to explain this "always": so, while his theory is right in so far as it is applied to certain individual cases, he is wrong in making it of universal application.* [Physics VIII.1]

However, the nature of probabilities in classical theory is quite different than those in quantum theory as emphasized by Born in his 1926 paper that first introduced probabilities into quantum mechanics: "The classical theory introduces the microscopic coordinates which determine the individual processes only to eliminate them because of ignorance by averaging over their values; whereas the new theory gets the same results without introducing them at all." [6] The intrinsic nondeterminism of the quantum succeeded the world of the deterministic clockwork but not before determinism reigned in the Newtonian worldview. These two types of randomness, *epistemic* randomness due to lack of complete knowledge of the state of the system and *ontic* randomness that is intrinsic even with complete knowledge of the state of the system, are the paradigms that would continue to compete for their roles within physical theory up to the present day [507]. Currently, the most stringent tests for the intrinsic randomness of quantum mechanics are from measurements showing violations of Bell inequalities, and a central ingredient in that conclusion is that the measurement settings could be freely chosen, at least to some degree [508].

Consciousness and Free Will

It also ought to be impossible to predict the future for a system that acts with free will. And free will is very different than simply having random occurrences. Free will is not yet defined by either deterministic theory or current probability theory as developed by Kolmogorov where each event in a sample space has an *a priori* probability measure assigned to it. If free will is different than random occurrences, even conventional quantum mechanics with Born indeterminacy may not be sufficient to explain free will. This is because the probabilities are determined by a law of causality, hence the conventional quantum theory would seem to be problematic in terms of free will. However, the insistent focus on freedom of choice in the measurement settings of the quantum Bell experiments naturally touches on the long contentious concepts of free will, consciousness, mind, and even the concept of the soul [509]. The concept of free will would allow two potential courses of action under identical initial circumstances. The identical circumstances would inevitably encompass both external conditions and the internal states of consciousness. The consciousness of inner mental life follows us during our waking hours, but experience suggests that it is banished during dreamless sleep, anesthesia, coma, and death [510]. The long-debated question is whether these phenomena are an illusion or alternatively are they our most intimate examples of genuine freedom of choice? After all, we appear to know almost from birth that while our mind is evident to itself, it is impervious to outside observers.

In what way can any of this be encompassed by physical theory, particularly without having first addressed the problem of measurement as discussed in Chapter 4? And how is it impacted by the questions of the nondeterminism versus determinism of particular physical processes? The so-called *hard problem* [511] of qualia of experiences—the redness of red, the painfulness of pain—is understanding how these could arise from physical processes [512], presumably including those associated with the living body and with the brain in particular. An inductive view of the universe has led some to dismiss this as insignificant; e.g., Gell-Mann has said that the hard problem consists only of the "redness of herrings" [513], which is contrary to the deductive approach argued in this book necessary to address problems of consciousness and measurement. There have long been attempts to locate within "thinking matter" the origins of consciousness, mind or soul, leading within cognitive neuroscience to the quest to identify the *neural correlates of consciousness* (NCC)

[510], the minimal neural mechanisms that are sufficient for any one conscious perception, thought or memory, given the necessary background conditions to maintain consciousness. There are currently both reductionist and holistic approaches to studying these aspects of consciousness, with the reductionist focusing on genetic, synaptic and cellular levels and the holistic attempting to understand large neural networks underlying cognition, actions, and emotion [514]. Neuroimaging methods, such as positron emission tomography and functional magnetic resonance imaging (fMRI), have recently allowed observation of neurons in human subjects under conditions where they attempt to focus on awareness, thoughts, and feelings. Craig Venter et al. in the paper presenting the first sequencing of the human genome commented on its future significance [515],

> *The real challenge for human biology, beyond the task of finding out how genes orchestrate the construction and maintenance of the miraculous mechanism of our bodies, will lie ahead as we seek to explain how our minds have come to organize thought sufficiently well to investigate our own existence.*
>
> J. Craig Venter et al., The Sequence of the Human Genome Science 291, 1304 (2001). Reprinted with permission from AAAS.

Those aspects of consciousness related to *awareness* and *volition* have benefitted from these neuroscience approaches, though much remains to be understood. The remaining hard problem of *subjectivity* or qualia can be expected to be resistant and inconclusive without a fundamental understanding and resolution of the measurement problem. However, unjustified conclusions have been drawn from these neuroimaging studies; e.g., Stephen Hawking's insistence on determinism in biological processes and that agency and free will are illusions [516, pp. 31-32],

> *...the molecular basis of biology shows that biological processes are governed by the laws of physics and chemistry and therefore are as determined as the orbits of the planets. Recent experiments in neuroscience support the view that it is our physical brain, following the known laws of science, that determines our actions and not some agency that exists outside those laws...so it seems that we are no more than biological machines and that free will is just an illusion.*

Neuroscience researchers have indeed interpreted some experiments as showing that the brain initiates conscious movements and free will before we are consciously aware of the will to move. The neuroscience experiment that initially raised suspicions about free will was a now classic paper by Libet et al. [517]. These experiments attempted to observe a person making a deliberate decision and use the timing of a physical event to capture the moment of the mental act and its associated brain activity. They studied participants using an electroencephalogram (EEG) and asked them to watch a clock face on which the point of a bright light was sweeping. The subjects were instructed to spontaneously but deliberately flex their wrist or finger as soon as they *felt the urge* to do so and note the light position. Significantly, their conscious decision to move was preceded by 550ms of a particular electrical change in the brain (*readiness potential*, RP) that originates in a region involved in motor preparation, the supplementary motor area (SMA). Since SMA activity preceded the conscious decision, it was argued that the subjects were found to exhibit *unconscious* activity via the EEG *before* reporting their conscious awareness. The subjects became aware of intention to move 350-400ms after the RP starts but 200 ms before the motor act. Apparently, their brains had already decided before the subjects were even aware of making a choice rather than the brain and body acting only after the mind has willed it!

How could this be possible? A neuroscientist might wonder how an underlying mechanics of precognition within the brain could play out in the situation of these experiments in terms of pulses released via synaptic vesicles toward the premotor cortex signaling the motor cortex and spinal cord to trigger muscle responses, and finally the emergence of a *decision to move.* Libet's experiments were widely discussed and criticized but also replicated with many variations in the experimental methods [518]. Does the actual decision to move occur before awareness? Libet has also reported that the RP was consistent with motor preparation even on occasions in which the subject decided to veto the prepared action and did not actually move [519] [520], some saying that this at least leaves the door open for "free won't," if not free will [518]. Actual human volition though can be described as *deliberating and acting on a dichotomy.* The world is divided into two possibilities and a deliberate choice of preference is made for one of them. There are two separate characteristics associated with free will: *willing* and *self-agency* [521]. The freely made decision for movement is the sense of willing, whereas the sense that "I am responsible for the movement" is the sense of self-agency. It has been argued that what Libet has called *self-paced voluntary acts* [520], where subjects are told to respond whenever they *feel an urge to act*, may not correspond to free will. Other Libet-type experiments have examined the difference between an intention to move in the future and an actual immediate movement and found an RP-like intention potential [522] associated with the intention to move. This would suggest that the brain is unconsciously active before a *thought*, and not just a movement. An RP-like potential has also been found associated with decision making, at the time of the selection of a letter without any associated movement [523].

These and related neuroscience experiments are a great beginning and may lead to a more fundamental understanding of consciousness and free will, but many issues still need to be resolved [521]. That is, it has not yet been shown convincingly at the neurophysiological level that a neural decision sufficient to cause movement occurs before the time of awareness of the decision to move [524]. Moreover, the discrimination of free will acts that require deliberation from those that are simply spontaneous or reflex is a key issue that has not been yet achieved by these experiments. It is of interest to note that the discrimination of actions that are reflex or spontaneous versus those that require deliberation can be traced to Aristotle [353]:

Thus an inanimate thing ... cannot do anything by chance, because it is incapable of choice ...The spontaneous on the other hand is found both in the lower beasts and in many inanimate objects. [Physics II.6]

As flicking a wrist or moving a body part is something a lower animal is capable of, the current experiments clearly have not sufficiently investigated free will in a manner consistent with Aristotle's view of deliberate intention in 350 BCE, let alone today. Before these details are understood, we should be cautious regarding statements in the neuroscience literature that appear to be drawing premature conclusions, such as [518]:

> *...modern neuroscience is shifting towards a view of voluntary action being based on specific brain processes...*
>
> Reprinted by permission from Macmillan Publishers Ltd: Nature Reviews | Neuroscience, P. Haggard, Human volition: towards a neuroscience of will, Vol. 9, p. 934 (2008).

As both the physics of consciousness and volition are in its infancy, conclusions of neuroscientists against nondeterminism may be seen as reflective of an inductive approach and that a problem of this fundamental importance, which has remained unsolved from the time of Aristotle, requires a deductive approach. Conclusions in this area that are reached using an inductive approach should be viewed with the utmost suspicion.

We are certainly far from being able to make a conclusion regarding the determinism of mental activities. The deductive researcher keeps all options in play, particularly when facing a problem within an unfamiliar terrain. Consider Niels Bohr, who as early as the 1930s was considering the prospects of how a physicist would confront the issues of phenomena in living organisms and the possibility that non-causality should not be ruled out [213][Kindle location 5185] (letter to Pascal Jordan 1931),

> *I am particularly concerned to emphasize that, epistemologically speaking, a natural place can be found in our world view for the laws obeyed by the phenomena of life. I agree of course with the point that it is above all the fundamental limitation of the applicability of the causality concept in the inorganic macrocosms which gives us the necessary freedom, and that in this sense acausality can be regarded as a characteristic of life. The peculiar reaction of the organism is, however, connected in the closest possible way with the laws of biology, which cannot be comprehended mechanically, and is therefore different in principle from the technical amplification devices used in the study of fluctuation phenomena. Just as the stability of the atomic phenomena is inseparably connected with the limitation of observation possibilities expressed by the uncertainty principle, so in my view the peculiarities of life phenomena are connected with the impossibility in principle of ascertaining the physical conditions under which life exists. Briefly, one could*

perhaps say that atomic statistics deals with the behavior of atoms under well-defined external conditions, whereas we cannot define the state of the organism on an atomic scale.

Bohr considered the comparison of measurement of atomic phenomena with the tinge of conscious inner experience [319, p. 100],

The unavoidable influence on atomic phenomena caused by observing them here corresponds to the well-known change of the tinge of the psychological experiences which accompanies any direction of the attention to one of their various elements.

Bohr did not draw final conclusions regarding the scale at which quantum phenomena might play a role in living or conscious systems. He did keep track of progress in biology, particularly from discussions with his colleague Max Delbrück (1906-1981) who had begun as a physicist but later pioneered the new area of molecular biology, finally winning the Nobel Prize in Physiology/Medicine for his work on the replication and genetics of viruses. It was the discovery of helical DNA by Crick and Watson in 1953 that convinced Bohr that at least DNA did not require knowledge of what happens at the atomic level in organisms. The mechanism of replication required no new laws and could be accounted for by physical and chemical explanations. Bohr then turned to the possibility that the fundamental unit of life might be the cell rather than the chromosomes and that the functions of the cell must be regulated by information from the organism as a whole.

Bohr believed that both the tinge or impressions felt during conscious experience as well as volition are contrary to a mechanistic deterministic description and as well, that the conscious process of the nervous system is not open to introspection by the external observer,

The fact that consciousness, as we know it, is inseparably connected with life ought to prepare us for finding that the very problem of the distinction between the living and the dead escapes comprehension in the ordinary sense of the word. That a physicist touches upon such questions may perhaps be excused on the ground that the new situation in physics has so forcibly reminded us of the old truth that we are spectators as well as actors in the great drama of existence.

Niels Bohr, Atomic Theory and the Description of Nature: Four essays with an Introductory Survey, Cambridge University Press, Reissue Edition 2011, p. 119.

From a biological point of view, we can hardly interpret the characteristics of psychical phenomena except by concluding that every conscious experience corresponds to a residual impression in the organism, amounting to an irreversible recording of the outcome

of processes in the nervous system which are not open to introspection and hardly adapted to exhaustive definition by mechanistic approach.

Reprinted by permission of Dover Publications, N. Bohr, Atomic Physics and Human Knowledge, Dover Publications 2010, p.77.

To illustrate such argumentation, we may refer to the old problem of the freedom of will. In an unrestricted deterministic approach this concept, of course, finds no place, but it is evident that the word volition is indispensable in an exhaustive description of psychical phenomena.

Reprinted by permission of Dover Publications, N. Bohr, Atomic Physics and Human Knowledge, Dover Publications 2010, p.89.

There have also been particular attempts to include consciousness within the framework of quantum measurement. Von Neumann (1932) demonstrated within his framework for measurement that it makes no difference whether the measurement projections are applied on the system immediately prior to the measurement, at the measurement apparatus, or anywhere within the brain of the observer [13]. Shortly afterward, London and Bauer (1939) developed this further with a measurement theory in which measurement occurs at the point when the observer becomes conscious of the particular outcome [525]. Eugene Wigner (1961) explored the role of consciousness with his *Wigner's Friend* thought experiment in which the example of Schrödinger's Cat is supplemented with a conscious *friend* who intervenes as Wigner attempts to carry out a measurement [217].

In David Deutsch's variant of Wigner's Friend, the friend answers in the affirmative when Deutsch asks the friend whether he sees a definite state of the cat but Deutsch does not ask the friend for the result [142]. Therefore, the state of the "cat and friend" has not yet been collapsed. As discussed in Chapter 4, these explorations have stimulated much discussion and alternative developments. However, ultimately measurement, consciousness, and free will must be understood on a fundamental basis before questions such as Wigner's Friend can be answered.

Search for the Soul, Mind and Consciousness

Near the end of the 19th century, when the discovery of the quantum was about to put an end to determinism in physical theory, the predominant view of biologists was that life was a machine. The biologist Thomas Huxley wrote in his "On the Hypothesis that Animals are Automata, and its History" [526],

...the feeling we call volition is not the cause of a voluntary act, but the symbol of that state of the brain which is the immediate cause of that act. We are conscious automata...

Philosopher William James responded to Huxley in "Are We Automata" [527] in favor of consciousness not being automata,

> *I think we can, and propose in the remainder of this article to show that this presumptive evidence wholly favours the efficacity of consciousness... Since the mere supernumerary depicted by the Conscious-Automaton-theory would be useless, it follows ... that if we can discover the utility of consciousness we shall overthrow that theory.*

However, despite the arguments and evidence he could present for consciousness, James was unable to link it to any known physical process,

> *You may, it is true, ascribe mind to a physical process. You may allow that the atom engaged in some present energy has a dreamlike consciousness of residual powers and a judgment which says, "Those are better than this". You may make the rain-drop flowing downhill posit an impossible ascent as its highest good. Or you may make the C, H, N and O atoms of my body knowingly to conspire in its construction as the best act of which they are capable. But if you do this, you have abandoned the sphere of purely physical relations.*

The early 20th century views of the mind were dominated by the psychologists with the constructs of Sigmund Freud (1900) and later the behaviorism of B. F. Skinner (1938), which eliminated any internal representations for mental process or consciousness. Gilbert Ryle (1948) eliminated the *ghost in the machine* and described mental processes merely in terms of transitions between states. Cognitive psychologists and linguists such as Noam Chomsky (1957) demonstrated that language acquisition cannot be explained by behavioral models alone and postulated internal mechanisms for processes that could not be directly observed. And all the while, philosophers of the mind continued to argue about the nature of the private and unique internal experiences of each of us [528] [529] leading to wide-ranging arguments and speculations across many disciplines; e.g., Nagel's influential "What is it Like to Be a Bat?" [530]. Philosophers also developed arguments at great length for whether our internal experiences are compatible with determinism; e.g., the counterfactual possibility of *acting otherwise* with free will cannot include violating natural laws; however, can free will still possibly be consistent with determinism by means of a "local miracle" as argued by David Lewis' "Are We Free to Break the Laws?" [531].

It has been tempting throughout history to match qualia or *tinge* with particular objects associated with some part of ourselves, be it of the site of the soul, mind, or consciousness, and these have become intertwined throughout the history of reason, understanding, and spiritual essence [509]. The Greek *psyche* and the Latin *anime* became the Old English term soul, first appearing in the 8th century Greek description of man's appearance in Hades. Originally designating agility and self-movement, it has been translated variously as appetite, desire, and passion, representing the power that animates physical movement as well as reflection and deliberation [532]. Like Democritus before him, Epicurus (341-270 BCE) argued that the soul comprised particles diffused throughout the body and these became separated at death. Centuries later, Dante went out of his way to make sure Epicurus was reserved a prime spot in the *Inferno* of *The Divine Comedy* (1320) for his demotion of the soul,

> *In this dark part are entombed*
> *Epicurus and all his followers*
> *Who made the soul die with the body.*

Aristotle (384-322 BCE) identified a segment responsible for movement and an immortal piece at the root of intellect. He was the first to identify the series of cavities in the brain now called cerebral ventricles as the anatomical location of the soul. Aristotle described the need for these distinctions for the soul in living beings [353],

> *Since some such originative sources are present in soulless things, and others in things possessed of soul, and in soul and in the rational part of the soul, clearly some potentialities will be non-rational and some will be accompanied by reason. This is why all arts, i.e. all productive forms of knowledge, are potentialities; they are principles of change in another thing or in the artist himself considered as other.* [Metaphysics IX.2]

> *Some things can exist apart and some cannot, and it is the former that are substances. And therefore, all things have the same causes, because, without substances, affections and movements do not exist. Further, these causes will probably be soul and body, or reason and desire and body.* [Metaphyics XII.5]

Aristotle appears to be contemplating something beyond substance called soul. The soul gives rise to non-rational potentialities and yet it is also possible for the potentialities to change via a rational formula. Aristotle might therefore have associated Schrödinger's equation with the rational change of the wave equation and the non-rational change due to a soul, would cause wave-function collapse which is probabilistic. Aristotle also developed particular detailed ideas regarding consciousness,

> *For our assumption is that things that are undergoing alteration are altered in virtue of their being affected in respect of their so-called affective qualities; for every body differs from another in possessing a greater or lesser number of sensible characteristics... But the alteration of that which undergoes alteration is also caused by the above-mentioned characteristics, which are affections of some*

> *particular underlying quality.* [Physics VII.2]
>
> *Thus the animate is capable of every kind of alteration of which the inanimate is capable; but the inanimate is not capable of every kind of alteration of which the animate is capable, since it is not capable of alteration in respect of the senses: moreover the inanimate is unconscious of being affected, whereas the animate is conscious of it, though there is nothing to prevent the animate also being unconscious of it when the process of the alteration does not concern the senses.* [Physics VII.2]

Had Aristotle been alive to see the developments in quantum mechanics, he may have argued that living systems are distinguished from inanimate systems by the existence of senses. That only animate systems measure, that is, consciousness is a necessary condition for measurement.

With the rise of anatomic research came specific proposals for the location of the soul or consciousness or other aspects of our being. For example, Galen (ca. 200 AD) used his anatomic research and his experience with spinal cord injuries as physician to the gladiators to conclude that the soul must reside in the brain (contrary to previous views that it should be in the heart), and this must also be the organ responsible for generating thoughts. Descartes divided the composition of human beings into a thinking substance or soul, the *res cogitans*, and a corporeal substance of physical matter, the *res extensa*. Descartes' view was so influential that from the late 17th century until the early 20th century, the predominant view in the West was some form of dualism, involving an immaterial soul interacting with a material brain. This dualistic view then required an explanation of the relation between mind and body. During his animal dissection work, Descartes identified a small cherry-sized structure called the pineal gland and insisted that it was the material site of his *cogito* and location of the rational soul. In his *De Anima Brutorum* (1672), physician Thomas Willis (1621-1675) made the identification that the corpus striatum was associated with imagination and the cerebral cortex with memory. The 17th and 18th centuries led to English empiricism with John Locke (1632-1704) arguing against Descartes and the other Rationalists, claiming that all knowledge results from experiences of the senses and reflection on those experiences, the mind's "reflecting on its own Operations within itself" with "internal Sense," from which we derive our ideas of "Perception, Thinking, Doubting, Believing, Reasoning, Knowing, Willing, and all the different actings of our own Minds." [533]. The famous anatomist Samuel Soemmerring (1755-1830) identified the fluid of the cerebral ventricles as the location of the soul. Isaac Newton (1643-1727), ever astute, avoided these issues altogether, stating that,

> *I can calculate the motion of Heavenly Bodies, but not the madness of people.*

However, these examples of proposals for the site of qualia, mind or the soul were based on theories or hypotheses that had little direct verification and would not lead to further understanding. An exception in this early era was the deductive work of Leonardo da Vinci (1452-1519) who used his anatomic studies, as well as other sources in a systematic way to make a judgment on the best candidate for the function and location of the soul to be at a specific site in the anterior portion of the third ventricle of the brain [534]. Evidence from his exploration of this problem can still be traced in his surviving papers. Leonardo's investigations into the structure of the nervous system, including the brain, spinal cord, and cranial nerves, was spread over about thirty years. Due to the high level of his artistic abilities, he was also able to develop a three-dimensional drawing format. His ability to develop these types of reconstructive drawing methods put him in a unique position to then use a series of deductive steps to fix a cerebral location of the soul, which he termed the *senso commun*. His conceptual reasoning for solving the problem of the *senso commun* can be followed in a series of particular experiments and vivisections.

An important early set of experiments involved the "pithing of a frog," which resulted in the destruction of the upper spinal cord and medulla oblongata. The frog remarkably remains living for several hours without its head, heart, interior organs, intestines or skin. However, a puncture of the spinal medulla causes it to spasm and die immediately. This could be understood as all of the nerves derive from the spinal cord. Leonardo's thinking can be followed from the words *sense of touch*, *cause of movement*, *origin of nerves*, and *transit of animal powers* next to the appropriate structures in the drawings for this study. With the identification of the spinal medulla, Leonardo had deduced evidence for the foundation of movement and life. The frog experiments were extended to many others, including that of damaging of a dog's brachial plexus, the network of nerves extending from the spinal cord (similar experiments were not carried out for another 250 years). These were supplemented by his many studies of the layers of the head of various animals as well as humans. These initial studies were later used to abstract and construct a three-dimensional synthesis of the nervous system of the human head based on his experimental investigations. Crucial to this was his ability to make three-dimensional drawings of the human skull, without the need for three-dimensional models, using techniques he had learned from architectural concepts for geometrizing space. These also included the techniques of cutaways and exploded views. In these drawings, he traced the optic nerves back to the optic chiasm and continued into the anterior cavity which he labeled *intelleto* (intellect) and *imprensiva* (sensory information). The olfactory and auditory nerves were bundled toward the middle ventricle labeled *volonta* (will) and *senso commun* as well as a posterior ventricle labeled *memoria.* Leonardo's deduction was that information from the sensory organs is integrated and acted upon by the middle ventricle, the *senso commun*, Figure 5.20.

However, models were also developed using his skill as a sculptor to outline the shape of the ventricular system by injecting wax with a syringe. He could dissect away brain tissue after the wax had set, resulting in an accurate representation of the ventricular system. Leonardo's use of deductive methods along with his highly

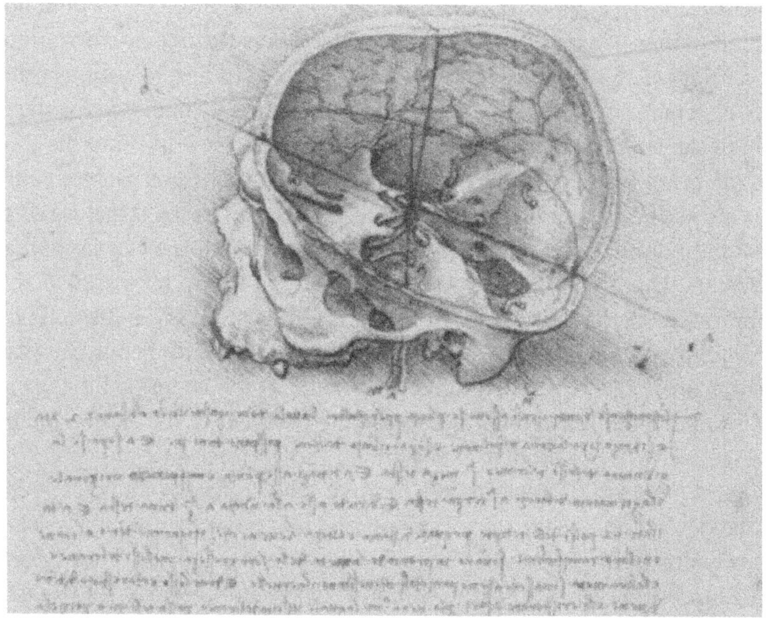

Figure 5.20 Leonardo da Vinci's deductive determination of the senso commun or site of the soul where all of the five senses come together, located in the anterior portion of the third ventricle above the optic chiasm, indicated by a series of intersecting lines he superimposed on drawings of the human cranium.

developed artistic skills enabled him to resolve quite complex anatomical structures. Leonardo determined the location of the *senso commun* to be just above the optic chiasm proximate to the anterior portion of the third ventricle. Current knowledge about brain structures indicates that this is a region which is vital to the way that humans perceive both their inner and outer worlds and these functions are quite sensitive to any damage in this area. Although the process by which information from the senses results in cognitive functions is still only partly understood, it is remarkable that Leonardo was able to narrow in on a crucial nexus of brain function using both deductive methods and the most sophisticated means of retrieving and representing information that were then at his disposal.

Leonardo's comments are revealing [535]:

> *The Common Sense, is that which judges of things offered to it by the other sense…And this name of Common Sense is given to it solely because it is the common judge of all the other five senses i.e. Seeing, Hearing, Touch, Taste and Smell. This Common Sense is acted upon by means of Sensation which is placed in a medium between it and the senses. Sensation is acted upon by means of the images of things presented to it by the external instruments…Surrounding things transmit their images to the senses and the senses transfer them to the*

Common Sense, and by it they are stamped upon by memory and are more or less retained according to the force of impression. [Richter no. 836]

The soul appears to reside in the seat of judgment, and the judicial part appears to be in that place where all the senses come together, which is called the "senso commun", and it is not all of it everywhere in the whole body as many believed, but all in this part. For if it were all in the whole and all in each part it would not have been necessary to make the instruments of the senses converge to one and the same concourse in one place only...How the sense gives to the soul and not the soul to the sense, and where the sensory function is missing from the soul, the soul in this life lacks information from the function of that sense, as appears in a mute or one born blind. [Richter no. 838]

It would take another 300 years so that by the 19th century, improvements in the design and function of microscopes led to understanding of the cellular organization in brain tissues [514] [534]. The first microscopic image of a nerve cell was obtained by Gabriel Valentin (1810-1883) in 1836, and Jan Evangelist Purkinje (1787-1869) identified the first nerve cell in 1837. Modern neuroscience as we know it began when Santiago Ramón y Cajal (1852-1934) gave critical evidence for the "neuron doctrine," the picture of neurons serving as the basis of signal functioning in the nervous system due to the precise interconnections of neurons [514]. This transformed the cellular view of the brain into our modern neuronal conception of it. Ramón y Cajal was attracted to the area by the hope of explaining the phenomenon of consciousness. He described his life-long exploration as an attempt to break into [536]:

...the utter darkness of the inner mechanism of psychic acts. The determination of the sequence of molecular processes that the neurons undergo ... during the productions of the concomitant phenomena of perception and thought, namely feelings, consciousness, and volition.

Rather in the mold of Leonardo, Ramón y Cajal was also an exceptional artist. It was his more than twenty years of producing meticulous ink drawings of brain cells in compressed detail, based on countless hours of viewing brain samples in the microscope, that transformed neurology and led to the breakthroughs for our current understanding of the brain based on the concept of the *neuron* as the basic building block of the nervous system. In addition, also similar to Leonardo, he developed the practice of making cumulative deductions from the large mass of details he had collected. His deductive skill and dedication gave him an uncanny ability to surmise the functional properties of neurons from his static drawings of brain cells. In addition to his neuron doctrine, Ramón y Cajal also proposed a principle of dynamic polarization by which electrical signaling is unidirectional within neurons, propagating from the receiving pole of the neuron to the dendrites and the cell body of the axon, and along the axon to the output pole of the neuron [514]. This allowed the picture of coherent neural circuits to develop with the brain's primary functions being information processing. This led to the later work of Sherrington and Eccles in the early 20th century to conclude that each neuron involves a competition between excitation and inhibition with a winner-take-all resolution leading to an integrative action of the brain in terms of signals at the level of individual neurons. Ramón y Cajal distinctively indicated the information flow between neurons that he had deduced by the use of arrows in his drawings and remarkably the directional arrows were later verified to be correct, Figure 5.21.

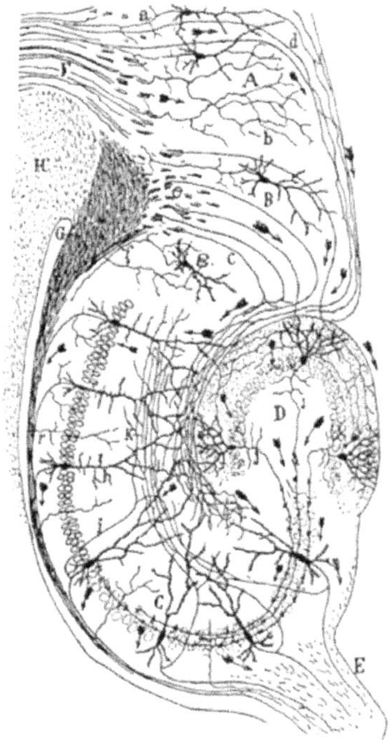

Figure 5.21: Ramón y Cajal's 1911 depiction of directionality of information flow in the nervous system, giving the first view of a coherent neural circuit. Image obtained by staining of a rodent hippocampus.

The state of neuroscience today, 100 years after Ramón y Cajal's work, still has not resolved the issues of consciousness and free will, and if Bohr is correct that consciousness and free will are non-mechanistic processes, it is logically possible that this must await a resolution of the measurement problem as a first step in this direction.

Scientific Methodology

Deductive versus Inductive Thought

Essential to the dark art of physics is determining whether the failing of a prediction of a theory is due to a minor anomaly or instead requires alteration reaching back to its foundations. By far, most research published each week in the physics journals is of the former type and progress in the knowledge base of physics, the *Nexus of Knowledge* (see Chapter 6), gradually grows as contributions continually expand and fine-tune our understanding, Figure 5.22.

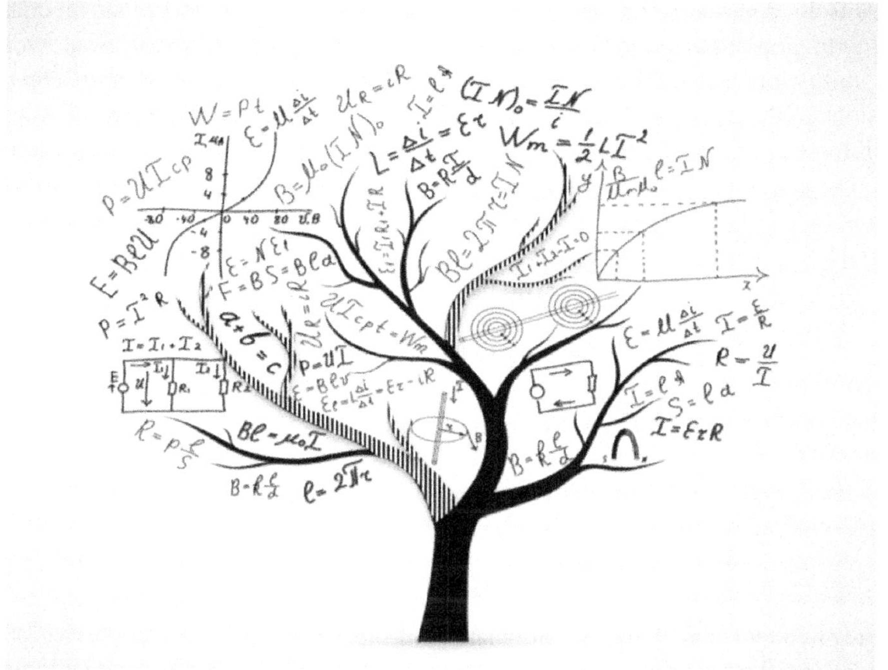

Figure 5.22: Nexus of Knowledge: knowledge base of physics grows outward from established results (trunk) by examining experimental results (branches) consistent with induction from current theoretical predictions within accepted error (black) and those not yet explained by existing theory (striped) which may require deduction.

Progress of this type can be achieved by using induction to extrapolate from existing theory (solid line branches). However, understanding of the quantum measurement problem is expected to be of the daring variety since it has resisted solution for so long and involves fundamental physics across a wide range of issues (striped branches). As discussed in Chapter 6, problems such as quantum measurement are expected to require a deductive approach rather than simply inductive extensions of existing theory. However, this clearly differs from a straightforward application of deductive logic since physics is further constrained both by the results of empirical observation and by the existing body of physical principles that have been historically established.

Due to the necessary consistency with these constraints, deduction within the foundations of physics is not an automatic process that can be easily mechanized, but rather an unruly enterprise requiring bold intuition and an ability to deal with paradox and uncertainty and contradiction. This is because, except for maintaining the established physical principles, all legitimate possibilities must remain on the table in the course of deduction and the investigator must be open to the circumstance found by Holmes that "when you have eliminated all which is impossible, then whatever remains, however improbable, must be the truth." [537]

Radical Conservatism

Niels Bohr, a master of the deductive construction of physical theory, had an uncanny ability to pinpoint and transform deficiencies within theories [9] [538]. Bohr evolved this ability into a process, wherein he would categorize the faults within different versions of the theory and play them off against each other revealing what J. L. Heilbron has called *partial truths* [539]. These were then used to make hypotheses that could be applied not only to the parts that agreed with experimental observation, but also used to isolate exceptions wherein he would begin another round of constructing partial truths [540]. This approach to theoretical physics requires great creativity and the ability to embrace and work with paradox and contradiction. However, deductive thinking also requires responsibility as well as a deep understanding of physics in order for such a process to successfully lead to a solid piece of science. John Wheeler, who regarded Bohr as his mentor and was part of his Copenhagen Institute in the 1930s, characterized him as a daring conservative [541],

> *Conservative against postulating any change in the battle-tested laws of physics, but in the application of them, daring.*

Inspired by Bohr, Wheeler would later call this principle *radical conservatism*, that of conservatively respecting great principles while pushing ideas into radically new directions and logically deducing the unexpected consequences and insights if still constrained by the great principles [542]. Wheeler's student Richard Feynman, who had discovered the sum-over-paths formulation of quantum mechanics, emphasized in an address at the 1961 Solvay Conference how the search for an *exception* or a *failure* is more prevalent in experiment than in theory [543, p. 89],

> *I now realize that there is much to be said for considering theoretically the possibility that Q.E.D. is exact, although incomplete. This assumption may be wrong, but it is precise and definite, and suggests many things to study theoretically, while the other negative assumption, (that it fails somehow) is not enough to suggest definite theoretical research. This is Wheeler's principle of "radical conservatism".*

Things are, of course, quite the other way for experimental research. One should look very hard for an "expected" failure. I have probably been converted from my prejudice that it must fail, just in time to be caught off base by an experiment next month showing that indeed it does.

The search for the exception is central to the process of deductive reasoning. Newton, Maxwell, Einstein, Bohr, and others used this strategy of deduction constrained by great principles to investigate the foundations of physics but it is seldom used today, particularly regarding the quantum measurement problem. Though Bohr's achievements are undeniable, his application of the deductive process to the problems in science has at times been surprisingly misunderstood. Deduction is explored at length in Chapter 6 where it is contrasted with the more commonly encountered practice of induction.

Besides Bohr's innate ability, he had been continually encouraged by the intellectual stimulation of a remarkable family life [9, p. 42] and by university teachers, such as family friend and philosopher Harald Høffding (1834-1931), who taught that the concept of a secure fact and the notion of a complete theory are at best ideals [540],

Neither [a secure fact nor a complete theory] is given in experience, nor can either be adequately supplied by our reason; so that, above and below, thought fails to continue, and terminates against an irrational.

What Bohr would insist upon in discussing a proposed conjecture, was whether he could draw arguments in favor of it from the available evidence: logical analysis was not for him a mere verification of consistency, which he regarded as trivial, but a potent and productive device for directing the mind. Wheeler recalled Bohr's Copenhagen Institute in the 1930's:

"The central idea of the institute was clear. No progress without a paradox... Explanation was never dry pedagogy, but a one-man tennis match in which Bohr hit the ball from one side of the court, then ran to the other fast enough to hit it back – the more volleys, the more enjoyable the game: Such-and-such an effect leads one to expect thus-and-so...Indeed one does see thus-and-such, but then so-and-so observed such-and-such ...That finding put us in immense difficulty. Just at this point so-and-so pointed out that the proper formulation of the principle is not what we thought, but thus-and-such...This discovery brought the whole subject in to order. But then

> *so-and-so realized that this extended principle stands in absolute contradiction to the stability of such-and-such...This discrepancy convinced us that we were absolutely lost. But just today we find that the new formulation itself is really completely nonsense...What fools we have all been! We have only to recognize such-and-such and we see at last that absolutely everything has to be exactly as it is."*
>
>

At a difficult point in a discussion, Bohr would sometimes make a joke based on the old saying about the two kinds of truths, one kind being so simple that the opposite was clearly false and the other so-called "deep truths" whose opposite also is a deep truth [503, p. 240].

Bohr's Atomic Model

A prime example of Bohr's constrained deductive process is his work on the constitution of the atom while in Ernest Rutherford's (1871-1937) laboratory in Manchester, the initial versions published as a trilogy entitled *On the Constitution of Atoms and Molecules* [544], which for the first time made atomic structure into a subject of scientific inquiry. Bohr had spent the year 1911 with J. J. Thomson (1856-1904) in Cambridge before joining Rutherford in Manchester in 1912. In 1897, Thomson had unraveled the mystery of the nature of cathode rays by discovering the electron and went on to consider atomic structure. The leading atomic theory that Thomson had been developing for several years was a *plum pudding* arrangement of positive charges sprinkled among rings of large numbers of electrons throughout the atom but which did not lead to a quantitative description of atoms. However, the experimental observation by Rutherford's group that α-particles were deflected as they pass through a thin gold foil led Rutherford instead to a planetary picture of the atom in which electrons orbit a massive nucleus according to classical mechanics. Bohr's initial examination found that the electrons could not be mechanically stable in Rutherford's planetary type of orbits. There would also be instability due to radiation of the charged electrons. What keeps the matter of such a configuration from collapsing in a fraction of a second, as it must since any orbiting electrons must radiate? Enforcing stability on the basis of classical physics would have led back to the unsatisfactory Thomson style of model.

From his earlier thesis work on electrons in metals, a detailed mathematical work, Bohr was convinced that Newtonian mechanics and Maxwell's electrodynamics were not adequate for a description of the atomic world. Classical orbits became unstable when populated with more than one electron and nothing in classical physics determines orbital radii or frequencies. Not only did Bohr find that the Rutherford atom was mechanically unstable, he also noticed that it had no characteristic radius to define the size of the atom. As Bohr saw it, this requires introduction of a quantity extraneous to the classical electrodynamics, i.e., Planck's elementary quantum of action, and Bohr noted [442, p. 51]

> *this constant [h] is of such dimensions and magnitude that it, together with the mass and the charge of the particles, can determine a length of the order of magnitude [of the atom's linear dimensions.*
>
> S. Petruccioli, Atoms, Metaphors and Paradoxes, Niels Bohr and the construction of a new physics, Cambridge University Press 1993.

This length is of the order of magnitude required to characterize a hydrogen atom and now known as the *Bohr radius* a_0,

$$a_0 = \frac{h^2}{(2\pi)^2 m_e e^2} \cong 0.53 \text{ Ångström} = 0.53x10^{-10} \text{ meter}. \quad (5.10)$$

However, he also relied heavily on the proposition that classical predictions still apply whenever quantum effects can be ignored, which led him to formulate and extensively exploit a *correspondence principle*. Bohr built up a series of alternative arguments, which were distinct but in part mutually contradictory in Part 1 of his trilogy, each containing a partial deductive truth, which led to a model with the principal feature that energy did not come out through continuous vibrations but discontinuously, in a transition from an orbit more distant to the one closer to the nucleus. As Bohr later recalled [214, p. 17]

> *A clue to the solution of this dilemma was, however, already provided by Planck's discovery of the elementary quantum of action, which was the outcome of a very different line of physical research. As is well-known, Planck was led to this fundamental discovery by his ingenious analysis of just such features of the thermal equilibrium between matter and radiation which, according to the general principles of thermodynamics, should be entirely independent of any specific properties of matter, and accordingly of any special ideas on atomic constitution.*
>
>

The principal postulates of the first model that emerged from Bohr's deductive path were:

- The dynamic equilibrium of the systems in stationary states can be discussed via ordinary classical mechanics while the passing of the systems between different stationary states cannot be treated on that basis.
- The transitions between the stationary states is followed by the emission of a homogeneous radiation whose frequency and energy are related by Planck's quantum relation, $E_n = nh\nu$.

The result was a new atomic theory that unexpectedly mixed classical and quantum ideas based on postulates that could be justified only by their empirical success. Bohr emphasized its preliminary and hypothetical character and admitted that [421, pp. 12-13]

> *I am by no means trying to give what might ordinarily be described as an explanation; nothing has been said here about how and why the radiation is emitted.*
>
> Niels Bohr, The Theory of Spectra and Atomic Constitution, Three Essays, Cambridge University Press, 1924.

However, application of Bohr's model was a spectacular success, giving quantitative agreement with the hydrogen atom spectral series seen in laboratory experiments and in stellar observations and it predicted other spectral series that were soon verified. It also identified the He^+ ion from stellar spectral lines that had been mistakenly attributed to atomic hydrogen. Bohr's theory gave us such concepts as the screening number and the self-consistent atomic field, and the order of building up of the elements in the periodic table. Bohr's approach had further success for about a dozen years with extensions by Arnold Sommerfeld and others to include elliptical orbits and relativistic effects, which produced fine structure in agreement with observations. And it was within the context of Bohr's struggle to understand atomic constitution via his models that: Wolfgang Pauli (1900-1958) in 1924 made his epic discovery of the exclusion principle, that no two electrons could exist in the same quantum state, and George Uhlenbeck (1900-1988) and Samuel Goudsmit (1902-1978) in 1925 discovered the electron spin. Bohr's was the first theory of atoms and molecules that addressed their structure in terms of the configurations of electrons. It initiated a quite new and productive development in the study of atomic, subatomic, molecular, and chemical phenomena along a path that eventually led to a consistent theory of quantum mechanics. The Nobel Prize in Physics was awarded to Bohr in 1922 "because of the assured results and because of the powerful stimulus which this theory has given to experimental as well as theoretical physics."

Despite these successes, many contemporary scientists objected to the apparent lack of foundation for Bohr's collection of postulates. And Bohr's approach could not easily be successfully extended to more complex atoms and to observations such as the response to an external magnetic field called the anomalous Zeeman effect. Understanding these would have to wait for the discovery of the full theories of quantum mechanics, Heisenberg's matrix mechanics in 1925 and Schrödinger's wave mechanics in 1926, which eventually became essential for a more complete understanding of atomic phenomena. However, several features of Bohr's insights were destined to become pervasive even with these developments: identification of the lowest-energy stationary state as the stable ground state which does not emit radiation; the attribution of spectra to radiation absorbed or emitted in transitions between stationary states; and the description of atoms in which an electron is in an excited state with very large quantum number n. Atoms existing in states given by high quantum numbers might indeed be quite large, with radii on the order of 0.01 mm. These large hydrogenic atoms were later called *Rydberg atoms* and were first observed in 1965 at the National Radio Astronomy Observatory when detected radiation from hydrogen atoms in interstellar space was found undergoing transitions between levels near $n = 100$ [545, p. 60]. These large atoms are well described by Bohr's picture and the full theory of quantum mechanics can actually justify the persistence of these several features of Bohr's early model. These aspects of the Bohr model have recently been demonstrated to be exact results of the Schrödinger equation examined in the limit of infinite dimensions where quantum mechanics morphs into classical mechanics [546], a remarkable testament to Bohr's deductive process.

Backstory to Deductive Thought

The example of Bohr's atomic work illustrates the innovative use of deduction applied to the subtle area of the foundations of physics. However, deductive logic as a fully systematic discipline can be traced to the ancient Greeks, beginning with Aristotle's (384-322 BCE) syllogistics and Chrysippus's (ca. 279-ca. 206 BCE) system of propositional logic but also included other categories of argument, demonstration and explanation [547] [548]. Greek thought was expansive and characterized by a relentless use of the power of logical thought, even if this led to paradox and contradiction [340, p. 54]. Although the senses may not be perfect, they also had a role as useful guides if employed with judgment. Aristotle's predecessors had already been assessing the criteria for proof and the rules of inference. Among the fundamental questions of most interest were: (1) the underlying material substance of the world including the origins of the atomic world-view, (2) whether or not and how change was possible, and most importantly for the future development of science, (3) the problem of knowledge (*epistemology*) and understanding the rules of reasoning, argumentation, and assessment of theories. The Greek inquiries into knowledge initiated the path that eventually led after much struggle to the methods of modern science.

The earliest scientific theories were formulated by the pre-Socratics, the Greek natural philosophers from the 6th and 5th centuries BCE who were among the first to explain phenomena in terms of naturalistic causes. These were not theories of physics in the modern sense of systematically constraining them by empirical observation. It has even been suggested that to understand these early Greeks, it is better to think of them not as physicists or scientists or even philosophers, but as poets [304, p. 12]. They were predecessors of modern science in that they pioneered the use of causal explanations in place of the tradition of myths and supernatural forces, even if the explanations were sometimes based on speculation and assumption. Imposing current values on the past, sometimes pejoratively labeled *Whig history* by historians, may not be out of place in judging the status of early physical theories [304]. Their importance becomes more apparent if we also compare them with their predecessors [549] [550]. Development of the sciences had their start at the beginning of the 5th century BCE with the Ionian natural philosophers who lived on the coast of Asia Minor, where the influences of the east and south were strongest. It then spread to the Pythagoreans and Eleatics (Parmenides and his school) of what is now southern Italy. Subsequently, philosophy experienced its peak in Athens with Socrates, Plato, and Aristotle. The 6th century BCE Greeks were the result of colonization of the Mediterranean shore and Aegean islands. In the 5th century BCE, the Greeks maintained contact only with the Persian empire. It was a time of questioning of the relation of man to the universe and the individual to society. In the 4th century BCE came the flourishing of Greek philosophy and science, Plato, and Aristotle. Alexander the Great, educated by Aristotle, set out in 334 BCE to overpower the Persian empire and conquer the world.

As a result, Greek culture spread to all conquered lands and Greece was enriched by contacts with Mesopotamia and Egypt, resulting in the important scientific developments of Hellenism and Alexandria in the 1st and 2nd centuries BCE [340, p. 45]. Aristotle's methods in logic were widely adopted following the translations of Boethius in the 6th century. By the 12th and 13th centuries, translations of the ancient texts became available and were assimilated across Europe. However, the Aristotelian model for scientific investigation had been severely challenged by the end of the 16th century. A competing view, in the form of the atomism of Democritus (ca. 460-ca. 370 BCE) and Epicurus (341-270 BCE), became known through the work of the Roman poet Lucretius (ca. 99-ca. 55 BCE), *On the Nature of Things* [506]. This view of the world as randomly moving indivisible atoms in an infinite void found its way into the 17th century to Galileo Galilei (1564-1642) in Italy, René Descartes (1596-1650) and Pierre Gassendi (1592-1655) in France, Robert Boyle (1627-1691) and Isaac Newton in England. By the end of the century this mechanistic philosophy became dominant although in a predominantly deterministic form. The scientific revolution may have appeared as a sharp transition focused on the 16th and 17th centuries when viewed from the scale of millennia but ways of thinking about the world had accumulated across a much larger scale. However, the disciplines of the deductive reasoning of Aristotle and the atomism of Democritus and Epicurus that trace back to ancient Greece are a vital part of science today and are foundational elements for understanding the measurement problem.

The intellectual transition to science in the time and place of ancient Greece is often attributed to favorable circumstances of geography, economics, religion, and politics [551]. The geography of mountains separating cities and the sea separating islands encouraged the formation of diverse city-states in which there was social freedom and democracy as well as no official religion requiring dogma or rituals to interfere with intellectual pursuits. In addition to these factors, Nicolaides has argued for the possible influence of the Greek language, citing a statement from the mathematical logician and historian Bertrand Russell (1872–1970): "The Greeks, borrowing from the Phoenicians, altered the alphabet to suit their language, and made the important innovation of adding vowels instead of having only consonants. There can be no doubt that the acquisition of this convenient method of writing greatly hastened the rise of Greek civilization." [551, p. 81] This development occurred around the 8th century BCE and spread rapidly. Having a phonetic language, where each sound has its own symbol, facilitated the clear formulation of abstract thoughts and encouraged discussion and refinement. It's a plausible and interesting possibility that the invention and flourishing of deductive reasoning, a necessary tool for understanding fundamental aspects of our world today, can be partly traced to the innovation of the vowel some 3000 years ago.

Aristotle's many treatises cover topics which could be seen as the continuation of a long history of inquiry by his predecessors. The exception is his development of logic, for which Aristotle emphasizes [547, p. 27],

> *When it comes to this subject, it is not the case that part had been worked out in advance and part had not; instead, nothing existed at all.* [Sophistical Refutations 34]
>
> R. Smith, Logic; pp. 27-65, In: The Cambridge Companion to Aristotle, J. Barnes (Editor), Cambridge University Press, 1999.

Aristotle was the first to conceive of a systematic treatment of inference and the *Prior Analytics* [353] is a complete exposition of his theory of syllogistics. There are also treatments of two types of arguments: *demonstration* which produces scientific proofs, the subject of his *Posterior Analytics,* and *dialectical argument* which focuses on debates between persons, the subject of his *Topics*. These are supplemented by *On Interpretation,* on the structure of propositions and their truth-conditions, and *Categories,* on the theory of meaning of the terms within propositions. The syllogism though is claimed to be the correct basis for all arguments; e.g., the universal affirmative "All Bs are Cs, all As are Bs, so all As are Cs" where each of the premises is a necessary truth. He goes on to explore many different variations and alternatives to this template. Aristotle recognizes two types of argument which lead to conclusions in fundamentally different ways, deduction and induction [547, p. 32],

> *A deduction is an argument in which, certain things being supposed, something else different from the things supposed follows of necessity because of their being so.*
>
> R. Smith, Logic; pp. 27-65, In: The Cambridge Companion to Aristotle, J. Barnes (Editor), Cambridge University Press, 1999.

Aristotle also recognized induction, which infers a general conclusion from a number of instances. This is a generalization of particulars to the universal, such as:

> *Socrates has red hair.*
> *Plato has red hair.*
> *Aristotle has red hair.*
> *All humans have red hair.*

However, a *single* counterexample can render an inductive argument false; e.g., Empedocles has black hair. By its very nature, inductive reasoning can dramatically fail for problems that cannot be solved using existing theory. Thus, induction would not be a good method to test the theory that all humans have red hair. Additional premises could be added in the attempt to make the inductive argument valid. However, as we continue to add more and more premises, we find that the inductive argument is being reworked into making it deductively valid. The large number of required supplementary premises typically required to make deduction valid illustrates how deduction applied to the physical world is an unruly process requiring boldness and intuition. However, deduction is the method that is appropriate for precisely those problems that demand the use of *exceptions* to the currently known theory for their respective solutions. Aristotle does not give a complete theory of inductive argument and his comments on it are less extensive than on deduction. We can recognize situations where induction can be useful; e.g., drawing conclusions by generalizing what is known [547, p. 32]:

> *Every A observed so far is B; therefore, every A without qualification is B.*
>
> R. Smith, Logic; pp. 27-65, In: The Cambridge Companion to Aristotle, J. Barnes (Editor), Cambridge University Press, 1999.

This is the situation in well-established areas of modern science where the inductive application of theory can be very productive.

Aristotle's work on logic was not by any means confined to the syllogisms and deduction. Throughout his works are extensive discussions on reasoning, language, definition, demonstration, dialectical argument, necessity, and possibility without any dependence on the method of the syllogism. Many of his views on reasoning, argument, and language are relatively independent of it. Sometimes these arguments can be thought of as logic supplemented with contingencies for *accidental occurrences* or qualifications along the lines of "*for the most part*" [548, p. 115]. Aristotle's scientific works can be seen to contain a great variety of informal arguments and a miscellany of more or less incredulous *truths about things* to incorporate into these arguments, without any hint of a worked-out syllogism [548, p. 113]. Commenters on Aristotle frequently attempted to restructure his arguments into syllogistic form. However, taking a system of logic outside of its purified framework and applying it to the world around us is a delicate task. This can be seen from the previous example of Bohr's atomic work and in the detailed discussion of deductive reasoning applied to the scientific approach in Chapter 6. Physical deductive reasoning must be constrained by the previously empirically established *great physical principles*. Resolving problems via deduction with these constraints is a form of *radical conservatism*, conservatively respecting great principles while pushing ideas into radically new directions and logically deducing the unexpected consequences and insights if still constrained by the great principles.

The absence of syllogistic argument throughout so many of Aristotle's scientific works may be a reflection of his attempts to carry out arguments in an era before the development of systematic empirical observation. The times of Aristotle lacked the empirical leverage to produce radical conservatism when applied to physical science. There has been a recent trend comparing physical theories of the pre-Socratics with fundamental results in modern physics; e.g., Thales' notion of sameness with the quest for a theory of everything. Although intriguing, such comparisons can seem strained and perhaps can be the result of revisionist cherry-picking when applied to modern empirical results. However, comparisons are less far-fetched when considering areas such as epistemology. When unencumbered by empirical issues, Aristotle's writings show him to be a deductive reasoner of the highest caliber. As a few examples in the area of epistemology will show, his thoughts appear relevant even for our discussions on quantum mechanics [353]:

> *All men by nature desire to know. An indication of this is the delight we take in our senses.* [Metaphysics I.1]

> *Further, if we admit in the fullest sense that something exists apart from the concrete thing, whenever something is predicated of the matter, must there, if there is something apart, be something corresponding to each set of individuals, or to some and not to others, or to none? If there is nothing apart from individuals, there will be no object of thought, but all things will be objects of sense, and there will not be knowledge of anything, unless we say that sensation is knowledge.* [Metaphysics III.4]

> *And, in general, if only the sensible exists, there would be nothing if animate things were not; for there would be no faculty of sense. The view that neither the objects of sensation nor the sensations would exist is doubtless true (for they are affections of the perceiver), but that the substrata which cause the sensation should not exist even apart from sensation is impossible. For sensation is surely not the sensation of itself, but there is something beyond the sensation, which must be prior to the sensation; for that which moves is prior in nature to that which is moved, and if they are correlative terms, this is no less the case.* [Metaphysics IV.5]

Had Aristotle lived to see the development of quantum mechanics, perhaps he would have argued that knowledge comes about from sensing or measurement. Aristotle states that sensation is knowledge. Furthermore, Aristotle indicates that something external must cause the sensation, other than the sensory system itself. Next consider Aristotle's thoughts on potentiality and actuality [353]:

> *For the same thing can be potentially at the same time two contraries, but it cannot actually.* [Metaphysics IV.5]

> *Further, one must observe that some causes can be expressed in universal terms, and some cannot. The primary principles of all things are the actual primary "this" and another thing which exists potentiality. The universal causes, then, of which we spoke do not exist.* [Metaphysics XII.5]

It seems reasonable that Aristotle might have associated the wave function as representing potentialities whereas actualities occur due to measurement. A similar comparison has been suggested by Heisenberg [552, pp. 9-10]. Aristotle may have anticipated the requirements of entanglement as shown in his statement:

> *One might suppose especially that the parts of living things and the corresponding parts of the soul are both, i.e. exist both actually and potentially, because they have sources of movement in something in their joints; for which reason some animals live when divided. Yet all the parts must exist only potentially, when they are one and continuous by nature.* [Metaphysics VII.16]

It is of interest to note that by associating an entangled state as being "one" and the parts or subsystems as existing only potentially, Aristotle's statement appears to parallel the theory of entangled states. This is because no subsystem of an entangled state can be specified in terms of a single pure state (which can be considered to be a single classical state), only via a mixture of pure states which requires more than one pure state for specification, so in this sense the subsystems would *only* exist potentially.

Aristotle's invention and development of logic remained essentially the main formulation of this subject until the mathematical treatments of logic by George Boole (1815-1864) and Augusta de Morgan (1806-1871) in the mid-19th century. This was followed by works on mathematical logic by Gottlob Frege (1848-1925) and Bertrand Russell to develop a set of logical axioms sufficient to contain all of mathematics. In the early 20th century, this culminated in Russell and Alfred North Whitehead's (1861-1947) system of logical axioms for mathematics published as the three volume opus *Principia Mathematica* and David Hilbert's program to prove the consistency of mathematics within axiomatic systems. These ambitions were shattered in 1931 by the publication of the incompleteness theorems of Kurt Gödel (1906-1978) that completely changed the nature and possibilities of logic. Gödel demonstrated the construction of a formula via Principia Mathematica that claims it is unprovable within the same system. This had the consequence that if it were provable it would be false and therefore that there will always be at least one true but unprovable statement in any sufficiently rich logical system. This was followed in 1936 by Alan Turing's (1912-1954) proof that the *halting problem* is undecidable: it is not always possible to decide whether a computational algorithm will complete its task. These works dramatically changed the view of how the nature of proof and truth can be embedded within logic and many repercussions followed. In particular, the mathematician and mathematical-physicist Roger Penrose, born the same year as Gödel's landmark paper, has argued that consciousness necessarily transcends the framework of Gödel's incompleteness theorem and Turing's halting theorem, and cannot be described by an algorithmically deterministic system [553].

The World as a Collection of Facts

Ludwig Wittgenstein (1889-1951), Bertrand Russell's one-time protégé in logic at Cambridge, may have been the closest counterpart to Aristotle as a deductive philosopher during the era when quantum mechanics was being developed, with the difference being [554, p. 357],

> *For Aristotle, the world is a collection of things: for Wittgenstein, the world is a collection of facts.*
>
> Aristotle Wittgenstein, Alias Isaac Newton, Between Fact and Substance, In: Newton's Scientific and Philosophical Legacy 1988, p. 357, G. DeBrock, © 1988 Kluwer Academic Publishers. With permission of Springer.

The ancient Greeks actually did not have a word for "fact" but could account for the inventory of things in their lives or imaginations: earth, water, fire, air, gods, angels, demons, the prime mover. In contrast, facts are relations between things or situational arrangements. Facts appear to have taken over our system of thought at about the same time that modern science developed during the 17th and 18th centuries. Instead of asking for the essence of material substances, science begins to ask how matter interacts with other matter. And Newton appears to be the first scientist to master this approach particularly with his universal theory of gravitation [554, p. 367]. However, not all scientists were up to the task, Descartes being a prime example. As we will see, the difference is between deductive and inductive reasoning. Deduction represents the approach of the most successful fundamental physical theories. This continued up to the development of quantum theory in the early 20th century with Bohr asking about the facts of a measurement and that observing the same system with different experimental arrangements can be mutually exclusive. Bohr was asking about the logic of facts and Wittgenstein attempted to address this at about the same time. Bohr later began using the term *phenomenon* to describe this [9, p. 425].

Ludwig Wittgenstein was perhaps the most influential philosopher of the 20th century though he was not in any sense a typical philosopher [555]. He inspired two important schools of thought, both of which he repudiated, i.e., logical positivism and linguistic philosophy. These influences stemmed from his early work *Tractatus Logico-Philosophicus* [556] and the posthumously published *Philosophical Investigations* [557] respectively, "the *early* and *late* Wittgenstein." Much of the focus of Wittgenstein, though approached very differently in the early and late works, dealt with the world, thought, and language and the inherent limitations in communicating aspects of these. The famous last line of the *Tractatus* is:

> *What we cannot speak about we must pass over in silence.*
>
> L. Wittgenstein, *Tractatus Logico-Philosophicus* Routledge, London 2000, p.74, translated by D.F Pears and B.F. McGuiness.

Although Wittgenstein did not have a direct interest in quantum mechanics, the trajectories of his ideas ran remarkably parallel to those of Bohr's with his complementarity view of the quantum. Interestingly, the geographical points and times of Wittgenstein's early seminal work also coincidentally overlap with those of Bohr's:

Manchester, Cambridge, and Como in the early 1900s. For Bohr: (1) Cambridge in 1911 with J.J. Thompson, (2) Manchester in 1911-12 with Rutherford for the atomic model, and (3) Como in 1927 for his announcement of complementarity. For Wittgenstein: (1) Manchester in 1908-11 for doctoral studies in aeronautical engineering which morphed into an obsession with the foundations of mathematics, (2) Cambridge in 1911-13 with Russell for work on logic, and (3) Como in 1918 as a prisoner of war captured at the Italian front with the draft of the *Tractatus* in his rucksack, which he completed in captivity. Wittgenstein was the son of an industrial magnate and grew up in the Palais Wittgenstein in Vienna, with frequent visitors such as Freud and Brahms, and surrounded by a precocious but intense family (three brothers committed suicide). From his inheritance, he was perhaps the wealthiest man in Europe at one point, but one of the first steps he took was to give away this fortune to simplify his life. Wittgenstein's plans to study physics with Boltzmann were ended by Boltzmann's unexpected death in 1906. As discussed in the section *Atomism Prevails*, Boltzmann's periods of depression may have been exacerbated by the intense opposition to his ideas on statistical irreversibility and atomism. As we have also seen in *The Fall of Classicality*, Planck's discovery of the quantum required him to overcome his previous rejection of Boltzmann's ideas. Wittgenstein's subsequent studies in aeronautical engineering in Manchester piqued an intense interest in the foundations of mathematics, and he was advised by Frege to join Russell in Cambridge. Turing later attended Wittgenstein's lectures on the foundations of mathematics.

Wittgenstein attempted in the *Tractatus,* published in 1921, to decipher the possible logic of the world in terms of facts. It was adopted by the logical-positivists as their playbook, though Wittgenstein could not take their efforts seriously. It takes the form of a seventy-five-page list of hierarchically numbered propositions in austere prose, mathematical logic, and Wittgenstein's innovation of the truth-table and begins,

1. The world is all that is the case.
1.1 The world is the totality of facts, not of things.
1.11 The world is determined by the facts, and by their being all the facts.
1.2 The world divides into facts.
2. What is the case - a fact – is the existence of states of affairs.
2.011 It is essential to things that they should be possible constituents of affairs.

L. Wittgenstein, *Tractatus Logico-Philosophicus* Routledge, London 2000, p.5, translated by D.F Pears and B.F. McGuiness.

During the development of quantum mechanics, Heisenberg had recalled Wolfgang Pauli saying [182, p. 206]:

> *It is part and parcel of the positivists' creed that facts must be taken for granted, sight unseen, so to speak. As far as I remember, Wittgenstein says: The world is everything that is the case. The world is the totality of facts, not of things. Now if you start from that premise, you are bound to welcome any theory representative of the*

case. The positivists have gathered that quantum mechanics describes atomic phenomena correctly, and so they have no cause for complaint.

However, as Bohr would have agreed, the constraints of formal logic are not subtle enough to encompass the quantum mechanics of measurement occurring under various experimental arrangements. Wittgenstein reportedly was first put off from the logical basis of his *Tractatus* while arguing with his friend, the Italian economist Piero Sraffa, that a proposition and that which it describes must have the same logical form. Sraffa made a Neapolitan gesture of contempt, brushing up underneath his chin, and asked,

What's the logical form of that?

Wittgenstein said that his discussions with Sraffa made him feel like a tree from which all branches had been cut [558, p. 15] and, like a true deductive reasoner, abruptly set out on a new path. This eventually led to *The Philosophical Investigations* and its exploration of facts through the meaning of concepts and activities within a multiplicity of what Wittgenstein called *language games*, by traveling with the word's uses through a complicated network of similarities, overlapping and criss-crossing. [557] In his survey of Bohr's contributions to the foundations of quantum theory, philosopher Edward MacKinnon summarizes [435, p. 115],

In his earlier conceptual analyses Bohr had focused on the proper way of extending classical concepts to, and restricting their usage in, quantum domains. His later work forced him to come to grips with the problem of how any concept has meaning. Though he never developed a systematic theory, he anticipated some of the key features later developed in Wittgenstein's Philosophical Investigations. The meaning of a word is determined by its usage in language, not by the objects it can or may denote. Bohr's analysis clarified the possible meanings of "particle" and "wave" within the contexts of experimental sources of information of atomic systems. Bohr also extended this approach to such higher order terms as "observe", "objective", "real", and "exist", analyzing the conditions of the possibility of unambiguous communication of information.

For Bohr, philosophical problems were about the general conditions for the communication of facts [559]:

What is it that we human beings ultimately depend on? We depend on

our words. We are suspended in language. Our task is to communicate experience and ideas to others. We must strive continually to extend the scope of our description, but in such a way that our messages do not thereby lose their objective or unambiguous character.

There is no quantum world. There is only an abstract quantum physical description. It is wrong to think that the task of physics is to find out how nature is. Physics concerns what we can say about nature.

The Philosophy of Niels Bohr, A. Petersen, Bulletin of the Atomic Scientists 19 (7), 8 (1963), Taylor & Francis, reprinted by permission of Taylor & Francis Ltd.

Newton's Hypotheses Non Fingo

Approaches for attempting to implement this task within theories of quantum measurement are discussed in detail in the section *The Rise of the Measurement Problem*. Bohr's requirement can be seen to actually be in accord with Newton's 17th century approach for communicating the facts of a world that happens not have a finite quantum of action. Newton does not make hypotheses about the material substance around us but instead focuses from the very start on the *facts*, the interactions that we experience, just as Bohr had focused on the experimental arrangements. The prime example is Newton's theory of gravitation in which a universal attractive force between masses would exert its influence instantly and extend over arbitrary distances of empty space. This violated the standard mechanical philosophy at that time, as dictated by Descartes and Boyle, in which explanation must be built up from hypothesized properties and motions of the materials filling space. In the second edition of the *Principia* of 1713, Newton introduced the General Scholium with its dictum *Hypotheses non fingo*, saying he would *not feign hypotheses* in place of sound explanation. It is enough that this force really exists and that it serves abundantly to account for the major phenomena of the heavens and our earth. Descartes viewed forces acting at a distance as something occult which cannot exist [340, p. 221]. However, Descartes' scenarios of vortices comprising materials filling space could not explain gravitation. Descartes' only secure *fact* was that he exists, the *Cogito*, the proposition that *if I think I am* as this requires an I that is thinking. From this, he proceeds to infer the reality of material substances of the world and their properties [304, p. 203].

Newton could be said to have a *top-down* approach whereas Descartes' approach is *bottom-up* [554, p. 367]. In private, Newton did make attempts to "explain gravity" [407, p. 40], but couldn't find a suitable explanation. He was satisfied in his place and time not to feign hypotheses. It would take almost 400 years of further development in mathematics and physics to explain gravity within Einstein's theory and gravitational interactions moving at the speed of light in order to justify why Newton's action at a distance can be an excellent approximation. In a 1672 letter to the Jesuit Father Ignace

Gaston Pardies, Newton expands on his approach in more detail [554, p. 369]:

> *For the best and safest method of philosophizing seems to be, first to enquire diligently into the properties of things, and to establish those properties by experiments and then to proceed more slowly to hypotheses for the explanation of them. For hypotheses should be employed only in explaining the properties of things, but not assumed in determining them; unless so far as they may furnish experiments. For if the possibility of hypotheses is to be the test of the truth and reality of things, I see not how certainty can be obtained in any science; since numerous hypotheses may be devised, which shall seem to overcome new difficulties. Hence it has been here thought necessary to lay aside all hypotheses, as foreign to the purpose.*
>
>

These requirements were later codified in the 1726 edition of the *Principia*, Vol. 2 as "Rules of Reasoning in Natural Philosophy," which might be regarded as guidelines for deductive reasoning in the physical sciences [308, pp. 202-205]:

Rule I. We are to admit no more causes of natural things than such as are both true and sufficient to explain their appearances.

Rule II. Therefore, to the same natural effects we must, as far as possible, assign the same causes.

Rule III. The qualities of bodies, which admit neither intension nor remission or degrees, and which are found to belong to all bodies within the reach of our experiments, are to be esteemed the universal qualities of all bodies whatsoever.

Rule IV. In experimental philosophy, we are to look upon propositions collected by general induction from phenomena as accurately or very nearly true, notwithstanding any contrary hypotheses that may be imagined, till such time as other phenomena occur, by which they may either be made more accurate, or liable to exceptions.

Deductive Reasoning Prevails

Another deductive reasoner was James Clerk Maxwell, who had brought classical physics to a close toward the end of the 19th century with his 1873 summation *A Treatise on Electricity and Magnetism.* He had completed the development of this branch of theory with his electromagnetic field equations, now referred to as

Maxwell's Equations. Maxwell revered Michael Faraday for the ingenuity of his historic experiments in revealing the deep subtleties of electromagnetism. Maxwell was inherently a deductive researcher and he also appreciated this trait in Faraday for his combination of conceptual imagination and systematically allowing for any possibility in the search for a solution. He also had a high opinion of the work of André-Marie Ampère (1775-1836) and called him the "Newton of electricity," however, Ampère leaned toward the inductive approach in his research. Despite the greatness of both Faraday and Ampère, Maxwell noted the difference in their approaches and presentation of results. Maxwell summarized in Part IV, Chapter III of [560]:

> *The method of Ampère, however, though cast into an inductive form, does not allow us to trace the formation of the ideas, which guided it. We can scarcely believe that Ampère really discovered the law of action by means of the experiments, which he describes. We are led to suspect, what, indeed, he tells us himself, that he discovered the law by some process which he has not shown us, and that when he had afterwards built up a perfect demonstration, he removed all traces of the scaffolding by which he had raised it.*
>
> *Faraday, on the other hand, shews us his unsuccessful as well as his successful experiments, and his crude ideas as well as his developed ones, and the reader, however inferior to him in inductive power, feels sympathy even more than admiration, and is tempted to believe that, if he had the opportunity, he too would be a discoverer.*

The reporting by Faraday of failures as well as successes in his experimental attempts is all too rare in scientific reporting, but it is an essential aspect of the deductive process. Faraday's law, the generation of an electromotive force in a coil due to a changing magnetic flux, is a fundamental result and is the working principle of electric motors, generators and transformers. Ampère had come close to discovering this law but without having Faraday's deductive thoroughness, he had ignored the crucial aspect of the changing magnetic flux and found no effect to his long-lasting regret.

On the hundredth anniversary of Maxwell's birth, Einstein described the deductive path Maxwell took in arriving at his celebrated equations, explaining that Maxwell began thinking in terms of mechanical models which would reproduce aspects of the known electrical and magnetic behaviors. But then he was able to extract the essence of these models, leaving behind what we now refer to as electric and magnetic fields described by differential equations [561],

> *If the idea of physical reality had ceased to be purely atomic, it still remained for the time being purely mechanistic; people still tried to explain all the events as the motion of inert masses; indeed, no other*

way of looking at things seemed conceivable. Then came the great change, which will be associated for all time with the names of Faraday, Maxwell, and Hertz. The lion's share in this revolution fell to Maxwell. He showed that the whole of what was then known about light and electromagnetic phenomena was expressed in his well-known double system of differential equations...Maxwell did, indeed try to explain, or justify, these equations by the intellectual construction of a mechanical model...But he made use of several constructions at the same time and took none of them really seriously, so that the equations alone appeared as the essential thing and the field strengths as the ultimate entities, not to be reduced to anything else.

A. Einstein, Maxwell's Influence on the Evolution of the Idea of Physical Reality, p. 266, In: James Clerk Maxwell: A Commemorative Volume on the one hundredth anniversary of his birth, Cambridge University Press, Cambridge 1931.

Red Flags for Deduction

As discussed in the section *Einstein's Quandary*, Einstein himself was a great deductive investigator but he did make different choices than Bohr on the issue of the completeness of quantum mechanics. What are possible extraneous factors that may influence one's choice along the deductive path? In a talk for Planck's sixtieth birthday in 1918, Einstein describes the significance for him of having a complete world-picture, perhaps suggesting that this may partly motivate such choices [562, p. 19]

Man tries to make for himself in the fashion that suits him best a simplified and intelligible picture of the world; he then tries to some extent to substitute this cosmos of his for the world of experience, and thus to overcome it. This is what the painter, the poet, the speculative philosopher, and the natural scientist do, each in his own fashion. Each makes this cosmos and its construction the pivot of his emotional life, in order to find in this way, the peace and security which he cannot find in the narrow whirlpool of personal experience.

The Cambridge Companion to Einstein, M. Janssen and C. Lehner (Editors), Cambridge University Press 2014.

In 1919, Einstein identified and contrasted two types of theory, *constructive* and *principle* [563],

We can distinguish various kinds of theories of physics. Most of them are constructive. They attempt to build up a picture of the more complex phenomena out of the materials of a relatively simple formal scheme from which they start out. Thus, the kinetic theory of gases seeks to reduce mechanical, thermal, and diffusional processes to movements of molecules—i.e., to build them up out of the hypothesis of molecular motion...Along with this most important class of theories there exists a second, which I will call "principle theories". These employ the analytic, not the synthetic, method. The elements which form their basis and starting-point are not hypothetically constructed but empirically discovered ones, general characteristics of natural processes, principles that give rise to mathematically formulated criteria which the separate processes or the theoretical representations of them have to satisfy. Thus, the science of thermodynamics seeks by analytical means to deduce necessary conditions, which separate events have to satisfy, from the universally experienced fact that perpetual motion is impossible. The advantages of the constructive theory are completeness, adaptability, and clearness, those of the principle theory are logical perfection and security of the foundations.

Note the similarities to the top-down versus bottom up theories of Newton and Descartes, respectively. In these terms, Einstein regarded the standard approach to quantum mechanics as a principle-theory and his disagreements with Bohr came in part from his insistence on a constructive theory. In developing quantum mechanics, Heisenberg claimed to base his approach on another criterion of Einstein's: only that which is directly observable should be introduced into a theory, as he thought Einstein had done in developing relativity. Heisenberg describes a conversation with Einstein on this issue where Einstein responds [182, p. 63]:

But you don't seriously believe that none but observable magnitudes must go into a physical theory? ... on principle, it is quite wrong to try founding a theory on observable magnitudes alone. In reality, the very opposite happens. It is the theory which decides what we can observe.

Einstein later illustrated how this occurs in the construction of theories with the case of Kepler's laws, the concept of an elliptical orbit had to exist first, to then inform the empirical data [564],

Now came the second and no less arduous part of Kepler's life work. The orbits were empirically known, but their laws had to be guessed from the empirical data. First, he had to make a guess at the mathematical nature of the curve described by the orbit, and then try it out on a vast assemblage of figures. If it did not fit, another hypothesis had to be devised and again tested. After tremendous search, the conjecture that the orbit was an ellipse with the sun at one of its foci was found to fit the facts. Kepler also discovered the law governing the variation in speed during one revolution, which is that the line sun-planet sweeps out equal areas in equal periods of time. Finally, he also discovered that the squares of the periods of revolution round the sun vary as the cubes of the major axes of the ellipses.

Another outside factor that can cause deviations from the deductive path is beliefs that are imposed or censored by authorities of church or state. The transition to an independent examination of the world began to take hold during the 17th century as part of the scientific revolution, though many exceptions have persisted. Though the scientific community needs to cautiously protect its knowledge base, Figure 5.22, and close ranks against speculative developments unsupported by observation, this trend can also move to become overly protective and only support *safe* research. This phenomenon results in the occasional insistence that "everything important is already known," and this is seen repeatedly throughout the history of science. In 1871, Maxwell commented on the state of science [399]

The opinion seems to have got abroad that in a few years all the great physical constants will have been approximately estimated, and that the only occupation which will then be left to men of science will be to carry out these measurements to another place of decimals...But we have no right to think thus of the unsearchable riches of creation, or the untried fertility of those fresh minds into which these riches will continue to be poured...

Indeed, in 1903, Albert Michelson (1852-1931), who had performed the landmark interferometer experiments in clarifying the basis of the theory of relativity, stated [565]

The more important fundamental laws and facts of physical science have all been discovered, and these are now so firmly established that the possibility of their ever being supplanted in consequence of new discoveries is exceedingly remote...Our future discoveries must be looked for in the 6th place of decimals.

Just a few years after the development of quantum theory, one of the pioneers of quantum mechanics and quantum electrodynamics, Paul Dirac (1902-1984) similarly proclaimed in 1929 that the underlying basis for a large part of physics and the whole of chemistry is completely known [566],

The underlying physical laws necessary for the mathematical theory of a large part of physics and the whole of chemistry are thus completely known, and the difficulty is only that the exact application of these laws leads to equations much too complicated to be soluble.

P.A.M. Dirac, Quantum Mechanics of Many-Electron Systems, Proceedings of the Royal Society of London, Series A 123 (792), 714 (1929).

This statement of Dirac's was widely quoted for decades afterwards and was interpreted as signifying the triumph of quantum theory. However, we will see in Chapter 6 that such statements signify ignorance of the measurement problem.

We have seen in this exploration of *Deductive versus Inductive Thought* through history that the method of physical deduction constrained by empirically established great principles repeatedly produces deep results in investigating the foundations of physics. Typical examples of great principles are the conservation laws, e.g., energy and momentum. Occasionally, there have been attempts to identify other more unconventional principles and explore the consequences of deduction constrained by these. A sometimes controversial example is the *Anthropic Principle*, that the laws of nature and parameters of the universe take on values that are consistent with conditions for life as we know it because we are here to observe them. Robert Dicke (1916-1997), a pioneer of quantum optics and gravitation, explored one of the first modern applications of this, though it is another idea that traces back to the ancient Greeks. Dicke's colleague John Wheeler gives a description of the scale of the universe, constrained by the fact that we are here [567]:

> *My Princeton colleague, Robert Dicke, expressed it this way: "What good is a universe without somebody around to look at it?" That, to be sure, was an old idea, going back not only to the Bishop Berkeley of the time of Newton, but all the way back to Parmenides, the precursor of Socrates and Plato. But it was new in the form that Dicke put it. He said if you want an observer around, you need life, and if you want life, you need heavy elements. To make heavy elements out of hydrogen, you need thermonuclear combustion. To have thermonuclear combustion, you need a time of cooking in a star of several billion years. In order to stretch out several billion years in its time dimension, the universe, according to general relativity, must be several billion years across in its space dimensions. So why is the universe as big as it is? Because we're here!*

From EPR to the Present

The Quantum Triumvirate

Bohr's response to the 1935 Einstein-Podolsky-Rosen (EPR) paper [474] was essentially the same as his view of Laplace's Demon. The EPR argument involves *the quantum triumvirate* of non-determinacy, entanglement and nonlocality. EPR considered an entangled state of two widely separated systems A and B, and considered how A is affected by a remote measurement on B. EPR used *locality* as a criterion to argue that the value of a local observable must be definite and not affected by a measurement at the remote location and gave a sufficient criterion for an *element of reality* of a physical quantity [474]:

> *If, without in any way disturbing a system, we can predict with certainty (i.e., with probability equal to unity) the value of a physical quantity, then there exists an element of reality corresponding to that quantity.*
>
> A. Einstein, B. Podolsky, and N. Rosen, Physical Review 47, 777 (1935), Copyright (1935) by the American Physical Society.

EPR demonstrated that the non-commuting variables of position and momentum could be measured more accurately than allowed by the uncertainty principle unless the measurement of the remote particle instantaneously affects the local particle. They concluded that quantum theory did not account for all elements of reality and must be incomplete. EPR expected that the inclusion of local hidden variables would make quantum theory complete and allow definite values to be given for all physical variables. Einstein later described the argument as forcing us to choose between the following two assertions [339]:

> (1) the description by means of the psi-function is complete.
>
> (2) the real states of spatially separate objects are independent of each other.

Bohr's response [469] focused on the ambiguity of EPR's *criterion of reality* and emphasized the different experimental arrangements required for unambiguously addressing complementary physical variables in quantum theory,

> *In fact, the renunciation in each experimental arrangement of the one or the other of two aspects of the description of physical phenomena—the combination of which characterizes the method of classical physics, and which therefore in this sense may be considered as complementary to one another—depends essentially on the impossibility, in the field of quantum theory, of accurately controlling the reaction of the object on the measuring instruments, i.e., the transfer of momentum in case of position measurements, and the displacement in case of momentum measurements.*
>
> N. Bohr, Physical Review 48, 696 (1935), Copyright (1935) by the American Physical Society.

Experimental tests of Bell's inequality eventually showed that quantum nonlocality implies that the local hidden variables sought by EPR do not exist. Bell's results implied that locality and realism in the sense of EPR are incompatible with quantum mechanics. The demands of no-signaling between space-like separated systems as well as a condition of locality imply that correlations comply with Bell's inequalities. However, the inequalities are violated by correlated entangled quantum particles. No-signaling requires that instantaneous communication is impossible. If no-signaling is accepted, Bell measurements cannot be deterministic, and hence they are random as long as the measurement settings are freely chosen. Nonlocal correlations cannot be revealed only by local descriptions of the systems and this instead requires an entangled state. Local observations will then generally reflect an intrinsic randomness in the presence of nonlocal correlations even if the entangled state is pure and completely known. Greenberger, Horne, and Zeilinger extended the EPR entangled states to three particles [568], which have additional subtle properties allowing an illustration of Bell's theorem in a very direct way, as shown in Appendix 5.A.

Some thirty years after EPR, a further result, the Kochen-Specker theorem [81] [569], reinforced the argument for nondeterminism in a different way than EPR by showing (in Hilbert space dimensions $d \geq 3$) that if we want to add hidden variables to avoid nondeterminism, the hidden variables have to be *contextual*; i.e., the outcome depends on the specific experimental arrangement used to measure the observable. Kochen-Specker does not refer to nonlocality but rather to the incompatibility of hidden variables associated with the system being measured. This incompatibility addresses the mutual exclusiveness of experimental arrangements and further strengthens Bohr's complementarity principle some thirty years earlier as a type of contexuality argument.

The tension between the concepts of non-determinacy, entanglement, and nonlocality also impacts how the quantum state or wave function is viewed. The question of the meaning of the wave function goes back to the beginning of quantum theory, with de Broglie and Schrödinger initially arguing that it was a real physical wave [4]. After Born's introduction of indeterminacy, the Copenhagen interpretation of Bohr, Heisenberg, and Pauli viewed the wave function as a wave of probability amplitude. Heisenberg further suggested that the quantum probability amplitude waves can be interpreted as a quantitative version of the concept of *potentia* from Aristotle's philosophy [552, pp. 9-10]. The tendency for an event to take place was recognized as an aspect of the world having an intermediate reality between object and idea. Potentiality and actuality were a dichotomy used by Aristotle to analyze a range of issues of the physical world related to motion, causality and physiology. Similarly, Heisenberg suggested that the quantum laws of nature determine the possibility of occurrence and not the phenomenon itself.

As discussed in Chapter 4, some recent interpretations and models for quantum theory such as many-worlds, de Broglie-Bohm, and spontaneous collapse regard the wave function as real, and are known as *ψ-ontic* theories (a state of reality). Other interpretations such as quantum Bayesian and quantum information-based approaches relate the wave function to representations of knowledge and are known as *ψ-epistemic* theories (a state of knowledge). More recently, Pusey, Barrett, and Rudolph (PBR) [570] proved a theorem showing that the quantum state must be ontic in a wide class of quantum theories. The main underlying assumption of the PBR theorem is that composite systems prepared in a product state should be independent of one another; i.e., the ontic distribution for $|\psi_a\rangle \otimes |\psi_b\rangle$ is the product of the ontic distribution for $|\psi_a\rangle$ and the ontic distribution for $|\psi_b\rangle$. Much effort has been spent on examining a variety of assumptions and approaches to determine under what conditions the wave function is allowed to be considered either ontic or epistemic [571].

However, a complete picture of the subtle roles played by indeterminacy, entanglement and nonlocality and how they are interrelated has emerged most fully from an understanding of a variety of Bell inequality experiments, beginning in the 1960s and up to the present day. For classical theories, randomness cannot be intrinsic but only a result of an incomplete description of the system. Quantum theory is known to give probabilistic predictions for particular experiments in which the preparation of the system is essentially perfect. Einstein claimed in the EPR paper that this could be explained by the incompleteness of quantum mechanics, and there should be a complete theory giving a deterministic result for every experiment. However, as discussed in Chapter 3, Bell's theorem indicated that local hidden variable theories are inconsistent with quantum mechanics. In hidden-variable theories with a local causal structure, correlations between space-like separated measurement events satisfy a variety of Bell inequalities. Numerous quantum systems have by now been experimentally shown to violate Bell inequalities. However, there are subtleties among the assumptions made in order to conclude that intrinsic nondeterminism in nature follows from quantum nonlocality. If no-signaling had not been assumed, then it would be possible to have deterministic descriptions of quantum correlations.

If no-signaling is assumed, the conclusion that Bell measurements imply nondeterminism relies on the further assumption that the Bell measurement settings are freely chosen and truly random. The theorem does not apply unless one has freedom to choose the detector's settings without modifying the state that is to be measured. However, it is never possible to completely certify the randomness of the settings and rule out scenarios, however implausible, to explain the initial random settings. An extreme example is *superdeterminism*, so-named by Bell as a possible loophole for his theorem, which hypothesizes extravagantly that all processes are completely predetermined by conditions in the past light-cone of both observers, so that each space-time point encodes the initial state of the universe. As Shimony, Horne, and Clauser had summarized [572]:

> *In any scientific experiment in which two or more variables are supposed to be randomly selected, one can always conjecture that some factor in the overlap of the backward light cones has controlled the presumably random choices. But, we maintain, skepticism of this sort will essentially dismiss all results of scientific experimentation. Unless we proceed under the assumption that hidden conspiracies of this sort do not occur, we have abandoned in advance the whole enterprise of discovering the laws of nature by experimentation.*
>
> Search for a Naturalistic World View, Volume II, Natural Science and Metaphysics, Abner Shimony, Cambridge University Press.

The implication is that if the initial conditions were to be somewhat different, the entire quantum mechanical theory would dissolve and not describe the full range of phenomena currently accounted for by quantum mechanics [573].

Though logically impossible to rule out, such conspiracy-type models become even further far fetched by other developments, which highlight the tension between quantum theory and determinism in a particularly striking way. These are methods for *Bell-certified randomness*, including quantum randomness expansion (generating and certifying a large number of random bits from a small seed of random bits) and quantum randomness amplification (generating uniformly random bits called *free* randomness from imperfectly random ones). These experimentally demonstrated protocols for certification, expansion and amplification use Bell measurements to produce outcomes that are fully unpredictable and achieve tasks that are not possible classically [574] [575] [576] [577] [578] [579]. For example, the scaling of an initial seed randomness has been demonstrated for protocols that give an exponential expansion of the randomness as strong as exponential, so that a seed of n-bits grows to 2^n-bits. The *Bell-certified* means that the resulting bits are close to uniformly random in terms of a quantitative relation between the amount of the Bell-inequality violation and the randomness observed that is also independent of the details of the measurement devices, i.e., *device-independent quantum randomness*. This is the strongest certification of nondeterminism that can be expected using quantum nonlocality. In a sense, this pushes any alternative possibilities of conspiracies in the measurement settings as tightly as possible into a corner so that they would have to be truly outlandish to be functioning in our universe. Such protocols also allow devising random number generators for which a Bell inequality violation guarantees that the output is random and immune to outside adversaries.

Although intrinsic quantum randomness can be certified in terms of nonlocality and entanglement, the relationships among the quantum triumvirate of nondeterminism, entanglement, and nonlocality are subtle and not as direct as might be expected; not simply, e.g., less Bell violation $\Longrightarrow$ less quantum randomness [580]. Instead, a maximum quantum randomness can be certified from arbitrarily small amounts of nonlocality or entanglement. The amount of randomness certified by nonlocal quantum correlations is inequivalent both to entanglement and nonlocality even though nonlocality is necessary for entanglement and nonlocality is necessary for certifying randomness. Probability distributions with maximal nonlocality do not necessarily contain maximal randomness. And contrariwise, distributions with arbitrarily small amounts of nonlocality may contain nearly maximal randomness. The quantitative relationships between these quantities require the challenging task of characterizing extremal properties of the boundaries of quantum correlations. As shown in [580], in the type of CHSH plot familiar from Chapter 3, the region of maximal randomness can be arbitrarily close to the CHSH bound of $S \leq 2$, below which local hidden variable models become possible. Violations of CHSH had been discussed in Chapter 3 as a crucial element in devising UMDT tests for distinguishing measurement from unitary evolution in terms of CHSH bounds. The generation of maximal certified quantum randomness, the signature of intrinsic nondeterminism, is intriguingly seen to emerge near this borderline of nearly product states with arbitrarily small entanglement.

The relation between the constraint of no-signaling and the randomly generated response to measurement exemplified by the Bell theorems was further extended by Conway (known outside of academia as inventor of the celebrated cellular automaton,

Game of Life) and Kochen (of the Kochen-Specker Theorem) in results they brazenly called the *Free Will Theorem* (FWT) and a strengthened version, the *Strong Free Will Theorem* [581] [582], which led to exchanges in the literature. These results in effect claimed that, in the absence of signaling, if Alice and Bob have the free will to measure their particles, then the particles have their own free will for how to respond. These results combined elements of EPR and Kochen-Specker along with the relativistic consequence of there being no preferred frame of reference determining the order of Alice and Bob's freely chosen space-like separated measurements to conclude that their measurements also cannot be determined by the prior state of the universe. Although Alice and Bob could be replaced by (pseudo) random number generators, Conway and Kochen contend that free will would still be required to choose the random number generators to avoid the possibility of them having been predetermined in the past. However, the use of the term *free will* here need not be quite the strong free will that haunts the psyche of philosophers. It could as well be the *free randomness* generated by the protocol of randomness amplification discussed previously [575]. With randomness amplification, sufficiently strong quantum correlations ensure that even if Alice and Bob cannot choose the measurements perfectly freely, the generated outputs are nonetheless perfectly free randomness. It has also been shown that with no-signaling and arbitrary freedom of choice of measurement settings, there are quantum processes with fully intrinsic randomness that are not mixed with apparent randomness due to the incompleteness of quantum mechanics [583]. In those cases, no alternative theories based on either signaling or no freedom of choice could give better predictions than quantum theory.

Quantum Demons

The name *Statistical Mechanics* actually came from the American physicist Josiah Willard Gibbs (1839-1903) in 1874. Inspired by the work of Maxwell and Boltzmann, he realized that this was an entirely new discipline and Gibbs developed his own system of statistical mechanics and defined an entropy in terms of ensembles [396]. He considered three types of ensembles: the micro canonical for systems that have definite energy, the canonical for systems in contact with heat baths and thus have fluctuating energies, and the grand canonical that has fluctuations in both energy and particle number. Bohr admired the ensemble approach of Gibbs and had said to Heisenberg "You read Gibbs and there are these chapters in Gibbs' book and that is really everything that can be said about thermodynamics." [398, p. 325] Gibbs' expression for the entropy in terms of states labeled by ν takes the form:

$$S = -k \sum\nolimits_{\nu} P_\nu \log P_\nu$$

where P_ν is the probability of the microstate and $\sum_\nu P_\nu = 1$. Gibbs' expression for the entropy has the same form later found for quantum systems with discrete energy levels by von Neumann. The founders of quantum mechanics had worried about the issues of thermodynamic irreversibility and entropy change during measurement. As an example of one of many discussions, here from a 1947 letter from Bohr to Pauli and Pauli's response [398, p. 325].

Bohr to Pauli:

> *The idea that any observation must necessarily involve an increase in entropy has been much discussed and I remember that already in the discussion... when you helped me with the "proofs" of the old paper on complementarity, I stressed the principle [of] irreversibility of the concept of observation. More specifically, any observation must make use of some registering device, whether through a photographic plate or directly by the retina of the eye, which involves processes of amplification by which free energy is spent.*

Pauli to Bohr:

> *The discussions...concerned the quantitative side of the connection of the concepts of entropy and of observation, a connection which, as we all agree, is of a very fundamental character. The problem arises whether there is a well-defined minimum of the increase in entropy, independent of the particular experimental arrangement in use, if a certain quantity ("observable") is measured. Our discussions seemed to indicate that this is actually the case, although we did not reach yet any final conclusion.*

The generalization of statistical mechanics to include quantum mechanics began with von Neumann in his 1927 paper, his 1929 paper on the quantum ergodic and H theorems, and his 1932 book, *Mathematical Foundations of Quantum Mechanics* [13]. Von Neumann had introduced his version of entropy based on considerations of the disorder of quantum states so that it has the same meaning as the thermodynamic entropy. The von Neumann entropy of the quantum density matrix ρ is

$$S(\rho) = -\mathrm{Tr}(\rho \log \rho) \ .$$

For a system with discrete energy levels and using the eigenstates of the Hamiltonian to compute the trace, the von Neumann entropy takes the same form as the Gibbs entropy. Von Neumann had established a procedure to determine the equilibrium distribution by requiring that the entropy becomes a maximum under certain subsidiary conditions. An important difference in quantum mechanics is the symmetries of the particles under permutations, which must be antisymmetric states for fermions or symmetric states for bosons. Von Neumann's 1929 paper "Proof of the ergodic theorem and the H-theorem in quantum mechanics," [584] revisited the problems of Boltzmann and Gibbs, and the issues of disorder and ergodicity, within the framework of quantum mechanics. As he summarizes in the introduction:

> *The object of the present paper is the clarification of the relations between the macroscopic and the microscopic point of view of complex systems...how the peculiar, seemingly irreversible behavior of entropy arises... And these questions shall be attacked with the means of quantum mechanics.*
>
> *In classical mechanics, it is known that these questions have led to the development of two elaborate theoretical systems: the statistical mechanics of Boltzmann and that of Gibbs. The former could not provide a final and satisfactory solution because it had to make essential use of so-called assumptions of disorder—and exactly to fathom the nature of this "disorder" is the real problem. The latter would basically be adequate for this program; however, it leads to a mathematical problem—the so-called quasiergodic problem ...the new quantum mechanics differs from the classical mechanics by being remarkably simple; it is due to this circumstance that in quantum mechanics, if we follow the Gibbsian path, we can reach the goal with relatively simple mathematical means. That is, it will be possible in what follows to prove the ergodic theorem and the H-theorem (which are the two questions mentioned above) without the need to recur to any assumption of disorder.*
>
>

Von Neumann's results have been newly revisited within the past decade with the renewed interest in quantum statistical mechanics [585]. Entropy production has been experimentally measured in a microscopic quantum system which breaks time-reversal invariance [586]. This utilized nuclear magnetic resonance of an ensemble of spin-1/2 particles from a Carbon-13 nucleus in a chloroform molecule in a liquid. The preparation in an initial thermal equilibrium state is what had singled out the preferred value of entropy, leading to irreversibility based on states that are most probable to occur, a microscopic quantum version of Boltzmann's arguments. Although ergodicity and time-averaging have often been used to justify entropy maximization in closed systems, ergodicity does not always appear to be applicable on the same regime over which statistical mechanics is successful in calculations [587]. This has generated recent attention in the fundamentals of quantum statistical mechanics [588], introducing the study of new concepts such as eigenstate thermalization of single quantum states, canonical typicality, and the relations between thermalization, integrability, and many-body localization. Can a closed quantum system actually behave as a thermal system? Thermalization of classical systems has been characterized in terms of concepts such as ergodicity and mixing, properties which enable systems to self-randomize. These properties reflect the statement that the most accessible microstates describe the same macroscopic state. However, the linear Schrödinger equation formally cannot exhibit such non-linear and chaotic behavior. A quantum many-body system initially in a pure state remains pure under unitary evolution with constant zero entropy. However, recent work has suggested that quantum systems can nevertheless thermalize, with indications that this is due to effectively thermal characteristics of the individual eigenstates so that expectation values are equal to their thermal values [587]. That this may happen for complex systems is called the *eigenstate thermalization hypothesis*. This was demonstrated experimentally using a Bose-Einstein condensate of Rubidium atoms in a two-dimensional optical lattice in which the local entropy from quantum entanglement plays the role of a thermal entropy [589]. For a pure state of two spatially separated entangled spins, such as in a Bell state, the local measurement of one of the spins is a statistical mixture. If this indeed behaves as a thermalized state, it is expected to grow extensively with subsystem volume and the experiments are consistent with such an extensive thermal entropy.

The extension of Maxwell's demon to quantum systems has been studied theoretically and experimental efforts on quantum demons have also begun with recent advances in techniques. However, Maxwell's demon exhibits new aspects quantum mechanically since measurement disturbs quantum states. More work can be extracted from a quantum demon than a classical demon [590] [591]. The indistinguishability of quantum particles affects the work that can be extracted [592]. Theoretical studies of Landauer's principle has been shown to apply in equilibrium quantum systems [593] [594] as well as non-equilibrium quantum systems [587]. The erasure of classical information encoded in quantum states by thermalization was considered by Lubkin [595]. For a system coupled to a thermal bath, the total entropy change on erasure of classical information encoded in quantum states by thermalization is the sum of the system and bath entropy changes with the minimum entropy change of the system given by [382]:

$$\Delta S_{sys} = k \ln 2 S(\rho)$$

where $S(\rho) = -\mathrm{Tr}(\rho \log \rho)$ is the von Neumann entropy of the density operator ρ. The factor of $k \ln 2$ is often customarily included for consistency with Landauer's principle. Landauer's principle has also been used to demonstrate the requirement of no-cloning for quantum states and the Holevo bound for accessible quantum information [596] [382].

Generalizing Maxwell's demon into the quantum realm has involved theoretical studies using quantum demons within classical engines [597] and also considering quantum demons within quantum engines [590], allowing a quantum treatment of the demon's entire cycle of operation including heat absorption, processing, decoherence, work generation and erasure. As an example, the energy stored in two-level systems can be utilized directly in scenarios where atoms are separated into different spatial paths and the energy is extracted from the atoms in the excited state and then recombining the paths. The which-way path information can then be erased isothermally and dissipated into the environment in preparation for the next cycle [598]. It has been suggested by Scully [599] [600] that quantum coherence of the ground state could be used to enhance the efficiency of the thermodynamic cycle beyond the Carnot efficiency by extracting work in the form of photons from a heat bath, quantum coherence playing the role of the demon in a quantum version of a Szilard one-molecule engine. Sorting occurs because cold atoms absorb less than they would otherwise in the absence of coherence whereas hot atoms emit photons. It has been proposed [601] that the information gained from measurements of the micro-levels could also extract additional work directly in form of mutual information I gained during the measurement, so that $W_{max} = -(\Delta E - T\Delta S) + k_B T\, I$. It has further been shown [602] that quantum correlations from separated entangled particles can be harnessed to produce mechanical work.

Experimental work on quantum Maxwell demons has also begun, including a superconducting qubit with a microwave cavity demon [603] and photonic light pulses with a demon in the form of a photodetector with feed-forward operation [604]. More recently, a quantum Maxwell demon has been demonstrated out of thermal equilibrium [605]. In the non-equilibrium domain, fluctuations of thermodynamic quantities become important, so additional entropy is produced, decreasing thermodynamic efficiency. Non-equilibrium effects will become increasingly important as devices approach the nanoscale and become more susceptible to thermal and quantum fluctuations. The understanding of fluctuations of systems in thermal equilibrium was pioneered by Maxwell, Boltzmann and Gibbs. Extending this to near-equilibrium systems was developed by Onsager, Green, and Kubo. However, in recent years, exact non-equilibrium quantum fluctuations results have been obtained by Jarzinski and Crooks [606] [607]. In addition to fabricated devices, biology offers systems that intrinsically operate as non-equilibrium systems. By ingesting nutrients, oxygen, and photons, living organisms maintain themselves by controlling energy and increasing the entropy of their surroundings.

In the equilibrium examples of open quantum systems, the system is in contact with a heat bath which establishes the temperature T. The heat bath is a source of stochasticity and irreversibility. More recent works on quantum Maxwell demons have attempted to replace the heat bath with the measurement process itself as the primary source of stochasticity, i.e., a "thermodynamics without bath" [608] [609] [610]. The motivation in these works has not been to understand the measurement problem but to determine the energy expenditure for measurements during quantum computing and error correction. Understanding the physical energy cost for projective measurements would determine the fundamental physical limitation to quantum computers in a way similar to the Landauer limit for classical computers. These works find that the system energy change incurs an additional expense due to the average entropy decrease. It might be thought that the energy cost for a measurement is just the energy difference.

$$\Delta E_S = \mathrm{Tr}[H_S(\acute{\rho}_S - \rho_S)]$$

where H_S is the system Hamiltonian and $\acute{\rho}_S = \sum p_k \acute{\rho}_{S,k}$ is the average post-measurement state and the p_k are the respective probabilities. However, a computational scenario will include steps similar to that of a Maxwell demon or a heat engine, involving storing the outcomes in a memory for readout and feedback. In addition, the same measurement device will be used repeatedly so that it needs to be restored to the initial state and reset. This will incur additional energy expenses in terms of the average entropy decrease. For projective measurements, the entropic decrease simply takes the form of a Shannon entropy:

$$E_{\text{proj}} = \Delta E_S - k_B T \sum_k p_k \ln p_k.$$

Any attempts by Maxwell demons to intervene in irreversible processes appear to be vanquished by Landauer's principle classically and at least in some cases quantum mechanically. Although the validity of Landauer's principle has been experimentally confirmed, perhaps it is not a necessary element in order to address Maxwell's demon in all situations. There have been arguments by Earman and Norton that there may be no need to exorcise the demon if it's assumed from the outset that the demon is governed by the laws of thermodynamics [611] [612]. Bennett, who had proposed using Landauer's principle to exorcize Maxwell's demon, states [613]:

> *One of the main objections to Landauer's principle, and in my opinion the one of greatest merit, is that raised by Earman and Norton, who argue that since it is not independent of the Second Law, it is either unnecessary or insufficient.*
>
>

If the demon is assumed to be a thermodynamic system already governed by the Second Law, no further consideration is needed to ensure that the Second Law is obeyed by the entire system and demon. Conversely, if the demon is not assumed to obey the Second Law, no supposition about the entropy cost of information processing can save the Second Law from the demon. Bennett has shown that with classical reversible information processing, i.e., where measurement is equivalent to copying, the act of information destruction in Landauer erasure has a cost exactly sufficient so that the Second Law is obeyed. Theoretical studies of the thermodynamic costs of quantum operations [614] find that the Landauer limit, at which computation becomes thermodynamically reversible, can be reached by classical systems but that for quantum systems there is an unavoidable excess heat generation that results in an inherent thermodynamic irreversibility. This quantum limit, however, applies to quantum computations that are represented as unitary operations that can always be run in reverse. The excess heat generation in this unitary case results from the no-cloning theorem, which prevents simply saving a copy of the output before reversing the computation. This prevents a quantum generalization of Bennett's procedure for the reversible unitary case. However, it has not yet been experimentally determined whether the measurements that Maxwell demons carry out in quantum systems are unitary or non-unitary. As discussed in Chapters 3 and 4 of this book, distinguishing unitary from non-unitary processes can be experimentally determined and is a key aspect of the quantum measurement problem. In particular, the understanding of Maxwell's demon at a more fundamental level requires solution of the measurement problem.

Other quantum phenomena might also be considered as quantum demons in the sense that they intervene in quantum measurements but can be distinguished from the Maxwell demons since they do not involve a thermodynamic cycle. One example is *Wigner's Friend*, in which an intelligent *friend* who intervenes as Wigner attempts to carry out a measurement [217]:

> *It is natural to inquire about the situation if one does not make the observation oneself but let someone else carry it out. What is the wave function if my friend looked... However, even in the case, in which the observation was carried out by someone else, the typical change in the wave function occurred only when some information (the yes or no of my friend) entered my consciousness. It follows that the quantum description of objects is influenced by impressions entering my consciousness.*
>
>

At what point in the measurement do the interference fringes of the wave function disappear, either for the friend or for Wigner: when the friend makes the observation or when he communicates the result to Wigner or when it is registered in Wigner's consciousness? And how does the thermodynamics of the friend intervene in the measurement? These are the typically recurring questions that require a theory of measurement in order to resolve them. If a demon requires measurement to open and close the hole in the diaphragm, then the resolution to the measurement problem can be expected to have an impact on the theory of thermodynamics.

Another process which could be viewed in terms of a quantum demon is the *delayed-choice quantum eraser* experiment. This originated in the continued attempt to understand the wave-particle duality as argued by Bohr and Einstein in the 1920-30s. An appropriate detector placed in one of the paths of a photon beam splitter is able to destroy the interference pattern, exemplifying Bohr's concept of complementarity. In some cases, this could often be explained as the result of uncontrollable momentum kicks to the quantum particle as quantified by Heisenberg's uncertainty principle. However, it was argued by Scully, Englert, and Walther [615] in 1991 that distinguishing paths could be accomplished solely using quantum entanglement even if no significant momentum kicks were present. The proposal resulted in heated debates in the literature. However, the concept was realized in experiments by Dürr, Nonn, and Rempe [314] in 1996 using beams of cold atoms diffracted by standing waves of light, which demonstrated that they could simply encode within the atoms the information as to which path was taken. This tagging of information is sufficient to cause the interference fringes to disappear entirely even without the presence of momentum kicks. The initially imposed atomic momentum distribution leaves an envelope pattern that was found not to be distorted at the location of the detector showing that momentum kicks cannot be responsible for destroying interference. Instead, the quantum entanglement between the atom's momentum and its internal state result in the attachment of a distinguishable atomic label to the path taken by the atom. The result is that the total atomic-plus-path wave function along one path is orthogonal to that along the others and so the paths can't interfere. However, the entanglement also results in an alternative way for distinguishing paths and destroying interference fringes.

The demonstrated experimental ability to utilize entanglement to distinguish interference paths led to a realization of the *delayed-choice quantum eraser* which had previously been proposed by Scully and Drühl in 1982 [616]. In the quantum eraser, the presence of information accessible to an observer and the subsequent *erasing* of this information, which we might view as being carried out by a demon, qualitatively changes the outcome of an experiment. Combined with John Wheeler's *delayed choice* [617] arrangement, the erasure by the demon can even take place long after the measurement has occurred. The quantum eraser is simply a device in which coherence is lost in a subset of the system but in which the coherence can be restored if a demon erases the tagging information which originally caused the interference to disappear. For a particle sent into a two-slit experiment in which one "tags" which slit the

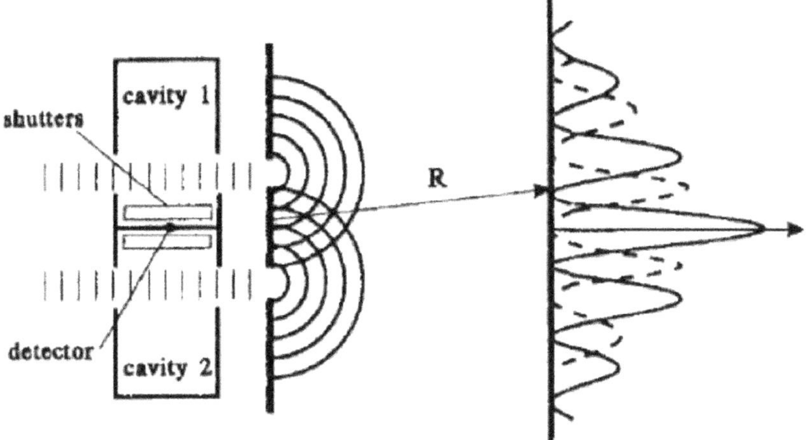

Figure 5.23: Delayed Choice Quantum Eraser. Shutters separate photons into two cavities. Detector wall absorbs photon and acts as a demon-like photodetector. Solid line with demon erasure, dashed line without [618].

particle goes through, then the interference pattern will disappear. But if one makes the which-slit tag information in-principle unobservable, the interference pattern can be restored. In an example proposed by Scully and Walther [618], Figure 5.23, atoms are detected one-by-one at the screen and interference is observed depending on the information provided by the detector in the center (i.e., the demon) between the microwave cavities. As long as this demon information still exists, the experimenter can choose to analyze his data in ways that will show fringes or not.

Preskill has proposed an explanation in [619] for which the quantum eraser is understood by realizing that Alice's state ρ_A is not the same as ρ_A accompanied by the information Alice has received from the demon. The information provided by the demon changes the physical description of Alice's particle. If Alice's state is the mixed state ensemble $\rho_A = I/2$, it is not possible for her to observe interference between states such as $|\uparrow_z\rangle_A$ and $|\downarrow_z\rangle_A$ (sum of the solid and dashed curves in Figure 5.23). However, if the demon provides her with the which-way information he has previously collected, Alice can select a sub-ensemble of her spins that are all in the pure state $|\uparrow_z\rangle_A$ which can now reveal an interference pattern (solid curve in Figure 5.23). The information from the demon can even be delayed an arbitrarily long time, resulting in a delayed-choice quantum eraser. If Alice and demon are space-like separated, there is no invariant meaning as in what order these events occurred. Alice could measure all of her spins long before the demon decides to intervene. This simply reveals that in cases where their correlations are due to quantum entanglement, the situation is completely symmetric between Alice and demon [303].

Alone in a Dark Wood

In his 1927 indeterminacy paper [7], Heisenberg describes the collapse of an electron wave function by repeated interaction with light of wavelength λ, so that "Each determination of position reduces therefore the wave packet back to its original size λ." In a letter to Bohr regarding the role played by wave function reduction in Heisenberg's 1927 indeterminacy paper, Pauli explains [4, p. 162],

> *This is precisely a point that was not quite satisfactory in Heisenberg; there the "reduction of the packets" seemed a bit mystical. Now however, it is to be stressed that at first such reductions are not necessary if one includes in the system all means of measurement. But in order to describe observational results theoretically at all, one has to ask what one can say about just a part of the whole system.*
>
> G. Bacciagaluppi and A. Valentini, Quantum Theory at the Crossroads: Reconsidering the 1927 Solvay Conference, Cambridge University Press, 2013.

Pauli's view is that an overall unitary evolution of the system coupled to the measurement device will lead to measurement. A statement by Heisenberg later in the 1950's from his book *Physics and Philosophy* suggests a similar viewpoint [620]:

> *It must be observed that the system which is treated by the methods of quantum mechanics is in fact a part of a much bigger system (eventually the whole world); it is interacting with this bigger system; and one must add that the microscopic properties of the bigger*

system are (at least to a large extent) unknown. This statement is undoubtedly a correct description of the actual situation...The interaction with the bigger system with its undefined microscopic properties then introduces a new statistical element into the description...of the system under consideration. In the limiting case of the large dimensions this statistical element destroys the effects of the "interference of probabilities" in such a manner that the quantum-mechanical scheme really approaches the classical one in the limit.

This is a point of view that has become almost commonplace in recent years as revealed in the survey of the different approaches to measurement presented in Chapter 4. This had begun to become prevalent in the decades following World War II after most research efforts became redirected and younger physicists lost touch with the views of the founders of quantum theory and the subtleties of the measurement problem. As additional examples of such views, consider Richard Feynman and Julian Schwinger, two of the most prominent theoretical physicists in the mid-20th century who, along with Shin'ichirō Tomonaga, had shared the Nobel Prize for their work in quantum electrodynamics [621]. Both viewed measurement as being a unitary phenomenon and therefore not an area of fundamental importance.

Feynman's colleague R.W. Hellwarth had said [622, p. 437],

Feynman thought that there was little of practical consequence left to be done in the theory of measurement. He certainly impressed upon his students that the so-called "collapse" of the wave function, said to occur upon measurement, follows the Schrödinger equation precisely. The subtlety in the process is the same as that in any irreversible process.

While a graduate student at Princeton, Feynman discussed these problems with von Neumann who was nearby at the Institute for Advanced Study but he was dissatisfied with von Neumann's approach. Attempting to formulate an objective statement, he had argued at the time [621, pp. 394-395],

If you could correlate the position of a piece of the world with other pieces, then if the other pieces (or their action or energy or something) would go to infinity and the correlation approaches a finite value, then that thing is measurable.

However, all of this could be addressed as an unexceptional unitary process.

Schwinger's views on measurement were presented at a Snowbird Conference, now in a record of the University of California Los Angeles archives, and he indicated that quantum measurement was a non-existent problem [623, pp. 368-369],

> *To me, the formalism of quantum mechanics is not just mathematics; rather it is a symbolic account of the realities of atomic measurement. That being so, no independent quantum theory of measurement is required—it is part and parcel of the formalism. This is not a universally held opinion, however. I quote from one recent paper: "Ordinary quantum mechanics is based on two distinct principles of evolution of the wave function. The first principle, to be applied in ordinary situations, is expressed by the Schrödinger equation, which provides a deterministic evolution of the wave function. The second principle, to be applied when a measurement takes place, is the reduction postulate, according to which the wave function undergoes a sudden stochastic evolution"... In my opinion, this is a desperate attempt to solve a non-existent problem, one that flows from a false premise, namely that of the von Neumann dichotomization of quantum mechanics. Surely physicists can agree that a microscopic measurement is a physical process, to be described as would any physical process, that is distinguished only by the effective irreversibility produced by amplification to the macroscopic level. Perhaps what has been lacking is a detailed analysis of the dynamics involved in some realistic measurements.*

The most significant development for the measurement problem following the period of the EPR discussions after 1935 were the *Bell inequalities* in 1964 and their experimental verification beginning with Aspect in 1982 [83]. Interest in the measurement problem waned with the disruption of World War II, and the subsequent focus was on new developments and applications of quantum theory including nuclear physics, elementary particles and relativistic quantum field theory, condensed matter, and low temperature physics, much of it beginning to be carried out in larger scale laboratories. The war also spurred developments of radar, missile technology, jet aircraft, computation, and nuclear technology. Governments invested in scientific research as they recovered economically from the war and support for research increased at universities and in systems of national laboratories. This affected the direction of research projects both due to the pressures of funding and due to the administrative organization of the new research institutions. The displacements of many researchers due to the war, including the founders of quantum theory, also necessarily affected research and collaborations [624, p. 1246]. Bohr escaped in 1943 from German-occupied Denmark and worked on the Manhattan Project in the United States, returning to Denmark in 1945 to continue work on quantum-theoretical problems. Schrödinger left Berlin in 1933 to Oxford and continued to migrate via Graz, Rome, Oxford, Belgium, and finally Ireland where Eamon de Valera installed him at the Dublin Institute for Advanced Studies. The pattern continued with scores of other scientists and collaborations inevitably were interrupted.

Interest in the foundations of quantum physics and the measurement problem faded with the rise of these new directions in research. Einstein and Bohr, the two deductive thinkers who had been at the focus of the measurement problem in the first half of the 20th century, had enormous stature, but young scientists coming into research were not motivated by their professors to study the foundations of quantum mechanics. Einstein died in 1955 and Bohr just seven years later in 1962, just before the Bell inequality developments. By the 1980s, the Copenhagen interpretation was summarized as [625]:

> *If I were forced to sum up in one sentence what the Copenhagen interpretation says to me, it would be "Shut up and calculate!"*
>
> Reproduced from N.D. Mermin, What's wrong with this pillow?, Physics Today, 42 (4), 9 (1989) with the permission of the American Institute of Physics.

This is largely the opposite of a rallying cry for students to investigate the measurement problem. The fundamental nondeterminism, discontinuities, and causal renunciation of Bohr, Born, and others continued to diminish in importance. New faces appeared who began to largely calculate and explain finer and finer effects, e.g., the theory of the Lamb shift for the measured shift due to the electromagnetic field interaction. Very few modern researchers appeared to fully grasp the measurement problem and often believed that the universe was described completely by Schrödinger's equation.

> *Midway in our life's journey, I went astray...*
> *and woke to find myself alone in a dark wood.*
> [Dante c.1320]

Midway in the 20th century, the state of the Quantum Measurement Problem indeed found itself alone in a dark wood. However, beginning in the 1950s there was a low-level resurgence of interest both in *interpretations* of quantum theory and in new approaches to the question of quantum measurement. The history of the period between 1950 and 1990 is well-described in Freire Junior's book, *The Quantum Dissidents* [320], giving the background of the theories of Bohm, Everett, Wigner, and others and the era of the development of Bell's theorem to the renewal of interest in quantum foundations. These efforts into measurement continued to heat up in the following decades, particularly after John Bell's (1928-1990) theorem in the 1960s allowing a test to discriminate the underlying workings of quantum theory, of Einstein's unholy choice between *hidden variables* and *spooky action at a distance*. However, the subject of measurement was also transformed in the last twenty years due to advances in experimental techniques which make it possible to perform repeated measurements, interaction-free measurements, quantum non-demolition, feedback and partial measurements on single continuously measured quantum systems. Implementation of these in physical systems has been made possible by the unprecedented level of control allowed by lasers, computers, and superconducting devices. The result is the ability to directly control and manipulate individual quantum systems and minimize undesired influences from the environment. In the process, thought-experiments are often now able to be realized involving single particle interferometers and cavities.

Starting with the resurgence of interest in measurement in the 1950s up to today, the collection of theories of quantum measurement have by now proliferated in great numbers. These include Bohmian mechanics, Relative-state or Many-worlds, GRW and Stochastic collapse theories, Consistent histories, Information-based, and Bayesian theories, among many others. These theories have been reviewed in detail in Chapter 4 along with a discussion of their capabilities for discerning the quantum measurement problem.

Are we now coming into the *Age of Quantum Measurement*? New developments devoted to measurement in both experiment and theory are constantly appearing in the journals, arXiv.org, and the proliferation of online organizations and social media platforms. Nevertheless, at this point in history, the *Quantum Measurement Problem* remains unresolved.

Appendix 5.A

Details from The Fall of Classicality

In his model of cavity oscillators, Planck was able to show a simple relation between the radiated energy density and the energy U for a single oscillator is

$$B(\nu,\mathrm{T}) = \frac{8\pi v^2}{c^3} U(\nu, T). \tag{A.1}$$

The function $U(\nu, T)$ was not known; however, Planck took a hint at how to proceed from the fundamental thermodynamic relation between entropy and energy changes in terms of the absolute temperature:

$$dS = dU/T. \tag{A.2}$$

Planck was inspired by Equations (A.1) and (A.2) to guess $U(\nu, T)$ by interpolating the second derivative d^2S/dU^2 between the Rayleigh-Jean and Wien laws via the equation

$$-\left(\frac{d^2S}{dU^2}\right)^{-1} = \alpha U + \beta U^2 \,. \tag{A.3}$$

The second term in Equation (A.3) corresponds to the Rayleigh-Jeans formula which implies $T \propto U$ at low frequencies and thus - $(d^2S/dU^2)^{-1} \propto U^2$ using Equation (A.2). The first term corresponds to the Wien formula which implies $1 / T \propto -\ln U$ and - $(d^2S/dU^2)^{-1} \propto U$ at high frequencies using Equation (A.2). Note that the full entropy obtained by integrating Equation (5.3) is of the form [412]:

$$S(U) = k\left[\log(1 + U/h\nu)^{1+U/h\nu} - \log(U/h\nu)^{U/h\nu}\right] \tag{A.4}$$

which suggested to Planck the combinatorial counting approach for the probabilities needed for Boltzmann's entropy formula with energies $E = h\nu$.

Einstein was able to show that high-frequency heat radiation carries a distinctive signature of finitely many spatially localized independent components. He confined his treatment to high frequencies since this corresponds to the Wien limit where he could treat them as independent. Firstly, he showed that the entropy change due a fluctuation process where N independent particles moving in a space of volume V_0 become confined to a sub-volume V is given by:

$$S - S_0 = kN\ln(V\,/\,V_0)\,. \tag{A.5}$$

Secondly, he compared this result with the entropy change from the Wien radiation formula for two systems of energy E occupying volumes V and V_0 of space:

$$S - S_0 = k(E/h\nu)\ln(V\,/\,V_0)\,. \tag{A.6}$$

It can be seen by comparing Equations (A.5) and (A.6) that the energy E of the heat radiation can be identified with N independent spatially localized quanta of magnitude $h\nu$, $E = Nh\nu$.

Issues in Bohr's Complementarity

Entanglement is responsible for the circumstance that "an unambiguous definition of the state of the system is naturally no longer possible." For Bohr, the "claim of causality" had meant a description in which energy and momentum are conserved. However, given Bohr's central role for the understanding of measurement, let us take a more careful look at a variety Bohr's statements regarding causality, closed and open systems, and other aspects of measurement from his statements in talks, papers and letters over the years. Though Bohr rarely invokes the "reduction of the wave-packet", he appears to relate this feature to the statistical nature of quantum mechanics in his 1927 Como lecture where had introduced complementarity [398, p. 94],

> *Due to the gradual spreading of the wave fields associated with the individuals, the statistical character of the description, however, is by no means limited to the inaccuracy expressed by Heisenberg's relations. Indeed, we are forced to contemplate a proper reduction of the spatial extension of the fields after every new observation.*
>
>

However, some thirty years later, Bohr is more explicit in his view that reduction of the wave packet is a necessary way of describing the situation due to the indivisibility of the quantum. In a letter to Pauli in 1955, Bohr explains [213, p. 568],

> *I take it for granted that, as regards the fundamental physical problems which fall within the scope of the present quantum mechanical formalism, we have the same view, but I am afraid that we sometimes use a different terminology. Thus, when speaking of the physical interpretation of the formalism, I consider such details of the procedure like "reduction of the wave packets" as integral parts of a consistent scheme conforming with indivisibility of the phenomena and the essential irreversibility involved in the very concept of observation. As stressed in the article, it is also in my view very essential that the formalism allows of well-defined applications only to closed phenomena, and that in particular the statistical description just in this sense appears as a rational generalization of the strictly deterministic description of classical physics.*
>
>

Note Bohr's emphasis here on "closed phenomena". In this passage, does Bohr imply causal behavior when he stresses "well-defined applications only to closed phenomena"? This would be supported by Bohr's use of "strictly speaking" when describing observation as a causal process in the following remark [214, p. 89]:

> *As all measurements thus concern bodies sufficiently heavy to permit the quantum to be neglected in their description, there is, strictly speaking, no new observational problem in atomic physics. The amplification of atomic effects, which makes it possible to base the account on measurable quantities and which gives the phenomena a peculiar closed character, only emphasizes the irreversibility characteristic of the very concept of observation. While, within the frame of classical physics, there is no difference in principle between the description of the measuring instruments and the objects under investigation, the situation is essentially different when we study quantum phenomena, since the quantum of action imposes restrictions on the description of the state of the systems by means of space-time coordinates and momentum-energy quantities.*
>
>

This is consistent with the possibility that Bohr believes that closed systems are causal, but because the interaction required during measurement is uncontrollable the results have an apparent non-deterministic appearance. No rigorous theoretical argument appears to be given by Bohr to justify fundamental nondeterminism versus unknowable determinism because of the necessity of measurement in forming what we know. Deterministic evolution is consistent with the requirement of loss of phase that would result due to an uncontrollable interaction, as Bohr explains in the following, [398, p. 62]:

> *Possibility for space-time description closely connected with conservation theorems. Measurement of energy or momentum with given accuracy implies loss of phase relations, which implies impossibility of interference by superposition.*

In a letter from Bohr to Pauli [398, p. 193], this concept is reaffirmed for which Bohr argued that the loss of phase that results from the measurement process limits the application of the superposition principle. It is well known that the loss of phase or a mixture also results in standard decoherence theory using deterministic evolution via Schrödinger's equation when considering system-measuring device interaction.

Does Bohr believe that the use of probability is fundamental or is being used because the underlying theory is deterministic but of such complexity that the use of probability is a good approximation? Consider Bohr's statement [494, p. 343],

> *However, it is most important to realize that the recourse to probability laws under such circumstances is essentially different in aim from the familiar application of statistical considerations as practical means of accounting for the properties of mechanical systems of great structural complexity. In fact, in quantum physics we are presented not with intricacies of this kind, but with the inability of the classical frame of concepts to comprise the peculiar feature of indivisibility, or "individuality," characterizing the elementary processes.*

Bohr claims that the use of probability is not to account for mechanical systems of great complexity. But why should quantum physics not be able to capture such intricacies? Bohr appears to believe that discontinuities were real and that nondeterminism was physical, but only when "open to direct observation" [439, p. 498],

> *Still it would appear that the essence of the theory may be expressed through the postulate that any atomic process open to direct observation involves an essential element of discontinuity or rather individuality completely foreign to the classical ideas and symbolized by Planck's quantum of action. This postulate at once implies resignation as regards the causal space-time coordination of atomic phenomena.*

The following argument appears to allow the possibility that closed systems obey

Schrödinger's equation but are fundamentally unknowable due to interaction during measurement in order to observe such systems [213][Kindle Locations 5430-5432],

> *A completely deterministic description of human conduct would require a knowledge of the so-called internal and external preconditions, which are in principle mutually exclusive and precisely in this situation should allow natural room for the immediate feeling of will.*
>
>

Similar arguments by Bohr center on the inability to determine the initial state of a system without affecting it due to the interaction with an external measuring device [310]:

> *The conception of a stationary state involves, strictly speaking, the exclusion of all interactions with individuals not belonging to the system. The fact that such a closed system is associated with a particular energy value, may be considered as an immediate expression for the claim of causality contained in the theorem of conservation of energy.*
>
>

This is an argument that again is consistent with causal or deterministic evolution for closed systems. Bohr's arguments appear to only require the condition of deterministic but unknowable evolution for closed systems. This survey of Bohr's statements therefore reveals a more nuanced view of how he may have viewed causality and other aspects of measurement in closed and open systems.

Details from The Quantum Triumvirate

The violation of Bell's theorems by quantum entangled particles can be illustrated in a very direct way by considering, instead of two nonlocal particles, three nonlocal two-level particles sharing entanglement via the special properties of the Greenberger-Horne-Zeilinger (GHZ) state [568], which in the z-basis takes the form:

$$|GHZ\rangle = \frac{1}{\sqrt{2}}(|1_z\rangle_A|1_z\rangle_B|1_z\rangle_C - |-1_z\rangle_A|-1_z\rangle_B|-1_z\rangle_C)\,.$$

These particles might be at the locations of spatially separated observers A, B, and C (i.e., Alice, Bob and Charlie) who each carry out measurements on the particle at their location using either Pauli matrix σ_x or σ_y , with the results being either +1 or -1. Following the description in Chapter 2 for treating polarized photons in terms of qubit states, these two measurements can be thought of as either representing position measurements of localized photon wave-packets (i.e., σ_x) or their counterparts after being sent through beam splitters (i.e. σ_y) [626]. A direct calculation using σ_x, σ_y and $|\mathrm{GHZ}\rangle$ finds that the results of measurements by A, B and C in the GHZ state always obey the following predictions of quantum theory:

$$\{\sigma_{Ax}\}\{\sigma_{Bx}\}\{\sigma_{Cx}\} = -1$$
$$\{\sigma_{Ax}\}\{\sigma_{By}\}\{\sigma_{Cy}\} = 1$$
$$\{\sigma_{Ay}\}\{\sigma_{Bx}\}\{\sigma_{Cy}\} = 1$$
$$\{\sigma_{Ay}\}\{\sigma_{By}\}\{\sigma_{Cx}\} = 1$$

where $\{\sigma_{Ax}\}$ represents the result of measurement of σ_x by A, etc. Note the startling implication that the measurements of A, B, and C cannot be reproduced by local hidden variables as long as they don't conspire beforehand. This can be seen by multiplying the left-hand sides of these equations with the result $\{\sigma_{Ax}\}^2\{\sigma_{Ay}\}^2\{\sigma_{Bx}\}^2\{\sigma_{By}\}^2\{\sigma_{Cx}\}^2\{\sigma_{Cy}\}^2$, which equals just +1 since each local variable must have a definite value of either +1 or -1 by EPR's argument. However, the product of the right-hand sides is -1, giving a contradiction with quantum theory. Therefore, there cannot be definite values for any of the variables $\{\sigma_{Ax}\}$, $\{\sigma_{Ay}\}, \{\sigma_{Bx}\}, \{\sigma_{By}\}, \{\sigma_{Cx}\}, \{\sigma_{Cy}\}$ before the measurement takes place even though they are all locally measureable. This result requires us to reject local determinism.

Kraus Operators

In the 1970-1980s, von Neumann's formalism of measurement was generalized from projection operators to POVM's (Positive Operator Value Measurement) [627] [49] in terms of Kraus operators M_i [14, p. 91]

$$|\psi\rangle \longrightarrow \frac{M_i|\psi\rangle}{\||M_i||\psi\rangle\|}$$

where completeness requires $\sum_i M_i^\dagger M_i = I$. For a mixed state density operator,

$$\rho \longrightarrow \frac{M_i\rho M_i^\dagger}{\mathrm{Tr}(M_i\rho M_i^\dagger)}.$$

The measurement operators M_i represent properties of the measuring device and describe its effect on the system when the i-th outcome is registered. These generalized measurements are often viewed formally in terms of a projected measurement on an ancilla Hilbert space which is unitarily coupled to the system as justified by Naimark's Theorem [46] [628]. POVM's have been particularly useful for work in quantum information as realized in quantum optical and atomic systems. Within this framework, it is also straightforward to include the effects of disorder or information loss by averaging over subsets of Kraus operators, i.e., decoherence. In this way, a variety of extensions to the formalism of measurement have been developed that enable applications to a variety of phenomena including continuously monitored quantum systems, weak measurements, partial measurements, quantum trajectories, quantum filtering, feedback control, quantum jumps and stochastic master equations [186].

Quantum Trajectories and Jumps

The term *quantum trajectory* was introduced by Carmichael [184] for the stochastic evolution of a quantum state, and this been particularly useful for various types of photodetection including homodyne detection. This allows consideration of a sequence of measurements with the time evolution of the state conditioned on the results. The limit where the system is subjected to continuous measurement can then obey a stochastic version of the Schrödinger equation. These developments were motivated in part by the refining of experimental techniques to the point of enabling quantum-limited measurements on individual quantum systems. One of the first demonstrations of the power of these methods was the observation and prediction of quantum jumps of a single trapped ion [629] [183] [630]. These modern refinements brought a realization of the early concepts of the quantum founders from the 1920-30s, allowing the possibility to actually monitor the reduction of the wave function (i.e., Schrödinger's *damned jumping)* by the measurement process on an oscilloscope screen. These experiments also reveal that quantum jumps depend on the way the atom is monitored and what they describe is the state of the atom conditioned on the local state of the measuring device. The experimenter's decision to switch from observing circular rather than linear polarization results in the photon being emitted all at once as a quantum jump rather than leaking out in a diffusive way. This is precisely Bohr's complementarity [631]. These generalizations began within von Neumann's framework but now allowing analysis that goes far beyond basic projection operators as required by the myriad of contemporary tools for probing measurement. Despite the usefulness of these later developments for measurement techniques, they still do not bring us any closer to resolving the physics of the measurement problem. As discussed in Chapter 4, the status of the measurement problem remains the same as it did in von Neumann's day except for the deeper understanding of entanglement via the Bell inequality experiments.

CHAPTER 6

Scientific Approach

The realization that Schrödinger's equation would not suffice to provide a complete description of nature did not take a long time. When Schrödinger visited Bohr after formulating his now-famous equation, Bohr argued immediately that such a deterministic equation could not be sufficient to describe the quantum of action. Quantum theory prior to the discovery of Schrödinger's equation was dominated by the quantum of action, which was generally agreed to incorporate a statistical aspect that was indicative of non-causality or nondeterminism at a fundamental level. Hence, the idea that there would be some problem with a completely deterministic equation that is devoid of any statistical formalism arose overnight. The precise problems were made much more concise by Schrödinger [119] in which the prediction of entanglement under unitary evolution was found to be different than the product states expected under measurement.

One might expect that with the rather rapid identification of this issue with Schrödinger's equation, the problem could then be rapidly resolved. However, this has not occurred. Over ninety years has passed since the problem was identified, without solution. With the mathematical and physical developments that have occurred in the twentieth century, as well as the improvements in experimental abilities to probe nature, one would not have expected such a problem to remain unsolved for so long.

The answer we believe lies partly in the approach that many have been undertaking to resolve this problem. There are two general methodologies to consider, inductive and deductive reasoning. In a deductive investigation, one eliminates the possibilities one by one, subject to known constraints, until only a single possibility is left. In an inductive investigation, one works from what is known and expands outwards. In a deductive investigation, one creates hypotheses and attempts to eliminate these based on fundamental facts.

If one examines resolutions of most major physical problems in the past, there appears to be an underlying similarity in how they have been solved. The use of deductive reasoning most often appears when some new physical theory is discovered. Often such theories are considered revolutionary. As has been discussed already in Chapter 5, most scientific revolutions were advanced through the use of scientific deduction.

Aristotle introduced the notion of deductive and inductive reasoning.

Aristotle states [353, p. 108]

> *For every belief comes either through deduction (syllogism) or from induction.* [Prior Analytics II.23]

Popper [632, p. 9] states,

> *According to the view that will be put forward here, the method of critically testing theories, and selecting them according to the results of tests, always proceeds on the following lines. From a new idea, put up tentatively, and not yet justified in any way—an anticipation, a hypothesis, a theoretical system, or what you will—conclusions are drawn by means of logical deduction.*

Newton determined his laws from observations and deduction. The same has been attributed to Maxwell. Einstein utilized deductive reasoning to find errors in the prior approaches that led to his discovery of relativity. As well, Bohr utilized deductive reasoning in many of his investigations. On the other hand, the use of inductive reasoning most often is found when there is a discovery based on current physics.

An example of an inductive discovery is Dirac's quantization of the electromagnetic field. Lamb's finding of the Lamb shift and Bethe's subsequent computation. Feynman's path integrals and the predictions of particles such as quarks are based on symmetries within the current formalism of quantum mechanics. The vast majority of recent advances in medicine, electronics, materials, chemistry, aeronautics, etc. have been using inductive methodologies. One could argue that nearly 100% of all scientific progress and discoveries after 1926 have largely been inductive.

There are many who have utilized induction in failed attempts at resolving the measurement problem. As discussed in Chapter 4, Rosenfeld's paper on utilizing Schrödinger's equation to explain measurement is one such example. We suggest that if this problem could have been solved by an inductive application of Schrödinger's equation, it would have been solved in short order after 1926. Historically, the process that has been utilized to reveal new physics has overwhelmingly been deduction. The question one might ask is then why do so many people attempt to resolve this problem using induction?

One possibility is that the process of induction is taught and engrained in today's society in schools and universities rather than deduction. The idea of attending lectures, reading textbooks, and working through exercises where an inductive outcome is expected, is a procedure that is designed to develop the abilities of induction. A student graduates with a degree in a given subject when the student demonstrates the ability to solve a problem that one might be confronted within the future in an inductive manner. That is, the graduate with a degree is expected to be able to be given an abstract problem and formulate the problem in a manner so that the graduate can apply the methodologies that have been learned in school. There is certainly nothing wrong with this approach for nearly all problems—except those that cannot be solved by the current formalism. Welcome to the Measurement Problem.

The following differences should be noted regarding the application of induction and deduction. If the problem can be resolved using inductive techniques, both induction and deduction will give the same result and lead to the resolution of the problem. If the resolution of the problem requires the development or discovery of new principles for which the current inductive theory fails, then deduction will lead to the resolution of the problem, and induction may or may not. Note that in both cases, deduction resolves the problem. In cases where the problem falls within the inductive class of problems for which a given theory is designed to handle, induction succeeds generally faster than the application of deduction, but outside such a class, induction can fail. However, deduction applied to fundamental physics clearly differs from a straightforward application of deductive logic since physics is further constrained both by the results of empirical observation and by the existing body of physical principles that have been historically established. Due to the necessary consistency with these constraints, deduction within the foundations of physics is not an automatic process that can be easily mechanized, but rather an unruly enterprise requiring bold intuition and an ability to deal with paradox, uncertainty, and contradiction.

Historically, inductive reasoning is not a good approach to apply to fundamental physics in order to discover new physical laws and/or principles. Here are potential examples that illustrate why:

1. The Greeks' belief before 700 BC that Zeus is responsible for lightning striking the earth.

2. The conclusion that the Earth is round rather than flat. If you look around you it seems flat. Generalize that and you will conclude that the Earth is flat.

3. The conclusion that because a feather falls slower than a lead weight on the Earth, that the acceleration solely due to gravity must be proportional to mass. If one simply drops a feather versus a lead weight, there is air resistance and the feather falls slower than the lead weight. Implying induction, one generalizes that this is always the case, i.e., that gravitation accelerates objects differently based on the mass. It was believed prior to the discovery of Galileo in the 1500s that gravity would accelerate an object with higher mass faster than an object with lower mass. It is said that Galileo dropped two cannonballs of different masses from the top of the leaning tower of Pisa, and to the amazement of all, both cannonballs hit the ground at the same time. The fact that gravity accelerates objects independently of their mass was contrary to the inductive conclusion that was believed prior to 1500, and Galileo utilized a deductive process to reach this conclusion.

4. The inductive generalization, that because we do not see matter converting to energy, it is impossible for it to occur. This was the belief via induction before Einstein's relativity theory. This was only accepted after the special theory of relativity was developed by Einstein with the formulation of new principles discovered via deduction.

5. The assumption that the conversion between different periodic elements is physically impossible, before the understanding of nuclear processes. Again, as this is not seen in standard chemical reactions, it is easy to incorrectly utilize induction to generalize this result to all physical processes. It was generally the belief in the 19th century that transmutation was impossible. However, this is false and it is possible via nuclear processes.

6. The assumption before the discovery of atoms, that all phenomena are due to a continuum, and individual atoms do not exist. Remarkably, this was the belief up until the turn of the twentieth century. It is interesting because different views were expressed at different times throughout history. Most philosophers took the status-quo continuum view and opposed new theoretical ideas such as Boltzmann's proposal based on the statistical mechanics of atoms. J. P. Perrin's 1908 experimental confirmation of Einstein's Brownian motion predictions along with the determination of Avogadro's constant confirmed atomic kinetic theory. At that point, most philosophers and other skeptics quickly switched sides.

7. The assumption before Einstein's explanation of the photoelectric effect, that such a threshold effect did not exist, and lowering the frequency but increasing the intensity will be able to cause photo-ionization of matter. This deductive prediction by Einstein and the experimental confirmation led the way for a particle interpretation versus a wave interpretation.

8. Because we have seen that we can predict fundamental constants with extreme precision using a combination of unitary evolution and measurement, then it must be that all physical processes obey unitary evolution.

9. Because we have not isolated any processes to date that are non-unitary, then all processes must be unitary

10. The conclusion that consciousness and free will eventually will be explained by currently known theory.

11. The reasoning that if many atomic configurations evolve via unitary evolution, that all configurations of matter are described by unitary evolution.

New physical principles may very well be discovered via the resolution of the measurement problem. Einstein believed that quantum mechanics is incomplete. In other words, Einstein believed that quantum mechanics needed to be completed. If Einstein worked at great length and could not figure out how to complete quantum mechanics, one ought to consider the use of deduction rather than the use of induction regarding this problem. Interestingly, Einstein used deductive methods for his earlier discoveries of relativity and light quanta, while his later efforts in interpreting quantum theory relied more and more on induction.

Essentially, induction is a process by which one defines a problem and then uses established principles to reach a conclusion. Insofar as the problem falls into the class of problems for which the particular theory that you were taught is correct, then induction will give the correct result. Induction is essentially a process whereby one takes an abstract problem and formulates a precise problem using mathematical or other methods for which a process exists for its solution.

Logically speaking, why not use a deductive approach? One will succeed in solving the problem if the problem can also be solved within an inductive formalism. And one will have the best chance of succeeding if the problem cannot be resolved within an inductive formalism.

Deduction is rather the reverse process. First one needs to make a set of assumptions that will be used in the investigation. Secondly, one needs to determine the logical possibilities that exist for the resolution of the problem, based on these fundamental assumptions. There generally will be a large number of possibilities. Then the possibilities need to be evaluated in terms of experimental and theoretical evidence in a manner to eliminate the possibilities until a single possibility remains. If no possibility remains, then the set of assumptions is wrong. One then needs to change the assumptions based on what has been learned and repeat the process.

Induction is an orthodox approach. One attends school and learns time-tested methodologies required to be applied in practice. In terms of physics, new results are obtained from laws already known. Essentially, the knowledge base is continually pushed outward and growth occurs. However, such a knowledge base has a fundamental limitation because it is based on a given set of axioms or fundamental laws. Suppose that a given phenomenon cannot be explained on the basis of the axioms and fundamental laws that are contained within the inductive approach. Instead suppose that the given phenomenon demands the development of new laws to explain it. Then such a problem is outside the scope of induction and is best approached using deduction.

Newton in Principia Mathematica proposed four rules regarding the pursuit of science [308, p. 205]. Rule 4 is relevant to the measurement problem:

> *Rule 4 In experimental philosophy we are to look upon propositions inferred by general induction from phenomena as accurately or very nearly true, notwithstanding any contrary hypothesis that may be imagined, till such time as other phenomena occur, by which they may either be made more accurate, or liable to exceptions.*

The phenomenon of measurement, by all means, does appear to be an exception to unitary evolution. Rule 4 of Newton states that induction should either be made more accurate or liable to exceptions. The process that is most applicable to Rule 4, when the current theory cannot be inferred by induction, is the process of deduction.

Figure 6.1: Initial configuration of Nexus of Knowledge.

Nexus of Knowledge

Many at this point perhaps have some sense of the differences between inductive and deductive reasoning, but still the concepts could be strengthened. The Nexus of Knowledge is introduced to further help illustrate several fine points of these concepts and as well the limitations and shortcomings of inductive reasoning for certain classes of problems. The Nexus of Knowledge is a graph whereby any point represents a unique configuration for an experiment. In quantum terms, any point represents a unique quantum state of all the particles that compose an individual experimental arrangement. A hypothetical example is shown in Figure 6.1 where the paths emanating from an initial condition represents a potential time evolution of the experiment. Each end point is a potential empirical outcome or result. The Nexus of Knowledge outcomes are ultimately empirically based, and there may or may not exist a theory that predicts state evolution from one state to another. One can consider any end point as a new initial condition for a branch. Paths are initially unknown and are used to designate state changes of phenomena that can physically occur, and have as yet no experimental or theoretical foundation for their prediction. As the paths are initially unknown, there may be a substantially larger number of possible paths that exist in terms of hypotheses in addition to the initial paths, and such additional

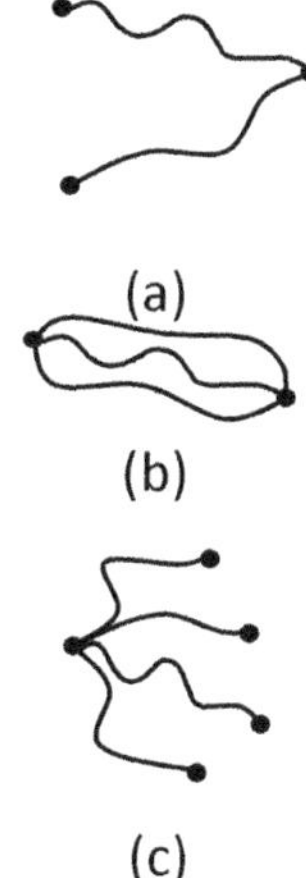

Figure 6.2: Types of state evolution (a) deterministic, (b) nondeterministic path with single final state, (c) nondeterministic outcome with multiple final states.

hypothetical paths will be eliminated as the correct paths are discovered. The additional hypothetical paths are left out, to simplify the graph.

A transition between states in a fully deterministic Nexus has a single deterministic final state but could have more than a single initial state as shown in Figure 6.2(a). A nondeterministic path Nexus illustrated in Figure 6.2(b) includes the case of

Figure 6.3: Verification of state evolution is indicated by the dashed lines.

bifurcations from a single initial state to a single final state but for which there are multiple possible paths, but only one of which is taken in any particular trial. A nondeterministic outcome Nexus is such in Figure 6.2(c) where there can be different outcomes for the same experiment beginning at a common state. Bifurcations represent nondeterministic outcomes.

When theoretical predictions, based on the current knowledge base and inductive reasoning, have been verified within some error ε that there has been observed state evolution from a given initial state to a final state, the path is changed to a dashed line as shown in Figure 6.3. Essentially dashed paths for which theoretical predictions and experimental predictions agree are indicative of a given theory being verified, or other theories that made different contrary predictions, being falsified. The fundamental scientific approach is based on such abilities and dashed paths of the Nexus reflect this, as stated in [633]:

> *Faced with difficulties in applying fundamental theories to the observed Universe, some researchers called for a change in how theoretical physics is done. They began to argue—explicitly—that if a theory is sufficiently elegant and explanatory, it need not be tested experimentally, breaking with centuries of philosophical tradition of defining scientific knowledge as empirical. We disagree. As the philosopher of science Karl Popper argued: a theory must be falsifiable to be scientific.*

Shown in Figure 6.4 is an example where the knowledge base is growing. The configurations of phenomena that are known to be understood within the current

Figure 6.4: Growing knowledge base.

framework grow with time and the dashed areas expand.

Figure 6.5: Deterministic knowledge base.

Figure 6.5 is an example of a fully deterministic knowledge base within a nondeterministic Nexus. It is known that the dashed evolution is possible although other nondeterministic possibilities have not been ruled out.

Paths represented by dash-dot pattern as illustrated in Figure 6.6 are used to denote

Figure 6.6: Theoretical predictions of state evolution that have not been empirically confirmed represented by the dash-dot pattern.

phenomena for which no exact experiment has taken place in order to test a particular arrangement, however, there exists general theory and confirmed experimental evidence for similar arrangements.

Double line paths, as illustrated in Figure 6.7, represent phenomena for which experiments or theory have called into question whether or not such phenomena can

Figure 6.7 Double line paths in the Nexus represent potential exceptions to the currently understood inductive framework.

be accurately described by the current inductive knowledge base. Such lines may actually be dashed and the experiments to date have been performed incorrectly or the theoretical assumptions did not match the experiment, for example if the theoretical modeling was too simplistic. Examples of such paths are similar to the Turing halting problem, the result of the Michelson–Morley experiment and the perihelion precession of mercury.

Paths with three lines are assigned to represent phenomena that have been verified at present to not be adequately described by proposed or current theory, as shown in Figure 6.9. Experiments are performed that prove within experimental error that a given theoretical prediction is wrong. An example of this is the Bohr-Kramers-Slater (BKS) theory which assumed that energy and momentum need not be conserved on every trial but rather on average. Compton found that experiments of electron-photon interaction conserve energy and momentum on every trial, even at the single quantum level. This invalidated the BKS theory. New theory was then needed to explain photon-matter interaction. Another historical example is the experimental Lamb Shift result showing that the weak coupling between the electromagnetic field and the atomic system needed to be taken into account in order to compute the levels of an atomic system.

Suppose there exists a theory that accurately predicts the Nexus in a particular

Figure 6.8: Dashed double line paths represent phenomena for which no a priori generalization of a theory can exist to make predictions, and one must take into account empirical evidence.

area. It may be possible that the theory can be generalized to apply to the Nexus states in a different area. On the other hand, it is possible that no generalization of a theory is applicable to other areas of the Nexus. In such a case, one must utilize empirical

Figure 6.9: Three-line paths have been verified at present to not be adequately described by proposed or current theory.

evidence to make predictions. In such cases, dashed double line paths are assigned as shown in Figure 6.8.

When there is no possibility of verifying predictions either theoretically or empirically, the path will be shown using the dash-dot-dot pattern as in Figure 6.10.

One might consider string theory as an example of a dash-dot-dot path. However, it is expected that string theory might eventually be testable in the future, see [634]. It is also possible that the amount of information and evidence available is not sufficient to narrow existing theories to a single theory or is underdetermined and multiple theories will often exist to explain a given phenomenon.

Consider the case whereby experimental results lie outside the region described by the current knowledge base. Phenomena would exist that cannot be described by the use of induction based solely on current theory. These phenomena are necessarily exceptions to what is currently understood. A new theory that explains such phenomena would be expected to be initially met with skepticism and resistance, particularly by those who prefer inductive reasoning and are unskilled in deductive reasoning. Scientists that use inductive reasoning would incorrectly classify such phenomena as those which can be understood by current theory, but the detailed predictions are generally too complex.

Let us analyze several examples of scientific processes in order to more fully understand the roles of inductive and deductive reasoning.

Example 1:

Figure 6.10: Paths are represented with the dash-dot-dot pattern when it is impossible to verify predictions either theoretically or empirically.

An engineer learns how to correctly design and accurately predict the response of the design for various electrical engineering circuits. He learns to build and test circuits in school and verifies their behavior up to a level of accuracy as shown in Figure 6.11.

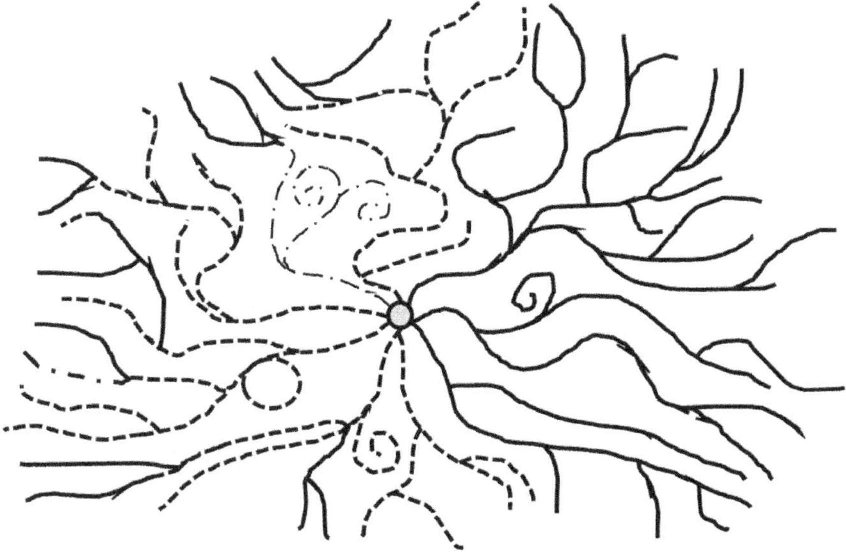

Figure 6.11: Engineer's learned knowledge base of circuits.

The engineer builds the circuit, tests it and measures the outputs, and confirms that his

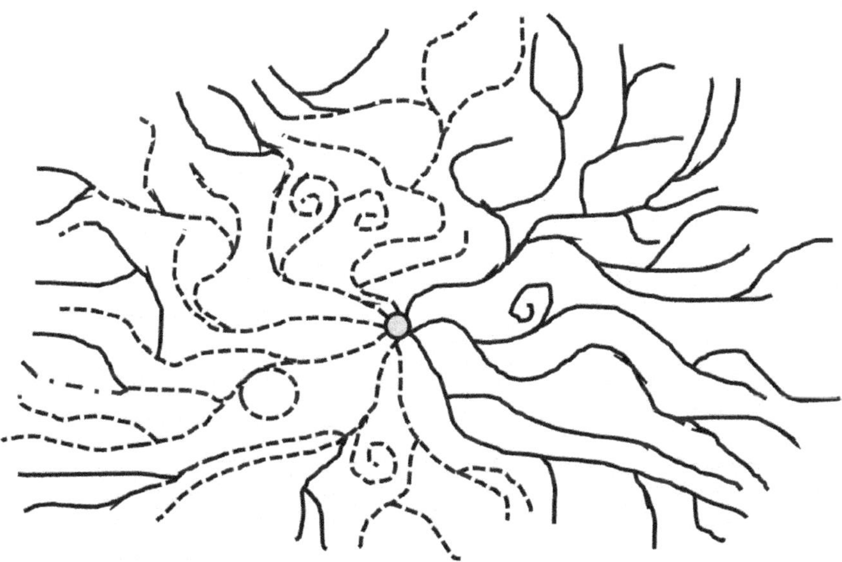

Figure 6.12: Confirmation and expansion of knowledge base when existing theory is applied to previously untested circuits and found to accurately reflect the phenomena.

predictions based on the current knowledge base are within an error ε. He is successful, and the path associated with this circuit is updated to a dashed pattern as seen in Figure 6.12.

Example 2: The same engineer is given a new circuit to design shown by the dash-dot pattern in Figure 6.13, one which has not been designed previously but is expected will work in conformity with the current knowledge base.

The engineer applies the theory and practices learned previously to design the desired circuit. It is found when the outputs of the circuit are measured, that it does not agree with expectations. These paths are changed to a double line pattern as shown in Figure 6.14.

The engineer is surprised by this finding and consults with others. They tell him that he must have done something wrong. Later, it is found that the engineer's modeling of the experiment was not sufficiently accurate. For this problem, more accurate modeling was needed and when this was done the results agreed with the theory. These paths are updated to a dashed pattern as seen in Figure 6.15. The general form of the Nexus is ultimately constrained by the principles of nature. It may be expected that a solution to the measurement problem will impose a similar structure to the physically realizable Nexus.

These examples indicate that an inductive reasoner sees the entire Nexus of Knowledge as dashed, even though it may not be. An inductive reasoner may be deceived unless the current state of knowledge is sufficiently advanced to where absolutely all phenomena can be predicted to within some error. Inductive reasoning is to largely compute or apply knowledge that one has learned.

There is generally a known methodology or approach that can be used to solve a

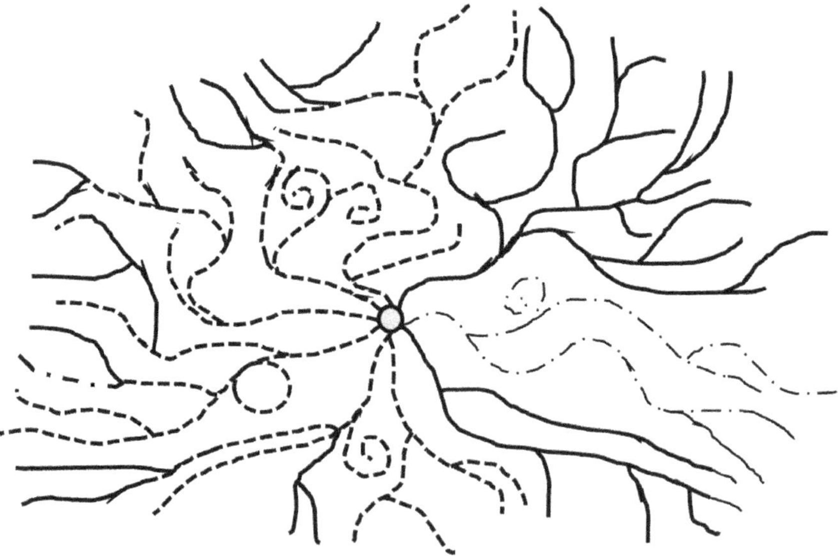

Figure 6.13: New circuit that has not been previously designed or tested is desired as shown by the dash-dot pattern.

problem inductively. Deductive reasoning can also be applied to problems that would be predicted correctly using the current knowledge base. For example, mathematical theorems are generally based on deductive reasoning and are guaranteed to be self-contained within dashed paths. In such cases, deductive reasoning will yield the same result as inductive reasoning.

When applying deductive reasoning to physical problems outside the knowledge base, the deductive solution may require rigor well beyond any inductive argument. One needs to be assured that the experimental conditions are correct to within the desired error ε needed to differentiate whether the problem is outside the current knowledge base or within the current knowledge base, but the modeling was not sufficient.

A new theoretical basis may be needed to explain the phenomena beyond anything known previously. It is necessary to verify experimentally that the new proposed theory is correct and explain why the previous theory was inadequate.

Inductive reasoning is successful when a problem can be solved or accurately approximated based on building upon existing theory and techniques. Deductive reasoning must be considered when a problem cannot be solved via induction using the current knowledge base. One can make additional assumptions but cannot be sure that the assumptions are correct. The use of deductive reasoning for cases for which the current knowledge base is insufficient, will necessarily yield a different result than what is predicted by current theory and thereby be initially considered improbable, surprising, and in certain cases even mystical.

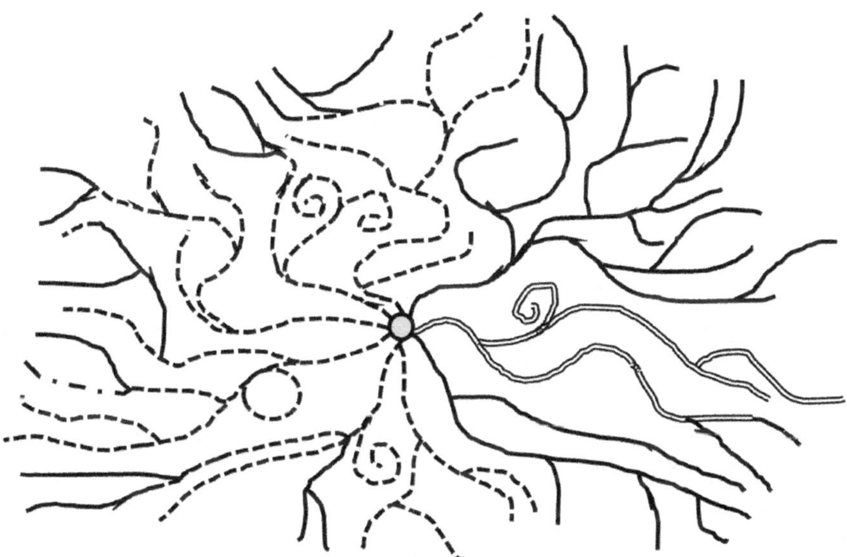

Figure 6.14 An engineer's perspective of the Nexus when it is found that a circuit has failed to produce the predicted outcome.

Deductive reasoning requires an approach in which many potential solutions are considered. Only the theory that cannot be eliminated is the correct answer. Any idea must remain "on the table" until eliminated In this case, the issues of consciousness and free will must be considered in any serious deductive attempt to solve the measurement problem.

Methodology of Deduction

So perhaps we have convinced the reader—deduction is the best approach to take regarding the resolution of the measurement problem. Now one may ask, "How does one best perform deduction?" In order to examine methods that have been used in deduction, consider comments from many of the practitioners of deduction.

Galileo [635, p. 225]:

> *Now you see how easy it is to be understood. So are all truths once they are discovered; the point is in being able to discover them.*
>
> Dialogue Concerning the Two Chief World Systems, Ptolemaic and Copernican, Second Revised edition, Galilei, Galileo, Translated by Stillman Drake, 1962, reprinted with permission of University of California Press Books.

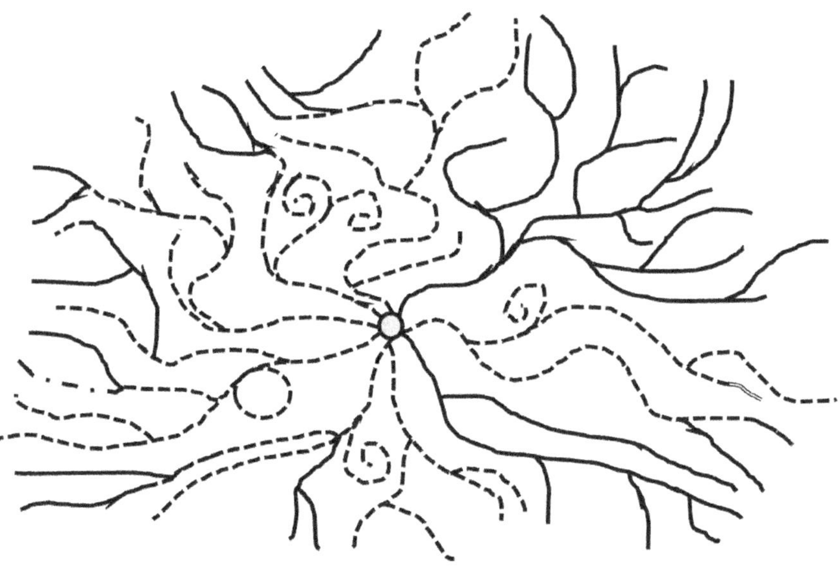

Figure 6.15: Final knowledge base of engineer after taking into account more accurate modeling.

Einstein [636, p. 221]:

> *... one comes closer to the distinguished scientific goals with a minimum of hypotheses or axioms to encompassing a maximum of empirical content by logical deduction*

Bohr [2]:

> *An expert is a man who has made all the mistakes which can be*

made, in a very narrow field.

Darwin [637, p. 28]:

I must begin with a good body of facts and not from a principle (in which I always suspect some fallacy) and then as much deduction as you please.

Maxwell [638, p. 157]:

It is by the use of analogies of this kind that I have attempted to bring before the mind, in a convenient and manageable form, those mathematical ideas which are necessary to the study of the phenomena of electricity. The methods are generally those suggested by the processes of reasoning which are found in the researches of Faraday and which, though they have been interpreted mathematically by Prof. Thomson and others, are very generally supposed to be of an indefinite and unmathematical character, when compared with those employed by the professed mathematicians.

Planck [639, p. 109]:

When the pioneer in science sets forth the groping feelers of his thought, he must have a vivid, intuitive imagination, for new ideas are not generated by deduction, but by an artistically creative imagination.

Popper [632, p. 278]:

We do not know: we can only guess. And our guesses are guided by the unscientific, the metaphysical (though biologically explicable) faith in laws, in regularities which we can uncover—discover.

Hoyle [640, p. 194]:

What happens in practice is that by intuitive insight, or other inexplicable inspiration, the theorist decides that certain features seem to him more important than others and capable of explanation by certain hypotheses. Then basing his study on these hypotheses the attempt is made to deduce their consequences. The successful pioneer of theoretical science is he whose intuitions yield hypotheses on which satisfactory theories can be built, and conversely for the unsuccessful.

Wheeler believed that the major results were a result of *radical conservatism* [542]:

> *"He was also developing an approach to physics that he called radical conservative-ism: Insist on adhering to well-established physical laws (be conservative), but follow those laws into their most extreme domains (be radical), where unexpected insights into nature might be found. He attributed that philosophy to his own revered mentor, Niels Bohr."*
>
>

Ampère [641, p. 159]:

> *By combining at random simple truths with each other, more complicated ones are deduced from them. This is the method of discovery, the special method of inventions, contrary to popular opinion.*

Ayrton [642, p. iii]:

> *The attempt to correlate all the known phenomena, and to bind them together into one consistent whole, led to the deduction of new facts, which, when duly tested by experiment, became parts of the growing body, and, themselves, opened up fresh questions, to be answered in their turn by experiment.*

Holmes [537]:

> *How often have I said to you that when you have eliminated the impossible, whatever remains, however improbable, must be the truth?*
>
> *Sir Arthur Conan Doyle, The Adventure of the Blanched Soldier, Strand Magazine, 1926.*

What can we learn from the statements from those who have successfully practiced deduction? Generally, there is agreement that the deductive method requires an approach that moves by proposing general hypotheses in a manner to test these by analogy, experiment, or otherwise and seeks to eliminate hypotheses to eventually reach a narrow or particular conclusion. The process is one of making errors and this is reaffirmed by Bohr's statement that, "An expert is one who has made all the mistakes which can be made."

It is certainly not desired to make *all* the errors that can be made, for there could be an astronomical number of such potential errors, but rather to develop a deductive process in which the number of errors is minimized. Unlike induction, there does not appear to be a concise methodology that has been laid out for successful application of deduction. Ampère recommends combining, at random, simple truths. According to

Hoyle, deduction requires intuition or other inexplicable inspiration and appears to Maxwell as a process of indefinite and unmathematical character. This is reaffirmed by Planck who says that artistically creative imagination is required. These statements indicate that hypotheses are developed through intuition and then combined in various manners and tested.

It is therefore desirable to determine what initial hypotheses should be used. Here Hoyle states, "The successful pioneer of theoretical science is he whose intuitions yield hypotheses on which satisfactory theories can be built, and conversely for the unsuccessful." Hence it appears that the initial hypotheses can be critical for the successful development of a theory. Suppose that one lacks intuition to go forward and proposed the initial hypotheses. The statement by Wheeler of radical conservatism, that is to be conservative by sticking to well-established physical principles, but probe them by exposing their most radical conclusions, should be considered. That is, consider initial constraints that are well-established principles and consider radical possibilities that can resolve the deductive problem for which the established principles are maintained. An example of a well-established principle is no-signaling whereby information cannot be transmitted faster than the speed of light.

A point regarding scientific revolutions noted by Kuhn [1, p. 52] is that,

> *Discovery commences with the awareness of anomaly, i.e., with the recognition that nature has somehow violated the paradigm-induced expectations that govern normal science. It then continues with a more or less extended exploration of the area of anomaly. And it closes only when the paradigm theory has been adjusted so that the anomalous has become the expected.*

The measurement problem does appear to be along the lines discussed by Kuhn, in that unitary evolution, while apparently sufficient to explain a large number of phenomena, is not sufficient to explain measurement and thereby lies the anomaly.

Mathematics is a primary tool in induction. In scientific deduction, mathematics is *not* a primary tool but a secondary tool. Hypotheses, formulation of Gedanken or thought experiments, and logic are primary tools in deduction. Once the problem is fully understood, and hypotheses are formed, one can begin to investigate and narrow the possibilities which would form the basis of a theory. Certainly, it is conceivable to consider all the possibilities and all the different theories, one-by-one. However, this is rather inefficient if there are a large number of possibilities, and in this respect, it is suggested that the problem be further broken down. Questions regarding the theory and/or properties that need to be tested should be defined for each possibility. Some properties will be common to more than one theory, and hence by testing or investigating properties that distinguish the most theories initially will lead to the fastest resolution of the problem. We denote a *deductive decision tree* by a diagram where each fork with two paths represents a binary question, and where the depth of the tree is m, as shown schematically in Figure 6.16. One sees that the number of

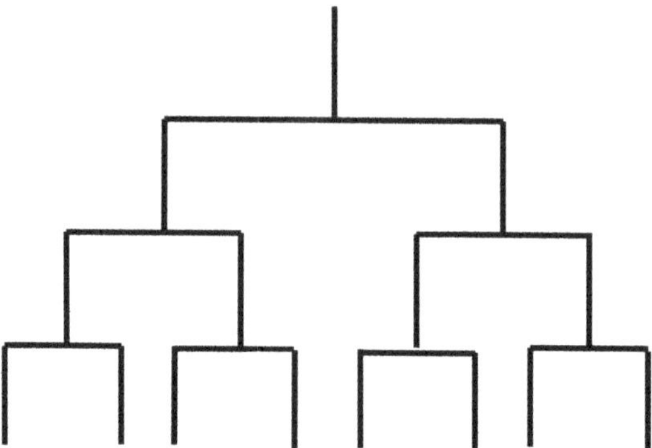

Figure 6.16 Deductive decision tree.

possibilities that would need to be considered from the outset increases exponentially with m. It would be typically more efficient to eliminate possibilities starting from the top and working downward. This can reduce 2^m possibilities to a single correct solution in only m steps. Hence it is desirable to begin by investigating properties or questions as high as possible on the tree, that can be evaluated by theoretical and experimental means in a reasonable time period. That is, if the question that defines the first fork is too difficult to currently answer, move down the tree until one can address those questions that are currently within reach. Each question that is answered in the lower parts of the tree will provide new results that had not been known in the initial stages of the investigation. One will then utilize all such new results to aid in moving up the tree.

In this sense, these first questions that are addressed might be considered as the most important as they eliminate the most possibilities. Formulation of the questions should proceed carefully in order to efficiently break the problem down. The good news is that if one gets an answer incorrect, then eventually there should be contradictions as one works down an incorrect path. This will eventually lead one back to the correct path. The bad news is that this is inefficient as in the worst case it would require to reach contradictions on all the incorrect paths in order to deduce where the incorrect path was taken. Another pitfall is that if one starts too high on the tree and reaches a wrong conclusion, then it may take a substantial amount of time to recover from this, if ever. Hence the saying, "don't bite off more than you can chew" is particularly important when addressing the first questions.

Summary

In summary, one might consider the following steps to aid in the solution of scientific problems:

1. Develop a comprehensive understanding of the problem before attempting to solve it.

2. If there is little or no reason to believe that there will be experimental contradiction when applying current theory, approach the problem via induction. If the problem is posed in terms of potential experimental contradiction with current theory, approach the problem via deduction.

3. Apply Bohr/Wheeler radical conservatism in developing a set of general assumptions from time-tested fundamental laws that constrain the solution of the problem. Do not consider assumptions that are particular to the problem at-hand, but rather have also been found to be applicable to a more general class of problems.

4. Develop a deductive decision tree for which the problem can be further broken down or subdivided into to a set of binary questions. Those questions that fall below a given fork are contingent upon that path being correct. If a given question can be answered, i.e., yes or no, then all other paths below the one that is incorrect are eliminated.

5. Based on the decision tree, develop Gedanken experiments to answer those questions as high as possible on the tree, but that also can be expected to be evaluated in a reasonable time based on current theoretical and experimental analysis. Utilize analogies with other phenomena and apply creativity, imagination, and insightfulness in developing hypotheses that would explain the Gedanken experiments.

6. Utilize new results obtained when resolving forks at a given level in the decision tree to aid in moving up the tree. Remove or eliminate possibilities as new results are obtained based on the formation of new Gedanken experiments and the application of theoretical and experimental analysis.

7. Always remain flexible in a deductive investigation—do not eliminate any particular proposal or idea from the solution set because it appears to be incorrect based on induction. All ideas remain on the table in a deductive investigation, until disproven.

Personality Traits

One might consider the personality traits of those who are expected to excel at induction versus deduction. Such traits could be applied to consider whether someone who excels at induction would also excel at deduction.

Induction tends to give answers and results much faster than deduction. If you are quick witted but quickly get inpatient when others are slower, then induction is for you. If you get angry if anybody would question your work, then induction is for you.

If you have trouble admitting to your errors, then induction is for you. Those who desire a hierarchical lifestyle whereby one is not interested in questioning orders but rather desires to follow orders, then induction is for you. If you prefer to work on problems that have a clear-cut methodology for solving them, then induction is for you. If you have a good amount of discipline to work through a problem for a short time, but after a while with no results prefer to give up and work on another problem, then induction is for you. If you are in a meeting and a large majority of the scientists express a particular opinion, you generally take the side associated with the majority inductive opinion, particularly if you know that you will be criticized if you don't, then induction is for you.

Deduction is a process of trial-and-error where intuition and creativity reign supreme. One goes down a road knowing that it most likely will lead nowhere. However, deductive thinking also requires responsibility as well as a deep understanding of physics in order for such a process to successfully lead to a solid piece of science. If you prefer to deeply think through problems before giving a definite answer, then deduction is for you. If you have excessive patience on working through problems, then deduction is for you. If you can't sleep until a problem is correctly solved, then deduction is for you. If you don't mind making and admitting to errors, then deduction is for you. If you desire at times to be excessively correct, rather than be considered by others to be correct, then deduction is for you. If you have the discipline to work on a given problem for years rather than weeks, then deduction is for you. If you don't allow what others believe or expect to stand in the way of your work, then deduction is for you. If you are in a meeting and a majority of the scientists express a particular inductive opinion, you generally take the side associated with what you believe, then deduction is for you.

Those who will take on inductive problems generally will be working through the problem in agreement with time-tested methodology and often with the support of the majority. Those that work on inductive problems, will generally have a highly acceptable manner of coping in society. A primary tool of inductive thinking is to formulate an abstract problem into mathematical equations and apply known methods for their resolution. Such methodology can be improved and perfected to a large degree.

Based on these personality traits, consider now whether someone who excels at induction would also excel at deduction. This seems doubtful as the personality traits of a highly successful inductionist are in many instances contrary to the useful personality traits of a highly successful deductionist. It is certainly conceivable that a person that is both a great inductionist and a great deductionist arises, but more likely an individual will excel at either induction or deduction, but not both. It is entirely possible that those who are the strongest at resolving problems via induction, find themselves at a complete loss when solving problems demanding deductive skills.

The Backlash of Society

Those that work on deductive problems will utilize a methodology that is foreign to the multitude of inductionists in society. Methods of Gedanken experiments rather than mathematics are a primary tool. Error, rather than success, is a typical outcome. The optimal methodology of deduction is illusive and is not easily mastered. One is generally working in a minority on a deductive problem, in nearly all stages of development except after the theory is either accepted or rejected. Those that work on deductive problems will generally have less than an acceptable manner of coping in society. History shows that one should be prepared and fully expect a substantial backlash from society when proposing a new theory when the methodology that was utilized is deductive. The status quo is dominated by inductive thinkers that you should expect will put forward stinging criticisms of any work that challenges the scientific foundations—as they understand them. Most who have worked on deductive problems have encountered this fact, including Bohr, Boltzmann, Einstein, Galileo, and Planck. Of course, these are some of the great names in science who had the necessary ability and deep understanding of physics required to carry out deduction correctly. These cases should not be confused with other forgotten names whose faulty attempts at deduction resulted in simply incorrect (or even 'crackpot') theories and who encountered stinging criticisms for very different reasons. Both the practitioner and the method must be held to very high standards to succeed in deduction.

Consider statements from some of the great deductionists.

Galileo stated [643, p. 134]:

> *It is surely harmful to souls to make it a heresy to believe what is proved.*

Einstein stated [644, p. 149]:

> *Great spirits have always encountered opposition from mediocre minds. The mediocre mind is incapable of understanding the man who refuses to bow blindly to conventional prejudices and chooses instead to express his opinions courageously and honestly.*

Planck stated [1, p. 151]:

> *A new scientific truth does not triumph by convincing its opponents and making them see the light, but rather because its opponents eventually die, and a new generation grows up that is familiar with it.*

Kuhn [1, p. 6]:

> *For the far smaller professional group affected by them, Maxwell's*

equations were as revolutionary as Einstein's, and they were resisted accordingly. The invention of other new theories regularly, and appropriately, evokes the same response from some of the specialists on whose area of special competence they impinge.

It has also been noted [645, p. 10] that Boltzmann was surrounded by critics that were opposed to his atomic hypothesis.

Note that Bohr, Boltzmann, Einstein, Galileo, Planck, and other deductionists have generally met with substantial resistance when proposing ideas that conflict with existing ideologies or dogmas. If you do not have the stomach for such societal backlashes, think twice before working on a problem that by all logical standards is best approached via deduction.

Exercises

6.1 Characterize the following in terms of best approached via an inductive versus deductive approach:

a. Discovery of Plank's quantum of action

b. An Architect's plans for a small single bedroom home

c. Discovery of Quarks by Gell-Mann

d. Atomic hypothesis by Boltzmann

e. Proof of Fermat's last theorem (1995)

f. An Architect's plans for the largest skyscraper ever built

6.2 Consider the following two statements by Maxwell in 1875 and Dirac in 1929:

Maxwell [399, p. 420]:

It is only a small part of the theory of the constitution of bodies which has as yet been reduced to the form of accurate deductions from known facts.

Dirac [566]:

The underlying physical laws necessary for the mathematical theory of a large part of physics and the whole of chemistry are thus completely known, and the difficulty is only that the exact application of these laws leads to equations much too complicated to be soluble.

a) Is Maxwell's statement an inductive or deductive statement?

b) Is Dirac's statement an inductive or deductive statement?

Assume that both Maxwell's statement and Dirac's statement are true. Conclude that the theoretical knowledge base of physics of mankind jumped from only a low level to knowing all physical laws in the span of 53 years.

c) How many experiments do you expect were conducted on the physics of the measurement process between 1875 and 1929?

6.3 It is stated in [633]:

> *Fundamentally, the multiverse explanation relies on string theory, which is as yet unverified, and on speculative mechanisms for realizing different physics in different sister universes. It is not, in our opinion, robust, let alone testable.*

How should quantum state transitions be diagrammatically assigned within a Nexus that are described by multiverses?

6.4 Consider the following questionnaire to determine if you are likely to perform well at deduction versus induction.

Assign numbers to all questions as follow:

0 = strongly agree, 5 = no particular opinion, 10 = strongly disagree

a) Consider the sequence 1, 3, 5, 7, 9, 11, 13, 15, x. From the sequence it follows that, x=17._______

b) The underlying physical laws of physics and chemistry are completely known, and the difficulty is only that the exact application of these laws leads to equations much too complicated to be soluble. _______

c) All physical theories are based on laws that have governing equations, otherwise they are not physical theories and should be relegated to philosophy. _______

d) Any resolution of the measurement problem cannot have any fundamentally new predictive power over current theory. _______

e) Rose purchased a pack of bottled water and placed it underneath the shopping cart. It was raining and when she got to her car, she hurriedly transferred the items from the cart to the car trunk and forgot the water. Jane arrived ten minutes later and found a cart with bottled water underneath a shopping cart. The bottled water she found belonged to Rose. _______

f) Unsolved problems such as the lack of an explanation for free will, the lack of a current physical theory that explains why conscious feelings exist and what they are, are problems that are very complex to model but of course will eventually be explained by current theory. _______

g) David and John headed two separate research groups and both were racing to discover a cure to a fast spreading disease. David's group discovered a cure and published it in February 1980 and John's group discovered a cure and published it in January 1980. In 2012 David was asked by a journal to write a general survey paper on this disease and cure. David cited many peoples' work including his own work in January 1980 but did not cite John's. One can conclude that David was aware of John's work but for some reason did not cite John's work. _______

h) Eventually computers may become so sophisticated that it becomes possible to very accurately model single particles and to simulate the motion of particles that compose conscious systems. Suppose that an experimenter is able to input identical queries to both such a simulation and the conscious system being simulated, with the task of distinguishing the simulation from the conscious system (Turing Test). Assume that the evolution of the particles is simulated in accordance with currently understood quantum theory and that the Hamiltonian of the conscious system and inputs received from the environment are known. The experimenter can never achieve his objective because it is known that the simulation and the conscious system are both governed by orthodox quantum theory for which statistical outcomes can always be simulated given the full Hamiltonian. _______

i) Sam was working in a construction project and fell off a scaffold twenty stories high. Sam was quickly taken to the hospital but he died three hours later. Sam would have lived had he been more careful and not fallen from the scaffold. _______

j) Because experiments to date have largely validated the use of current theory, there is no need for investigation of theories that would contradict current theory. _______

Tally up your responses. If you scored between:

100 Seasoned Deductionist, Sherlock Holmes
70-99 Assistant Deductionist, Dr. Watson
50-69 Greenhorn Deductionist, Inspector Lestrade
<50 Dyed in the Wool Inductionist—Danger. You are going to create confusion and spread incorrect dogma. For heaven's sakes, do us all a favor and STAY AWAY FROM THE MEASUREMENT PROBLEM!

CHAPTER 7

Closed and Open System Approaches

One of the reasons that the measurement problem has not yet been solved is that the number of potential solutions to the problem is large. This is because there is a lack of known constraints on the problem, which essentially allows a large number of theories to be put forward. One should strive to reduce the class of potential solutions through the discovery of new constraints that the solution to the measurement problem obeys.

The application of potential constraints depends partly on whether or not the solution is ultimately an approach for which a closed system of particles constitutes a measurement device, or instead the solution is an open systems approach for which an external action is required on a set of particles in order for the particles to function as a measurement device. For both cases, constraints will be examined beginning with those constraints that can be considered regardless of whether or not a measurement device constitutes a closed system.

In those physical instances for which a large number of physical constraints were to exist that narrow state evolution to a single evolution law, then unitary evolution would be a reasonable law to consider. We begin by examining why Schrödinger's equation does provide a compelling rationale to explain evolution when a given system is overly constrained in such a manner by physical laws.

Schrödinger's Equation

We know that Schrödinger unitary evolution evolves an eigenstate $|\psi_e\rangle$ of the Hamiltonian by a time varying phase change given by $e^{i\lambda_e t}|\psi_e\rangle$ where λ_e is the eigenvalue of the Hamiltonian corresponding to the eigenstate $|\psi_e\rangle$. λ_e is the energy when the state is in such an eigenstate and the Schrödinger equation dictates that such eigenstates evolve with a complex phase that progresses proportional to $\lambda_e t$.

There is a substantial body of theoretical and experimental evidence that indicates many systems of interest, when they do evolve unitarily, evolve according to the specific unitary evolution found via Schrödinger's equation. This subject will be further discussed in substantially more detail within the next volume of the Progress on the Physics of Quantum Measurement series.

Nonlinear Wave Function Theory

There are many constraints that can be considered regardless of whether or not a closed system constitutes a measurement device. Consider the imposition of the results of Chapter 3 for which unitary Schrödinger evolution cannot account for the results of measurement.

From the results of Chapter 3, we know that a measurement process exists and this process is necessarily non-Schrödinger. The reasoning behind the result of Chapter 3 for which unitary Schrödinger evolution cannot account for the results of measurement was due to the dichotomy of the entanglement prediction under unitary Schrödinger evolution versus the product state under measurement. Hence a unitary operation that is not a Schrödinger unitary operation would have to be capable of resolving this dichotomy. However, it is easy to see that any linear map, be it unitary or non-unitary, cannot resolve the dichotomy required to result in a single product state for all superposition input states. That is, for any linear map, the same issue of entanglement versus product state will arise as in Chapter 3, and this will again lead to a measurement problem. Hence there only remains the consideration of non-linear wave function operations.

Non-Linear Wave Function, Linear Density Operator Evolution

Generally, the solution to the measurement problem lives in the class of non-linear evolution in wave function. However, linearity in density operator is preserved in conventional quantum mechanics in both the deterministic unitary postulate of von Neumann as well as the nondeterministic measurement postulate. In both cases of conventional quantum mechanics and as well due to the no-signaling arguments by Wald in [646], Gisin in [251], and further extended in [647] [648], there is good rationale for taking quantum dynamics to be described by linear density operator evolution. This will be further discussed in a future volume in the Progress in Physics of Quantum Measurement Series.

Completely Positive Maps

Consider a Hilbert space $\mathcal{H}_A$ and a linear map $\mathcal{E}(\varrho)$ where ϱ is a density operator on $\mathcal{H}_A$. If $\mathcal{E}(\varrho)$ is positive for all density matrices ϱ, then $\mathcal{E}$ is defined as a positive map.

Suppose a density matrix ρ_{AB} of a bipartite system A, B with Hilbert space $\mathcal{H}_A \otimes \mathcal{H}_B$ is acted upon locally on A, i.e. via an operation $\hat{\rho}_{AB} = (\mathcal{E} \otimes I)\rho_{AB}$. A positive map $\mathcal{E}$ defined on $\mathcal{H}_A$ is a completely positive (CP) map for which $\hat{\rho}_{AB}$ is also positive on $\mathcal{H}_A \otimes \mathcal{H}_B$ for all ρ_{AB}. A trace preserving linear map has a Kraus sum-operator representation if and only if it is completely positive (CP). If a trace preserving linear map $\mathcal{E}(\varrho)$ is also completely positive, then $\mathcal{E}$ is a quantum channel.

Note that when the initial state is a tensor product $\rho_{AB} = \varrho_A \otimes \varrho_B$, it is seen that

$(\mathcal{E}\otimes I)\rho_{AB} = \mathcal{E}(\varrho_A)\otimes\varrho_B$ is always positive since $\mathcal{E}(\varrho_A)$ is positive. Additionally, if one considers a unitary operation $U\rho_{AB}U'$, it is found that if one restricts initial states to be tensor product initial states, such restricted initial state maps also have a Kraus representation for the reduced dynamics corresponding to Markovian dynamics. One might also consider initially correlated states. In the presence of initial correlations, such a map may not exist and the dynamics may not be Markovian. However, state dependent Kraus operators always can be found as a function of the initial condition and the unitary evolution operator [649]. Further issues of non-Markovian evolution have been explored in [650] [651].

For certain correlated initial states, non-CP maps may be physically relevant, but they found additional restrictions were needed [652]. The authors state:

> *Correlated initial states, and therefore non-CP maps, may be expected to be needed as the methods of quantum information theory are applied to systems as diverse as excitons in the condensed phase nuclear spins in semiconductors, and quantum optical systems, all exhibiting non-Markovian behavior. When such open system evolutions can be modeled by linear subsystem dynamical maps, these maps will not typically be CP.*
>
> Quantum Information Processing, Beyond complete positivity, Vol. 15, 2016, p. 1349 , J. M. Dominy and D.I A. Lidar
>

It has also been shown in [653] that there exist linear density maps that describe the evolution of a subsystem that is initially entangled with another system for which the linear map is not completely positive. Hence it appears that certain relevant physical open system maps can be linear, and may not be CP.

A point argued in [647] is that the general form of quantum state evolution, whether unitary or measurement, is completely positive. The argument is essentially that if one allows entangled states on $\mathcal{H}_A \otimes \mathcal{H}_B$ and a non-CP local operation on B, that there would be cases for which the overall state would be not positive, which would contradict the basic tenants of positive probabilities upon which von Neumann theory, as well as probability theory, is predicated. This is true, if all states are allowed to occur in a Hilbert space.

On the other hand, a characteristic of measurement is that macroscopic superpositions are not always seen. The measurement problem requires one to explain why the entangled states that are predicted under unitary evolution are not occurring. It may be a valid induction that entangled states can occur, but at least from what has been established at this point, it is not a valid deduction based on the evidence of the characteristics of the problem for which entangled superpositions are not seen. One might argue that such states could occur, but that they are impeded by some physical reason such as Penrose's gravitational reduction and such impedance reduces such states to a product state. That is certainly one possibility. Another logical possibility that must be considered in a deductive investigation is that certain states can never occur. And it has been shown in [648] that if one assumes linear density operator evolution and that density matrices are mapped one-to-one and onto density matrices, then the resulting evolution is unitary. Hence it seems the inability of certain states to ever occur is a feature that is directly related to von Neumann measurement and the existence of a process that is different from unitary evolution. Due to the fact that measurement is not a one-to-one and onto process, certain states are excluded from occurring by the formalism itself, and hence the issue of whether or not all operations are CP, particularly those in the process of measurement, does not appear to have been properly addressed, as of yet.

There still exist some questions in the literature as noted in [652]:

> *Despite much recent attention and progress, the situation for more general initial conditions (e.g., for families of thermal states) is not well understood. Is completely positivity indispensable? Is there a consistent framework for the dynamics arising from general initial conditions?*

It had been claimed [654] that the equation for wave function collapse is necessarily CP under the assumption of linear density operator and Markov evolution. If this were the case, then CP would certainly form a natural addition to linear density operator and the Lindblad CP form could be taken as a deductive assumption for the general form of quantum evolution. However, the claim in [654] was shown by counterexample to be false by Diosi [655] who exhibited a Markov process that is not CP.

As in the case of linear density operator, the case of CP evolution at this time does appear to be a compelling argument when restricted to theories in which all superposition states can occur. Until such time as it is ruled out, CP maps appear to be a potential candidate to describe quantum evolution in the case of theories that strictly rule out certain superposition states from physically ever occurring. In the case of the latter, it is known that there are cases where superpositions of certain eigenstates never occur, and the imposition of superselection rules is required. For example, pure superposition states of different charge are prohibited. In such cases, the possibility of non-CP maps should also be considered until such time as the matter is more thoroughly understood, and before reaching firm conclusions in a deductive investigation that could very well rule out a non-CP theory that could turn out to be the correct theory.

Theory and Classification of Measurement Operations

There exist numerous categories of measurement operations [656] [186] that have been considered in the literature. Whether or not such operations are possible in terms of actual measurement is unknown, and the successful resolution of the measurement problem is expected to provide further insight into this problem. In this section, some of these classes will be reviewed. It is sometimes found that different authors use the same term for different effects. These will be differentiated when found.

A projection operator P is a linear idempotent transformation, i.e., one for which $P^2 = P$. A projection operator has the property that repeated application of P to $P|\psi\rangle$ does not alter $P|\psi\rangle$. Note that all rank 1 operations of the form $|\psi\rangle\langle\psi|$ are projection operators, but a projection operator need not be rank 1.

A POVM of N possible outcomes is specified by a set of positive self-adjoint (Hermitian for finite dimensional matrices) operators of the form $E_k = {M_k}^\dagger M_k$, $k \in \Omega = \{1, \cdots, N\}$ with the properties

$$\sum E_k = I,$$

where $\langle\psi|E_k|\psi\rangle \geq 0 \ \forall \, |\psi\rangle \in \mathcal{H}_A, E_k \in \mathcal{B}(\mathcal{H})$, and $\mathcal{B}(\mathcal{H})$ denotes the set of bounded linear operators on the Hilbert space $\mathcal{H}$. When the k th result occurs, the density operator ϱ is transformed according to

$$\mathcal{E}_k(\varrho) = \frac{M_k \varrho {M_k}^\dagger}{p_k} \tag{7.1}$$

with probability $p_k = \mathrm{Tr}\left[M_k \varrho {M_k}^\dagger\right] = \mathrm{Tr}[\varrho E_k]$. We define POVM observables via a set of positive operators $\{E_k\}$ satisfying $\sum E_k = I$, $E_k \in \mathcal{B}(\mathcal{H})$. Aside from the possibility of non-Hermitian decompositions M_k, $E_k = {M_k}^\dagger M_k$ one can always find a decomposition $E_k = {A_k}^2$, where A_k is any square root of E_k. Note that A_k are not necessarily unique and the map $\mathcal{E}(\varrho)$ may not be unique for a given set of operators E_k. When the A_k are further constrained to be positive semi-definite, A_k can also be shown to be unique and will be denoted $A_k = {E_k}^{1/2}$.

We note that there exists a more general representation of $\mathcal{E}_k(\varrho)$ for which $\mathcal{E}_k(\varrho) = \sum_i p_k \mathcal{E}_{k,i}(\varrho)$ that are defined as *inefficient* in [186, p. 32]. In such cases a pure state is generally transformed to a mixed state for a given outcome. Although this is possible mathematically, the authors are not aware of any physical measurement that has been shown to be necessarily caused by an inefficient non-unitary action, and hence the question of whether or not physical measurement operations allow such measurements is an open question. On the other hand, experimental evidence does support the form of Equation (7.1). Henceforth, all measurements will be assumed to be of the form of Equation (7.1). The density matrix that is averaged over all measurement results is often referred to as Kraus decomposition or operator-sum form:

$$T(\varrho) = \sum M_k \varrho {M_k}^\dagger. \tag{7.2}$$

Note that for a given set of POVM elements E_k, the transformation of the density matrix is not unique, until a particular decomposition is considered of the E_k elements that are to comprise the operators in the transformation of the density matrix. Hence there can be different implementations of a given set of POVM elements E_k. Projection operators P_k that have a non-trivial degenerate subspace transform the density matrix according to an Ansatz by Lüders that corrected a deficiency in von Neumann's original treatment. A Lüders [657] transformation occurs when the density matrix transforms according to:

$$\mathcal{E}_k^L(\varrho) \equiv \frac{P_k\, \varrho P_k}{P(k)}. \tag{7.3}$$

A general transformation of ϱ may or may not be a Lüder's transformation. We denote the average density matrix given a Lüder's instrument as $T^L(\varrho)$. Consider the following which can be used to consider a von Neumann Hermitian observable A (which is not necessarily a positive matrix) as a POVM. A Hermitian matrix has a spectral decomposition. $A = \sum_i \lambda_i P_i$ where the P_i are projection operators and λ_i are the unique eigenvalues of A. The probability of the k th outcome of the observable A is given by $\mathrm{Tr}(\varrho A_k)$. A POVM with elements $\{A_k\}$ would yield the correct probabilities of the observable. However, not only does the probability of an observable need to coincide with von Neumann's postulate, but the state transformation needs to coincide with the projection onto the eigenstates of an observable. The POVM defined via Lüder's instrument of Equation (7.3) suffices to implement a von Neumann Hermitian observable.

It has been noted previously that in defining POVM positive operators $\{E_k\}$ satisfying $\sum E_k = I$, $E_k \in \mathcal{B}(\mathcal{H})$, that the map $\mathcal{E}(\varrho)$ may not be unique for a given set of operators E_k. When applying the decomposition $E_k = {E_k}^{1/2}{E_k}^{1/2}$, we will refer to the resulting implementation as a generalized Lüder's instrument. The density matrix when the k th outcome occurs is given by

$$\mathcal{E}_k^L(\varrho) = \frac{{E_k}^{1/2}\, \varrho {E_k}^{1/2}}{P(k)}, \tag{7.4}$$

where $P(k)$ is the probability of the k th outcome. An adjoint map of $\mathcal{E}_k(\varrho)$ that maps observables, denoted $\mathcal{E}_k^*(A)$: $\mathcal{B}(\mathcal{H}) \to \mathcal{B}(\mathcal{H})$, $A \in \mathcal{B}(\mathcal{H})$, can be defined via the representation for which the operator is transformed and the state ϱ is constant. In analogy to the unitary case for which the $\mathcal{E}_k^*(A)$ is the Heisenberg representation and $\mathcal{E}_k(\varrho)$ the Schrödinger representation, the probability of measurement must be the same, which is given by the duality requirement that $\mathrm{Tr}(A_k\mathcal{E}_k(\varrho)) = \mathrm{Tr}(\mathcal{E}_k^*(A_k)\varrho)$.

A POVM observable $A = \{A_k\}$ will be said to be commutative if $[A_i, A_j] = 0$, $i \neq j \in \Omega_A$ where Ω_A is the outcome space for A. Two POVMs $A = \{A_k\}$, $B = \{B_k\}$ with outcomes in Ω_A and Ω_B respectively are said to commute if $[A_i, B_j] = 0$, for all $i, j \in \Omega_A \times \Omega_B$, where $[A, B]$ denotes the commutator between A and B. A POVM observable A has rank 1 if $\mathcal{R}(A_i) = 1, \forall i$, where $\mathcal{R}(B)$ denotes the rank of the matrix B.

A joint measurement in its most general form is a single measurement $C = \{C_{i,j}\}$ that has outcomes in the product space $i, j \in \Omega_A \times \Omega_B$. Note that sequential measurements such as A followed by B is a joint measurement, but not all joint measurements need be sequential measurements.

Given a joint measurement C, the marginal observables are computed as i.e. $A = \sum_j C_{i,j}$, and $B = \sum_i C_{i,j}$. Two POVM observables A and B are said to be jointly measurable if there exists a measurement C on $\Omega_A \times \Omega_B$ that returns A and B as marginal POVM observables. POVM observables A and B that commute, are also jointly measurable [229].

Sharp Measurement

A sharp measurement is defined by Busch et al. [658] [659] as one for which the M_k in Equation (7.1) are projection operators P_k, including projectors that have rank greater than one. Wiseman et al [186, p. 37] define sharp measurement for which the M_k are rank 1 projection operators. Herein, we will adopt the more general definition of sharp measurement proposed by Busch.

In the case of unsharp observables, there exist examples of joint measurable observables that are not commutative. However, if either A or B are sharp measurements, then joint measurability also implies commutation [229].

Informationally Complete Measurement

An informationally complete measurement is one in which the set of the probabilities of outcomes $\{\mathrm{Tr}[\varrho E_k]\}$ can be inverted to uniquely determine the state ϱ. The construction of bases that implement informationally complete measurements is discussed in [660].

Repeatable Measurement

Suppose a measurement of a system is made initially in state ρ via a POVM observable A that is implemented via $\mathcal{E}_k(\varrho)$. If the outcome of the measurement is $i \in \Omega_A$, a repeatable measurement gives the same outcome $i \in \Omega_A$. After the first measurement, the probability of the ith outcome $P(i)$ is unity if the measurement is repeatable, that is $P(i) = \mathrm{Tr}[\mathcal{E}_i(\varrho)\, E_i] = 1$, provided $\mathrm{Tr}[\varrho\, E_i] > 0$.

A measurement is repeatable only if it is not continuous [661]. An equivalent condition for repeatability is that the dual map $\mathcal{E}_l^*(A_m) = 0$ for all $l \neq m$ [229]. There exists an implementation $\mathcal{E}_k(\varrho)$ of a POVM observable A that is a repeatable measurement if and only if each A_i has an eigenvalue of unity [662].

If A is a repeatable POVM observable and $\mathcal{D}(\mathcal{H}) = d < \infty$ has an outcome space with cardinality given by $|\Omega_A| = d$, then A is sharp, where the dimension of the Hilbert space $\mathcal{H}$ is denoted by $\mathcal{D}(\mathcal{H})$. If $|\Omega_A| \geq d - 1$ then A is commutative [229].

Minimally Disturbing Measurement

Minimally disturbing measurement, back-action evading measurement (BAE), non-disturbing measurement, and non-demolition measurement, are all classes of measurements that are designed to minimize in some manner the disturbance of the measurement.

As noted previously, a POVM given by elements $E_k \equiv {M_k}^\dagger M_k$ also could have the same elements as a second POVM E_k but of the form $E_k = {\widehat{M}_k}^\dagger \widehat{M}_k$where $\widehat{M}_k \not\equiv M_k$. That is, it is possible that the measurement instruments are different corresponding to the idea that there exist multiple ways of performing a measurement with the same POVM elements.

Consider a generalized Lüder's implementation $\mathcal{E}_k^L(\varrho)$ whereby $M_k = {E_k}^{1/2}$. In such a case $E_k = {M_k}^{\dagger} M_k = {\widehat{M}_k}^{\dagger} \widehat{M}_k$. Wiseman et al. [186] refers to the generalized Lüder's measurement as a minimally disturbing measurement since one can always decompose an arbitrary operator $\widehat{M}_k$ via the polar decomposition theorem according to $\widehat{M}_k = U({\widehat{M}_k}^{\dagger} \widehat{M}_k)^{1/2}$, whereby $({\widehat{M}_k}^{\dagger} \widehat{M}_k)^{1/2}$ is unique. Hence $\widehat{M}_k = U M_k$ .The fidelity between the initial state and final average state is also found to be maximized when $M_k = {E_k}^{1/2}$ [663]. $\widehat{M}_k$ requires an operation U that, provides additional back-action beyond that required to implement ${E_k}^{1/2}$.

Back-Action Evading Measurement

Suppose that, given an observable A there is no disturbance on-average, i.e. $T_A(\varrho) = \varrho$ for all $\varrho \in \mathfrak{M}(\mathcal{H})$. In such a case, $\mathcal{E}_k(\varrho)$ is proportional to the identity map for all $k \in \Omega_A$, and there is no information gain on any outcome [661] [664]. This shows that there must be some disturbance of the system being measured in order for there to be information gain.

Given a measurement that implements observable A and an implementation that has average mapping $T_A(\varrho)$, the measurement of A is a back-action evading (BAE) measurement if for all $\varrho \in \mathfrak{M}(\mathcal{H})$ and all $j \in \Omega_A$ the probability of measuring A_j is the same whether or not $T_A(\varrho)$ is previously applied. The condition for back-action evading measurement is $\mathrm{Tr}[A_j T_A(\varrho)] = \mathrm{Tr}[A_j \varrho]$ [186]. A back-action evading measurement is one for which the probability of measurement is unaffected whether or not the average mapping of A is implemented before the second measurement of A.

The condition for non-disturbing measurement can be rewritten equivalently in terms of the dual map of A, $T_A^*(A_j) = A_j$ and $j \in \Omega_A$ [229]. BAE measurement is also referred to in the literature as *measurement of the first kind* [229]. A Lüder's implementation of a commutative POVM observable is also BAE (see Exercise 7.3). A sufficient condition for a BAE measurement is that each A_i has an eigenvalue of unity.

Non-Disturbing Measurement

The concept of back-action evading measurement can be extended to include two measurements. Given two observables A and B and an implementation that has average mapping $T_A(\varrho)$, A does not disturb B if for all $\varrho \in \mathfrak{M}(\mathcal{H})$ and all $j \in \Omega_B$ the probability of measuring B_j is the same whether or not A is measured previously to B_j. The condition for non-disturbing measurement is $\mathrm{Tr}[B_j T_A(\varrho)] = \mathrm{Tr}[B_j \varrho]$.

The condition for non-disturbing measurement can be rewritten in terms of the dual map of A, $T_A^*(B_j) = B_j$ and $j \in \Omega_B$ [229]. Note by setting B=A, one obtains a BAE measurement.

It has been proven in [659] that for finite dimensional Hilbert spaces A does not disturb *B* only if A and B commute. It has been noted [229] that in an infinite dimensional space there exist non-commuting observables for which *A* does not disturb *B*.

The following results are established in [229]:

1. If *A* does not disturb *B*, then *A* and *B* are jointly measurable. However, there exist examples for which *A* and *B* are jointly measurable but *A* does disturb *B*.

2. Given a POVM observable *B* for which $B_i^2 \in \overline{\text{span}}(B), \forall i \in \Omega_B$, there exist implementations of *A* that do not disturb *B* if and only if *A* and *B* commute, where span(*B*) denotes the set of linear combinations of the vectors in set *B*.

3. Given the generalized Lüder's implementation for A, $\mathrm{T}_\mathrm{A}^\mathrm{L}(\varrho)$, if *A* and *B* commute then *A* does not disturb *B* (See Exercise 7.5)

4. Let *B* be a POVM observable such that $\mathcal{D}(\text{span}(B)) \geq (d-1)^2 + 1$, where $\mathcal{D}(A)$ denotes the dimension of the linear vector space *A*, and $\mathcal{D}(\mathcal{H}) = d$. There exist implementations of *A* that do not disturb *B* if and only if *A* and *B* commute. In the case where $\mathcal{D}(\mathcal{H}) = 2$, there exist implementations of *A* that do not disturb *B*, and there exist implementations of *B* that do not disturb *A*, if and only if *A* and *B* commute. In the case when $\mathcal{R}(B)=1$, where $\mathcal{R}(B)$ denotes the rank of *B*, there exist implementations of *A* that do not disturb *B* if and only if *A* and *B* commute.

Non-Demolition Measurement

Consider a system coupled to an apparatus initially in the state $\rho_S \otimes \rho_A$. A measurement operator *X* operates on the system during a time period $[t_{0,}t_1]$ for which the operator correctly correlates with the input state and is a repeatable measurement. The measurement operator *X* is a quantum non-demolition (QND) observable if the system is projected into an eigenspace corresponding to a distinct eigenvalue of the observable and not absorbed or removed in the process of measurement [665].

A sufficient condition for a QND [186, p. 39] is that the operator *X* in the Heisenberg representation is a constant of motion i.e. assuming a time-invariant Hamiltonian that couples the system and apparatus,

$$X \otimes I = U(t)^\dagger (X \otimes I) U(t).$$

However, it is not necessary that the observable be a constant of motion in order to qualify as a QND measurement as shown in [666]. Rather the condition given in [666] is that the observable *X* commute with itself at times when the observable is measured, i.e., $[X(t), X(t')] = 0$, where $t, t' \in [t_0, t_1]$.

The theory of continuous non-demolition measurement has been under development by several groups. The Belavkin equation is a stochastic approach that is often referred to as a quantum filter and has been applied to continuous non-demolition measurement [667]. Additionally, Korotkov and others have been active developing and applying theory to describe continuous measurement, see for example [668].

Indirect Measurement

The concept of indirect measurement is to couple a probe to a system via a unitary interaction and then perform a measurement on the probe. In this manner, one can learn information regarding the system, yet the information is learned indirectly by measuring only the probe. This is similar to the measurement procedure of von Neumann described in [13, pp. 441-445]. Ozawa showed that even though the indirect measurement models are only a subclass of all the possible quantum measurements, every measurement is statistically equivalent to an indirect measurement model [669].

Additionally, one can generalize this scheme to more than one probe, particularly when more than a single quantity is desired to be known regarding the system. For example, one might consider the measurement of non-commuting observables by coupling two probes to the system, one probe designed to extract information regarding the position of the system and the other probe designed to extract information regarding momentum. The use of two probe "meters" to extract both position and momentum information was proposed by Arthurs and Kelly [670]. An additional characteristic of this scheme is that the positions of the two probe meters commute in [670], hence ideal simultaneous measurements can be made of the positions of the two meters.

It was found in [671] that the use of correlated probes can provide more precise joint measurement outcomes than if the same probes were individually applied. However, Heisenberg's uncertainty relationship was found not to be violated.

Weak Measurement

Weak measurement, as developed by Aharonov et al. [672, p. 233], is actually a unitary operation for which systems weakly interact assuming the measurement-system unitary coupling formalism considered by von Neumann [13, pp. 441-445] in concert with a pre-selected state and post-selected state. The post-selected state occurs via bona fide measurement. Weak measurement and the weak values that result have been used to analyze and shed light on a number of quantum phenomenon.

It is claimed in an experiment in [673] that Heisenberg's uncertainty can be violated by weak measurement. Further analysis however indicates that the latter experiment still conforms with the spirit of Heisenberg's principle via error measures that are quantified by distances between observables [674].

In any case, other than the possibilities of pre and post selection, the intervening weak interaction is unitary and not a measurement operation.

Protective Measurement

It might be surprising to learn that it is believed possible to measure a system and determine a quantum wave function precisely so long as the wave function is an eigenstate of the system Hamiltonian. Aharonov et al. [672, pp. 214-216] showed that this can be accomplished using weak adiabatic measurement assuming the measurement-system unitary coupling formalism considered by von Neumann [13, pp. 441-445].This means for example that an entire atomic wave function that represents a given energy orbital can be determined without disturbing the state of the atom. Moreover, the measurement is claimed to protect the system from changing state compared to the non-adiabatic case of strong interaction, so long as the initial state of the system is an eigenstate of the system Hamiltonian.

When the system is initially in a superposition of eigenstates, an entangled superposition of measurement devices is predicted under unitary evolution [672, p. 218] which is claimed will collapse to one of the two readouts that is associated with an eigenstate of the system Hamiltonian.

Protective measurement is a rather remarkable finding. Protective measurement has been recently implemented in [675].

Non-Local Measurement

Consider two spin ½ particles for which two parties, Alice and Bob separated by a distance *d*, each have a single particle at their disposal. Suppose that it is possible for Alice and Bob to perform arbitrary measurements on the two particles. An interesting question is whether or not it is possible for Alice and Bob to make a non-local measurement of these particles, faster than would allow the speed of light to propagate between the two parties, i.e., faster than d/c. It is shown in [672, p. 197] that the majority of non-local measurements are impossible to make without violating causality, which is consistent with a result established in 1931 by Landau and Peierls [676]. On the other hand, Aharonov et al. [672, p. 198], assuming the measurement-system unitary coupling formalism considered by von Neumann [13, pp. 441-445], found a loophole for which it may be possible to measure a particular state faster than d/c --that being a bipartite superposition with precisely equal coefficients:

$$\frac{1}{\sqrt{2}}(|\uparrow\rangle|\uparrow\rangle + |\downarrow\rangle|\downarrow\rangle).$$

Closed System Approaches

Consider a system that is desired to be measured and a large set of particles (including atoms, molecules, and other fundamental constituents of matter) with Hamiltonian H that are in a configuration defined by the quantum state $|\Psi\rangle$. The configuration space consists of both spatial coordinates for the particles and their wave functions as well as the energies and couplings between all sets of particles, allowing the formation of molecules, dissociation, etc.

The closed system approach posits that under certain conditions given by $|\Psi\rangle \in \mathcal{R} \subset \mathcal{H}$, the set of particles forms a δ efficient measurement device for the system, that is, the probability of measurement when the particle completely impinges on the detector, is greater than δ. The set $\mathcal{R}$ defines those states of particles for which the device will be considered efficient. Other states $|\Psi\rangle \notin \mathcal{R}$ may form measurement devices, but are not δ efficient.

In the closed system approach, it is further stipulated that the physical reason that the device is capable of measuring the system when $|\Psi\rangle \in \mathcal{R}$ is related only to the internal physics of the set of particles. That is, if one had available $\mathcal{R}$, one would be expected to see different physics in terms of why the particles function as a δ efficient detector, while other configurations that are less efficient as well as those configurations that act unitarily.

Note that the requirement that a device function non-unitarily is related only to the physics of the particles in the closed system approach. That is, the goings-on of the rest of the universe are irrelevant. The internal physics of the closed system is significant in the closed system approach, not the environment. The physics of the closed system when the state $|\Psi\rangle \in \mathcal{R}$ is the reason that the closed system constitutes a bona fide measurement device.

Historical statements during the formation of the Copenhagen interpretation by Bohr, Heisenberg, Dirac, and Pauli indicate the framers believed that closed systems evolve mechanically and deterministically. In their view, it is the inability to be capable of knowing exactly the preconditions that allows for a seemingly nondeterministic action. Moreover, if an attempt is made to determine the state evolution of a closed system, a device is required that interacts with the closed system. Such an interaction was believed to create a sufficient disturbance due to the Heisenberg uncertainty relationship for which an uncontrollable action occurs leading to the destruction of coherence.

We have seen however that this explanation is not sufficient to be considered as a resolution of the measurement problem, as long as one can initialize the state of a detector, as the entanglement predicted under unitary evolution is inconsistent with any particular outcome occurring in the closed system hypothesis.

In 1927 this would not have been fully understood, as the quantifiable differences between entanglement and product states were not known. The paradoxes of unitary evolution were further illustrated by Schrödinger in 1935, for which the concept of entanglement was first introduced [119]. The quantifiable difference between product states and entanglement became clear with the advent of Bell's inequality in 1964 and the era of quantum information that began in the mid-1990s.

In closed system approaches, there is physical rationale for which non-unitary evolution occurs that can be traced to the internal properties of a given closed set of particles. One example of a closed system approach is whereby a set of particles exists in a configuration that has a sufficient measurement latency, bandwidth, etc., and a significantly amplified signal when the particle impinges on the closed system that is typical of a detector. The act of amplification may be claimed to be non-unitary, irreversible, and be the direct cause of measurement.

Another example of a closed system approach is the proposition that a conscious system is a measurement device because a conscious system has the property of becoming aware of external phenomenon upon measurement. In such a theory, it could be surmised that the interaction between a system of particles and an external quantum particle, for which no conscious tinge within the closed system of particles can result, would constitute a deterministic unitary interaction.

Continuous stochastic theories are based on an underlying continuous stochastic process that affects the evolution of quantum particles. Such theories suggest that the existence of such a continuous stochastic process would provide a mechanism to resolve the measurement problem, although to date the existence of such a process has not been reported. Whether such theories are closed or open depends ultimately on the physics as to why such effects are occurring. If the effect always occurs, and is a property of individual particles that can be considered irrespective of the environment, then the effect could be considered to be a closed system effect. Moreover, Adler in [100] makes the important point that only environmental particles that are within a distance $c\tau_l$, where c is the speed of light and τ_l is the measurement latency time, can causally affect the experimental outcome without violating no-signaling.

Considerations in Closed Systems

Major considerations in a closed system deductive approach are the formation of the initial assumptions that will constitute and drive the investigation. An approach has been suggested in Chapter 6, in which the Bohr/Wheeler principle of radical conservatism provides the guide of the selection of conservative time-tested assumptions as well as the radical probing of such assumptions during the investigation. The importance of the selection of time-tested assumptions cannot be overly stressed, as the violation of this rule could easily lead to the dismissal of the correct solution and lead to significant wasted effort. A logical and rigorous foundation is desired in order to properly proceed.

We propose the following Ten Commandments of measurement in closed system investigation:

The Ten Commandments

(of measurement)

1) Thou shalt conserve energy

   Total energy of the system plus measuring device plus space-time surrounding environment is conserved in the process of measurement.

2) Thou shalt conserve momentum

   Total momentum of the system plus measuring device plus space-time surrounding environment is conserved in the process of measurement.

3) Thou shalt conserve charge

   The net charge of the system plus measuring device plus space-time surrounding environment is conserved in the process of measurement.

4) Measurement shalt be gauge invariant

   The measurement results and statistics of occurring are independent of the particular electromagnetic gauge.

5) Thou shalt not signal

   Measurement theory cannot lead, in principle, to information transmitted faster than the speed of light.

6) Thou shalt have no other preferred reference frames before Me

   There is no preferred frame that can result in the theory of measurement for which the theory is different than in other frames.

7) Thou shalt obey the equivalence principle

   The equivalence principle due to Einstein is that it is impossible to locally distinguish if one is being accelerated by a gravitational force or by a local force. Any measurement theory that violates such distinguishability is invalid.

8) Thou shalt not create perpetual motion machines

   Work from a fixed source cannot be delivered perpetually to drive a separate system. Any measurement theory that violates this premise is invalid.

9) Thou shalt make measurements obeying $|\Psi|^2$

   Measurement theory must respect Born's law.

10) Ye who covet the theory of measurement shalt be exalted amongst the heathen Unitarians.

These Ten Commandments and their implications on the theory of measurement will be expanded upon by the authors in future publications regarding the theory of closed system measurement.

Open System Approaches

An open system approach is one whereby the physical reason that a given configuration of particles $|\Psi\rangle$ undergoes measurement is that the action of the particles when going into a superposition causes a phenomenon external to the closed system to exert a non-unitary back-action on the closed system that takes the form of measurement. For example, consider a unitary action for which the particles $|\Psi\rangle$ evolve into a macroscopic superposition of distinct positions. One could posit that there is some fundamental principle that reacts to such a superposition by inducing a back-action on the system, in a manner that opposes the system superposition, and is of the form of a non-unitary measurement.

The idea that external interaction is responsible for measurement is related to the hypothesis that a closed system by itself will always evolve according to Schrödinger's equation. Such a view was indeed held by many of the progenitors of the Copenhagen interpretation of quantum theory. In such a view, something external must intervene in order for a system to collapse. According to Bohr, the uncontrollable interaction of a system with a macroscopic device causes collapse. Whether or not such an interaction was actually evolving according to Schrödinger's equation or not did not seem to be of much consequence or interest to Bohr. Rather, due to the uncontrollable interaction and Heisenberg's uncertainty relationships, the evolution was not predictable and could never be predictable, and therefore the best one could ever hope for would be a quantum mechanical measurement theory that provided a statistical account. That is, statistics are not just a convenient way of computing unitary evolution in the Copenhagen interpretation, it is imposed on the theory due to a fundamental impossibility of precisely knowing the entire particle behavior in the interaction of a system and macroscopic device.

Hence the rise of the development of open-system approaches can be identified with the proposition that all closed systems evolve via Schrödinger's equation. However, as we have seen in our Chapter 3 development, unitary evolution by itself is not a sufficient explanation of measurement. Hence open-system approaches must inevitably provide some additional non-unitary external mechanism or action that acts on the closed system to affect a measurement.

In the case of many-worlds theory, a system that evolves into a superposition evokes a reaction that causes a splitting of worlds and hence a back-action on the system for which a non-unitary measurement occurs. One might also consider that if a body with substantial mass is put into a stationary superposition of distinct locations, that there is a general relativistic back-action that modifies the non-relativistic quantum mechanical laws so that the superposition is no longer stationary. Such a back-action might conceivably effect a collapse similar to that proposed by Penrose and others.

One could also consider that there exist Category 2 type theories that are open system approaches. For example, suppose the environment when sufficiently coupled to a system in a superposition is simply impossible to reverse. That is, it is not only for all practical purposes (FAPP) reversible, but there is some yet to be discovered law for which a particular unitary evolution would be intrinsically irreversible. One could postulate that in such a case, certain unitary state evolutions would be physically irreversible in principle, and the point that this would occur would be described by non-unitary evolution rather than unitary evolution. Related theories have been advocated, with various differences by Omnès [295] and others.

Considerations in Open Systems

How are energy and other fundamental quantities conserved if a particle collapses due to an external reaction that opposes a particular system superposition? If the external source passes energy or momentum onto the system, then one would need to include the external phenomenon in order to consider conservation of energy and momentum. The consideration of no-perpetual motion as well no-decrease of entropy would also generally require taking the external source into consideration.

On the other hand, there are several considerations in open systems that can be directly applied to the system that is measured. Gauge invariance of measurement results should hold regardless of the reduction mechanism. No-signaling also must always apply. That there is no preferred frame and the equivalence principle are also valid considerations in open systems. Furthermore, Born's rule and Lüder's rule should also be met by open system solutions.

Another consideration in theories that utilize Everett's many-worlds interpretation (MWI) is the possibility of communication between universes, such as found in [677]. If a new universe occurs each time a measurement occurs, what is to prevent communication between distinct universes? A proposal in [678] suggests a technique to test signaling between worlds:

> *For the MWI it has been shown in the previous sections that inter-world communication on a time scale of minutes should be possible with state of the art quantum-optical equipment.*
>
> Foundation of Physics, On a possibility to find experimental evidence for the many-worlds interpretation of quantum mechanics, Vol. 27, 1997, P. 559, R. Plaga © Plenum Publishing Corporation 1997. With permission of Springer.

Measurement, in open system theory, is often explained when decoherence occurs. MWI does not provide a well-defined rationale for choosing one basis versus another for the act of splitting or branching. The use of decoherence to supplement Everett's theory does provide an approach whereby the splitting into different branches occurs for which the subsystem density matrix becomes in the process of decoherence diagonal or non-coherent in a particular (or preferred) basis.

Conservation laws are violated on any single trial in MWI, but are upheld on average. In terms of interaction-free measurement, Elitzur and Vaidman [107] also state that conservation laws are restored when all collapse possibilities are considered in the MWI:

> *This experiment gives a clear demonstration of the violation of conservation laws by the collapse of a quantum state. For example, if the potentials in regions A and B are different, then the expectation value of the particle's energy changes though no change happens to the photon. Conservation laws are restored when all branches of the quantum state are considered together.*
>
> Foundation of Physics, Quantum mechanical interaction-free measurements, *Vol. 23 1993, p. 987, A. C. Elitzur, L.* Vaidman © Plenum Publishing Corporation 1993. With permission of Springer.

If there is a violation of energy and momentum on individual trials, then this should be observable if experiments were conducted to establish whether or not energy and momentum are strictly upheld.

An issue that is often vague in open system approaches is precisely the conditions under which a measurement occurs. In open system theory, is complete decoherence required before a splitting of the worlds, or is partial decoherence allowed? If all the branches of a measurement can occur, one must explain how the Born rule comes about, while taking into account the possibility of re-combination of previous split worlds.

Energy-driven stochastic Schrödinger equations can be considered closed systems for the purposes of energy, momentum, and other fundamental considerations [255], one can also consider such solutions as open system solutions if there is an external source that causes the system to undergo stochastic non-unitary evolution. In such a case, the external source could contribute resources such as energy or entropy in order for the system plus external source to satisfy conservation laws. For example, to explain stochastic Schrödinger equations, an actual noise source is hypothesized to exist by Pearle in [232]. Using such a noise source, it was shown by Pearle that energy is conserved. With an external noise source, such stochastic reduction models are open systems insofar as measurement occurs on the system. In such a case, the source of such a noise source would need to be identified and established. If an infinite energy source simply is posited to exist that feeds the measurement process, then the question arises as to whether or not this poses a violation of the non-existence of perpetual motion machines.

Exercises

7.1 Classify whether or not GRW meets the definition for each of the following categories of measurement: 1) a sharp measurement, 2) informationally complete measurement, 3) repeatable measurement, 4) minimally disturbing, 5) BAE, 6) QND.

7.2 Suppose that a POVM observable A consists of two outcomes. Show that A is commutative.

7.3 Show that the Lüder's implementation of a commutative POVM observable is also BAE.

7.4 Show that a repeatable POVM is also BAE. Find an example of a BAE that is not repeatable.

7.5 Given the generalized Lüder's implementation for A, $\mathrm{T}_{\mathrm{A}}^{\mathrm{L}}(\varrho)$, show that if A and B commute then A does not disturb B.

7.6 Show that a sufficient condition for a BAE measurement is that each A_i has an eigenvalue of unity.

7.7 Find an example of a repeatable POVM that is non-commutative. Hint: consider measurement operators of the form of a direct sum of two measurement operators i.e. $A_i = A_{i,1} \oplus A_{i,2}$ for which the $A_{i,1}$ are orthogonal projectors.

7.8 Prove that an informationally complete measurement is not a BAE measurement. Hint: Consider that a measurement for which there is no disturbance on average $T_A(\varrho) = \varrho$ cannot be an informationally complete measurement.

7.9 Prove that a QND is BAE.

7.10 Given two jointly measurable POVM observables A and B for which A is sharp, show that A and B commute.

7.11 Show that the measurement operators underlying GRW are commutative, hence GRW is BAE.

CHAPTER 8

Conclusions

Current Situation

In the quantum mechanics textbook by Messiah [679, pp. 47-48] published in 1961, it is stated:

> *The interpretation under discussion here is the statistical interpretation of the Copenhagen school. It is the one we are developing in this book. After violent controversies, it has finally received the support of the great majority of physicists. However, it had (and still has) a number of die-hard opponents, among which should notably list Einstein, Schrödinger, and de Broglie. The controversy has finally reached a point where it can no longer be decided by any further experimental observations; it henceforth belongs to the philosophy of science rather than to the domain of physical science proper.*

Messiah's last statement is false; we have shown that the measurement problem is open to further investigation, both experimentally and theoretically. It may have been thought to be impossible to conduct mesoscopic entanglement experiments in 1961 when Messiah's statement was published, but consider the statement by Aspelmeyer and Schwab in 2008 [122]

> *The last five years have witnessed an amazing development in the field of nano- and micromechanics. What was widely considered fantasy ten years ago is about to become an experimental reality: the quantum regime of mechanical systems is within reach of current experiments.*

Certain experimental tools that appear useful to investigate the measurement problem are those that discriminate entangled systems from non-entangled systems, and there have already been a large number of publications that have shown how this can be accomplished on both microscopic systems, for example photon entanglement experiments in [201] [85], and the mesoscopic to macroscopic regimes [22] [26] [27] [125] [126] [127] [128] [129] [272] [275].

The issue of non-separability that Einstein disliked and is predicted by Schrödinger's equation was not generally appreciated until after 1964 in the wake of the discovery of Bell's inequality. The fact that such nonlocality has been confirmed experimentally adds to the strength of Einstein's arguments with Bohr. If these two great men were still alive after such developments, it is likely that Einstein would have made new arguments regarding the incompleteness of quantum theory—that Bohr would not have dismissed.

Future Work

Experimental investigation of the measurement problem will require an investment. Mesoscopic to macroscopic experiments will need to be designed and probed at various temperatures, including low temperatures. Although significant experimental progress has been made in quantum information, experiments needed to test the limitations of superpositions are only beginning.

In our view and seemingly in a purely logical sense, a generalization of the Copenhagen interpretation aiming toward completion as desired by Einstein, is the primary approach that should first be considered for resolution of the measurement problem.

All theories remain on the table in a deductive investigation unless disproven. Currently this includes theories that advocate dual-descriptions of matter as well as non-dual or universal descriptions. There is little doubt that the measurement problem is best approached in this manner as has been discussed in Chapter 6.

Gisin in [79] considers several approaches for future work. We agree with Gisin regarding one of the approaches considered:

> *Assume there is only one sort of stuff, but certain arrangements of this stuff make it special. For example, assume everything is made out of elementary particles, but certain arrangements of atoms and photons make them act as measuring apparatuses. Hence the question: Which configurations of atoms and photons characterize measurement setups?*

Experimental steps that make sense at present in regard to furthering progress on the measurement problem include determining experimentally whether or not energy/momentum conservation holds on average or strictly for all trials for mesoscopic to macroscopic systems. Prior experimental evidence from scattering experiments in the 1930s indicates that those interactions are described by unitary interaction, but whether or not this is extendable to mesoscopic to the macroscopic regime is unknown. Numerous potential violations of strict energy conservation have already been suggested in various theories put forward to explain measurement, yet there has been no experimental confirmation. Experimentally, bona fide detection devices are needed in order to properly investigate this issue. At least a mesoscopic to macroscopic device in the sense that Bohr would have agreed constituted a detection device should be considered, for example a photo-avalanche detector. A two-state detection device can be utilized and the interaction-free measurement case can be investigated to determine if the energy of a particle is changed. As discussed in Chapter 4, there are many physical measurement theories that could be immediately removed from consideration if it is established that energy/momentum conservation strictly holds. For example, decoherence theories that incorporate many-world splitting would be ruled out and various closed-system stochastic Schrödinger equations would be ruled out. Such experimental results would also be a major aid to theoretical work, as it would allow one to concentrate on classes of theories that satisfy either strict or average conservation laws, depending on what is found experimentally.

On the other hand, if it is found that energy/momentum does not strictly hold on every trial, then many current theories that only assume on-average conservation, would certainly gain and become further accredited. Furthermore, such an experimental outcome would certainly fuel further research into theories that require conservation be violated on any particular trial and only conserved on average.

Based on our own research into the measurement problem, we make the following prediction regarding the outcome of such energy/momentum experiments:

> *Energy and momentum are always conserved in Nature, during unitary evolution and as well on all individual trials of the measurement process.*

Overall, it is recommended for future work that the following areas of investigation be emphasized regarding the measurement problem:

1. The use of deductive scientific methodologies when addressing the resolution of the problem of entanglement predicted under Schrödinger unitary evolution versus the product state predicted under measurement.
2. The theoretical investigation of the physics of the measurement process. Theoretical development of Category I or II type models as defined in Chapter 4, that address the respective requirements of either R1.1-R1.4 for Category I models or R2.1-2.3 for Category II models.
3. Experimentation at the mesoscopic to macroscopic regime:
    a. Investigation to determine how well proposed theories meet the respective requirements R1.1-R1.4 for Category I models or R2.1-2.3 for Category II models.

   b. Determine via experimental investigation whether there are specific characteristics that can be experimentally identified with the measurement process.
   c. Experimental confirmation of a bona fide measurement event.
   d. Experimental determination of configurations of atoms that constitute a bona fide measuring device.
   e. Experimental determination of the characteristic time and length scales due to the measurement process.
   f. Experimental confirmation of relationships between fundamental parameters and any fundamental constants that are needed to relate parameters.
4. Experimentation at the mesoscopic to macroscopic regime to determine if energy/momentum are conserved on average versus on every measurement trial.

Summary

After we finished writing, we imagined two extreme readers and their reactions to this book. On the one hand, we considered a strongly inductive and recalcitrant reader who resists our intended message. On the other hand, we considered a strongly deductive reader who embraces, with diligent effort, each argument presented...

On the one hand, if you read this book with a strong inductive mindset, then you most likely have learned very little regarding the measurement problem. Perhaps you found the discussion regarding the wave-particle duality and the Schrödinger cat paradox correct, but old hat. You found the chapter on the characteristics of unitary evolution pedestrian and lacking in imagination to explain real-world phenomena. You expect that the demonstration in Chapter 3 that is capable of distinguishing unitary evolution from non-unitary evolution independent of the number of particles, must have some error because of the extraordinary success of the predictions of quantum mechanics down to the umpteenth digit, even if you yourself cannot pinpoint the error due to your lack of background in quantum information. Because of this, you reject outright the claims in Chapter 4 as preposterous, particularly regarding the inclusion of the hypothesis of consciousness as related to quantum mechanics. You do not believe that Niels Bohr had ever seriously considered consciousness and free will as related to quantum mechanics, and even if so, you heard he had changed his ideas in later life. You agree with Dirac that the basic scientific principles are largely known and what remains is only limited by the problem's complexity and lack of sufficient computer power to currently solve it. Because of this, you believe that induction is the proper manner to pursue the measurement problem, if indeed such a problem even exists, which you doubt. You feel your expenditure purchasing this book was a waste of money, and you are going to see if you can get a refund so that you can instead purchase, "The Memoirs of Richard Nixon." In our view, you are toxic to major scientific progress. We would like to proceed to our next book on our approach to solving the measurement problem. It's rather difficult when there are many old curmudgeons still around, perpetuating false notions, suppressing the free-flow of scientific information, and working to choke progress on this problem with your expectations that all of Nature can be reduced to a trivial pseudo-classical resonance equation.

On the other hand, if you have read this book with a strong deductive mindset, then we expect that you feel that you have learned a substantial amount regarding the measurement problem. You learned the basics of wave-particle duality and the Schrödinger cat paradox in Chapter 1. You learned in Chapter 2 the many characteristics of unitary evolution such as its complete reversibility and constant entropy. In Chapter 3 you learned how to judge whether or not unitary evolution can suffice to explain measurement phenomena via an experimental manner that provides different outcomes for the predictions of unitary evolution versus measurement. You understand that entanglement is not measurement. You see how the deductive scientific process leads to the rejection that Nature evolves strictly according to Schrödinger's equation. In Chapter 4 you learned more precisely the definition of the quantum measurement problem as well as its misconceptions. You learned of our proposed requirements of what any theory should demonstrate before being accepted in any serious scientific deductive effort. In Chapter 5 you learned the fascinating *unabridged* history of the measurement problem that was not taught to you in school, neither in elementary school, nor in high school, nor at the university where you may have majored in quantum mechanics. And you now have a rather altered view of quantum and scientific history compared to your views before you read this book. You learned from Chapter 6 that deduction has overwhelmingly been the scientific standard utilized by great scientists when investigating areas in which truly new scientific discoveries were made. You see clearly now the rationale as to why the unresolved issues of the measurement problem should be investigated via a deductive approach and not via strict application of induction. You now understand from Chapter 7 the difference between closed and open system approaches and appreciate the difference. As well you learned several new categories of measurement that you were not aware of previously. You also learned the major considerations in developing a new theory of measurement and feel armed and ready now to attack the measurement problem.

Stay tuned— No doubt the reaction of the typical reader will fall somewhere in between these two extremes presented. We welcome hearing your actual reaction at our companion website to this book, theQMP.com. Numerous additional resources are also available including videos and a follow-on FAQ discussion section.

The authors have been working on the theory of closed system measurement for a rather long time. So be sure to utilize the measurement process and keep your non-unitary senses alert for our sequel.

Abbreviations

BAE	Back-action evading
BBO	Beta-Barium-Borate
BCE	Before Common Era
BEC	Bose-Einstein Condensate
BKS	Bohr-Kramers-Slater
CHSH	Clauser-Horne-Shimony-Holt
CMB	Cosmic Microwave Background
CNOT	Controlled Not
CP	Completely Positive
CSL	Continuous Spontaneous Localization
DNA	Deoxyribonucleic acid
EEG	Electroencephalography
EM	Electromagnetic
EPR	Einstein-Podolsky-Rosen
ER	Einstein-Rosen
FAPP	For All Practical Purposes
FAQ	Frequently Asked Questions
fMRI	Functional Magnetic Resonance Imaging
GHZ	Greenberger-Horne-Zeilinger
GRW	Ghirardi-Rimini-Weber
IAS	Institute for Advanced Study
IFM	Interaction-Free Measurement
LOCC	Local Operations and Classical Communication
MWI	Many Worlds Interpretation
NCC	Neural Correlates of Consciousness
NMR	Nuclear Magnetic Resonance
NOON	Quantum state consisting of a superposition of *N* particles in Mode 1 with zero particles in Mode 2 and vice-versa.
OPF	Operational Probabilistic Framework
PBR	Pusey-Barrett-Rudolph
POVM	Positive Operator Valued Measure
QED	Quantum Electrodynamics
RP	Readiness Potential
SL	Spontaneous Localization
SMA	Supplementary Motor Area

SP	Surface Plasmon
SPDC	Spontaneous Parametric Down Conversion
SQUID	Superconducting Quantum Interference Device
SVD	Singular Value Decomposition
TI	Transactional Interpretation
QB	Quantum Bayesian
QED	Quantum Electrodynamics
QND	Quantum Nondemolition
QMP	Quantum Measurement Problem
UMDT	Unitary versus Measurement Discrimination Test
WPM	Wave-Particle Model
YAI	Yet Another Interpretation.

Bibliography

[1] T. S. Kuhn, *The structure of scientific revolutions, International Encyclopedia of Unified Science, vol. 2, no. 2,* Chicago: The University of Chicago Press, 1970.

[2] Wikiquote, "Niels Bohr --- Wikiquote," 2017. [Online]. Available: https://en.wikiquote.org/w/index.php?title=Niels_Bohr&oldid=2297980.

[3] D. Howard, "Nicht Sein Kann was Nicht Sein Darf, or the Prehistory of EPR, 1909-1935: Einstein's Early Worries about the Quantum Mechanics of Composite Systems," in *Sixty-Two Years of Uncertainty*, A. I. Miller, Ed., New York, Plenum Press, 1990.

[4] G. Bacciagaluppi and A. Valentini, Quantum Theory at the Crossroads: Reconsidering the 1927 Solvay Conference, Cambridge University Press, 2013.

[5] P. Holland, The Quantum Theory of Motion: An Account of the de Broglie-Bohm Causal Interpretation of Quantum Mechanics, Cambridge University Press, 1995.

[6] M. Born, "On the Quantum Mechanics of Collisions," in *Quantum theory and measurement*, Princeton University Press (Orig. article published in 1927), 1983, p. 52.

[7] W. Heisenberg, "The physical content of quantum kinematics and mechanics," in *Quantum Theory and Measurement*, Princeton University Press, 1983 (Orig. article published in 1927), pp. 62-84.

[8] L. I. Mandelstamm and I. E. Tamm, "The Uncertainty Relation between Energy and Time in Non-relativistic Quantum Mechanics," *Journal of Physics, USSR,* vol. 9, pp. 249-254, 1945.

[9] A. Pais, Niels Bohr's Times in Physics, Philosophy, and Polity, Oxford: Oxford University Press, 1991.

[10] M. Born, "The statistical interpretation of quantum mechanics," 1954. [Online]. Available: www.nobelprize.org/nobel_prizes/physics/laureates/1954/born-lecture.pdf.

[11] M. H. Stone, "On One-Parameter Unitary Groups in Hilbert Space," *Annals of*

Mathematics, vol. 33, no. 2, pp. 643-648, 1932.

[12] E. Wigner, "On Unitary Representations of the Inhomogeneous Lorentz Group," *The Annals of Mathematics, Second Series,* vol. 40, no. 1, pp. 149-204, 1939.

[13] J. von Neumann, Mathematical Foundations of Quantum Mechanics, Princeton, NJ: Princeton University Press, 1955.

[14] M. A. Nielsen and I. L. Chuang, Quantum Computation and Quantum Information, Cambridge: Cambridge University Press, 2010.

[15] A. M. Gleason, "Measures on the Closed Subspaces of a Hilbert Space," *Journal of Mathematics and Mechanics,* vol. 6, no. 6, pp. 885-894, 1957.

[16] B. Opanchuk, L. Rosales-Zárate, R. Y. Teh and M. D. Reid, "Quantifying the mesoscopic quantum coherence of approximate NOON states and spin-squeezed two-mode Bose-Einstein condensates," *Physical Review A,* vol. 94, p. 062125, 2016.

[17] R. Y. Teh, L. Rosales-Zárate, B. Opanchuk and M. D. Reid, "Signifying the nonlocality of NOON states using Einstein-Podolsky-Rosen steering inequalities," *Physical Review A,* vol. 94, p. 042119, 2016.

[18] I. Afek, O. Ambar and Y. Silberberg, "High-NOON States by Mixing Quantum and Classical Light," *Science,* vol. 328, no. 5980, pp. 879-881, 14 May 2010.

[19] B. Pepper, R. Ghobadi, E. Jeffrey, C. Simon and D. Bouwmeester, "Optomechanical superpositions via nested interferometry," *Physical Review Letters,* vol. 109, p. 023601, 2012.

[20] N. Brunner, C. Branciard and N. Gisin, "Possible entanglement detection with the naked eye," *Physical Review A,* vol. 78, p. 052110, 2008.

[21] M. Arndt and K. Hornberger, "Testing the limits of quantum mechanical superpositions," *Nature Physics,* vol. 10, pp. 271-277, 1 April 2014.

[22] J. I. Korsabakken, F. K. Wilhelm and K. B. Whaley, "The size of macroscopic superposition states in flux qubits," *European Physical Letters,* vol. 89, p. 30003, February 2010.

[23] M. Arndt, O. Nairz, J. Vos-Andreae, C. Keller, G. van der Zouw and A. Zeilinger, "Wave-particle duality of C-60 molecules," *Nature,* vol. 401, pp. 680-682, 1999.

[24] S. Gerlich, L. Hackermüller, K. Hornberger, A. Stibor, H. Ulbricht, M. Gring, F. Goldfarb, T. Savas, M. Müri, M. Mayor and M. Arndt, "A Kaptitza-Dirac-Talbor-Lau interferometer for highly polarizable molecules," *Nature Physics,* vol. 3, pp. 711-715, 19 August 2007.

[25] P. Haslinger, N. Dörre, P. Geyer, J. Rodewald, S. Nimmrichter and M. Arndt, "A universal matter-wave interferometer with optical ionization gratings in the time domain," *Nature Physics,* vol. 9, pp. 144-148, 10 February 2013.

[26] S. Eibenberger, S. Gerlich, M. Arndt, M. Mayor and J. Tüxen, "Matter–wave

interference of particles selected from a molecular library with masses exceeding 10 000 amu," *Physical Chemistry Chemical Physics,* vol. 15, p. 14696, 2013.

[27] T. Kovachy, P. Asenbaum, C. Overstreet, C. A. Donnelly, S. M. Dickerson, A. Sugarbaker, J. M. Hogan and M. A. Kasevich, "Quantum superposition at the half-metre scale," *Nature,* vol. 528, pp. 530-533, 12 2015.

[28] D. M. Stamper-Kurn, G. E. Marti and H. Müller, "Verifying quantum superpositions at metre scales," *Nature,* vol. 537, pp. E1-E2, 2016.

[29] T. Kovachy, P. Asenbaum, C. Overstreet, C. A. Donnelly, S. M. Dickerson, A. Sugarbaker, J. M. Hogan and M. A. Kasevich, "Kovachy et al. reply," *Nature,* vol. 537, pp. E2-E3, 1 September 2106.

[30] M. Zawisky, M. Baron, R. Loidl and H. Rausch, "Testing the world's largest monolithic perfect crystal neutron interferometer," *Nuclear Instruments and Methods in Physics Research A,* vol. 481, pp. 406-413, 2002.

[31] M. Zawisky, J. Springer, R. Farthofer and U. Kuetgens, "A large-area perfect crystal neutron interferometer optimized for coherent beam-deflection experiments: Preparation and performance," *Nuclear Instruments and Methods in Physics Research A,* vol. 612, pp. 338-344, 2010.

[32] D. N. Spergel, "The dark side of cosmology: Dark matter and dark energy," *Science,* vol. 347, no. 6226, pp. 1100-1102, 2015.

[33] P. de Bernardis et al, "A flat Universe from high-resolution maps of the cosmic microwave background radiation," *Nature,* vol. 404, pp. 955-959, 27 April 2000.

[34] J. B. Hartle, "Theories of everything and Hawking's wave function of the universe," in *The Future of Theorectical Physics and Cosmology, Celebrating Stephen Hawking's 60th Birthday*, G. W. Gibbons, E. P. S. Shellard and S. J. Rankin, Eds., Cambridge, Cambridge University Press, 2003, pp. 38-49.

[35] J. B. Hartle and S. W. Hawking, "Wave function of the Universe," *Physical Review D,* vol. 28, no. 12, pp. 2960-2975, 1983.

[36] G. F. R. Ellis, "Physics, complexity and causality," *Nature,* vol. 435, p. 743, 9 June 2005.

[37] M. Reed and B. Simon, Methods of Modern Mathematical Physics I: Functional Analysis, vol. I, New York, NY: Academic Press, 1972.

[38] D. F. Walls and G. J. Milburn, Quantum Optics, Berlin: Springer, 1995.

[39] C. Cohen-Tannoudji, J. Dupont-Roc and C. Wunderlich, Atom-Photon Interactions: Basic Processes and Applications, New York, NY: Wiley, 1998.

[40] R. Sorkin, "Quantum Mechanics as Quantum Measure Theory," *Modern Physics Letters A,* vol. 9, no. 33, pp. 3119-3127, 1994.

[41] U. Sinha, C. Couteau, T. Jennewein, R. Laflamme and G. Weihs, "Ruling Out Multi-Order Interference in Quantum Mechanics," *Science,* vol. 329, pp. 418-

421, 2010.

[42] R. F. Werner, "Quantum states with Einstein-Podolsky-Rosen correlations admitting a hidden-variable model," *Physical Review A,* vol. 40, no. 8, pp. 4277-4281, 15 October 1989.

[43] R. Horodecki, P. Horodecki, M. Horodecki and K. Horodecki, "Quantum Entanglement," *Reviews of Modern Physics,* vol. 81, no. 2, pp. 865-942, April-June 2009.

[44] V. Vedral, "The role of relative entropy in quantum information theory," *Reviews of Modern Physics,* vol. 74, pp. 197-234, January 2002.

[45] S. Weinberg, Lectures on Quantum Mechanics, Cambridge: Cambridge University Press, 2013.

[46] M. A. Naimark, *Izv. Akad. Nauk. SSSR Ser. Mat.,* vol. 4, p. 277, 1940.

[47] W. F. Stinespring, "Positive functions on C*-algebras," *Proceedings of the American Mathematical Society,* vol. 6, no. 2, p. 1, 1955.

[48] T. Heinosaari, T. Miyadera and M. Ziman, "An invitation to quantum incompatibility," *Journal of Physics A: Mathematical and Theoretical,* vol. 49, p. 123001, 2016.

[49] K. Kraus, A. Böhm, J. Dollard and W. Wootters, States, effects, and operations: fundamental notions of quantum theory : lectures in mathematical physics at the University of Texas at Austin, Springer-Verlag, 1983.

[50] H. Araki and E. H. Lieb, "Entropy Inequalities," *Communications in Mathematical Physics,* vol. 18, pp. 160-170, 1970.

[51] G. Weihs and A. Zeilinger, "Photon statistics at beam splitters: an essential tool in quantum information and teleportation," in *Coherence and Statistics of Photons and Atoms*, Wiley and Wiley, 2001.

[52] C. K. Hong, Z. Y. Ou and L. Mandel, "Measurement of subpicosecond time intervals between two photons by interference," *Physical Review Letters,* vol. 59, no. 18, pp. 2044-2046, 1987.

[53] B. Schumacher and M. D. Westmoreland, "Locality and Information Transfer in Quantum Operations," *Quantum Information Processing,* vol. 4, no. 1, pp. 13-34, February 2005.

[54] W. Wootters and W. Zurek, "A Single Quantum Cannot be Cloned," *Nature,* vol. 299, pp. 802-803, 28 October 1982.

[55] A. K. Pati and S. L. Braunstein, "Impossibility of Deleting an Unknown Quantum State," *Nature,* vol. 404, pp. 164-165, 2000.

[56] H. Barnum, C. M. Caves, C. A. Fuchs, R. Jozsa and B. Schumacher, "Noncommuting Mixed States Cannot Be Broadcast," *Physical Review Letters,* vol. 76, p. 2818, 1996.

[57] D. Mayers, *Physical Review Letters,* vol. 78, p. 3414, 1997.

[58] H. K. Lo and H. F. Chau, *Physical Review Letters,* vol. 78, p. 3410, 1997.

[59] G. Chiribella, G. M. D'Ariano, P. Perinotti and B. Valiron, "Quantum computations without definite causal structure," *Physical Review A,* vol. 88, p. 022318, 2013.

[60] G. Rubino, L. A. Rozema, A. Feix, M. Araüjo, J. M. Zeuner, L. M. Procopio, Č. Brukner and P. Walther, "Experimental verification of an indefinite causal order," *Science Advances,* vol. 3:e, p. 1602589, 2017.

[61] P. G. Kwiat, E. Waks, A. G. White, I. Appelbaum and P. H. Eberhard, "Ultrabright source of polarization-entangled photons," *Physical Review A,* vol. 60, no. 2, pp. R773-R776, August 1999.

[62] P. Hariharan and B. Sanders, *Progress in Optics,* vol. 36, p. 49, 1996.

[63] J. D. Franson, *Physical Review Letters,* vol. 62, p. 2205, 1989.

[64] Y. H. Shih, "Entangled biphoton source - property and preparation," *Reports on Progress in Physics,* vol. 66, p. 1009, 2003.

[65] X. Y. Zou, L. J. Wang and L. Mandel, "Induced Coherence and Indistinguishability in Optical Interference," *Physical Review Letters,* vol. 67, p. 318, 1991.

[66] L. J. Wang, X. Y. Zou and L. Mandel, "Induce coherence without induced emission," *Physical Review A,* vol. 44, p. 4614, 1991.

[67] M. Krenn, A. Hochrainer, M. Lahiri and A. Zeilinger, "Entanglement by Path Identity," *Physical Review Letters,* vol. 118, p. 080401, 2107.

[68] A. B. Klimov, J. L. Romero, J. Delgado and L. L. Sánchez-Soto, "Master equations for effective Hamiltonians," *Journal of Optics B: Quantum and Semiclassical Optics,* vol. 5, pp. 34-39, 2003.

[69] R. A. Millikan, "A Direct Photoelectric Determination of Planck's "h"," *The Physical Review,* vol. 7, pp. 355-388, 1 March 1916.

[70] A. H. Compton, "A Quantum Theory of the Scattering of X-Rays by Light Elements," *The Physical Review,* vol. 21, no. 5, pp. 483-502, May 1923.

[71] W. Bothe and H. Geiger, "Über das Wesen des Comptoneffeckts: ein experimenteller Beitrag zur Theories der Strahlung," *Zeitschrift fur Physik,* vol. 32, no. 9, pp. 639-663, 1925.

[72] P. Grangier, G. Roger and A. Aspect, "Experimental Evidence for a Photon Anticorrelation Effect on a Beam Splitter: A New Light on Single-Photon Interferences," *Europhysics Letters,* vol. 1, no. 4, pp. 173-179, 15 Febsruary 1986.

[73] T. Weidlich, Appointment denied: the inquisition of Bertrand Russell, Prometheus Books, 2000.

[74] J. W. E. Lamb and M. O. Scully, "The photoelectric effect witlhout photons," in *Polarization, Matière et Rayonnement*, Paris, Presses University de France, 1969, pp. 363-369.

[75] J. F. Clauser, "Experimental distinction between the quantum and classical

field-theoretic predictions for the photoelectric effect," *Physical Review D,* vol. 9, pp. 853-860, 1974.

[76] H. J. Kimble, M. Dagenais and L. Mandel, "Photon antibunching in resonance fluorescence," *Physical Review Letters,* vol. 39, pp. 691-695, 1977.

[77] A. Bassi and G. Ghirardi, "A general argument against the universal validity of the superposition principle," *Physics Letters A,* vol. 275, pp. 373-381, 10 2000.

[78] A. J. Leggett, "The quantum measurement problem," *Science,* vol. 307, pp. 871-872, 2005.

[79] N. Gisin, "Collapse. What else?," *ArXiv e-prints : 1701.08300v2,* 1 2017.

[80] G. S. Agarwal, Quantum Optics, Cambridge: Cambridge University Press, 2013.

[81] J. S. Bell, "On the Problem of Hidden Variables in Quantum Mechanics," *Reviews of Modern Physics,* vol. 38, pp. 447-452, 1966.

[82] J. F. Clauser and A. Shimony, "Bell's theorem. Experimental tests and implications," *Reports on Progress in Physics,* vol. 41, p. 1881, 1978.

[83] A. Aspect, P. Grangier and G. Roger, "Experimental Realization of Einstein-Podolsky-Rosen-Bohm Gedankenexperiment: A New Violation of Bell's Inequalities," *Physical Review Letters,* vol. 49, pp. 91-94, 7 1982.

[84] Z. Y. Ou and L. Mandel, "Violation of Bell's Inequality and Classical Probability in a Two-Photon Correlation Experiment," *Physical Review Letters,* vol. 61, pp. 50-53, 7 1988.

[85] Y. H. Shih and C. O. Alley, "New Type of Einstein-Podolsky-Rosen-Bohm Experiment Using Pairs of Light Quanta Produced by Optical Parametric Down Conversion," *Physical Review Letters,* vol. 61, pp. 2921-2924, 12 1988.

[86] L. K. Shalm, E. Meyer-Scott, B. G. Christensen, P. Bierhorst, M. A. Wayne, M. J. Stevens, T. Gerrits, S. Glancy, D. R. Hamel, M. S. Allman, K. J. Coakley, S. D. Dyer, C. Hodge, A. E. Lita, V. B. Verma, C. Lambrocco, E. Tortorici, A. L. Migdall, Y. Zhang, D. R. Kumor, W. H. Farr, F. Marsili, M. D. Shaw, J. A. Stern, C. Abellán, W. Amaya, V. Pruneri, T. Jennewein, M. W. Mitchell, P. G. Kwiat, J. C. Bienfang, R. P. Mirin, E. Knill and S. W. Nam, "Strong Loophole-Free Test of Local Realism," *Physical Review Letters,* vol. 115, p. 250402, 18 December 2015.

[87] B. G. Christensen, K. T. McCusker, J. B. Altepeter, B. Calkins, T. Gerrits, A. E. Lita, A. Miller, L. K. Shalm, Y. Zhang, S. W. Nam, N. Brunner, C. C. W. Lim, N. Gisin and P. G. Kwiat, "Detection-Loophole-Free Test of Quantum Nonlocality, and Applications," *Physical Review Letters,* vol. 111, p. 130406, 9 2013.

[88] M. A. Clauser, A. Shimony and R. A. Holt, "Proposed Experiment to Test Local Hidden-Variable Theories," *Physical Review Letters,* vol. 23, pp. 880-884, 10 1969.

[89] B. S. Cirel'son, "Quantum Generalization of Bell's Inequality," *Letters in Mathematical Physics,* vol. 4, pp. 93-100, 1980.

[90] N. Gisin, "Bell's inequality holds for all non-product states," *Physics Letters A,* vol. 154, no. 5,6, pp. 201-202, 1991.

[91] S. Popescu and D. Rohrlich, "Which states violate Bell's inequality maximally?," *Physics Letters A,* vol. 169, pp. 411-414, 1992.

[92] N. Brunner, D. Cavalcanti and S. Pironio, "Bell nonlocality," *Reviews of Modern Physics,* vol. 86, pp. 419-478, April-June 2014.

[93] M. Giustina, M. A. Versteegh, S. Wengerowsky, J. Handsteiner, A. Hochrainer, K. Phelan, F. Steinlechner, J. Kofler, J.-Å. Larsson, C. Abellán, W. Amaya, V. Pruneri, M. W. Mitchell, J. Beyer, T. Gerrits, A. E. Lita, L. K. Shalm, S. W. Nam, T. Scheidl, R. Ursin, B. Wittmann and A. Zeilinger, "Significant-Loophole-Free Test of Bell's Theorem with Entangled Photons," *Physical Review Letters,* vol. 115, p. 250401, 18 December 2015.

[94] B. Hensen, H. Bernien, A. E. Dreau, A. Reiserer, N. Kalb, M. S. Blok, J. Ruitenberg, R. F. L. Vermeulen, R. N. Schouten, C. Abellan, W. Amaya, V. Pruneri, M. W. Mitchell, M. Markham, D. J. Twitchen, D. Elkouss, S. Wehner, T. H. Taminiau and R. Hanson, "Loophole-free Bell inequality violation using electron spins separated by 1.3 kilometres," *Nature,* vol. 526, pp. 682-686, 29 October 2015.

[95] S. M. Roy, "Multipartite Separability Inequalities Exponentially Stronger than Local Reality Inequalities," *Physical Review Letters,* vol. 94, p. 010402, 2005.

[96] M. Seevinck and J. Uffink, "Local commutativity versus Bell inequality violation for entangled states and versus non-violation for separable states," *Physical Review A,* vol. 76, p. 042105, 2007.

[97] A. K. Biswas, G. Compagno, G. M. Palma, R. Passante and F. Persico, "Virtual photons and causality in the dynamics of a pair of two-level atoms," *Physical Review A,* vol. 42, p. 4291, 1990.

[98] J. León and C. Sabín, "Generation of atom-atom correlations inside and outside the mutual light cone," *Physical Review A,* vol. 79, p. 012304, 2009.

[99] R. Passante, F. Persico and L. Rizzuto, "Nonlocal field correlations and dynamical Casimir--Polder forces between one excited-and two ground-state atoms," *Journal of Physics B: Atomic, Molecular and Optical Physics,* vol. 40, p. 1863, 2007.

[100] S. L. Adler, "Why decoherence has not solved the measurement problem: a response to P.W. Anderson," *Studies in History and Philosophy of Science Part B: Studies in History and Philosophy of Modern Physics* , vol. 34, no. 1, pp. 135-142, 2003.

[101] S. J. van Enk, "Single-particle entanglement," *Physical Review A,* vol. 72, pp. 064306-1-3, 2005.

[102] S. L. Braunstein, A. Mann and M. Revzen, "Maximal Violation of Bell

Inequalities for Mixed States," *Physical Review Letters,* vol. 68, pp. 3259-3261, 1992.

[103] W. K. Wootters, "Entanglement of Formation of an Arbitrary State of Two Qubits," *Physical Review Letters,* vol. 80, no. 10, pp. 2245-2248, 1998.

[104] G. Vidal, "On the characterization of entanglement," *J. Mod. Opt.,* vol. 47, p. 355, 2000.

[105] S. Popescu and D. Rohrlich, "Generic quantum nonlocality," *Physics Letters A,* vol. 166, pp. 293-297, 1992.

[106] M. Renninger, "Messungen ohne Storung des MeßobJekts," *Zeitschrift fur Physik,* vol. 158, pp. 417-421, 1960.

[107] A. C. Elitzur and L. Vaidman, "Quantum mechanical interaction-free measurements," *Foundations of Physics,* vol. 23, pp. 987-997, 7 1993.

[108] Y. Feng, R. Duan and M. Ying, "Unambiguous discrimination between mixed quantum states," *Physical Review A,* vol. 70, p. 012308, Jul 2004.

[109] M. B. Plenio and S. Virmani, "An Introduction to entanglement measures," *Quantum Information and Computation,* vol. 7, pp. 1-51, 2007.

[110] J. S. Bell, "On the Einstein-Podolsky-Rosen paradox," *Physics,* vol. 1, pp. 195-200, 1964.

[111] D. Howard, "Revisiting the Einstein-Bohr Dialogue," *The Jerusalem Philosophical Quarterly,* vol. 56, pp. 57-90, January 2007.

[112] Wikipedia, "Ontology --- Wikipedia, The Free Encyclopedia," 2017. [Online]. Available: https://en.wikipedia.org/w/index.php?title=Ontology&oldid=795548207.

[113] Wikipedia, "Epistemology --- Wikipedia, The Free Encyclopedia," 2017. [Online]. Available: https://en.wikipedia.org/w/index.php?title=Epistemology&oldid=795821247.

[114] P. Lewis, "How Bohm's Theory Solves the Measurement Problem," *Philosophy of Science,* vol. 74, pp. 749-760, 2007.

[115] R. Feynman, R. Leighton and M. Sands, Feynman Lectures on Physics, Volume I, California Institute of Technology, 1963.

[116] J. S. Bell, "Against Measurement," *Physics World,* vol. 3, pp. 33-40, 1990.

[117] L. Rosenfeld, "The Measuring Process in Quantum Mechanics," *Progress of Theoretical Physics Supplement,* vol. E65, pp. 222-231, 1 1965.

[118] A. Daneri, A. Loinger and G. M. Prosperi, "Quantum theory of measurement and ergodicity conditions," *Nuclear Physics,* vol. 33, pp. 297-319, 1962.

[119] E. Schrödinger, "The present situation in quantum mechanics," *Naturwissenshaften (English translation in Proceedings of the American Philosophical Society vol 124),* vol. 23, pp. 802-812, 1935.

[120] J. Bub, Bananaworld: Quantum Mechanics for Primates, Oxford University

Press, 2012.

[121] S. Haroche, J. M. Raimond and J. P. Dowling, "Exploring the Quantum: Atoms, Cavities, and Photons.," *American Journal of Physics,* vol. 82, pp. 86-87, 1 2014.

[122] M. Aspelmeyer and K. Schwab, "EDITORIAL: Focus on Mechanical Systems at the Quantum Limit," *New Journal of Physics,* vol. 10, p. 095001, 9 2008.

[123] I. Lovchinsky, A. O. Sushkov, E. Urbach, N. P. Leon, S. Choi, K. De Greve, R. Evans, R. Gertner, E. Bersin, C. Müller, L. McGuinness, F. Jelezko, R. L. Walsworth, H. Park and M. D. Lukin, "Nuclear magnetic resonance detection and spectroscopy of single proteins using quantum logic," *Science,* vol. 351, pp. 836-841, 2 2016.

[124] J. Millen, P. Z. G. Fonseca, T. Mavrogordatos, T. S. Monteiro and P. F. Barker, "Cavity Cooling a Single Charged Levitated Nanosphere," *Physical Review Letters,* vol. 114, p. 123602, 3 2015.

[125] M. Aspelmeyer, P. Meystre and K. Schwab, "Quantum optomechanics," *Physics Today,* vol. 65, pp. 29-35, 2012.

[126] S. Haroche, "Controlling photons in a box and exploring the quantum to classical boundary," *International Journal of Modern Physics A,* vol. 29, p. 1430026, 4 2014.

[127] C. U. Lei, A. J. Weinstein, J. Suh, E. E. Wollman, A. Kronwald, F. Marquardt, A. A. Clerk and K. C. Schwab, "Quantum Nondemolition Measurement of a Quantum Squeezed State Beyond the 3 dB Limit," *Physical Review Letters,* vol. 117, p. 100801, 9 2016.

[128] F. Lecocq, J. B. Clark, R. W. Simmonds, J. Aumentado and J. D. Teufel, "Quantum Nondemolition Measurement of a Nonclassical State of a Massive Object," *Physical Review X,* vol. 5, p. 041037, 10 2015.

[129] R. Riedinger, S. Hong, R. A. Norte, J. A. Slater, J. Shang, A. G. Krause, V. Anant, M. Aspelmeyer and S. Görblacher, "Non-classical correlations between single photons and phonons from a mechanical oscillator," *Nature,* vol. 530, pp. 313-316, 2 2016.

[130] J. Preskill, "Lecture Notes for Ph219/CS219: Quantum Information, Chapter 3," 2015. [Online]. Available: http://www.theory.caltech.edu/people/preskill/ph219/chap3_15.pdf.

[131] M. Ozawa, "An operational approach to quantum state reduction," *Annals of Physics,* vol. 259, pp. 121-137, 1997.

[132] G. Chiribella, G. M. D'Ariano and P. Perinotti, "Informational derivation of quantum theory," *Physical Review A,* vol. 84, p. 012311, 2011.

[133] B. Cruikshank and K. Jacobs, "The role of quantum measurements in physical processes and protocols," *Quantum Science and Technology,* vol. 2, no. 3, p. 033001, 2017.

[134] R. B. Griffiths, Consistent quantum theory, Cambridge University Press, 2003.

[135] J. N. McElwaine, "Approximate and exact consistency of histories," *Physical Review A,* vol. 53, no. 4, pp. 2021-2032, 4 1996.

[136] R. B. Griffiths, "Consistent histories and quantum reasoning," *Physical Review A,* vol. 54, p. 2759, 1996.

[137] F. Dowker and A. Kent, "On the consistent histories approach to quantum mechanics," *Journal of Statistical Physics,* vol. 82, p. 1575, 1 3 1996.

[138] R. B. Griffiths, "Choice of consistent family, and quantum incompatibility," *Physical Review A,* vol. 57, pp. 1604-1618, Mar 1998.

[139] H. Everett, ""Relative State" Formulation of Quantum Mechanics," *Reviews of Modern Physics,* vol. 29, no. 3, pp. 454-462, 1957.

[140] B. DeWitt and N. Graham, The Many-Worlds Interpretation of Quantum Mechanics, Princeton University Press, 1973.

[141] J. Barrett, The Quantum Mechanics of Minds and Worlds, Oxford University Press, 2003.

[142] D. Deutsch, "Quantum theory as a universal physical theory," *International Journal of Theoretical Physics,* vol. 24, pp. 1-41, 1985.

[143] F. Schmidt-Kaler, H. Haffner, M. Riebe, S. Guide and others, "Realization of the Cirac-Zoller controlled-NOT quantum gate," *Nature,* vol. 422, p. 408, 2003.

[144] T. Monz, K. Kim, W. Hänsel, M. Riebe, A. S. Villar, P. Schindler, M. Chwalla, M. Hennrich and R. Blatt, "Realization of the quantum Toffoli gate with trapped ions," *Physical Review Letters,* vol. 102, p. 040501, 2009.

[145] C. Monroe, D. M. Meekhof, B. E. King, W. M. Itano and D. J. Wineland, "Demonstration of a fundamental quantum logic gate," *Physical Review Letters,* vol. 75, p. 4714, 1995.

[146] L. Isenhower, E. Urban, X. L. Zhang, A. T. Gill, T. Henage, T. A. Johnson, T. G. Walker and M. Saffman, "Demonstration of a neutral atom controlled-NOT quantum gate," *Physical Review Letters,* vol. 104, p. 010503, 2010.

[147] C. J. Broadbent, R. M. Camacho, R. Xin and J. C. Howell, "Preservation of energy-time entanglement in a slow light medium," *Physical Review Letters,* vol. 100, p. 133602, 2008.

[148] H. Everett, *The Theory of the Universal Wave Function,* Ph.D. Thesis, Princeton University, 1956.

[149] W. H. Zurek, "Environment-induced superselection rules," *Physical Review D,* vol. 26, pp. 1862-1880, 1982.

[150] W. H. Zurek, "Decoherence and the transition from quantum to classical," *Physics Today,* vol. 44, pp. 36-44, 1991.

[151] D. Bohm and B. J. Hiley, "Measurement understood through the quantum potential approach," *Foundations of Physics,* vol. 14, pp. 255-274, 3 1984.

[152] D. Dürr, S. Goldstein and N. Zanghì, "Quantum Equilibrium and the Role of Operators as Observables in Quantum Theory," *Journal of Statistical Physics,* vol. 116, pp. 959-1055, 2004.

[153] X. Oriols and J. Mompart, "Overview of Bohmian Mechanics," in *Applied Bohrmian Mechanics From Nanoscale Systems to Cosmology*, Pan Stanford, 2012, p. 15.

[154] T. Maudlin, "Why Bohm's Theory Solves the Measurement Problem," *Philosophy of Science,* vol. 62, pp. 479-483, 1995.

[155] D. Dürr, S. Goldstein and N. Zanghì, "Bohmian Mechanics and Quantum Theory: An Appraisal," J. T. Cushing, A. Fine and S. Goldstein, Eds., Springer, 1996, pp. 21-44.

[156] E. Schrödinger, "Discussion of Probability between Separated Systems," *Proceedings of the Cambridge Physical Society,* vol. 31, no. 4, pp. 555-563, 1935.

[157] C. A. Brasil, F. F. Fanchini and R. Jesus Napolitano, "A simple derivation of the Lindblad equation," *Revista Brasileira de Ensino de Fisica,* vol. 35, 2013.

[158] J. Cramer, "An overview of the transactional interpretation of quantum mechanics," *International Journal of Theoretical Physics,* vol. 27, pp. 227-236, 2 1988.

[159] J. S. Bell, Speakable and unspeakable in quantum mechanics: Collected papers on quantum philosophy, Cambridge University Press, 2004.

[160] A. Zeilinger, Dance of the Photons: From Einstein to Quantum Teleportation, Farrar, Straus and Giroux, 2010.

[161] G. 't Hooft, "The Fate of the Quantum," in *{Time and Matter 2013 (TAM2013) Venice, Italy}*, 2013.

[162] G. 't Hooft, The Cellular Automaton Interpretation of Quantum Mechanics, Springer International Publishing, 2016.

[163] J. G. Cramer, "The transactional interpretation of quantum mechanics," *Reviews of Modern Physics,* vol. 58, no. 3, pp. 647-687, 7 1986.

[164] J. G. Cramer, "The Transactional Interpretation of Quantum Mechanics and Quantum Nonlocality," *ArXiv e-prints arXiv :1503.00039,* 2015.

[165] J. G. Cramer, "The Quantum Handshake: A Review of the Transactional Interpretation of Quantum Mechanics," 2005.

[166] E. T. Jaynes and F. W. Cummings, "Comparison of quantum and semiclassical radiation theories with application to the beam maser," *IEEE Proc.,* vol. 51, pp. 89-109, 1963.

[167] D. Meschede, H. Walther and G. Müller, "One-Atom Maser," *Physical Review Letters,* vol. 54, no. 6, pp. 551-554, 2 1985.

[168] F. Bernardot, P. Nussenzveig, M. Brune, J. M. Raimond and S. Haroche, "Vacuum Rabi Splitting Observed on a Microscopic Atomic Sample in a

Microwave Cavity," *EPL (Europhysics Letters),* vol. 17, p. 33, 1992.

[169] M. G. Raizen, R. J. Thompson, R. J. Brecha, H. J. Kimble and H. J. Carmichael, "Normal-mode splitting and linewidth averaging for two-state atoms in an optical cavity," *Physical Review Letters,* vol. 63, no. 3, pp. 240-243, 7 1989.

[170] L. S. Bishop, J. M. Chow, J. Koch, A. A. Houck, M. H. Devoret, E. Thuneberg, S. M. Girvin and R. J. Schoelkopf, "Nonlinear response of the vacuum Rabi resonance," *Nature Physics,* vol. 5, pp. 105-109, 2 2009.

[171] J. H. Eberly, N. B. Narozhny and J. J. Sanchez-Mondragon, "Periodic Spontaneous Collapse and Revival in a Simple Quantum Model," *Physical Review Letters,* vol. 44, no. 20, pp. 1323-1326, 5 1980.

[172] G. Rempe, H. Walther and N. Klein, "Observation of quantum collapse and revival in a one-atom maser," *Physical Review Letters,* vol. 58, no. 4, pp. 353-356, 1 1987.

[173] S. Haroche, "Entanglement, Mesoscopic Superpositions and Decoherence," *Physica Scripta,* vol. T76, pp. 159-164, 1998.

[174] L. Marchildon, "Causal Loops and Collapse in the Transactional Interpretation of Quantum Mechanics," *Physics Essays,* vol. 19, p. 422, 2006.

[175] B. G. Englert, J. Schwinger and M. Scully, "Is Spin Coherence Like Humpty-Dumpty? I: Simplified Treatment," *Foundation of Physics,* vol. 18, pp. 1045-1056, 10 1988.

[176] J. Schwinger, M. O. Scully and B. G. Englert, "Is spin coherence like Humpty-Dumpty? - II. General theory," *Zeitschrift fur Physik,* vol. 10, pp. 135-144, 6 1988.

[177] M. O. Scully, B.-G. Englert and J. Schwinger, "Spin coherence and Humpty-Dumpty. III. The effects of observation," *Physical Review A,* vol. 40, no. 4, pp. 1775-1784, 8 1989.

[178] A. E. Allahverdyan, R. Balian and T. M. Nieuwenhuizen, "A sub-ensemble theory of ideal quantum measurement processes," *Annals of Physics,* vol. 376, no. Supplement C, pp. 324-352, 2017.

[179] A. E. Allahverdyan, R. Balian and T. M. Nieuwenhuizen, "Understanding quantum measurement from the solution of dynamical models," *Physics Reports,* vol. 525, pp. 1-166, 2013.

[180] C. A. Fuchs, N. D. Mermin and R. Schack, "An introduction to QBism with an application to the locality of quantum mechanics," *American Journal of Physics,* vol. 82, pp. 749-754, 8 2014.

[181] M. Born, "Quantenmechanik der Stossvorgänge," *Zeitschrift fur Physik,* vol. 38, pp. 803-827, 1927.

[182] W. Heisenberg, Physics and Beyond, New York: Harper Torchbooks, 1972.

[183] W. Nagourney, J. Sandberg and H. Dehmelt, "Shelved optical electron

amplifier: Observation of quantum jumps," *Physical Review Letters,* vol. 56, p. 2797, 1986.

[184] H. J. Carmichael, An open systems approach to quantum optics, Berlin: Springer, 1993.

[185] C. W. Gardiner, A. S. Parkins and P. Zoller, "Wave-function quantum stochastic differential equations and quantum-jump simulation methods," *Physical Review A,* vol. 46, p. 4363, 1992.

[186] H. M. Wiseman and G. J. Milburn, Quantum Measurement and Control, Cambridge: Cambridge University Press, 2010.

[187] H. M. Wiseman and J. M. Gambetta, "Are Dynamical Quantum Jumps Detector Dependent?," *Physical Review Letters,* vol. 108, no. 22, may 2012.

[188] P. A. M. Dirac, "The Quantum Theory of the Emission and Absorption of Radiation," *Proceedings of the Royal Society of London A: Mathematical, Physical and Engineering Sciences,* vol. 114, pp. 243-265, 1927.

[189] H. M. Wiseman, S. J. Jones and A. C. Doherty, "Steering, Entanglement, Nonlocality, and the Einstein-Podolsky-Rosen Paradox," *Physical Review Letters,* vol. 98, p. 140402, 2007.

[190] E. G. Cavalcanti, C. J. Foster, M. Fuwa and H. M. Wiseman, "Analog of the Clauser-Horne-Shimony-Holt inequality for steering," *Journal of the Optical Society of America B,* vol. 32, no. 4, p. A74, mar 2015.

[191] E. G. Cavalcanti and R. Lal, "On modifications of Reichenbach's principle of common cause in light of Bell's theorem," *Journal of Physics A: Mathematical and Theoretical,* vol. 47, p. 424018, 2014.

[192] A. M. Steinberg, P. G. Kwiat and R. Y. Chiao, "Measurement of the single-photon tunneling time," *Physical Review Letters,* vol. 71, p. 708, 1993.

[193] R. Y. Chiao and A. M. Steinberg, "VI: Tunneling Times and Superluminality," *Progress in Optics,* vol. 37, pp. 345-405, 1997.

[194] R. Chiao, "Superluminal phase and group velocities: A tutorial on Sommerfeld's phase, group, and front velocities for wave motion in a medium, with applications to the ``instantaneous superluminality'' of electrons," *ArXiv e-prints : 1111.2402,* 11 2011.

[195] M. Renninger, "Zum Wellen-Korpuskel-Dualismus," *Zeitschrift für Physik A ,* vol. 136, pp. 251-261, 1953.

[196] R. Hughes, Quantum Mechanics, Harvard University Press, 1989.

[197] B. d'Espagnat, Conceptual foundations of quantum mechanics, Perseus Books, 1999.

[198] N. Bohr, "Atomic Theory and Mechanics," *Nature,* vol. 116, pp. 845-852, 1925.

[199] A. C. Elitzur and S. Dolev, "Nonlocal effects of partial measurements and quantum erasure," *Physical Review A,* vol. 63, p. 062109, 6 2001.

[200] E. Altewischer, M. P. Exter and J. P. Woerdman, "Plasmon-assisted transmission of entangled photons," *Nature,* vol. 418, pp. 304-306, 7 2002.

[201] L. Mandel and E. Wolf, Optical Coherence and Quantum Optics, Cambridge University Press, 1995.

[202] B. Misra and E. C. G. Sudarshan, "The Zeno's paradox in quantum theory," *Journal of Mathematical Physics,* vol. 18, pp. 756-763, 4 1977.

[203] W. M. Itano, J. Heinzen, J. J. Bollinger and D. J. Wineland, "Quantum Zeno effect," *Physical Review A,* vol. 41, pp. 2295-2300, 3 1990.

[204] V. Frerichs and A. Schenzle, "Quantum Zeno effect without collapse of the wave packet," *Physical Review A,* vol. 44, no. 3, pp. 1962-1968, 8 1991.

[205] D. Home and M. A. B. Whitaker, "A Conceptual Analysis of Quantum Zeno; Paradox, Measurement, and Experiment," *Annals of Physics,* vol. 258, pp. 237-285, 1997.

[206] W. L. Power and P. L. Knight, "Stochastic simulations of the quantum Zeno effect," *Physical Review A,* vol. 53, no. 2, pp. 1052-1059, 2 1996.

[207] T. P. Spiller, "The Zeno effect: measurement versus decoherence," *Physics Letters A,* vol. 192, pp. 163-168, 1994.

[208] C. Balzer, R. Huesmann, W. Neuhauser and P. E. Toschek, "The quantum Zeno effect - evolution of an atom impeded by measurement," *Optics Communications,* vol. 180, pp. 115-120, 6 2000.

[209] G. C. Ghirardi, C. Omero, T. Weber and A. Rimini, "Small-time behaviour of quantum nondecay probability and Zeno's paradox in quantum mechanics," *Il Nuovo Cimento A (1971-1996),* vol. 52, pp. 421-442, 1979.

[210] P. G. Kwiat, A. G. White, J. R. Mitchell, O. Nairz, G. Weihs, H. Weinfurter and A. Zeilinger, "High-efficiency quantum interrogation measurements via the quantum Zeno effect," *Physical Review Letters,* vol. 83, p. 4725, 1999.

[211] W. M. Itano, "Perspectives on the quantum Zeno paradox," *Journal of Physics: Conference Series,* vol. 196, p. 012018, 2009.

[212] F. London and E. Bauer, "La théorie de l'observation en mécanique quantique," in *Quantum Theory and Measurement*, Princeton University Press, 1983 (article first published in 1939), p. 217.

[213] N. Bohr, Niels Bohr: Collected Works, Complementarity Beyond Physics (1928-1962), vol. 10, D. Fahvrholdt, Ed., North Holland, 1999.

[214] N. Bohr, Atomic Physics and Human Knowledge, Dover Publications, 2010.

[215] N. Bohr, The Philosophical Writings of Niels Bohr: Essays 1958-1962 on atomic physics and human knowledge, vol. 3, Ox Bow Press, 1995.

[216] W. Heisenberg, Physics and philosophy: the revolution in modern science, Allen & Unwin St. Leonards, Australia, 1959.

[217] E. P. Wigner, "Remarks on the Mind-Body Question," in *Philosophical Reflections and Syntheses*, J. Mehra, Ed., Berlin, Heidelberg, Springer Berlin

Heidelberg, 1995, pp. 247-260.

[218] B. d'Espagnet, "The Quantum Theory and Reality," *Scientific American,* vol. 241, pp. 158-181, Nov. 1979.

[219] J. C. Eccles, How the self controls its brain, Springer, 1994.

[220] H. P. Stapp, Mindful universe: Quantum mechanics and the participating observer, Springer, 2011.

[221] H. Stapp, Quantum Theory and Free Will: How Mental Intentions Translate Into Bodily Actions, Springer, 2017.

[222] S. Hameroff and R. Penrose, "Consciousness in the universe: A review of the 'Orch OR' theory," *Physics of Life Reviews,* vol. 11, pp. 39-78, 2014.

[223] L. E. Marks, The unity of the senses: Interrelations among the modalities, Academic Press, 1978.

[224] G. C. Ghirardi, A. Rimini and T. Weber, "Unified dynamics for microscopic and macroscopic systems," *Physical Review D,* vol. 34, no. 2, pp. 470-491, 1986.

[225] J. S. Bell, "Are there quantum jumps?," in *Speakable and Unspeakable in Quantum Mechanics: Collected Papers on Quantum Philosophy*, 2 ed., Cambridge University Press, 2004, p. 201–212.

[226] G. C. Ghirardi, P. Pearle and A. Rimini, "Markov processes in Hilbert space and continuous spontaneous localization of systems of identical particles," *Physical Review A,* vol. 42, no. 1, pp. 78-89, 7 1990.

[227] A. Bassi, G. Ghirardi and D. G. M. Salvetti, "The Hilbert-space operator formalism within dynamical reduction models," *Journal of Physics A: Mathematical and Theoretical,* vol. 40, p. 13755, 2007.

[228] K. Jacobs and D. A. Steck, "A straightforward introduction to continuous quantum measurement," *Contemporary Physics,* vol. 47, pp. 279-303, 2006.

[229] T. Heinosaari and M. M. Wolf, "Nondisturbing quantum measurements," *Journal of Mathematical Physics,* vol. 51, p. 092201, 2010.

[230] A. Bassi, E. Ippoliti and B. Vacchini, "On the energy increase in space-collapse models," *Journal of Physics A: Mathematical and General,* vol. 38, p. 8017, 2005.

[231] W. Feldmann and R. Tumulka, "Parameter diagrams of the GRW and CSL theories of wavefunction collapse," *Journal of Physics A: Mathematical and Theoretical,* vol. 45, no. 6, p. 065304, 2012.

[232] P. Pearle, "Completely quantized collapse and consequences," *Physical Review A,* vol. 72, p. 022112, 2005.

[233] S. L. Adler, "Gravitation and the noise needed in objective reduction models," *ArXiv e-prints : 1401.0353,* 2014.

[234] S. Adler, "Lower and upper bounds on CSL parameters from latent image formation and IGM heating," *Journal of Physics A: Mathematical and*

Theoretical, vol. 40, p. 2935, 2007.

[235] A. Vinante, M. R., P. Falferi, M. Carlesso and A. Bassi, "Improved Noninterferometric Test of Collapse Models Using Ultracold Cantilevers," *Physical Review Letters,* vol. 119, p. 110401, 2017.

[236] K. J. McQueen, "Four tails problems for dynamical collapse theories," *Studies in History and Philosophy of Science Part B: Studies in History and Philosophy of Modern Physics,* vol. 49, pp. 10-18, 2015.

[237] G. C. Hegerfeldt, "Remark on causality and particle localization," *Physics Review D,* vol. 10, no. 10, pp. 3320-3321, 11 1974.

[238] T. D. Newton and E. P. Wigner, "Localized states for elementary systems," *Reviews of Modern Physics,* vol. 21, p. 400, 1949.

[239] I. Bialynicki-Birula, "Exponential localization of photons," *Physical Review Letters,* vol. 80, p. 5247, 1998.

[240] N. Barat and J. C. Kimball, "Localization and causality for a free particle," *Physics Letters A,* vol. 308, pp. 110-115, 2003.

[241] G. C. Hegerfeldt, "Violation of causality in relativistic quantum theory?," *Physical Review Letters,* vol. 54, p. 2395, 1985.

[242] D. Z. Albert and B. Loewer, "Tails of Schrödinger's cat," in *Perspectives on quantum reality*, Springer, 1996, pp. 81-92.

[243] S. L. Adler and A. Bassi, "Is Quantum Theory Exact?," *Science,* vol. 325, pp. 275-276, 2009.

[244] A. Bassi and G. Ghirardi, "Dynamical reduction models," *Physics Reports,* vol. 379, pp. 257-426, 2003.

[245] P. Mörters and Y. Peres, "Brownian Motion," 2012. [Online]. Available: http://yuvalperes.com/brbook.pdf.

[246] Y.-J. Lee and K.-G. Yen, "Analysis of complex Brownian motion," *Communications on Stochastic Analysis,* vol. 2, p. 7, 2008.

[247] A. Barchielli and M. Gregoratti, Quantum trajectories and measurements in continuous time: the diffusive case, vol. 782, Springer, 2009.

[248] R. L. Hudson and K. R. Parthasarathy, "Quantum Ito's formula and stochastic evolutions," *Communications in Mathematical Physics,* vol. 93, no. 3, pp. 301-323, September 1984.

[249] V. Gorini, A. Kossakowski and E. C. G. Sudarshan, "Completely Positive semigroups of N-level systems," *Journal of Mathematical Physics,* vol. 17, no. 5, p. 821.

[250] G. Lindblad, "On the generators of quantum dynamical semigroups," *Communications in Mathematical Physics,* vol. 48, pp. 119-130, 1976.

[251] N. Gisin, "Stochastic quantum dynamics and relativity," *Helv. Phys. Acta,* vol. 62, p. 363, 1989.

[252] N. Gisin and I. C. Percival, "Quantum state diffusion, localization and quantum dispersion entropy," *Journal of Physics A: Mathematical and General,* vol. 26, p. 2233, 1993.

[253] K. R. Parthasarathy and A. R. Usha Devi, "Asymptotic spectral stability of the Gisin-Percival state diffusion," *ArXiv e-prints : 1707.08157,* 2017.

[254] K. R. Parthasarathy and A. R. Usha Devi, "From quantum stochastic differential equations to Gisin-Percival state diffusion," *Journal of Mathematical Physics,* vol. 58, p. 082204, 2017.

[255] D. C. Brody and L. P. Hughston, "Quantum State Reduction," *ArXiv e-prints : 1611.02664v2,* 2016.

[256] D. C. Brody and L. P. Hughston, "Quantum noise and stochastic reduction," *Journal of Physics A Mathematical General,* vol. 39, pp. 833-876, 1 2006.

[257] P. Pearle, "Problems and aspects of energy-driven wave-function collapse models," *Physical Review A,* vol. 69, p. 042106, 2004.

[258] L. Acardy, Y. G. Lu and I. Volovich, Quantum theory and its Stochastic Limit, Springer, 2002.

[259] N. Gisin and I. C. Percival, "The quantum-state diffusion model applied to open systems," *Journal of Physics A: Mathematical and General,* vol. 25, p. 5677, 1992.

[260] S. Adler, Quantum Theory as an Emergent Phenomenon, Cambridge University Press, 2004.

[261] D. J. Bedingham and O. J. E. Maroney, "Time symmetry in wave-function collapse," *Physical Review A,* vol. 95, p. 042103, 4 2017.

[262] S. Gerlich, S. Eibenberger, M. Tomandl, S. Nimmrichter, K. Hornberger, P. J. Fagan, J. Tüxen, M. Mayor and M. Arndt, "Quantum interference of large organic molecules," *Nature Communications,* vol. 2, p. 263, 3 2011.

[263] O. Romero-Isart, A. C. Pflanzer, F. Blaser, R. Kaltenbaek, N. Kiesel, M. Aspelmeyer and J. I. Cirac, "Large Quantum Superpositions and Interference of Massive Nanometer-Sized Objects," *Physical Review Letters,* vol. 107, no. 2, p. 020405, 7 2011.

[264] O. Romero-Isart, M. L. Juan, R. Quidant and J. I. Cirac, "Toward quantum superposition of living organisms," *New Journal of Physics,* vol. 12, p. 033015, 2010.

[265] T. Li and Z.-Q. Yin, "Quantum superposition, entanglement, and state teleportation of a microorganism on an electromechanical oscillator," *Science Bulletin,* vol. 61, pp. 163-171, 2016.

[266] S. Nimmrichter and K. Hornberger, "Macroscopicity of mechanical quantum superposition states," *Physical Review Letters,* vol. 110, p. 160403, 2013.

[267] C. M. DeWitt and D. Rickles, "The role of gravitation in physics: report from the 1957 Chapel Hill Conference," Edition Open Sources, 2011.

[268] G. Baym and T. Ozawa, "Two-slit diffraction with highly charged particles: Niels Bohr's consistency argument that the electromagnetic field must be quantized," *Proceedings of the National Academy of Sciences,* vol. 106, pp. 3035-3040, 2009.

[269] C. W. Misner, K. S. Thorne and J. A. Wheeler, Gravitation, 1973.

[270] C. A. Mead, "Possible connection between gravitation and fundamental length," *Physical Review,* vol. 135, p. B849, 1964.

[271] W. Marshall, C. Simon, R. Penrose and D. Bouwmeester, "Towards quantum superpositions of a mirror," *Physical Review Letters,* vol. 91, p. 130401, 2003.

[272] M. Arndt and K. Hornberger, "Testing the limits of quantum mechanical superpositions," *Nature Physics,* vol. 10, pp. 271-277, 4 2014.

[273] M. Scala, M. S. Kim, G. W. Morley, P. F. Barker and S. Bose, "Matter-Wave Interferometry of a Levitated Thermal Nano-Oscillator Induced and Probed by a Spin," *Physical Review Letters,* vol. 111, p. 180403, 11 2013.

[274] A. Belenchia, D. M. T. Benincasa, S. Liberati, F. Marin, F. Marino and A. Ortolan, "Testing Quantum Gravity Induced Nonlocality via Optomechanical Quantum Oscillators," *Physical Review Letters,* vol. 116, no. 16, p. 161303, 4 2016.

[275] Kaltenbaek, Rainer et al, "Macroscopic quantum resonators (MAQRO): 2015 Update," *EPJ Quantum Technology,* vol. 3, p. 5, 2016.

[276] N. Bohr and L. Rosenfeld, "On the question of the measurability of electromagnetic field quantities," in *Selected papers of Léon Rosenfeld*, D. Reidel Publishing Co., 1979, pp. 357-400.

[277] A. Peres and N. Rosen, "Quantum limitations on the measurement of gravitational fields," *Physical Review,* vol. 118, p. 335, 1960.

[278] Y. Ng and H. v. Dam, "Limitation to Quantum Measurements of Space-Time Distances," *Annals of the New York Academy of Sciences,* vol. 755, pp. 579-584, 1995.

[279] X. Calmet, M. Graesser and S. D. H. Hsu, "Minimum length from quantum mechanics and classical general relativity," *Physical Review Letters,* vol. 93, p. 211101, 2004.

[280] R. Penrose, "On gravity's role in quantum state reduction," *General relativity and gravitation,* vol. 28, pp. 581-600, 1996.

[281] L. Diósi, "A universal master equation for the gravitational violation of quantum mechanics," *Physics letters A,* vol. 120, pp. 377-381, 1987.

[282] L. Diósi, "Intrinsic time-uncertainties and decoherence: comparison of 4 models," *Brazilian Journal of Physics,* vol. 35, pp. 260-265, 2005.

[283] A. Bassi, K. Lochan, S. Satin, T. P. Singh and H. Ulbricht, "Models of wave-function collapse, underlying theories, and experimental tests," *Reviews of Modern Physics,* vol. 85, p. 471, 2013.

[284] S. Bera, S. Donadi, K. Lochan and T. P. Singh, "A comparison between models of gravity induced decoherence," *Foundations of Physics,* vol. 45, pp. 1537-1560, 2015.

[285] L. Diósi, "Gravitation and quantum-mechanical localization of macro-objects," *Physics Letters A,* vol. 105, no. 4-5, pp. 199-202, 1984.

[286] A. D. K. Plato, C. N. Hughes and M. S. Kim, "Gravitational effects in quantum mechanics," *Contemporary Physics,* vol. 57, no. 4, 2016.

[287] A. Peres, "Interaction of quantized and unquantized systems," *Nuclear Physics,* vol. 48, pp. 622-624, 1963.

[288] V. Husain and D. R. Terno, "Dynamics and entanglement in spherically symmetric quantum gravity," *Physical Review D,* vol. 81, no. 4, p. 044039, 2 2010.

[289] V. Husain and O. Winkler, "Quantum Hamiltonian for gravitational collapse," *Physical Review D,* vol. 73, p. 124007, 2006.

[290] C. Marletto and V. Vedral, "Witness gravity's quantum side in the lab," *Nature,* vol. 547, pp. 156-158, 2017.

[291] N. Altamirano, P. Corona-Ugalde, R. B. Mann and M. Zych, "Gravity is not a pairwise local classical channel," *ArXiv e-prints:1612.07735,* 2016.

[292] D. Kafri, J. M. Taylor and G. J. Milburn, "A classical channel model for gravitational decoherence," *New Journal of Physics,* vol. 16, p. 065020, 2014.

[293] B. G. Englert, "Fringe visibility and which-way information: An inequality," *Physical Review Letters,* vol. 77, p. 2154, 1996.

[294] M. Gell-Mann and J. B. Hartle, "Classical equations for quantum systems," *Physical Review D,* vol. 47, p. 3345, 1993.

[295] R. Omnes, The interpretation of quantum mechanics, Princeton University Press, 1994.

[296] R. Omnes, "A quantum approach to the uniqueness of Reality," *ArXiv e-prints : 1302.1750,* 2013.

[297] T. Maudlin, "Three measurement problems," *Topoi,* vol. 14, no. 1, pp. 7-15, mar 1995.

[298] T. Maudlin, Quantum non-locality and relativity, 3. ed. ed., Chichester: Wiley-Blackwell, 2011.

[299] H. R. Brown and D. Wallace, "Solving the Measurement Problem: De Broglie--Bohm Loses Out to Everett," *Foundations of Physics,* vol. 35, pp. 517-540, 2005.

[300] M. Esfeld, "Collapse or no collapse? What is the best ontology of quantum mechanics in the primitive ontology framework?," *ArXiv e-prints 1611.09218,* 2016.

[301] S. Goldstein, "Quantum theory without observers," *Physics Today,* vol. 51, pp. 42-47, 1998.

[302] Wikipedia, "De Broglie–Bohm theory --- Wikipedia, The Free Encyclopedia," 2017. [Online]. Available: https://en.wikipedia.org/w/index.php?title=De_Broglie-Bohm_theory&oldid=801054438.

[303] X. S. Ma, J. Kofler and A. Zeilinger, "Delayed-choice gedanken experiments and their realizations," *Reviews of Modern Physics,* vol. 88, no. 1, p. 015005, January-March 2016.

[304] S. Weinberg, To Explain the World, New York: Harper Collins Publishers, 2015.

[305] R. Sorabji, Matter, Space and Motion: Theories in Antiquity and Their Sequel, Cornell University Press, 1988.

[306] C. Flammarion, L'atmosphere meterologie populaire, Paris: Hachette, 1888.

[307] E. Nagel, "Measurement," *Erkenntnis,* vol. 2, no. 1, pp. 311-355, 1931.

[308] I. Newton, The Mathematical Principles of Natural Philosophy Vol I-III, London: Benjamin Motte, 1729.

[309] G. Holton and S. Brush, Physics, the Human Adventure, New Brunswick: Rutgers University Press, 2006.

[310] N. Bohr, "The Quantum Postulate and the Recent Development of Atomic Theory," *Nature,* vol. 121, no. (Suppl.), pp. 580-589, 1928.

[311] N. Bohr, "Quantum Mechanics and Physical Reality," *Nature,* vol. 136, pp. 1025-1026, 1935.

[312] L. Rosenfeld, "Niels Bohr in the Thirties, Consolidation and extension of the conception of complementarity," in *Niels Bohr, His Life and Work as Seen by HIs Friends and Colleagues*, S. Rozental, Ed., New York, Interscience Publishers, 1967, pp. 114-136.

[313] A. Plotnitsky, Niels Bohr and Complementarity, New York: Springer-Verlag, 2012.

[314] S. Dürr, T. Nonn and G. Rempe, "Origin of quantum-mechanical complementarity probed by a `which-way' experiment in an atom interferometer," *Nature,* vol. 395, pp. 33-37, 3 September 1998.

[315] P. Busch and C. Shilladay, "Complementarity and uncertainty in Mach–Zehnder interferometry and beyond," *Physics Reports,* vol. 435, pp. 1-31, 2006.

[316] S. Haroche and J.-M. Raimond, "Bohr's Legacy in Cavity QED," in *Bohr 1913-2013*, O. Darrigol, B. Duplantier, J. Raimond and V. Bivasseau, Eds., Basel, Birkhauser Verlag, 2016, pp. 103-146.

[317] A. Petersen, "The Philosophy of Niels Bohr," in *Niels Bohr A Centenary Volume*, A. P. a. K. P. J. French, Ed., Cambridge, MA: Harvard University Press, 1985, pp. 299-310.

[318] L. Rosenfeld, "Niels Bohr's contribution to epistemology," *Physics Today,* pp.

47-54, October 1963.

[319] N. Bohr, The Philosophical Writings of Niels Bohr: Atomic theory and the description of nature, vol. 1, Ox Bow Press, 1987.

[320] O. F. Junior, The Quantum Dissidents, Rebuilding the Foundations of Quantum Mechanics (1950-1990), Heidelberg: Springer, 2015.

[321] I. Newton, Opticks: or, A Treatise on the Reflections, Refractions, Inflections and Colours of Light, Fourth, corrected ed., London: William Innys, 1730.

[322] J. Brooke, Science and Religion: Some Historical Perspectives, Cambridge: Cambridge University Press, 1991.

[323] B. Gower, "Probability, Planets, and Newton's Methodology," Dortrecht, Kluwer Academic Publishers, 1988.

[324] M. Vogelsberger, S. Genel, V. Springel, P. Torrey, D. Sijacki, D. Xu, G. Snyder, S. Bird, D. Nelson and L. Hernquist, "Properties of galaxies reproduced by a hydrodynamic simulation," *Nature,* vol. 509, no. 7499, pp. 177-182, 7 May 201.

[325] P. F. Hopkins, "Supernovae, supercomputers, and galactic evolution," *Physics Today,* pp. 70-71, April 2017.

[326] P. Laplace, A Philosophical Essay on Probabilites, New York: Dover Publications, 1951.

[327] G. H. von Wright, Causality and Determinism, New York, 1924.

[328] M. A. Deakin, "Nineteenth century anticipations of modern theory of dynamical systems," *Archive for History of Exact Sciences,* vol. 39, no. 2, pp. 183-194, 1988.

[329] R. Lipschitz, "De explicatione per series trigonometricas insttuenda functionum unius arbitrariarum, et praecipue earum, quae per variablis spatium finitum valorum maximorum et minimorum numerum habent infintum disquisitio," *J. Reine Angew. Math,* vol. 63, pp. 296-308, 1864.

[330] S. C. Fletcher, "What Counts as a Newtonian System? The View from Norton's Dome," *European Journal for Philosophy of Science,* vol. 2, no. 3, pp. 275-297, October 2012.

[331] A. S. Helps, Realmah, London: MacMillan and Company, 1868.

[332] I. Pitowsky, "Laplace's Demon Consults an Oracle: The Computational Complexity of Prediction," *Studies in the History and Philosophy of Modern Physics,* vol. 27, no. 2, pp. 161-180, 1996.

[333] S. Lloyd, "A Turing test for free will," *Philosophical Transactions of the Royal Society A,* vol. 370, pp. 3597-3610, 2012.

[334] A. M. Turing, "On computable numbers, with an applications to the Entscheidungsproblem," *Proceedings of the London Mathematical Society,* vol. 42, pp. 230-265, 1936.

[335] C. Moore, "Unpredictability and Undecidability in Dynamical Systems,"

Physical Review Letters, vol. 64, no. 20, pp. 2354-2357, 1990.

[336] B. Russell, "On the Notion of Cause," *Proceedings of the Aristotelian Society, New Series,* vol. 13, pp. 1-26, 1912-1913.

[337] G. M. D'Ariano, F. Manessi and P. Perinotti, "Determinism without causality," *Physica Scripta,* vol. T163, p. 014013, 2014.

[338] L. Campbell and W. Garnett, The Life of James Clerk Maxwell, with selections from his correspondence and occasional writings, London: MacMillan and Co., 1884.

[339] A. Einstein, "Autobiographical Notes," in *Albert Einstein: Philosopher-Scientist*, vol. One, P. A. Schillp, Ed., London, Cambridge University Press, 1949, pp. 2-94.

[340] K. Simonyi, A Cultural History of Physics, Boca Raton: CRC Press, 2012.

[341] C. N. Yang, "The conceptual origins of Maxwell's equations and gauge theory," *Physics Today,* vol. 67, no. 11, pp. 45-51, 2014.

[342] J. C. Maxwell, "On Physical Lines of Force," *Philosophical Magazine,* vol. 90, pp. 11-23, 1861.

[343] H. Kubbinga, "Newton's Theory of Matter," in *Newton's Scientific and Philosophical Legacy*, Dortrecht, Kluwer Academic Publishers, 1988, pp. 321-341.

[344] H. E. B. Viana and P. A. Porto, "The Development of Dalton's Atomic Theory as a Case Study in the History of Science: Reflections for Educators in Chemistry," *Science and Education,* vol. 19, pp. 75-90, 2010.

[345] A. J. Rocke, "In Search of El Dorado: John Dalton and the Origins of the Atomic Theory," *Social Research,* vol. 72, no. 1, pp. 125-158, Spring 2005.

[346] M. J. Nye, "The Nineteenth-Century Atomic Debates and the Dilemma of an 'Indifferent Hypothesis'," *Studies in History and Philosophy of Science,* vol. 7, no. 3, pp. 245-268, 1976.

[347] J. C. Maxwell, "Molecules," *Nature,* pp. 437-441, September 1873.

[348] M. Bächtold, "Saving Mach's View on Atoms," *Journal for General Philosophy of Science,* vol. 41, no. 1, pp. 1-19, June 2010.

[349] H. R. Post, "Atomism 1900 I," *Physics Education,* vol. 3, pp. 225-232, 1968.

[350] S. Seth, "Crisis and the Construction of Modern Theoretical Physics," *The British Journal for the History of Science,* vol. 40, no. 1, pp. 25-51, March 2007.

[351] D. Wegener, "De-anthropomorphizing energy and energy conservation: The case of Max Planck and Ernst Mach," *Studies in the History and Philosophy of Modern Physics,* vol. 41, pp. 146-159, 2010.

[352] J. P. Heilbron, The Dilemmas of an Upright Man. Max Planck as Spokesman for German Science, Los Angeles: University of California Press, 1986.

[353] Aristotle, The Complete Works of Aristotle - The Revised Oxford Translation, J. Barnes, Ed., Princeton: Princeton University Press, 1984.

[354] A. Kekulé, "On the Existence of Chemical Atoms," *The American Journal of Science and Arts,* vol. XLIV, no. 130, 131, 132, July, September, November 1867.

[355] E. Broda, "Philosophical Biography of L. Boltzmann," in *Acta Physica Austriaca (Supplementum X)*, Vienna, 1973.

[356] E. N. Hiebert, "Boltzmann's Conception of Theory Construction: The Promotion of Pluralism," in *Probabilistic Thinking, Thermodynamics and the Interactions of the History and Philosophy of Science*, vol. 146, J. Hintikka, D. Gruender and E. Agazzi, Eds., Dortrecth, Springer, 1980.

[357] C. Cercignani, Ludwig Boltzmann, The Man Who Trusted Atoms, Oxford: Oxford University Press, 1998.

[358] L. Meitner, "Looking Back," *Bulletin of the Atomic Scientists,* vol. 20, no. 9, pp. 4-12, 1964.

[359] H. Minkowski, "Space and Time," in *The Principle of Relativity*, New York, Dover Publications, 1952, pp. 75-91.

[360] C. A. Gearhart, "Planck, the Quantum. and the Historians," *Physics in Perspective,* vol. 4, pp. 170-215, 2002.

[361] H. Kragh, "The origina of radiactivity: from solvable problem to unsolved non-problem," *Archive for History of Exact Sciences,* vol. 50, pp. 331-358, 1997.

[362] K. Svozil, "Physical Unknowables," in *Kurt Gödel and the Foundations of Mathematics, Horizons of Truth*, M. B. Baaz, C. H. Papadimitriou, H. W. Putnam, D. S. Scott and C. L. J. Harper, Eds., Cambridge, Cambridge University Press, 2011, pp. 213-251.

[363] G. Gamow, "Zur Quantentheorie des Atomkernes," *Zeitschrift der Physik,* vol. 51, pp. 204-212, 1928.

[364] R. W. Gurney and E. U. Condon, "Wave mechanics and radioactive disintegration," *Nature,* vol. 122, p. 439, 1928.

[365] M. Born, The Born-Einstein Letters, Correspondence between Albert Einstein and Max and Hedwig Born from 1916 to 1955 with commentaries by Max Born, London: The Macmillan Press LTD, 1971.

[366] A. Einstein, *Verh. Deutsch. Phys. Ges.,* vol. 18, p. 318, 1916.

[367] A. Einstein, *Mitt. Phys. Ges. Zürich,* vol. 16, p. 47, 1916.

[368] A. Einstein, *Phys. Zeitschr.,* vol. 18, p. 121, 1917.

[369] M. Jammer, Einstein and Religion: Physics and Theology, Princeton: Princeton University Press, 1999.

[370] K. Hentschel, "Quantum Jumps," in *Compendium of Quantum Physics*, D. Greenberger, K. Hentschel and F. Weinert, Eds., Berlin, Heidelberg, Springer, pp. 599-601.

[371] T. S. Kuhn, L. Rosenfeld, E. Rundinger and Aage Petersen, "American Institute of Physics Oral History Interviews," 31 October 1962. [Online]. Available: https://www.aip.org/history-programs/niels-bohr-library/oral-histories/4517-5. [Accessed 30 August 2017].

[372] D. Favrholdt, "Niels Bohr and Realism," in *Niels Bohr and Contemporary Philosophy*, J. Faye and H. J. Folse, Eds., Springer-Science + Business Media, B.V., 1994, pp. 77-96.

[373] J. A. Wheeler, "A Septet of Sibyls: Aids in the Search for Truth," *American Scientist,* pp. 360-377, 4 October 1956.

[374] H. P. Robertson, "The uncertainty principle," *Physical Review,* vol. 34, pp. 163-164, 1929.

[375] E. Schrödinger, "Zum Heisenbergschen Unschärfeprinzip," *Sitzungberichten der Preussischen Akademie der Wissenschaften,* vol. 19, pp. 296-323, 1932.

[376] P. J. Coles, M. Berta, M. Tomamichel and S. Wehner, "Entropic uncertainty relations and their applications," *Reviews of Modern Physics,* vol. 89, p. 015002, 19 December 2017.

[377] H. S. Karthik, A. U. Devi and A. K. Rajagopal, "Joint measurability, steering and entropic uncertainty," *Physical Review A,* vol. 91, p. 012115, 2015.

[378] H. S. Karthik, A. U. Devi, J. P. Tej, A. K. Rajagopal, Sudha and A. Narayanan, "N-term pairwise-correlation inequalities, steering, and Joint measurability," *Physical Review A,* vol. 95, p. 052105, 2017.

[379] C. G. Knott, Life and Scientific Work of Peter Guthrie Tait, Cambridge: Cambridge University Press, 1911.

[380] W. Thomson, "Kinetic Theory of the Dissipation of Energy," *Nature,* pp. 441-444, 9 April 1874.

[381] J. Joule, "On Matter, Living Force, and Heat. The Kinetic Theory of Gases," in *The Scientific Papers of James Prescott Joule*, London, The Physical Society, 1884.

[382] K. Maruyama, F. Nori and V. Vedral, "Colloquium: The physics of Maxwell's demon and information," *Reviews of Modern Physics,* vol. 81, no. 1, pp. 1-23, January-March 2009.

[383] M. v. Smoluchowski, "Experimentell Nachweisbare der Üblichen Thermodynamik Widersprechende Molekularphânomene," *Physikalische Zeitung,* vol. 13, pp. 1069-1080, 1912.

[384] L. Szilard, "Über die Entropieverminderung in einem Thermodynamischen System bei Eingriffen Intelligenter Wesen," *Zeitschrift für Physik,* vol. 53, pp. 840-856, 1929.

[385] R. Landauer, "Irreversibility and Heat Generation in the Computing Process," *Advances in Chemical Physics,* vol. 5, pp. 183-191, 1961.

[386] L. Brillouin, "Maxwell's Demon Cannot Operate: Information and Entropy. I,"

Journal of Applied Physics, vol. 22, pp. 334-337, 1951.

[387] D. Gabor, "Light and Information," *Progress in Optics,* vol. 1, pp. 111-153, 1964.

[388] C. H. Bennett, "The thermodynamics of computation - a review," *International Journal of Theoretical Physics,* vol. 21, pp. 905-940, 1982.

[389] A. Bérut, *Nature,* vol. 483, p. 187, 2011.

[390] A. O. Orlov, *Japanese Journal of Applied Physics,* vol. 51, p. 06FE10, 2012.

[391] Y. Jun, *Physical Review Letters,* vol. 113, p. 190601, 2014.

[392] G. L. Snider, E. P. Blair, C. C. Thorpe, B. T. Appleton, G. P. Boechler, A. O. Orlov and C. S. Lent, in *12th IEEE Conference on Nanotechnology (IEEE-NANO)*, 2012.

[393] H. S. Leff and A. F. Rex, Maxwell's Demon: Entropy, Information, Computing, Princeton: Princeton University Press, 1990.

[394] J. Bub, "Maxwell's Demon and the Thermodynamics of Computation," *Studies in History and Philosophy of Modern Physics,* vol. 32, no. 4, pp. 569-579, 2001.

[395] J. Uffink, "Compendium of the foundations of classical statistical physics," in *Handbook of the philosophy of science - philosophy of physics*, J. Butterfield and J. Earman, Eds., Amsterdam, Elsevier, 2007.

[396] M. v. Strien, "The nineteenth century conflict between mechanism and irreversibility," *Studies in History and Philosophy of Modern Physics,* vol. 44, pp. 191-205, 2013.

[397] D. Flamm, "Life and Personality of Ludwig Boltzmann," in *Acta Physica Austriaca (Supplementum X)*, Vienna, 1973.

[398] N. Bohr, Niels Bohr: Collected Works, Foundations of Quantum Physics I (1926-1932), vol. 6, J. Kalckar, Ed., North Holland, 1985.

[399] J. C. Maxwell, The Scientific Papers of James Clerk Maxwell, Vol. II, Cambridge University Press, 1890.

[400] J. L. Lebowitz, "Boltzmann's Entropy and Time's Arrow," *Physics Today,* vol. 46, no. 9, pp. 32-38, September 1993.

[401] R. Feynman, The Character of Physical Law, Cambridge, MA: MIT Press, 1967.

[402] R. Penrose, The Road to Reality, A Complete Guide to the Laws of the Universe, New York: Alfred A. Knopf, 2006.

[403] P. Ehrenfest and E. Tatiana, The Conceptual Foundations of the Statistical Approach in Mechanics, Mineola, NY: Dover Publications, 2002.

[404] J. Mehra, The Solvay Conferences on Physics, Aspects of the Development of Physics Since 1911, Dordrecht-Holland: D. Reidel Publishing Company, 1975.

[405] M. O. Scully and M. S. Zubairy, Quantum Optics, Cambridge University Press,

1997.

[406] C. Huygens, Treatise on Light, London, 1912 (originally published in 1690).

[407] J. Cohen, "Newton's Third Law and Universal Gravity," in *Newton's Scientific and Philosophical Legacy*, P. D. a. G. Scheurer, Ed., Dortrecht, Kluwer Academic Publishers, 1988, pp. 25-53.

[408] C. Hakfoort, "Newton's 'Optiks' and the Incomplete Revolution," in *Newton's Scientific and Philosophical Legacy*, P. B. Scheurer and G. Debrock, Eds., Dortrecht, Kluwer Academic Publishers, 1988, pp. 99-112.

[409] G. Garratt, The Early History of Radio: From Faraday to Marconi, London: The Institute of Engineering and Technology, 2006.

[410] M. Jammer, The Conceptual Development of Quantum Mechanics, New York: McGraw-Hill, 1966.

[411] R. Rilke, The Notebooks of Malte Laurids Brigge, Dalkey Archive Press, 2008.

[412] L. Boya, "The Thermal Radiation Formula of Planck (1900)," *Rev. Real Academia de Ciencias Zaragoza,* vol. 58, pp. 91-114, 2003.

[413] A. Babin and A. Figotin, "The History of Views on Charges, Currents and the Electromagnetic Field," in *Neoclassical Theory of Electromagnetic Interactions. Theoretical and Mathematical Physics*, London, Springer, 2016.

[414] M. J. Klein, "Thermodynamics and Quanta in Planck's Work," *Physics Today,* pp. 294-302, November 1966.

[415] J. L. Heilbron, "Max Planck's compromises," in *Quantum Mechanics at the Crossroads, New Perspectives from History, Philosophy and Physics*, J. Evans and A. S. Thorndike, Eds., Berlin, Springer, 2007, pp. 21-37.

[416] J. L. Heilbron, "Max Planck's compromises on the way to and from the Absolute," in *Quantum Mechanics at the Crossroads, New Perspectives from History, Philosophy and Physics*, J. Evans and A. S. Thorndike, Eds., Berlin, Springer-Verlag, 2007, pp. 21-37.

[417] O. Darrigol, "The Historians' Disagreement over the Meaning of Planck's Quantum," in *Revisiting the Quantum Discontinuity*, Berlin, Max-Planck-Institut für Wissenschaftsgeschichte., 2000, pp. 3-21.

[418] T. Kuhn, Black-Body Theory and the Quantum Discontinuity, Chicago: University of Chicago Press, 1987.

[419] O. Darrigol, From c-Numbers to q-Numbers, The Classical Analogy in the History of Quantum Theory, Berkeley: University of California Press, 1993.

[420] L. Rosenfeld, "The Wave-Particle Dilemma," in *The Physicist's Conception of Nature*, J. Mehra, Ed., Dortrecht, Springer, 1973, pp. 251-263.

[421] N. Bohr, The Theory of Spectra and Atomic Collisions, Three Essays, Cambridge: Cambridge University Press, 1924.

[422] N. Bohr, "Zur Frage der Polarization der Strahlung in der Quantentheorie," *Zeitschrift fur Physik,* vol. 6, pp. 1-9, 1921.

[423] A. Whitaker, Einstein, Bohr and the Quantum Dilemma: From Quantum Theory to Quantum Informations, Cambridge: Cambridge University Press, 2006.

[424] E. Schrödinger, "Are There Quantum Jumps? Part I," *The British Journal for the Philosophy of Science,* vol. III, pp. 109-123, 1952.

[425] E. Schrödinger, "Are There Quantum Jumps? Part II," *British Journal for the Philosophy of Science,* vol. 3, pp. 233-242, 1952.

[426] M. Bitbol, "Schrödinger Against Particles and Quantum Jumps," in *Quantum Mechanics at the Crossroads, New Perspectives from History, Philosophy and Physics*, J. Evans and A. S. Thorndike, Eds., Berlin, Springer-Verlag, 2007, pp. 81-106.

[427] J. Stachel, "The Other Einstein: Einstein Contra Field," *Science in Context,* vol. 6, pp. 275-290, 1993.

[428] N. Mott, "The Wave Mechanics of alpha-Tracks," *Proceedings of the Royal Society,* vol. A124, pp. 81-84, 1929.

[429] W. Heisenberg, The Physical Principles of the Quantum Theory, Dover Publications, 2013.

[430] R. Feynman, QED: The strange theory of light and matter, Princeton, New Jersey: Princeton University Press, 2006.

[431] J. A. Wheeler, "The Young Feyman," *Physics Today,* pp. 24-28, February 1989.

[432] A. Einstein, "Über die von der molekularkinetischen Theorie der Wärme geforderte Bewegung von in ruhenden Flüssigkeiten suspendierten Teilchen," *Annalen der Physik,* vol. 17, no. 8, pp. 549-560, 1905.

[433] J. D. Norton, "Atoms, entropy, quanta: Einstein's miraculous argument of 1905," *Studies in the History and Philosophy of Modern Physics,* vol. 37, pp. 71-100, 2006.

[434] J. Mehra, Einstein, Hilbert, and the Theory of Gravitation: Historical Origins of General Relativity Theory, Dortrecht: D. Reidel Publishing Company, 1974.

[435] E. MacKinnon, "Bohr on the Foundation of Quantum Theory," in *Niels Bohr: A Centerary Volume*, A. P. French and P. J. Kennedy, Eds., Cambridge, MA: Harvard University Press, 1985, pp. 101-120.

[436] J. Stachel, Einstein from 'B' to 'Z', Boston: Birkhäuser, 2002.

[437] A. Pais, Subtle is the Lord, The Science and Life of Albert Einstein, Oxford: Oxford University Press, 1982.

[438] D. W. Belousek, "Einstein's 1927 Unpublished Hidden-Variable Theory: Its Background, Context and Significance," *Studies in History and Philosophy of Modern Physics,* vol. 27, no. 4, pp. 437-461, 1997.

[439] J. Mehra, "Niels Bohr's Discussions with Albert Einstein, Werner Heisenberg and Erwin Schrödinger: The Origins of the Principle of Uncertainty and

Complementarity," *Foundations of Physics,* vol. 17, pp. 461-506, 1987.

[440] L. Gilder, The Age of Entanglement: When Quantum Physics was Reborn, New York: Vintage Books, 2008.

[441] N. Bohr, Niels Bohr Collected Works The Emergence of Quantum Mechanics, vol. 5, K. Stolzenburg, Ed., Elsevier, 1984.

[442] S. Petruccioli, Atoms, Metaphors and Paradoxes, Niels Bohr and the construction of a new physics, Cambridge: Cambridge University Press, 1993.

[443] W. Pauli, "Die allgemeinen Prinzipien der Wellenmechanik," in *Handbuch der Physik*, vol. 24, Berlin, Springer, 1933.

[444] M. D. Eisaman, J. Fan, A. Migdall and S. V. Polyakov, "Invited Review Article: Single-photon sources and detectors," *Review of Scientific Instruments,* vol. 82, p. 071101, 2011.

[445] R. Heilmann, J. Sperling, A. Perez-Leija, M. Gräfe, M. Heinrich, S. Nolte, W. Vogel and A. Szameit, "Harnessing click detectors for the genuine characterization of light states," *Scientific Reports,* vol. 6, p. 19489, 14 January 2016.

[446] P. A. M. Dirac, The Principles of Quantum Mechanics, Clarendon-Press, 1930, p. x + 257.

[447] R. L. Pfleegor and L. Mandel, "Interference of Independent Photon Beams," *Physical Review,* vol. 159, no. 5, pp. 1084-1088, 25 July 1967.

[448] A. Shimony, "Metaphysical Problems in the Foundations of Quantum Mechanics," *International Philosophical Quarterly,* vol. 18, no. 1, pp. 3-17, 1978.

[449] E. Schrödinger, "Quantisation as a Problem of Proper Values (Part I)," in *Collected Papers on Wave Mechanics*, Providence, RI: AMS Chelsea Publishing, 2010, pp. 1-12.

[450] E. Schrödinger, "Quantisation as a Problem of Proper Values (Part IV)," in *Collected Papers on Wave Mechanics*, Providence, RI: AMS Chelsea Publishing, 2010, pp. 102-123.

[451] S. Perovic, "Schrödinger's interpretation of quantum mechancis and the relevance of Bohr's experimental critique," *Studies in History and Philosophy of Modern Physics,* vol. 37, pp. 275-297, 2006.

[452] R. S. Shankland, "An apparent failure of the photon theory of scattering," *Physical Review,* vol. 49, pp. 8-13, 1936.

[453] N. Bohr, "Conservation Laws in Quantum Theory," *Nature,* pp. 25-26, 4 July 1936.

[454] J. C. Jacobsen, "Correlation between Scattering and Recoil in the Compton Effect," *Nature,* p. 25, 4 July 1936.

[455] W. G. Cross and N. F. Ramsey, "The Conservation of Energy and Momentum in Compton Scattering," *The Physical Review,* vol. 80, no. 6, pp. 929-936,

1950.

[456] R. Hofstadter and J. A. McIntyre, "Simultaneity in the Compton Effect," *Physical Review,* vol. 78, no. 1, pp. 24-28, 1950.

[457] E. Maier, "Particle Billiards, Captured on Film," [Online]. Available: https://www.mpg.de/4444655/S005_Flashback_092-093.pdf. [Accessed 6 September 2017].

[458] W. Bothe, "Nobelprize.org," [Online]. Available: https://www.nobelprize.org/nobel_prizes/physics/laureates/1954/bothe-lecture.html. [Accessed 6 September 2017].

[459] A. Ádám, L. Jánossy and P. Varga, *Acta Physica Hungerica,* vol. 4, p. 301, 1955.

[460] A. Ádám, L. Jánossy and P. Varga, *Annalen der Physik,* vol. 16, p. 408, 1955.

[461] J. F. Clauser, "Early History of Bell's Theorem," in *Quantum [Un]speakables, From Bell to Quantum Information*, R. A. Bertlmann and A. Zeilinger, Eds., Berlin, Springer-Verlag, 2002, pp. 61-98.

[462] J. F. Clauser, "Experimental distinction between the quantum and classical field-theoretic predictions for the photoelectric effect," *Physical Review D,* vol. 9, no. 4, pp. 853-860, 1974.

[463] T. Guerreiro, B. Sanguinetti, H. Zbinden, N. Gisin and A. Suarez, "Single-photon space-like antibunching," *Physics Letters A,* vol. 376, pp. 2174-2177, 2012.

[464] E. Schrödinger, "Probability relations between separated systems," *Proceedings of the Cambridge Philosophical Society,* vol. 32, pp. 446-452, 1936.

[465] K. D. Thomas, "The Advent and Fallout of EPR, An IAS teatime conversation in 1935 introduces an ongoing debate over quantum physics," *The Institute Letter,* Fall 2013.

[466] A. Fine, The Shaky Game: Einstein, Realism and the Quantum Theory, Chicago: University of Chicago Press, 1986.

[467] W. J. Moore, Schrödinger: Life and Thought, Cambridge: Cambridge University Press, 1989.

[468] D. Bohm, Quantum Theory, Englewood Cliffs, NJ: Prentice-Hall, Inc., 1950.

[469] N. Bohr, "Can Quantum-Mechanical Description of Physical Reality be Considered Complete?," *Physical Review,* vol. 48, pp. 696-702, 15 October 1935.

[470] W. H. Furry, "Note on the Quantum-Mechanical Theory of Measurement," *Physical Review,* vol. 49, pp. 393-399, 1 March 1936.

[471] W. H. Furry, "Remarks on Measurements in Quantum Theory," *Physical Review,* vol. 49, p. 476, 1 March 1936.

[472] A. Einstein and N. Rosen, "The Particle Problem in the General Theory of

Relativity," *Physical Review,* vol. 48, pp. 73-77, 1 July 1935.

[473] J. Maldacena and L. Susskind, "Cool horizons for entangled black holes," *Forschritte der Physik, Progress of Physics,* vol. 61, no. 9, pp. 781-811, September 2013.

[474] A. Einstein, B. Podolsky and N. Rosen, "Can Quantum-Mechanical Description of Physical Reality be Considered Complete?," *Physical Review,* vol. 47, pp. 777-780, 1935.

[475] C. G. Darwin, "A collision problem in the wave mechanics," *Proceedings of the Royal Society A,* vol. 124, pp. 375-394, 1929.

[476] F. J. Duarte, "The origin of quantum entanglement experiments based on polarization measurements," *The European Physical Journal H,* vol. 37, pp. 311-318, 2012.

[477] P. A. Dirac, "On the annihilation of electrons and protons," *Cambridge Philosophical Society,* vol. 26, pp. 361-375, 1930.

[478] J. A. Wheeler, "Polyelectrons," *Annals of the New York Academy of Sciences,* vol. 48, pp. 219-238, 1946.

[479] M. H. Pryce and J. C. Ward, "Angular correlation effects with annhilation radiation," *Nature,* vol. 160, p. 435, 1947.

[480] H. S. Snyder, S. Pasternack and J. Hornbostel, "Angular correlation of scattered annihilation radiation," *Physical Review,* vol. 73, pp. 440-448, 1948.

[481] E. Bleuler and H. L. Bradt, "Correlation between the states of polarization of the two quanta of annihilation radiation," *Physical Review,* vol. 73, p. 1398, 1948.

[482] R. C. Hanna, "Polarization of annihilation radiation," *Nature,* vol. 162, p. 332, 1948.

[483] C. S. Wu and I. Shaknov, "The angular correlation of scattered annihilation radiation," *Physical Review,* vol. 77, p. 136, 1950.

[484] D. Bohm and Y. Aharonov, "Discussion of experimental proof for the paradox of Einstein, Rosen and Podolsky," *Physical Review,* vol. 108, pp. 1070-1076, 1957.

[485] Yin, Juan et al., "Satellite-based entanglement distribution over 1200 kilometers," *Science,* vol. 356, pp. 1140-1144, 2017.

[486] K. W. Chan, C. K. Law and J. H. Eberly, "Quantum entanglement in photon-atom scattering," *Physical Review A,* vol. 68, p. 022110, 2003.

[487] J. von Neumann, John von Neumann: Selected Letters, M. Rédei, Ed., Providence, RI: The American Mathematical Society, 2005.

[488] H. Reichenbach, The direction of time, M. Reichenbach, Ed., Berkeley - Los Angeles, CA: University of California Press, 1956.

[489] A. Shimony, Search for a naturalistic world view, vol. II, Cambridge: Cambridge University Press, 1993.

[490] P. Adcroft and E. Datlow, Eds."Interview: John Bell, Interviewee: Mann, Charles; Crease, Robert," *Omni Magazine,* vol. 10, no. 8, p. 84, May 1988.

[491] J. Bub, "Von Neumann's 'No Hidden Variables' Proof: A Re-Appraisal," *Foundations of Physics,* vol. 40, pp. 1333-1340, 2010.

[492] M. Stöltzner, "What John von Neumann Thought of the Bohm Interpretation," in *Epistemological and Experimental Perspectives on Quantum Physics*, D. e. a. Greenberger, Ed., Dortrecht, Kluwer Academic Publishers, 1999, pp. 257-262.

[493] J. von Neumann, in *New Theories in Physics*, Paris, International Institute of Intellectual co-operation, 1939, pp. 30-45.

[494] N. Bohr, Niels Bohr: Collected Works, Foundations of Physics II (1933-1958), vol. 7, J. Kalckar, Ed., North Holland, 1996.

[495] J. von Neumann, "Wahrscheinlichkeitstheoretischer Aufbau der Quantenmechanik," *Königliche Gesellschaft der Wissenschaften zu Göttingen. Mathematisch-physikalische Klasse,* pp. 245-272, 1927.

[496] A. Duncan and M. Janssen, "(Never) Mind your p's and q's: Von Neumann versus Jordan on the foundations of quantum theory," *The European Physical Journal H,* vol. 38, pp. 175-259, 2013.

[497] J. D. Sneed, "Von Neumann's Argument for the Projection Postulate," *Philosophy of Science,* vol. 33, no. 1/2, pp. 22-39, 1966.

[498] G. Lüders, "Concerning the state-change due to the measurement process," *Annalen der Physik (Leipzig),* vol. 15, no. 9, pp. 663-670, 2006.

[499] D. Howard, "Who Invented the "Copenhagen Interpretation"? A Study in Mythology," *Philosophy of Science,* vol. 71, no. 5, pp. 669-682, December 2004.

[500] L. Rosenfeld, "Misunderstandings about the Foundations of Quantum Theory," in *Observation and Interpretation: A Symposium of Philosophers and Physicists*, S. Körner, Ed., New York, Academic Press, 1957, pp. 41-45.

[501] H. Zinkernagel, "Niels Bohr on the wave function and the classical/quantum divide," *Studies in History and Philosophy of Modern Physics,* vol. 53, pp. 9-19, 2016.

[502] G. Bacciagaluppi and E. Crull, "Heisenberg (and Schrödinger, and Pauli) on hidden variables," *Studies in History and Philosophy of Modern Physics,* vol. 40, pp. 374-382, 2009.

[503] N. Bohr, Albert Einstein: Philosopher-Scientist : Discussion with Einstein on epistemological problems in atomic physics, Cambridge University Press, 1949, pp. 200-241.

[504] A. Einstein, "Remarks to the Essays Appearing in this Collective Volume," in *Albert Einstein: Philosopher-Scientist*, P. A. Schilpp, Ed., London, Cambridge University Press, 1949, p. 663.

[505] C. C. W. Taylor, "The Atomists," in *The Cambridge Companion to Early Greek Philosophy*, A. A. Long, Ed., Cambridge, Cambridge University Press, 1999, pp. 181-204.

[506] S. Greenblatt, The Swerve, New York: W.W. Norton & Company, 2011.

[507] M. N. Bera, A. Acin, M. Kus, M. Mitchell and M. Lewenstein, "arXiv.org/quant-ph," 17 August 2017. [Online]. Available: https://arxiv.org/abs/1611.02176v2. [Accessed 19 August 2017].

[508] M. J. Hall, "Local Deterministic Model of Singlet State Correlations Based on Relaxing Measurement Independence," *Physical Review Letters,* vol. 105, p. 250404, 2010.

[509] G. Makari, Soul Machine, The Invention of the Modern Mind, New York: W.W. Norton & Company, 2015.

[510] C. Koch, The Quest for Consciousness: A Neurobiological Approach, Englewood, CO: Roberts & Company, 2004.

[511] D. J. Chalmers, "Facing up to the Problem of Consciousness," *Journal of Consciousness Studies,* vol. 2, no. 3, pp. 200-219, 1995.

[512] F. Crick and C. Koch, "A framework for consciousness," *Nature Neuroscience,* vol. 6, no. 2, pp. 119-126, 2003.

[513] M. Gell-Mann, "Consciousness, Reduction, and Emergence," *Annals of the New York Academy of Sciences,* pp. 41-49, April 2001.

[514] T. D. Albright, T. M. Jessell, E. R. Kandel and M. I. Posner, "Progress in the Neural Sciences in the Century after Cajal (and the Mysteries That Remain)," *Annal of the New York Academy of Sciences,* vol. 929, pp. 11-40, April 2001.

[515] J. Craig Venter et al, "The Sequence of the Human Genome," *Science,* vol. 291, pp. 1304-1351, 2001.

[516] S. Hawking and L. Mlodinow, The Grand Design, New York: Bantam Books, 2010.

[517] B. Libet, C. A. Gleason, E. W. Wright and D. K. Pearl, "Time of Conscious Intention to Act in Relation to Onset of Cerebral Activity (Readiness-Potential), The Unconscious Initiation of a Freely Voluntary Act," *Brain,* vol. 106, pp. 623-642, 1983.

[518] P. Haggard, "Human volition: towards a neuroscience of will," *Nature Reviews | Neuroscience,* vol. 9, pp. 934-945, December 2008.

[519] B. Libet, "Conscious subjective experience vs. unconscious mental functions: A theory of the cerebral processes involved," in *Experimental and Theoretical Studies of Consciousness*, Ciba Foundation Symposium #174 ed., Chichester: Wiley, 1993.

[520] B. Libet, "Do We Have Free Will?," *Journal of Consciousness Studies,* vol. 6, no. 8-9, pp. 47-57, 1999.

[521] M. Hallett, "Physiology of Free Will," *Annals of Neurology,* vol. 80, no. 1, pp.

5-12, July 2016.

[522] M. C. Vinding, M. Jensen and M. Overgaard, "Distinct electrophysiological potentials for intention in action and prior intention for action," *Cortex,* vol. 50, pp. 86-99, January 2014.

[523] P. Alexander, A. Schlegel and W. e. a. Sinnott-Armstrong, "Readiness potentials driven by non-motonic processes," *Consciousness and Cognition,* vol. 39, pp. 38-47, 2016.

[524] P. G. H. Clarke, "The Libet experiment and its implications for conscious will," *The Faraday Papers,* pp. 1-4, 2013.

[525] F. London and E. Bauer, La Theorie de l'Observation en Mecanique Quantique., Paris: Hermann & Cie, 1939.

[526] T. H. Huxley, "On the Hypothesis that Animals are Automata, and its History," in *Collected Essays*, vol. I, London, Macmillan, 1893, pp. 199-250.

[527] W. James, "Are We Automata," *Mind,* vol. 4, no. 13, pp. 1-22, January 1879.

[528] T. Nagel, "What is the mind-brain problem?," *Experimental and Theoretical Studies of Consciousness,* vol. 174, pp. 1-13, 1993.

[529] J. R. Searle, "The problem of consciousness," *Experimental and Theoretical Studies of Consciousness,* vol. 174, pp. 61-80, 1993.

[530] T. Nagel, "What Is It Like To Be a Bat?," *Philosophical Review,* vol. 83, pp. 435-450, 1974.

[531] D. K. Lewis, "Are We Free to Break the Laws?," *Theoria,* vol. 47, pp. 113-121, 1981.

[532] G. Santoro, M. D. Wood, L. Merlo, G. P. Anastasi, F. Tomasello and A. Germanò, "The Anatomic Location of the Soul From the Heart, Through the Brain, to the Whole Body, and Beyond: A Journey Through Western History, Science, and Philosophy," *Neurosurgery,* vol. 64, no. 4, pp. 633-643, October 2009.

[533] V. Caston, "Aristotle on Consciousness," *Mind,* vol. 111, pp. 751-815, October 2002.

[534] R. F. Del Maestro, "Leonardo da Vinci: the search for the soul," *Journal of Neurosurgery,* vol. 89, pp. 874-887, 1998.

[535] J. P. Richter, The Literary Works of Leonardo Da Vinci, London: Sampson Low, Marston, Searle & Rivington, 1883.

[536] A. Tozzi, J. F. Peters, A. A. Fingelkurts, A. A. Fingelkurts and P. C. Marijuán, "Brain projective reality: Novel clothes for the emperor," *Physics of Life Reviews,* vol. 21, pp. 46-55, 2017.

[537] A. C. Doyle, "The Adventure of the Blanched Soldier," *Strand Magazine,* November 1926.

[538] F. Aaserud and J. L. Heilbron, Love, Literature, and the Bohr Model of Atomic Structure, 1913-1925, Cambridge: Cambridge University Press, 2013.

[539] J. Heilbron, "Bohr's First Theories of the Atom," in *Niels Bohr: A Centenary Volume*, Cambridge, Harvard University Press, 1985, p. 352.

[540] J. L. Heilbron, "The Mind that Created the Bohr Atom," in *Niels Bohr; 1913-2013*, vol. 68, Springer International Publishing, 2016, pp. 57-98.

[541] J. Wheeler, "Niels Bohr, the man," *Physics Today,* vol. 38, no. 10, p. 66, 1985.

[542] C. W. Misner, K. S. Thorne and W. H. Zurek, "John Wheeler, relativity, and quantum information," *Physics Today,* pp. 40-46, April 2009.

[543] R. Feynman, "The Present Status of Quantum Electrodynamics," in *The Quantum Theory of Fields, Proceedings of the Twelfth Conference on Physics at the Univeristy of Brussels*, New York, 1961.

[544] N. Bohr, "Part 1: On the Constitution of Atoms and Molecules: Part 2: On the Constitutions of Atoms and Molecules; Part 3: Systems Containing Several Nuclei," *Philosophical Magazine,* vol. 26, pp. 1-25; 476-502; 857-875, 1913.

[545] H. Kragh, Niels Bohr and the Quantum Atom, Oxford: Oxford University Press, 2012.

[546] A. Svidinsky, M. Scully and D. Hershback, "Bohr's molecular model, a century later," *Physics Today,* pp. 33-39, January 2014.

[547] R. Smith, "Logic," in *The Cambridge Companion to Aristotle*, J. Barnes, Ed., Cambridge, Cambridge University Press, 1999, pp. 27-65.

[548] R. J. Hankinson, "Philosophy of Science," in *The Cambridge Companion to Aristotle*, J. Barnes, Ed., Cambridge, Cambridge University Press, 1999, pp. 109-139.

[549] G. Lloyd, Early Greek science: Thales to Aristotle, London: Chatto and Windus, 1970.

[550] D. C. Lindberg, The Beginnings of Western Science: The European Scientific Tradition in Philosophical, Religious, and Institutional Context, 600 B.C. to A.D. 1450, Chicago: The University of Chicago Press, 2009.

[551] D. Nicolaides, In the Light of Science: Our Ancient Quest for Knowledge and the Measure of Modern Physics, Amherst: Prometheus Books, 2014.

[552] W. Heisenberg, "Planck's discovery and the philosophical problems of atomic physics," in *On Modern Physics*, New York, Clarkson N. Potter, Inc., 1961, pp. 3-19.

[553] R. Penrose, The Emperor's New Mind, Oxford: Oxford University Press, 1989.

[554] G. Debrock, "Aristotle Wittgenstein, Alias Isaac Newton Between Fact and Substance," in *Newton's Scientific and Philosophical Legacy*, P. S. a. G. Debrock, Ed., Dortrecht, Kluwer Academic Publishers, 1988, pp. 355-377.

[555] R. Monk, The Duty of Genius, New York: The Free Press, 1990.

[556] L. Wittgenstein, Tractatus Logico-Philosophicus, London: Routledge, 2000.

[557] L. Wittgenstein, Philosophical Investigations, New York: The Macmillan

Company, 1953.

[558] N. Malcom, Ludwig Wittgenstein, London: Oxford University Press, 1975.

[559] A. Peterson, "The Philosophy of Niels Bohr," *Bulletin of the Atomic Scientists,* vol. 7, no. 19, pp. 8-14, 1963.

[560] J. C. Maxwell, A Treatise on Electricity and Magnetism, Oxford: Clarendon Press, 1873.

[561] A. Einstein, "Maxwell's Influence on the Evolution of the Idea of Physical Reality," in *James Clerk Maxwell: A Commemerative Volume on the one hundredth anniversary of his birth*, Cambridge, Cambridge University Press, 1931, pp. 266-270.

[562] M. Janssen and C. Lehner, "Introduction," in *The Cambridge Companion to Einstein*, Cambridge, Cambridge University Press, 2017, pp. 1-37.

[563] A. Einstein, "What is the theory of relativity?," *London Times,* 28 November 1919.

[564] A. Einstein, "On the occasion of the three hundredth anniversary of Kepler's death," *Frankfurter 'Zeiturig (Germany),* 9 November 1930.

[565] A. A. Michelson, Light Waves and Their Uses, The University of Chicago Press, 1903.

[566] P. Dirac, "Quantum Mechanics of Many-Electron Systems," *Proceedings of the Royal Society of London; Series A,* vol. 123, no. 792, pp. 714-733, 1929.

[567] J. Wheeler, "From the Big Bang to the Big Crunch," *Cosmic Search,* vol. 1, no. 4, p. 2, Oct, Nov, Dec 1979.

[568] D. M. Greenberger, M. A. Horne and A. Zeilinger, "Going Beyond Bell's Theorem," in *Bell Theorem, Quantum Theory and Conceptions of the Universe*, Dordrecht, Kluwer Academic, 1989, pp. 69-72.

[569] S. Kochen and E. P. Specker, "The problem of hidden variables in quantum mechanics," *Journal of Mathematics and Mechanics,* vol. 17, pp. 59-87, 1967.

[570] M. F. Pusey, J. Barrett and T. Rudolph, "On the reality of the quantum state," *Nature Physics,* vol. 8, pp. 475-478, 2012.

[571] M. S. Leifer, "Is the Quantum State Real? An Extended Reviews of the Psi-ontology Theorems," *Quanta,* vol. 3, no. 1, pp. 67-155, 2014.

[572] A. Shimony, M. A. Horne and J. F. Clauser, "Comment on "The Theory of Local Beables"," *Epistemological Letters,* vol. 9, p. 1, 1976.

[573] S. Weinstein, "Nonlocality Without Nonlocality," *Foundations of Physics,* vol. 39, pp. 921-936, 2009.

[574] R. Colbeck, "Quantum and Relativistic Protocols for Secure Multi-Party Computation," University of Cambridge, Cambridge, 2006.

[575] R. Colbeck and R. Renner, "Free randomness can be amplified," *Nature Physics,* vol. 8, pp. 450-453, 2012.

[576] S. Pironio, A. Acin, S. Massar, A. B. d. l. Giroday, D. N. Matsukevich, P. Maunz, S. Olmschenk, D. Hayes, L. Luo, T. A. Manning and C. Monroe, "Random numbers certified by Bell's theorem," *Nature,* vol. 464, pp. 1021-1024, 15 April 2010.

[577] U. Vazirani and T. Vidick, "Certified quantum dice," *Philosophical Transactions of the Royal Society,* vol. 379, pp. 3432-3448, 2012.

[578] R. Gallego, L. Masanes, G. De la Torre, C. Dhara, L. Aolita and A. Acin, "Full randomness from arbitrarily deterministic events," *Nature Communications,* vol. 4, p. 2654, 30 October 2013.

[579] A. Acin, "Randomness in quantum physics," *Euresis journal,* vol. 6, pp. 35-48, Winter 2014.

[580] A. Acin, S. Massar and S. Pironio, "Randomness versus Nonlocality and Entanglement," *Physical Review Letters,* vol. 108, p. 100402, 2012.

[581] J. Conway and S. Kochen, "The Free Will Theorem," *Foundations of Physics,* vol. 36, no. 10, pp. 1441-1473, 10 October 2006.

[582] J. H. Conway and S. Kochen, "The Strong Free Will Theorem," *Notices of the American Mathematical Society,* vol. 56, no. 2, pp. 226-232, 2009.

[583] C. Dhara, G. d. l. Torre and A. Acin, "Can Observed Randomness Be Certified to Be Fully Intrinsic?," *Physical Review Letters,* vol. 112, p. 100402, 2014.

[584] J. von Neumann, "Proof of the ergodic theorem and the H-theorem in quantum mechanics," *The European Physical Journal H,* pp. 201-237, 35.

[585] S. Goldstein, J. L. Lebowitz, C. Mastodonato, R. Tumulka and N. Zanghi, "Normal typicality and von Neumann's quantum ergodic theorem," *Proceedings of the Royal Society A,* vol. 466, pp. 3202-3224, 2010.

[586] T. B. Batalhão, "Irreversibility and the Arrow of Time in a Quenched Quantum System," *Physical Review Letters,* vol. 115, p. 190601, 2015.

[587] J. Goold, M. Huber, A. Riera, L. d. Rio and P. Skrzypczyk, "The role of quantum information in thermodynamics - a topical review," *Journal of Physics A: Mathematical and Theoretical,* vol. 49, pp. 1-52, 2016.

[588] P. Riemann, "Typical fast themalization processes in many-body systems," *Nature Communications,* vol. 7, p. 10821, 1 March 2016.

[589] A. M. Kaufman, M. E. Tai, A. Lukin, M. Rispoli, R. Schittko, P. M. Preiss and M. Greiner, "Quantum thermalization through entanglement in an isolated many-body system," *Science,* vol. 353, no. 6301, pp. 794-800, 19 August 2014.

[590] S. Lloyd, "Quantum-mechanical Maxwell's demon," *Physical Review A,* vol. 56, no. 5, pp. 3374-3382, November 1997.

[591] W. H. Zurek, "Quantum discord and Maxwell's demons," *Physical Review A,* vol. 67, p. 012320, 2003.

[592] S. Kim, T. Sagawa, S. De Liberato and M. Ueda, "Quantum Szilard engine," *Physical Review Letters,* vol. 106, p. 70401, 2011.

[593] S. Hilt, S. Shabbir, J. Anders and E. Lutz, *Physical Review E,* vol. 83, p. 030102, 2011.

[594] B. Piechocinska, *Physical Review A,* vol. 61, p. 062314, 2000.

[595] E. Lubkin, *International Journal of Theoretical Physics,* vol. 26, p. 523, 1987.

[596] M. B. Plenio and V. Vitelli, "The physics of forgetting: Landauer's erasure principle and information theory," *Contemporary Physics,* vol. 42, no. 1, pp. 25-60, 2001.

[597] W. H. Zurek, H. S. Leff and A. F. Rex, in *Frontiers of Nonequilibrium Statistical Physics*, 1986, pp. 151-161.

[598] M. O. Scully, *Physical Review Letters,* vol. 87, p. 220601, 2001.

[599] M. O. Scully, Y. Rostovtsev, Z.-. E. Sariyanni and M. S. Zubairy, *Physica E,* vol. 29, p. 29, 2005.

[600] H. T. Quan, P. Zhang and C. P. Sun, *Physical Review E,* vol. 73, p. 036122, 2006.

[601] T. Sagawa and M. Ueda, "Second law of thermodynamics with discrete quantum feedback," *Physical Review Letters,* vol. 100, p. 080403, 2008.

[602] L. Del Rio, J. Åberg, R. Renner, O. Dahlsten and V. Vedral, "The thermodynamic meaning of negative entropy," *Nature,* vol. 474, pp. 61-63, 2011 June 2011.

[603] N. Cottet, S. Jezouin, L. Bretheau, P. Campagne-Ibarcq, Q. Ficheux, J. Anders, A. Auffèves, R. Azouit, P. Rouchon and H. Benjamin, "Observing a quantum Maxwell demon at work," *Proceedings of the National Academy of Sciences of the United States of America,* vol. 114, no. 29, pp. 7561-7564, 2017.

[604] M. D. Vidrighin, O. Dahlsten, M. Barbieri, M. S. Kim, V. Vedral and I. A. Walmsley, "Photonic Maxwell's Demon," *Physical Review Letters,* vol. 116, p. 050401, 2016.

[605] P. A. Camati, J. P. Peterson, T. B. Batalhão, K. Micadei, A. M. Souza, R. S. Sarthour, I. S. Oliveira and R. M. Serra, "Experimental Rectification of Entropy Production by Maxwell's Demon in a Quantum System," *Physical Review Letters,* vol. 117, p. 240502, 2016.

[606] C. Jarzynski, "Diverse phenomena, common themes," *Nature Physics,* vol. 11, pp. 105-107, 2015.

[607] P. Hänggi and P. Talkner, "The other QFT," *Nature Physics,* vol. 11, pp. 108-110, 2015.

[608] C. Elouard, D. Herrera-Martí, B. Huard and A. Auffèves, "Extracting Work from Quantum Measurement in Maxwell's Demon Engines," *Physical Review Letters,* vol. 118, p. 260603, 2017.

[609] K. Abdelkhalek, Y. Nakata and D. Reeb, "Fundamental energy cost for quantum measurement," *ArXiv e-prints : 1609.06981v2,* 23 June 2017.

[610] K. Brandner, M. Bauer, M. T. Schmid and U. Seifert, "Coherence-enhanced

efficiency of feedback-driven quantum," *New Journal of Physics,* vol. 17, p. 065006, 2015.

[611] J. Earman and J. D. Norton, "Exorcist XIV: The Wrath of Maxwell's Demon. Part I. From Maxwell to Szilard," *Studies in History and Philosophy of Science Part B: Studies in History and Philosophy of Modern Physics,* vol. 29, no. 4, pp. 435-471, June 1998.

[612] J. Earman and J. D. Norton, "Exorcist XIV: The Wrath of Maxwell's Demon. Part II. From Szilard to Landauer and Beyond," *Studies in History and Philosophy of Science Part B: Studies in History and Philosophy of Modern Physics,* vol. 30, no. 1, pp. 1-40, 1999.

[613] C. H. Bennett, "Notes on Landauer's principle, reversible computation, and Maxwell's Demon," *Studies in History and Philosophy of Modern Physics,* vol. 34, pp. 501-510, 2003.

[614] D. J. Bedingham and O. J. E. Maroney, "The thermodynamic cost of quantum operations," *New Journal of Physics,* vol. 18, p. 113050, 2016.

[615] M. O. Scully, B.-G. Englert and H. Walther, "Quantum optical tests of complementarity," *Nature,* vol. 351, pp. 111-116, 9 May 1991.

[616] M. O. Scully and K. Drühl, "Quantum eraser: A proposed photon correlation experiment concerning observation and "delayed choice" in quantum mechanics," *Physical Review A,* vol. 25, no. 4, pp. 2208-2213, April 1982.

[617] J. A. Wheeler, in *Problems in the Formulations of Physics*, G. T. d. Francia, Ed., Amsterdam, North-Holland, 1979.

[618] M. O. Scully and H. Walther, "An Operational Analysis of Quantum Eraser and Delayed Choice," *Foundations of Physics,* vol. 28, no. 3, pp. 399-413, 1998.

[619] J. Preskill, "Lecture Notes for Ph219/CS219: Quantum Information, Chapter 2," 2017. [Online]. Available: http://www.theory.caltech.edu/~preskill/ph219/chap2_15.pdf. [Accessed 2017].

[620] K. Camilleri and M. Schlosshauer, "Niels Bohr as philosopher of experiment: Does decoherence theory challenge Bohr's doctrine of classical concepts?," *Studies in History and Philosophy of Modern Physics,* vol. 49, pp. 73-83, 2015.

[621] S. Schweber, QED and the Men Who Made It; Dyson, Feynman, Schwinger, and Tomonaga, Princeton: Princeton University Press, 1994.

[622] J. Mehra, The Beat of a Different Drum, The Life and Science of Richard Feynman, Oxford: Oxford University Press, 2000.

[623] J. Mehra and M. Kimball, Climbing the Mountain: the Scientific Biography of Julian Schwinger, Oxford: Oxford University Press, 2000.

[624] J. Mehra and H. Rechenberg, The Historical Development of Quantum Theory, The Completion of Quantum Mechanics 1926-1941, vol. 6, New York: Springer-Verlag, 2001.

[625] N. D. Mermin, "What's wrong with this pillow?," *Physics Today,* vol. 42, no. 4, p. 9, April 1989.

[626] L. Vaidman, "Variations on the Theme of the Greenberger-Horne-Zeilinger Proof," *Foundations of Physics,* vol. 29, no. 4, pp. 615-630, 1999.

[627] E. B. Davies, Quantum theory of open systems, London: Academic Press, 1976.

[628] J. B. Conway, A course in functional analysis, Second ed., New York: Springer, 1990.

[629] R. J. Cook and H. J. Kimble, "Possibility of direct observation of quantum jumps," *Physical Review Letters,* vol. 54, p. 1023, 1985.

[630] J. C. Bergquist, R. G. Hulet, W. M. Itano and D. J. Wineland, "Observation of quantum jumps in a single atom," *Physical Review Letters,* vol. 57, p. 1699, 1986.

[631] T. J. Hollowood, "Copenhagen quantum mechanics," *Contemporary Physics,* vol. 57, no. 3, pp. 289-308, 2015.

[632] K. Popper, The logic of scientific discovery, Routledge, 2005.

[633] G. Ellis and J. Silk, "Scientific method: Defend the integrity of physics," *Nature,* vol. 516, 2014.

[634] R. Dawid, String theory and the scientific method, Cambridge University Press, 2013.

[635] G. Galilei, Dialogue concerning the two chief world systems, Ptolemaic & Copernican, Univ of California Press, 1967.

[636] A. Einstein, Mein Weltbild, Amsterdam: Querido Verlag, 1934.

[637] P. McDonald, Oxford dictionary of medical quotations, Oxford Medical Publications, 2004.

[638] J. C. Maxwell, The Scientific Papers of James Clerk Maxwell, Vol. I, vol. 1, Cambridge University Press, 1890.

[639] M. Planck, Scientific autobiography: And other papers, Open Road Media, 2014.

[640] H. Kragh, Cosmology and controversy: The historical development of two theories of the universe, Princeton University Press, 1999.

[641] J. R. Hofmann, André-Marie Ampére: Enlightenment and Electrodynamics, vol. 7, Cambridge University Press, 1996.

[642] H. Ayrton, The electric arc, Cambridge University Press, 2012.

[643] J. A. Hammerton and A. Mee, The World's Greatest Books, vol. 13, McKinlay, Stone, Mackenzie, 1910.

[644] A. Calaprice, The Ultimate Quotable Einstein, Princeton University Press, 2011.

[645] G. Gallavotti, W. L. Reiter and J. Yngvason, Boltzmann's legacy, European

Mathematical Society, 2008.

[646] R. M. Wald, "Quantum gravity and time reversibility," *Physical Review D,* vol. 21, pp. 2742-2755, May 1980.

[647] C. Simon, V. Bužek and N. Gisin, "No-signaling condition and quantum dynamics," *Physical Review Letters,* vol. 87, p. 170405, 2001.

[648] T. F. Jordan, "Assumptions that imply quantum dynamics is linear," *Physical Review A,* vol. 73, p. 022101, Feb 2006.

[649] D. M. Tong, L. C. Kwek, C. H. Oh, J.-L. Chen and L. Ma, "Operator-sum representation of time-dependent density operators and its applications," *Physical Review A,* vol. 69, p. 054102, 2004.

[650] I. Vega and D. Alonso, "Dynamics of non-Markovian open quantum systems," *Reviews of Modern Physics,* vol. 89, p. 015001, 2017.

[651] H. P. Breuer, E. M. Laine, J. Piilo and B. Vacchini, "Colloquium: Non-Markovian dynamics in open quantum systems," *Reviews of Modern Physics,* vol. 88, p. 021002, 2016.

[652] J. M. Dominy and D. A. Lidar, "Beyond complete positivity," *Quantum Information Processing,* vol. 15, pp. 1349-1360, 2016.

[653] T. F. Jordan, A. Shaji and E. C. G. Sudarshan, "Dynamics of initially entangled open quantum systems," *Physical Review A,* vol. 70, p. 052110, Nov 2004.

[654] A. Bassi, D. Dürr and G. Hinrichs, "Uniqueness of the equation for quantum state vector collapse," *Physical Review Letters,* vol. 111, p. 210401, 2013.

[655] L. Diósi, "Comment on "Uniqueness of the Equation for Quantum State Vector Collapse"," *Physical Review Letters,* vol. 112, p. 108901, 2014.

[656] P. Busch, P. J. Lahti and J. P. Y. K. Pellonpää, Quantum Measurement, Springer, 2016.

[657] G. Lüders, "Translation of Lüders', Uber die Zustandsanderung durch den Messprozess," *Annalen der Physik (Leipzig) 15, No. 9, 663 - 670 (2006),* 29 2 2004.

[658] P. Busch, M. Grabowski and J. L. Pekka, Operational Quantum Physics, Springer, 1995.

[659] P. Busch and J. Singh, "Lüders theorem for unsharp quantum measurements," *Physics Letters A,* vol. 249, pp. 10-12, 1998.

[660] G. M. D'Ariano, P. Perinotti and M. F. Sacchi, "Informationally complete measurements and group representation," *Journal of Optics B: Quantum and Semiclassical Optics,* vol. 6, p. S487, 2004.

[661] P. Busch, "``No Information Without Disturbance": Quantum Limitations of Measurement," in *Quantum Reality, Relativistic Causality, and Closing the Epistemic Circle*, W. C. Myrvold and J. Christian, Eds., 2008, p. 229.

[662] A. Arias, A. Gheondea and S. Gudder, "Fixed points of quantum operations," *Journal of Mathematical Physics,* vol. 43, pp. 5872-5881, 2002.

[663] K. Banaszek, "Information gain versus state disturbance for a single qubit," *Open Systems and Information Dynamics,* vol. 13, pp. 1-16, 2006.

[664] M. M. Wolf, "Quantum channels & operations: Guided tour," *Lecture notes available at http://www-m5. ma. tum. de/foswiki/pub M,* vol. 5, 2012.

[665] T. C. Ralph, S. D. Bartlett, J. L. O'Brien, G. J. Pryde and H. M. Wiseman, "Quantum nondemolition measurements for quantum information," *Physical Review A,* vol. 73, p. 012113, Jan 2006.

[666] M. Tsang and C. M. Caves, "Evading Quantum Mechanics: Engineering a Classical Subsystem within a Quantum Environment," *Physical Review X,* vol. 2, p. 031016, Sep 2012.

[667] V. P. Belavkin, "A new wave equation for a continuous nondemolition measurement," *Physics Letters A,* vol. 140, pp. 355-358, 1989.

[668] A. N. Korotkov, "Selective quantum evolution of a qubit state due to continuous measurement," *Physical Review B,* vol. 63, p. 115403, 2001.

[669] M. Ozawa, "Uncertainty relations for noise and disturbance in generalized quantum measurements," *Annals of Physics,* vol. 311, no. 2, pp. 350-416, 2004.

[670] E. Arthurs and J. Kelly, "B.S.T.J. Brief: On the Simultaneous Measurement of a Pair of Conjugate Observables," *Bell System Technical Journal brief,* vol. 44, no. 4, pp. 725-729, 1964.

[671] T. J. Bullock and P. Busch, "Focusing in Arthurs-Kelly-type Joint Measurements with Correlated Probes," *Physical Review Letters,* vol. 113, p. 120401, 2014.

[672] Y. Aharonov and D. Rohrlich, Quantum paradoxes: quantum theory for the perplexed, John Wiley & Sons, 2008.

[673] L. A. Rozema, A. Darabi, D. H. Mahler, A. Hayat, Y. Soudagar and A. M. Steinberg, "Violation of Heisenberg's Measurement-Disturbance Relationship by Weak Measurements," *Physical Review Letters,* vol. 109, no. 10, p. 100404, 9 2012.

[674] P. Busch and N. Stevens, "Direct tests of measurement uncertainty relations: what it takes," *Physical Review Letters,* vol. 114, p. 070402, 2015.

[675] Piacentini et al, "Determining the quantum expectation value by measuring a single photon," *Nature Physics,* vol. 13, pp. 1191-1194, 2017.

[676] L. Landau and R. Peierls, "Erweiterung des Unbestimmtheitsprinzips für die relativistische Quantentheorie," *Zeitschrift für Physik A Hadrons and Nuclei,* vol. 69, pp. 56-69, 1931.

[677] J. Polchinski, "Weinberg's nonlinear quantum mechanics and the Einstein-Podolsky-Rosen paradox," *Physical Review Letters,* vol. 66, pp. 397-400, Jan 1991.

[678] R. Plaga, "On a possibility to find experimental evidence for the many-worlds interpretation of quantum mechanics," *Foundations of Physics,* vol. 27, pp.

559-577, 4 1997.

[679] A. Messiah, Quantum Mechanics Vol. 1, Interscience, 1961.

Index

Made in the USA
Columbia, SC
27 September 2021